Spacecraft Maneuver with Performance Guaranteed

Yufeng Gao · Dongyu Li

Spacecraft Maneuver with Performance Guaranteed

Control, Games, and Cooperation

Springer

Yufeng Gao
Department of Control Science and Engineering
Harbin Institute of Technology
Harbin, China

Dongyu Li
School of Cyber Science and Technology
Beihang University
Beijing, China

ISBN 978-981-99-4652-5 ISBN 978-981-99-4653-2 (eBook)
https://doi.org/10.1007/978-981-99-4653-2

This Springer imprint is published by the registered company Springer Nature Singapore Pte Ltd.
The registered company address is: 152 Beach Road, #21-01/04 Gateway East, Singapore 189721, Singapore

To my family and friends.

Preface

Facing future-oriented aerospace applications, large-scale space construction and on-orbit services have rapidly developed. In such emerging and increasingly complex spacecraft maneuvering and control tasks, more precise control accuracy and higher performance guaranteed need to be fully considered.

This book is dedicated to solving the performance guaranteed spacecraft control challenges with the expected transient and strict steady-state constraints, namely *spacecraft performance guaranteed control*. It is designed so that the aerospace closed-loop system can theoretically meet the pre-defined or prescribed performance requirements with the simple parameter selection. Furthermore, the expected performance constraints or indicators of the aerospace system time-domain performance response, such as the settling time, overshoot, steady-state error, and state amplitude, will be articulated and prescribed in the control design.

This book systematically proposes a series of spacecraft performance guaranteed control algorithms based on the practical situations of the aerospace system. For a single spacecraft, relevant control algorithms are proposed with the considerations of settling time constraints, transient performance normalization, input command constraints, and optimization indexes faced by the on-orbit spacecraft during complex space missions. For the pre-combination and post-combination control of multiple spacecraft, game algorithms based on performance guaranteed are proposed and analyzed. For spacecraft formations, control schemes are proposed to drive the network spacecraft to the desired attitude and orbit configurations, where practical issues such as full-state constraints, nonlinear uncertainties, output feedback, and collision avoidance are likewise discussed.

This book provides the theoretical basis and simulation experience for researchers and engineers to develop high-performance, high-precision spacecraft control algorithms.

Harbin, China
November 2022

Yufeng Gao

Contents

Chapter 1
Introduction

1.1 Research Background

The concept of spacecraft On-Orbit Service (OOS) was first proposed in the 1960s and has been widely practiced in subsequent practical missions [1, 2]. The spacecraft on-orbit servicing tasks typically include inspections, repairs, refueling, assembly or upgrades of target spacecraft in orbit, etc. [3]. In completing these tasks, the on-orbit service spacecraft usually needs to rendezvous and dock, fly around, accompany or hover with the target spacecraft. Since some interaction behaviors need to be maintained between spacecraft during these short-range relative motions, more accurate relative motion control that considers performance constraints is of great significance for spacecraft short-range on-orbit service missions.

The study of the spacecraft's relative motion process focuses on analyzing the spacecraft's time domain response and behavior during the control process. The time domain analysis method of the control system is a method to analyze the system in the time domain directly. The evaluation of control system performance is divided into two categories: transient (dynamic) performance indicators and steady-state performance indicators. The transient performance indicators include the settling time (delay time, rise time, peak time, adjustment time, etc.) and the overshoot. The time parameter is used to evaluate the response speed of the closed-loop system, and the overshoot is used to evaluate the damping degree of the system. The steady-state performance index mainly refers to the steady-state error of the system. The performance of the closed-loop system can be further decomposed through the time-domain response performance indicators [4]. The constraints on the time domain response indicators of the closed-loop system can be regarded as the performance constraints of the closed-loop system, which can also be called *performance guaranteed*. In order to achieve precise control for the spacecraft system, the control methods that add constraints on transient and steady-state performance indicators can be studied and applied.

Y. Gao and D. Li, *Spacecraft Maneuver with Performance Guaranteed*,
https://doi.org/10.1007/978-981-99-4653-2_1

From the point of view of control theory, the traditional control methods, such as proportional-integral-derivative (PID) control, adaptive control, sliding mode control, and robust control, can be proved the stability of closed-loop systems through Lyapunov theory and other stability theories. The transient and steady-state performance of the closed-loop systems, including the settling time, can be adapted by parameter adjustment [5]. These control methods are suitable for simpler space missions. However, these methods lack theoretical strict constraints and guarantees for the transient and steady-state performance, including the settling time for the closed-loop system state to converge to the origin (or its neighborhood) under complex operating conditions [6]. Specifically, since the constraints of the closed-loop system time domain response are not considered in these control methods, in the transient process of system convergence, there may be uncertain settling time, uncertain steady-state error, uncertain transient trajectory, undesired overshoot, and so on. These are not conducive to the complex on-orbit service missions of spacecraft with higher precision requirements. Although related tasks can be accomplished through the elaborate design of these algorithms and complex experiments, the uncertain design brings hidden dangers to more complex tasks. Faced with a spacecraft that may have irregular shapes and uncertain motion, the relative motion of the spacecraft in close service usually has strict constraints. In such cases, traditional unconstrained control will bring defects that cannot cope with complex constraints in practical applications. Therefore, this research hopes to achieve the relative motion control objectives of the on-orbit service spacecraft through performance guaranteed constraints, including the transient performance and behavior are guaranteed and constrained; the steady-state error is controlled with high precision. That is, through the performance guaranteed control method, the artificial active constraints of performance parameters such as settling time, overshoot, and steady-state error in the performance response of the closed-loop system are realized. This means that the maximum settling time, overshoot, and steady-state error of the on-orbit service spacecraft relative motion can all meet the practical mission requirements set artificially. The control accuracy of the closed-loop system can even be improved through tighter performance constraints.

From the spacecraft engineering point of view, in the conventional attitude or orbit control problems of a single spacecraft or multiple spacecraft with relatively long distances, the control constraints for the time domain response are only determined by the mission objectives. Therefore, the main objectives of attitude control are tracking or regulation with considering the control accuracy. The corresponding strict constraints are lacking. However, in the on-orbit service missions, two or more spacecraft perform attitude or relative orbital maneuvers or tracking within relative proximity. Considering the task execution window period, the control accuracy and complexity, and the physical configuration of the target spacecraft and the service spacecraft for close rendezvous, the attitude or relative orbital control of the service spacecraft will face the following practical problems:

(1) The closed-loop system settling time needs to meet the expected time or time window requirements *to guarantee the closed-loop system settling time*.

(2) The control accuracy is required to be higher. That is, a smaller steady-state error is required *to guarantee the steady-state performance* of the closed-loop system.

(3) In the motion control process, the state of the closed-loop system must always satisfy strict constraints. These constraints are used *to guarantee the transient performance*. That is, the transient performance needs to meet the desired constraints, which means that excessive overshoot needs to be strictly limited (this constraint can be converted into a constraint function of the states).

To sum up, for the increasingly complex on-orbit service missions involving dual or even multiple spacecraft, the practical constraints on the transient and steady-state responses of the spacecraft system are more complex. The control system significantly increases the demands on the desired relative motion control accuracy. It is worth noting that the spacecraft motion system can be transformed into the general second-order system or Euler–Lagrange system, etc. [7, 8]. After the system model is transformed, the spacecraft motion control or relative motion control problem can be further abstracted into the control problem based on the spacecraft dynamics and kinematics models. The same is true for the attitude motion of the on-orbit service spacecraft relative to the target spacecraft. On the other hand, the algorithms designed with this general model can also be transferred to the precise motion control design of other motion systems.

This book focuses on the precise control of the spacecraft's relative motion, conducts in-depth research on related issues, and presents theoretical results. These include spacecraft performance guaranteed control, performance guaranteed filter control, performance guaranteed optimal control, performance guaranteed differential game control, and the cooperative problem with state constraints. In this book, the transient and steady-state performance requirements and the maximum settling time of the closed-loop systems are emphasized.

1.2 On-Orbit Service Spacecraft Control

1.2.1 On-Orbit Service Spacecraft

In the 1960s, the concept of spacecraft on-orbit service gradually took shape. At that time, the focus of on-orbit service was to provide necessary maintenance for on-orbit spacecraft to extend their life or expand their size and function. During the Skylab mission in 1973, the astronauts performed a series of on-orbit operations, including repairing the super-energy battery array and microwave antenna, which actually proved the availability of the on-orbit service concept [2]. Subsequent space shuttle projects have given an enormous impetus to the development of on-orbit service technology, including astronauts using the Space Shuttle Challenger to carry out the maintenance of the attitude control system. The replacement of electronic components on the SMM satellite (the Solar Maximum Satellite) launched in 1980, using the Space Shuttle Discovery conducted similar maintenance tasks on the Palapa

B2 and Westar 6 satellites [9]. Since then, astronauts have successfully performed five on-orbit service missions to the Hubble Space Telescope through the space shuttle, making outstanding contributions to the spacecraft's on-orbit service missions [10].

In 1997, the successful launch of ETS-VII equipped with space robots was followed by a series of on-orbit service missions, such as autonomous rendezvous and docking (AR&D), satellite monitoring, module replacement, fueling, and truss assembly, which also made it the first autonomous on-orbit service spacecraft [2]. In 1998, the successful launch of Zarya, the first module of the International Space Station (ISS), marked the beginning of the space station's construction. The space station's construction further enriched the tasks of on-orbit services, especially large-scale on-orbit assembly and equipment maintenance [11]. In 2007, the development and on-orbit verification of the Orbital Express further promoted the development of the spacecraft's on-orbit service technology. Its mission components include the Autonomous Space Transfer and Robotic Orbiter (ASTRO) and the Next generation of Satellite (NEXTSat). Its mission objectives include autonomous rendezvous and capture, refueling, and testing of the Orbital Replacement Unit (ORU) [12, 13].

In the 21st century, with the increasing demand for space resources, and the rapid development of space technology, the technical framework of the on-orbit services has continued to improve. The practices of the on-orbit technology have continued to develop, such as the National Mission plan [9], the Robotic Servicing of Geostationary Satellites (RSGS) [14], Phoenix [15], Dragonfly [16], the Deutsche Orbital Servicing Mission (DEOS) [17], the Intelligent Building Blocks for On-Orbit Satellite Servicing (iBOSS) [2], the Space Infrastructure Servicing (SIS) [18], etc. In 2021, China successfully launched the Tianhe core module of the space station into the predetermined orbit with the Long March-5B carrier rocket in Wenchang, Hainan. The on-orbit assembly and construction of the China Space Station are in full swing.

The on-orbit service mission has experienced the development from the on-orbit service of conventional-scale spacecraft to large-scale spacecraft. The related technologies for the on-orbit service missions mainly include the system strategy, payloads, astronauts and manned services, space robots, orbital design and planning, spacecraft construction and layout, connection and separation systems, attitude and orbital control, propulsion, thermal control, energy management, data processing, and transmission, tracking telemetry, command, and communication, etc. [2]. Facing the future, on-orbit service missions will face more opportunities and challenges. As reported in [19], the on-orbit service technology and satellite service are essential technologies for future space applications.

This section reviews the on-orbit service application and control technology involved in this book. The theoretical development of related aspects in recent years is emphatically summarized. The attitude control design is mainly considered in Part I and Part II of this book. Part III considers some cooperative relative position control for multiple spacecraft. Since most of this book is about spacecraft attitude control, this section focuses on the results of attitude research. The relative position/orbit control problem is reviewed in the corresponding chapters.

1.2.2 Spacecraft Relative Motion Control

The attitude control problem of the on-orbit serving spacecraft mainly considers the relative attitude control and attitude control after combining the serving and target spacecraft. Among them, for the relative attitude control problem of the close-range service spacecraft, if the attitude maneuver information of the target spacecraft is known, the problem can be regarded as an attitude tracking problem considering some constraints. The attitude control problem with constraints will be described in detail later. The traditional spacecraft attitude tracking control has been studied for many years. Here is a brief list of the latest research results on spacecraft attitude control that have emerged in recent years, such as the finite-time attitude control [20, 21], the fuzzy control [22], the performance constraint control (detailed later), the observer-based attitude control [23–26], the event-driven control [27, 28], the adaptive control [29], the optimal control [30, 31], the filter-based control [32], the multi-model predictive control [33], the neural-network-based adaptive control [34], etc. These control methods are often used in related spacecraft attitude control tasks, such as the spacecraft formation flight attitude control [35, 36], the liquid-filled spacecraft attitude control [37], the attitude control with the unknown inertial [38], the fault-tolerant control [21, 23, 29, 39], the magnetic torque control [40], the multiple-fault diagnosis [34], etc.

Compared with the attitude control problem of a single spacecraft, the attitude control problem of the on-orbit service spacecraft faces more diverse engineering backgrounds. It faces more complex constraints, such as more precise control accuracy, more complex input and output constraints, motion safety zones, obstacle avoidance, precise pointing control, and more. Here, theoretical research results on practical problems related to attitude control of spacecraft serving in orbit in recent years are classified. For example, attitude control of serving spacecraft with considering the performance constraints and the physical constraints (detailed later), attitude control of non-cooperative targets [41, 42], takeover control [43, 44], hybrid actuator attitude control [45], attitude control for combined spacecraft or postcapture spacecraft [41, 46–52], fault-tolerant control [41, 51], game-based attitude control [53, 54], control allocation for spacecraft attitude control [55, 56], attitude control using line-of-sight observations [57, 58], spacecraft autonomous proximity [59] and so on. For the above-mentioned various control problems, the control methods often used in recent years include finite-time control [60], event-driven control [42, 46, 50, 61], neural-network-based adaptive control [46, 61], adaptive control [41], model-free control [47, 49] and so on.

In spacecraft attitude control engineering, there are many practical problems, such as control performance constraints, attitude angular velocity constraints, flexible vibration, input signal saturation, signal hysteresis, signal noise, excessive fuel or energy consumption, external disturbances, and so on. In the following sections, combined with the research content of this book, the research status of spacecraft attitude performance constraints, input signal processing, optimal or suboptimal control, differential games, and cooperative control problems are introduced.

1.2.3 Spacecraft Constraint Control

First, we introduce the constraints on the state/output of the spacecraft attitude system. In the spacecraft attitude control problem, it is often necessary to consider the motion boundary or the corresponding physical constraints. The physical constraints can be transformed into performance constraints of the system state, including various time-domain performance response indicators. The research status of control theory about performance constraints will be introduced in detail later. For spacecraft engineering, the performance constraints can be transformed into the state constraints of the spacecraft system, such as the attitude angle constraints and the angular velocity constraints during the attitude maneuvers. There are many research results on the constraint control design of the spacecraft attitude motion, as follows.

Reference [62] considers the actual physical constraints in spacecraft attitude control, that is, the attitude angular velocity constraints and the control input constraints, and designs the PID-like spacecraft attitude optimal controller. Reference [63] considers the state constraints of spacecraft attitude and proposes a spacecraft attitude constraint controller based on the reference governors and nonlinear model predictive control. Reference [64] considers the spacecraft attitude motion limited range and designs the spacecraft attitude controller based on quadratic programming in the constrained area. Reference [65] considers the spacecraft attitude state constraints and designs the model predictive controller with fixed-point constraints. Reference [66] considers the constraints of the inclusion and exclusion areas of the spacecraft attitude motion and proposes the constrained spacecraft attitude control algorithm based on the fast nonlinear model predictive control method. Reference [67] considers the spacecraft's attitude is constrained by certain boundaries and designs the desaturation attitude-constrained model predictive control algorithm. References [68, 69] consider attitude angular velocity constraints, control input constraints, attitude motion forbidden area, and attitude-related strict constraints, and respectively propose the time-optimal spacecraft attitude maneuver path planning and control algorithms under boundary and pointing constraints, and the energy-optimal spacecraft attitude maneuver path planning and control algorithm. Reference [70] considers the angular velocity constraints and designs the time-suboptimal control algorithm for the spacecraft's real-time attitude agile maneuvering. Reference [71] considers the attitude angular velocity constraints and designs the attitude tracking control algorithm for flexible spacecraft. Reference [72] considers the state constraints in the attitude maneuvering and designs the model predictive sliding mode control algorithm for the finite-time three-axis spacecraft attitude tracking. Reference [73] considers the multiple exclusion cone constraints so that the attitude motion cannot enter the excluded cone area. It considers the control actuator saturation and proposes the explicit reference governor approach for the spacecraft attitude control with nonconvex constraints.

Then, we consider the spacecraft attitude system constraints on the input signals. The actuators of the actual spacecraft attitude control system often face certain constraints, such as input amplitude constraints, frequency constraints, and so on. There are many research results on input saturation of the spacecraft attitude control.

Based on the variable structure control, Ref. [74] solves the problem of spacecraft attitude input saturation by designing a matrix that is left multiplied by the maximum control input and proves the stability of the closed-loop system through the Lyapunov stability theorem. Reference [75] solves the problem of control input saturation by directly designing the coefficients before the input upper bound. Simultaneously, it also analyzes the problem of controller saturation on the control rate. Reference [76] solves the saturation problem of the spacecraft attitude control actuator through the anti-windup control method. Reference [77] introduces the hyperbolic tangent function into the controller to limit the control input amplitude and solves the problem of control input saturation. Reference [78] directly applies the saturation function to deal with the control input's upper and lower boundaries and proves the closed-loop system's stability and the controller's effectiveness through boundedness analysis. Reference [79] considers the spacecraft attitude control problem with a variable input saturation upper bound and applies a non-singular normalization function with modulus constraints to solve the input saturation problem. For the problem of the flexible spacecraft attitude control with the unknown input saturation, Ref. [80] proposes intelligent proportional differential control, which effectively solves the problems of attitude control and input saturation. Reference [81] designs the normalized gain parameter, which can effectively deal with the problem of control input saturation. In Ref. [82], the controller is the product of the saturation upper bound value and the time-varying gain. This gain is a linear combination of normalized-function-like terms with variable parameters such that the control input always satisfies the constraints. Reference [83] proposes a robust attitude control method for spacecraft based on extended state observers, using quadratic programming techniques to overcome the actuator saturation problem. Reference [21] uses the hyperbolic tangent function to actively constrain the controller and designs the independent variable of the hyperbolic tangent function so that the spacecraft attitude closed-loop system can be stabilized within the finite time, and the algorithm also considers fault-tolerant control. Reference [84] proposes a robust backstepping attitude controller for spacecraft based on disturbance observers considering input saturation and measurement uncertainty. Reference [85] designs the anti-disturbance and actuator saturation spacecraft fault-tolerant attitude controller, in which the anti-saturation technique is applied to deal with the actuator saturation problem. Reference [86] considers both the control input amplitude and the rate saturation constraints and designs a spacecraft attitude control algorithm based on a disturbance observer. Reference [87] considers the control input amplitude and rate constraints simultaneously and proposes a flexible spacecraft attitude controller based on the disturbance observer. Considering the input saturation problem, Ref. [88] proposes an adaptive fault-tolerant attitude controller for satellite redirection. Reference [89] considers control input saturation and designs a finite-time attitude control algorithm for rigid body spacecraft. Reference [90] considers the control input saturation and attitude constraints and proposes an attitude control algorithm suitable for rigid-body spacecraft. Reference [91] designs a constrained dual-loop controller for attitude control of rigid spacecraft with inertial uncertainty, specified angular velocity, and control constraints, in which a command filter is applied so that the controller effectively satisfies the control input constraints.

1.2.4 Spacecraft Optimal Control

The optimal control of spacecraft attitude has received significant attention since the beginning of the aerospace industry. Due to the limited fuel or energy carried by spacecraft, in the process of spacecraft attitude control, most of the spacecraft energy consumption is expected to be optimal. Considering specific and/or unique task requirements, time-optimal control [92, 93] or time-fuel optimal control may also be desired in some tasks. Reference [94] designs the quadratic performance index on the control input and defines the Hamiltonian function and the Lagrange multiplier. By solving the differential equation, the solution that makes the performance index function optimal is obtained.

In Ref. [95], the quadratic performance index about the system state and control input is designed. As in the previous references, the optimal control solution of the spacecraft attitude maneuver is obtained. Reference [96] designs the quadratic performance index about attitude parameters and attitude angular velocity and solves the optimal control solution through the Hamilton-Jacobi equation. In Ref. [97], the quasi-quadratic performance index considering the terminal time is designed, and the penalty function is added to the index. Based on this performance index, the spacecraft attitude optimal controller is proposed. In Ref. [98], the related theory of Lie group optimal control is proposed, and it is applied to spacecraft attitude control. Reference [99] defines the quadratic performance index, considering the state at the terminal time, control input, and related variables, and designs the spacecraft adaptive attitude optimal controller. Reference [100] defines the quadratic performance index about the system state and the control input and obtains the optimal control solution of the spacecraft attitude tracking by solving the Hamilton-Jacobi equation. Reference [101] uses explicit model predictive control to design spacecraft attitude control with the system state minimization. In Ref. [102], the performance index of the state and the control input is designed, and the optimal controller for spacecraft attitude tracking based on sliding mode is designed. Based on the nonlinear optimal control, Ref. [103] designs an attitude-orbit optimal controller during the approaching process of the spacecraft. In Ref. [104], the global stability of the spacecraft attitude control based on quaternion is considered, in which the quadratic index function of the attitude parameters and attitude angular velocity is considered. Reference [105] considers the energy optimization problem of spacecraft attitude control constrained by the convergence rate. Reference [106] studies the nonlinear optimal control problem when the spacecraft approaches the rolling target. In Ref. [62], the quadratic performance index function about spacecraft attitude, attitude angular velocity, and control input is designed, and the constrained optimal PID-like control method for spacecraft attitude is proposed. Reference [107] uses the control Lyapunov function (CLF) method to design the spacecraft attitude optimal sliding mode controller. In

Ref. [108], the optimal output feedback controller for spacecraft attitude tracking is designed based on the observer. Reference [109] designs the null-space-based optimal control reallocation scheme for spacecraft under input saturation. Based on the extended state observer and adaptive dynamic programming, Ref. [110] designs the spacecraft output feedback attitude optimal controller, in which the adaptive neural network is used to fit the solution of the Hamilton-Jacobi-Bellman equation to obtain the optimal control input. Similarly, [111] also applies adaptive dynamic programming to solve the corresponding optimal control issues.

Most existing optimal controllers only consider the control itself and can achieve the energy-optimized control objective. However, in these optimal control designs, the closed-loop systems are generally not considered to satisfy the strict constraints. This means that, for optimality, transient or steady-state performance may not be considered. For the spacecraft attitude maneuver in the orbit service mission, we prefer to achieve the optimal control that satisfies the desired constraint conditions. This is also a design motivation in this book.

1.2.5 Spacecraft Differential Games

Game theory provides us with fundamental concepts and elements such as strategy, game value, pure strategy, or mixed strategy for multiple players [112, 113]. Finding the optimal strategy for a multi-player system becomes the main objective of the game problems. Due to the pursuit of the optimal strategy of multi-participants in game theory, the application prospect of game theory in multi-spacecraft missions has been brought.

Faced with the relative motion of two or more spacecraft, game theory can provide optimal decision-making or strategy for multi-spacecraft control rather than simply tracking control. This means that in the on-orbit service missions, the service spacecraft is no longer simply tracking the cooperative or non-cooperative target spacecraft. Instead, a set of optimal or suboptimal strategies for the players themselves is devised through game theory. Furthermore, the expected relative motion objectives of the multi-spacecraft system can be realized, and the system's overall energy consumption can be reduced.

For the continuous-time multi-spacecraft relative motion system and the control system of each spacecraft, Differential Game theory [113, 114] provides the game strategy solution method in the continuous-time domain. On the other hand, based on the Nash Equilibrium theory [115], the differential game theory helps to construct the Nash equilibrium state of the continuous system among the multiple spacecraft [116–118]. Under Nash equilibrium, each individual/participant can achieve the current optimal solution for its control strategy under strong coupling with other individuals/participants. The control input of each individual/participant is optimal under the current conditions for the cooperative system. References [119–127], etc., provide us with the game problem-solving methods based on adaptive dynamic programming and the basic frameworks for the game control design.

For the spacecraft differential game, the research results in recent years are mainly divided into two aspects: non-cooperative games and cooperative games. The spacecraft non-cooperative game is mainly researched in the space non-cooperative debris removal and non-cooperative target relative motion control. Among them, there are many studies on spacecraft pursuit-evasion games. Reference [128] addresses the optimal interception problem of an optimally evasive orbital target by a pursuing spacecraft. Reference [129] studies the basic spacecraft pursuit-evasion game problem. Reference [130] solves the problem of spacecraft pursuit-evasion trajectory optimization under the Euler–Hill reference frame. Reference [131] designed the hybrid spacecraft on-orbit pursuit-evasion game algorithm. Reference [132] studies control algorithms for pursuit-evasion games using only angular navigation. Reference [133] considers the problem of the pursuit-evasion game between two spacecraft in low orbit and gives the numerical solution. Reference [134] applies the state-dependent Riccati equation method to solve the spacecraft pursuit-evasion game problem. Reference [135] uses reachable sets to study the problems of spacecraft rendezvous and pursuit. Reference [136] addresses the dimension-reduction solution of free-time differential games for spacecraft pursuit-evasion. Reference [137] addresses the saddle point of the orbital pursuit-evasion game under J2-perturbed dynamics. Reference [138] addresses spacecraft pursuit-evasion game with different thrust configurations. Reference [139] studies the switching strategies in the pursuit-evasion game problem. Reference [140] designs the escape strategy under incomplete information. Based on the differential game theory, the H_∞ terminal rendezvous algorithm for spacecraft in an elliptical orbit is designed in [141]. Reference [142] considers the target behavior uncertainty and designs a non-cooperative game control algorithm. Reference [143] provides the non-cooperative game algorithm design for spacecraft attitude output regulation. Reference [144] further considers the fault-tolerant control problem for the zero-sum differential games. For the spacecraft actuator, [145] designs the fault-tolerant control algorithm based on the zero-sum game. Reference [146] is oriented to engineering practice, considering ΔV and spacecraft orbit analysis. The pursuit-evasion game algorithm based on ΔV is designed based on the game theory and objective function analysis.

The cooperative spacecraft game, also often called the spacecraft non-zero-sum game, is primarily studied in the optimal decision of spacecraft formation flight, cooperative attitude control, cooperative relative motion, etc. The relevant research results are briefly introduced here. For the spacecraft rendezvous and motion synchronization, Ref. [147] uses the state-dependent Riccati equation (SDRE) method to solve the linear quadratic differential game problem and obtains the corresponding control algorithm. Under this control framework, the multi-spacecraft systems achieve optimal rendezvous and motion synchronization. Reference [148] proposes the takeover control for failed spacecraft based on the linear quadratic differential game approach. Among them, multiple small satellites participate in the cooperative game to jointly complete the attitude takeover control of the failed satellite. Reference [149] designs the distributed cooperative game attitude takeover control algorithm. Reference [150] designs the cooperative game algorithm for modular spacecraft. Based on the cooperative game method, Ref. [151] designs the optimal angular momentum management algorithm for the hybrid spacecraft attitude control actuator.

In the above research results, the algorithms mainly focus on the design of the game strategy and the control algorithm itself. The performance of the closed-loop system mainly depends on the parameter design of the game algorithm. The performance guaranteed or motion constraints of the closed-loop systems should be discussed more. In this book, this problem is concerned. While the performance is guaranteed, this book focuses on the relative motion cooperative game between the multiple spacecraft and the cooperative control of the assembled/combined spacecraft.

1.2.6 Spacecraft Cooperation

For multi-spacecraft or related service missions, the issue of cooperation between multiple spacecraft needs to be considered. More generally, the multi-spacecraft system can be abstracted as the multi-agent system. Research on multi-agent systems is more extensive and adaptable. Moreover, most of these research results can be applied to spacecraft control.

The multi-agent systems (or networked agent systems, etc.) [152, 153] can well explain the behavior of biological groups in nature and has a wide range of applications. Since the multi-agents can emerge the group collaborative intelligence at the system level that surpasses the intelligence capabilities of a single agent, in practical applications, it has many advantages and characteristics, such as:
(1) Low cost: due to the relatively simple structure and single function of a single agent, it is often challenging to complete specific complex tasks, and the multi-agents can replace the single complex agent to achieve functions that it cannot have;
(2) Strong survivability: when one or part of the agents in the multi-agent system fails, the configuration of the multi-agent system can be optimized and reconstructed to continue to maintain the operation of the system without affecting the completion of the networked system tasks and overall functionality;
(3) High flexibility: it can flexibly build and adjust the team and structure of the multi-agent system in real-time to complete different tasks and realize differentiated functions.

The distributed cooperative control of the multi-agent system refers to the formation of regular and orderly coordinated movements of multiple agents through distributed cooperation, which contains many meaningful topics, such as consensus and synchronization [152, 154–157], flocking [158–160], rendezvous [161, 162], tracking [163–165], formation [166–169], containment [170, 171], circumnavigation [172–174], distributed problem solving [175, 176], distributed learning control [177, 178], etc.

For multi-spacecraft systems, multi-agent systems can reduce the cost of a single spacecraft, improve fault tolerance, expand application range, and perform multi-spacecraft formations. In this book, we focus on the problem of multi-agent formation constraint and containment control applicable to spacecraft or multi-agent systems.

The distributed formation control of the multi-agent system refers to the design of the communication topology network and the distributed controller so that the state of the networked agent can follow the desired formation configuration, and maintain or adjust the configuration over time, to achieve the needs of the practical multi-agent tasks, such as environment exploration, obstacle avoidance, etc. We can classify most of the existing formation control problems by defining the desired formation configuration, such as Displacements constraints, Distances constraints, Ratio of Distances constraints, Bearings constraints, Barycentric Coordinates constraints, Complex Laplacian Matrix constraints, Affine Laplacian Matrix constraints, etc. The desired configuration can be fixed or dynamically changed. As described in [179], each constraint method can maintain the 'invariance'. The most common configuration is defined by relative positional information between agents, also known as 'displacement' or 'bias'. This kind of constraint method can maintain its own 'invariance' when the formation shifts, will not cause the degradation or deterioration of the maneuver formation configuration, and can be applied to multi-agent tasks where the formation moves freely in space. For example, Ref. [180] proposes the potential function-based formation controller for a group of mobile agents with limited sensing ranges, which can avoid collisions while driving multi-agents to achieve the desired configuration. In this reference, a method to analyze the stability of the closed-loop systems is also proposed instead of Barbalat's lemma. Similar to the tracking control of the multi-agent systems, scholars also apply formation behavior to Euler–Lagrange systems. Reference [181] studies the multi-agent formation control problem described by the Euler–Lagrange systems. The adaptive controller proposed in this reference guides the agents to form the desired configuration through the potential functions and combines the good transient characteristics and robustness of the sliding mode control technique. The definition method based on the distance constraints can maintain its own 'invariance' when the formation is translated and rotated. For example, the distance-based controller proposed in [182] can make the multi-agent systems form the desired formation configuration based on the two-dimensional space body coordinate system. In [183], the authors use the single-view distance estimation method to propose the practical formation movement control scheme for a group of quadcopters. Among them, the agents neither need to communicate with each other nor need global information. The definition method based on the orientation constraints can maintain its own 'invariance' when the formation is translated and scaled. For example, the algorithm in [184] can realize the formation control of the dual-integral multi-agent systems in any dimension and solve practical problems such as input disturbance and obstacle avoidance. In addition to the above definition methods, there are also methods based on the distance ratio [185], the barycentric coordinates [186], the complex Laplace [168, 187], the affine Laplace [179, 188] and so on to describe the constraints of the formation configurations. They can guarantee the 'invariance' of the constraints in various formation transformations.

There is a special kind of multi-agent formation problem, also known as the 'containment problem', that is, first control some agents to form the desired formation configuration, which we call these agents the leaders, and then control the rest of the agents to remain within the formation structure formed by the leaders, which we call these agents the followers. Among them, the navigator usually has strong decision-making and planning capabilities. Since the follower is in the formation area, it can be equipped with less navigation equipment to achieve specialized functions. In this problem, the actual leader can also be replaced by a virtual dynamic trajectory or area constraints to form the desired formation area. From this perspective, the containment problem can also be considered a motion constraint or performance guaranteed problem.

At first, Refs. [189, 190] adopt the 'Stop-Go' strategy to control the followers, which can drive the followers to move into the formation structure surrounded by the leaders. Reference [191] introduces the region-based shape control for a swarm of robots. The swarm robots move as a whole in the desired area. Simultaneously, the interaction of distributed information is used to maintain the relative minimum distance between each agent. In the time-varying undirected topology, Ref. [192] designs the control strategy so that the followers can all converge to the formation area formed by the static leaders. The system's stability is analyzed based on LaSalle's invariance principle of the switched system. Considering the dynamic navigator, Ref. [193] designs the distributed containment control algorithm under the undirected graph for a group of unmanned vehicles modeled by double-integral dynamics under the condition of missing measurement information and only using local information of the agent. Subsequently, scholars have analyzed such formation problems from different perspectives. Under the directed communication graph, Refs. [194, 195] design the distributed containment controllers based on the output feedback for the linear time-invariant systems. Reference [196] considers the communication delay of the second-order multi-agent system in the formation-containment problem. Reference [197] proposes the formation-containment control algorithm for second-order multi-agent systems under undirected graphs under the condition that only sampling position information is used. Considering the coupled heterogeneous linear systems, [198] proposes the output formation-containment controller under intermittent communication. The formation-containment flight problem of the multiple unmanned aerial vehicles is studied in [199], and the proposed controller can be applied to both fixed and time-varying topological graphs.

For more research progress on the formation coordination problems of the multi-agent systems in recent years, related reviews can be referred to, such as [200, 201].

1.3 Performance Guaranteed Control

1.3.1 Problem Formulation of Performance Guaranteed Control

The time domain analysis of the control system is the method to directly analyze the system properties in the time domain, which has the advantages of intuition and accuracy. It can provide all the information on the system's time response, which reflects the control system's performance. The control system performance evaluation is divided into transient (dynamic) performance and steady-state performance indicators. The *transient performance indicators* include settling time (delay time, rise time, peak time, adjustment time, etc.) and overshoot. These time parameters are used to evaluate the response rate of the closed-loop system, and the overshoot is used to evaluate the damping of the system. The *steady-state performance indicators* mainly refer to the steady-state errors of the system. The performance of the closed-loop system can be further revealed by the time-domain performance response indicators of the system [4].

In the traditional control design, after the controller is designed, it is usually necessary to use typical input signals (such as unit step function, unit ramp function, unit acceleration function, etc.) to measure the performance indicators. In order to obtain the time-domain response performance indicators of the closed-loop system under specific task conditions, the closed-loop system performance indicators under the action of the designed controller are usually obtained through computer numerical simulation and physical experiments. However, such a control design requires extensive parameter selections, simulations, and physical tests and may require more complex designs to improve the robustness of the closed-loop system.

Compared with verifying the controller's performance by parameter selections, simulations, and calculation, if the controller can be designed after adding constraints on the performance indicators, the practical performance of the closed-loop system can theoretically meet the performance constraints. Then, such the closed-loop system satisfies the expected performance constraints, that is, the practical performance indicators meet the expected performance indicators, the performance also can be further improved through parameter design, and the *performance guarantees* of the closed-loop system is theoretically given. Then, as long as reasonable performance indicators and constraints are designed according to the practical task requirements, the closed-loop system performance always satisfies the expected performance constraints. Simultaneously, different constraints can be imposed on the time domain response of the closed-loop system state or error, and the transient and steady-state performance indicators of the closed-loop system can be actively adjusted. The controller or control algorithm that meets the above characteristics can be regarded as the *performance guaranteed control* method.

The model reference adaptive control problem was proposed in [202]: forcing the system error to be less than an (arbitrarily small) prespecified constant after an (arbitrarily short) prespecified period of time, with an (arbitrarily small) prespecified upper bound on the amount of overshoot. This achievement provides the method to actively constrain the transient and steady-state performance of the system by designing performance indicators. Simultaneously, the closed-loop system settling time can also be indirectly constrained by setting piecewise constraints. Relying on strong performance constraints on system states or errors, the performance guaranteed design methods depend only on desired control metrics. They are not directly related to system and controller parameters. That is to say, if the closed-loop system under the action of the performance guaranteed controller can satisfy the performance constraints at all times, then no matter how the controller parameters or system parameters change within a reasonable range, the closed-loop system's settling time, overshoot and steady-state errors can satisfy the desired performance.

Broadly, the method of adding strict constraints on system states or outputs, called *state constraints* or *output constraints*, is also a class of performance guaranteed methods. These methods adopt non-decreasing constraints, pay more attention to the maximum feasible boundary of the system state, and relax the strict guarantee on the convergence performance. Better performance guaranteed about system state or output ranges can be achieved. More generally, the formation behavior of networked systems also has the requirement of performance guaranteed control. Specifically, looking for a control method that guarantees the boundary of formation configuration can achieve the desired performance guaranteed formation motion. Some individuals in the formation form the desired configuration, and other individuals move within this configuration and are guaranteed not to escape from the range of this configuration. Such a control method in the above scenario is called *formation-containment control*. According to different scenarios, the formation configuration responds to changes, and other formation individuals move in the configuration, which can not only meet the performance guaranteed and performance constraints but also avoid obstacles or regional boundaries. Therefore, formation-containment control can also be considered a multi-constraint control and performance guaranteed control for networked formation systems. This method has been mentioned in Sect. 1.2.6 and is analyzed and designed in detail in Part III of this book.

In the following sections, an overview of the research results is given based on the development of performance guaranteed class methods.

1.3.2 Performance Constraint Control with Constant Value Constraints

Based on the idea of the performance guaranteed control, some research results using constant-valued-constraints to limit the closed-loop system errors gradually appeared [203–209].

The research result [203] proposed the robust adaptive control with considering the constant constraint of steady-state error for a typical multi-input-multi-output nonlinear system. Specifically, the predefined small constant is used as the steady-state error boundary of the closed-loop system, and the switching function about the steady-state error constraint is designed. When the closed-loop system state converges to the steady-state error interval, it always remains within the steady-state error constraint interval. The control objective of making the steady-state error more diminutive than the predefined constraint parameters can be achieved by the controller. Based on [203], the robust adaptive controller based on a neural network is designed for the single-input-single-output nonlinear system in [204]. The same as [203], the switching function is applied in this work so that the steady-state error of the closed-loop system can converge to the predefined steady-state error range. References [205, 206] design the adaptive robust sliding mode controllers for multi-input-multi-output systems with uncertain nonlinearity so that the steady-state errors of the closed-loop system can converge to the predefined constant constraint ranges. However, these methods only consider the constraints of the steady-state errors, that is, the boundary guarantees for the steady-state errors, and does not fully consider the transient performance constraints. The transient performance of closed-loop systems often has substantial uncertainties. Therefore, in the case of more precise control requirements, these methods are insufficient to constrain the states or errors. References [207, 208] employ constant-value constraints to ensure that the system output is always within the constraints. However, the constraints in these references are too broad, so it can only be ensured that the system output does not escape the constraints.

Reference [209] applies piecewise constraints. That is, the state of the closed-loop system must satisfy the constant constraints within the time T when the control starts. When the time is greater than T, the closed-loop system state needs to satisfy another smaller constant value constraint so that there can be multiple constant constraint segments. Different from the Refs. [203, 204], such a constant piecewise constraint makes the closed-loop system state always need to satisfy the constraints in the entire control process, and the constraint range for the error is decreasing. Therefore, both transient and steady-state performance can be guaranteed within a specific decreasing range. In addition, such constraints can be customized. Therefore, the results in [203, 204] that can only constrain the steady-state error are extended to the method that can constrain both transient and steady-state performance. The piecewise/segmented constant performance constraints may cause control abrupt input changes. In addition, the existence of a switching function may also easily cause signal chattering. Considering the above problems, the constant value constraints of piecewise decreasing can be continuously refined and transformed into constraints of time-varying continuous decreasing functions. In this way, the restraint ability and control performance of the controller can be improved.

The state-constrained control based on the barrier Lyapunov function (BLF) is also a performance guaranteed method [210–221]. Its purpose is to provide constant constraints on the system states or errors so that the states or errors always operate within the predefined constraint region.

Different from the aforementioned error constraint control, the constant value constraint range adopted by the BLF method is large, and it is not necessary to set a small range to constrain the steady-state error. In addition, the constant value in the constraint condition needs to be larger than the initial errors, and the constraint guarantees that the error does not exceed a predefined range. Reference [210] considers the constrained control problem of nonlinear systems based on the symmetry barrier functions proposed by [211]. The barrier Lyapunov function is in the form of $V = \frac{1}{2}\log\frac{k_b^2}{k_b^2-e^2}$, where e represents the variable to be constrained, the positive constant k_b is the constraint range of the variable e, $k_b \geq ||e(0)||$, and $\log(\cdot)$ is the natural logarithmic function. Then, its derivative with respect to time is $\dot{V} = \frac{e\dot{e}}{k_b^2-e^2}$. By $\dot{e}$, the candidate Lyapunov functions can be associated with the system to design performance-constrained controllers. There are related function items in the controller similar to V. From the barrier Lyapunov function form, it is not difficult to find that when the norm of e is close to k_b, V will increase. Therefore, the control effect will increase so that the error tends to be far away from the constraint boundary. Reference [212] applies the barrier Lyapunov function to design a full-state constrained controller for higher-order nonlinear systems. Correspondingly, the Ref. [213] proposed an adaptive neural network output feedback performance-constrained controller based on the barrier Lyapunov function. Reference [214] considers some state constraints and proposes a nonlinear controller based on the barrier Lyapunov function, which partially relaxes the performance constraints. Simultaneously, the guaranteed/constrained-performance control technique based on the barrier Lyapunov function has been further developed, such as the constraint control for nonlinear switching systems [215, 216], the adaptive fuzzy control with input and output constraints [217]. In addition, the performance guaranteed control based on the barrier Lyapunov function has also been applied in many aspects, such as the ships[218], the hypersonic vehicles [219, 220], the circuit systems [221], etc.

1.3.3 Funnel Control

Considering the sufficient constraints on the closed-loop system states or tracking errors and the continuously decreasing function constraints, related efforts have emerged that use time-varying decreasing functions to strongly constrain the system states or tracking errors, such as the funnel control [222–243], the prescribed performance control [6, 244–246], etc. These are all guaranteed/constrained-performance control methods that impose constraints on the states or tracking errors of the closed-loop system through a continuous function so that the states or tracking errors are always within the constrained ranges. In this subsection, we first introduce the basic principles of funnel control.

Specifically, the funnel control constrains and guarantees the transient and steady-state performance of the closed-loop system states or tracking errors through an arbitrary time-varying constraint function. The constraint range described by this function needs to contain the initial value of the state or error. In the funnel control framework, the distance from the system state or error to the constraint boundary is taken as the primary function term, which causes the constraint condition to be shaped like a funnel in the coordinate domain, called funnel control. The basic framework of the funnel control is designed in [222] for chemical reaction models. Specifically, considering the task objective that both transient and steady-state performance are constrained, the control algorithm is proposed in [222] as $u(t) = \text{sat}_{[u_{\min}, u_{\max}]}(-k(t)e + u^*)$, where e denotes the error to be stabilized, $\text{sat}_{[u_{\min}, u_{\max}]}(\cdot)$ represents the saturation function of the control input with upper and lower bounds $[u_{\min}, u_{\max}]$, u^* is the auxiliary controller, which needs to be designed according to the specific situation, and $k(t)$ is the time-varying funnel control gain related to the error e and the distance of the error to the funnel boundary. As a simple case, the design $k(t) = \text{dist}(e, F_t)$ can be considered, where F_t represents the predefined funnel boundary condition. Considering that F_t consists of two boundaries $\phi_1(t) = \phi_2(t) = \phi(t) > 0$, then the feedback control can be written as $u(t) = \text{sat}_{[u_{\min}, u_{\max}]}(-e/(\phi - |e|) + u^*)$. It can be seen from the controller form that when the error e is close to the boundary value ϕ, the norm of the function term $-e/(\phi - |e|)$ will increase, and the control effect will increase accordingly to achieve stronger of negative feedback control. The controller makes the closed-loop system reduce the error and keep the controlled error variable away from the boundary value. When the error value is far from the boundary value, the norm of the function term $-e/(\phi - |e|)$ decreases, and the controller makes the closed-loop system gradually stabilize. Therefore, the closed-loop system error can be constrained by the funnel formed in space by the distance between the errors and the constraints so that the errors can always remain within the constraint range. This approach constrains not only steady-state performance but also transient performance.

Reference [223] considers parameter uncertainties and designs the state feedback control with funnel constraint gain $u(t) = -k(t)e(t)$. Among them, the time-varying gain $k(t)$ is similar to the aforementioned funnel control gain, which also considers the relationship between the errors and the constraint boundary distances. When the distance becomes smaller, the gain becomes larger, and when the distance becomes larger, the gain decreases. Simultaneously, the issue of closed-loop system settling time is further mentioned in this reference by crossing fixed points related to time and error values. Reference [224] designed a funnel control for nonlinear multi-mass flexible systems. As in the previous references, the design focuses on selecting the time-varying gain, which is still related to the error and the distance of the constraint boundary. In [224], a constraint function as $\partial F_{exp}(t) = \varphi_{exp,0} \exp(-t/T_{exp}) + \varphi_{exp,\infty}$ is designed, where $\partial F_{exp}(t)$ represents the time-varying constraint function, $\varphi_{exp,0}$ and $\varphi_{exp,\infty}$ are the initial and steady-state error constraints, respectively, T_{exp} is the expected settling time constraint. Reference [225] improved the performance of the original funnel control. Specifically, in the original funnel control, the distance between the error and the boundary is the difference between the positive definite constraint value and the error norm. After the improvement, the distance between the error and the boundary is redefined as the distance obtained by the error point at the current moment as a vertical line orthogonal to the constraint boundary. In this way, a more reasonable control gain can be obtained, and the system performance can be improved. Reference [226] summarizes the main achievements of the funnel control above and designs a proportional-integral (PI) funnel control for a dual-mass system, in which the time-varying gain associated with the funnel constraint ensures that the system error always satisfies the constraints. References [227, 228] design the funnel controllers considering input saturation for linear multiple-input multiple-output systems and nonlinear single-input single-output systems, respectively. The constraining effect of the funnel boundary formed by the performance constraints is still reflected in the time-varying gain.

Regarding funnel control, it has been continuously deepened and expanded in theory, such as funnel control for linear differential-algebraic systems [229], funnel control for systems with relative degree two, higher relative degree systems or arbitrary relative degree systems [230–236], adaptive fuzzy funnel control [237–239], event-driven adaptive funnel control [240]. In addition, funnel control is also used in practical applications such as rigid rotary joint manipulators [241], motor servo systems [242], underactuated multibody systems [243], etc.

1.3.4 Prescribed Performance Control

The aforementioned traditional funnel control mainly considers the distance between the error and the constraint boundary. That is, the control is related to the error norm. On the other hand, for the primary funnel control, since the error-constraint norm is used, the positive and negative constraints on the error tend to be symmetric. Therefore, it is difficult to solve the problem of constraining the overshoot of the closed-loop system. Of course, if the positive and negative constraints of the system error are designed separately, the problem of overshoot constraints can be solved. However, the judgment conditions and switching functions may need to be included. In order to solve the complex performance constraint problem, the prescribed performance control (PPC) is proposed by scholars [6, 244]. In the prescribed performance control, the positive and negative constraints of each element of the system error are designed separately, and different parameters are used to deal with the constraint problems of the convergence rate, overshoot, and steady-state error. These constraint-related parameters provide great flexibility for performance constraints. On the other hand, through the error constraint mapping, the inequality constraint problem can be transformed into the stability problem of the auxiliary system described by the equation. Therefore, traditional control design methods are more convenient for performance-constrained control design. In the prescribed performance control algorithm, the performance constraints and parameter design, as well as the error transformation method, give the algorithm a more flexible application mode and a more complex description ability of constraints.

References [6, 244–246], as the first essential results on the prescribed performance control method, provide the basic technique of the PPC method. Like the aforementioned steady-state error constraint control, BLF-based constraint control, funnel control, etc., the PPC also needs to design the performance constraint function first. This performance constraint function needs to absorb the advantages of the aforementioned performance constraint methods. Furthermore, the customizable design for complex constraints is fully considered in PPC. The performance constraint function is defined as a monotonically decreasing positive definite time-varying function, which can be chosen as $\rho(t) = (\rho_0 - \rho_\infty)\exp(-lt) + \rho_\infty$. In this formula, ρ_0 represents the initial value of the performance constraint function; ρ_∞ represents the expected performance constraint function value when $t \to \infty$, which means that this parameter is adjusted to actively set the custom steady-state error range; l is a constant coefficient, reflecting the slope of the performance function, which means that this term is related to the convergence rate of the closed-loop system state or error. The performance constraint function exemplified here is an exponentially convergent monotonically decreasing function that can be used to constrain and decrease a variable. For system state or error variables, according to the performance constraint function, it is easy to write the relevant performance constraints, such as when $e(0) > 0$, $-\delta\rho(t) < e(t) < \rho(t)$, when $e(0) < 0$, $-\rho(t) < e(t) < \delta\rho(t)$. In the constraints, $e(t)$ represents the system variable to be constrained, such as state or error, and $\delta \in [0, 1]$ is a constant parameter. It is easy to know that when $\delta = 0$, the system is not allowed to overshoot according to the constraints. The reason for distinguishing the sign of the initial value of the constrained variable is mainly to limit the overshoot. If the initial value of a variable element to be constrained is positive, the overshoot direction is in the negative direction. The overshoot can be limited by setting the constraint range in the negative direction by δ. For multi-input-multi-output systems, the constraints described by the performance constraint function can be imposed on each system state or error element. This means that each element of the system state or error can be individually bounded by performance constraints, which is a great advantage of PPC.

On the other hand, the error transformation method is adopted in PPC, which maps the constraint range to the entire state space through a bounded function. Specifically, considering that the constraints are a set of inequalities, it is challenging to directly exploit the inequalities to design the controller. Therefore, the bounded time-varying function with the same value range as the inequality is designed to be equal to the variable constrained, such as the system state or error. The independent variables of the time-varying function can be obtained by calculating the inverse function of the time-varying function. It is easy to know that when the independent variable is bounded, the value of the time-varying function is always within the range of constraints. When the independent variable parameters are unbounded, the time-varying function approaches the constraint boundary. Therefore, the original inequality constraint problem is transformed into a new problem of the boundedness of independent variables. The traditional Lyapunov method can deal with the variable boundedness problem and the controller design problem. Therefore, the complex problem of performance-constrained controller design is further simplified.

Hereinafter, the basic principle of error transformation in PPC will be explained. Specifically, taking the initial state or error is greater than 0 as an example, divide both sides of the inequality of the constraint condition by the non-zero and positive definite function $\rho(t)$, and we can get $-\delta < e(t)/\rho(t) < 1$. Take $S(\varepsilon) = e(t)/\rho(t)$, then if $-\delta < S(\varepsilon) < 1$ is satisfied, then the system state or error satisfies the performance constraint, where ε is the independent variable of the function $S(\cdot)$. Considering the value range of the function $S(\varepsilon)$, the function can be designed as

$S(\varepsilon) = (\exp(\varepsilon) - \delta \exp(-\varepsilon))/(\exp(\varepsilon) + \exp(-\varepsilon))$. Next, find the independent variable ε, and one has $\varepsilon = S^{-1}(e(t)/\rho(t))$, then the derivative of ε with respect to time is $\dot{\varepsilon} = \frac{\partial S^{-1}(e(t)/\rho(t))}{\partial e(t)} \dot{e}(t) + \frac{\partial S^{-1}(e(t)/\rho(t))}{\partial \rho(t)} \dot{\rho}(t)$. Through $\dot{e}(t)$, the system can be substituted into the previous formula to obtain an auxiliary system with an explicit control input. Design the controller such that ε is bounded and the closed-loop system is stable. Furthermore, the closed-loop system state or error satisfies the performance constraints described by the performance function $\rho(t)$ and the overshoot parameter δ. The closed-loop system state or error has the desired convergence rate, overshoot, and steady-state error. When the initial state or error value is less than 0, the design process is similar. For the definition of specific parameters, please refer to the Refs. [6, 244–246].

Since then, the theoretical research on PPC has continued to deepen and expand. Reference [247] further expands the relevant theory of the PPC method and designs the robust prescribed performance controller with partial state feedback for cascaded systems with uncertain nonlinearities. Reference [248] designs the unobserved adaptive prescribed performance output feedback control based on the neural networks for the uncertain systems. Reference [249] designs an output feedback controller based on an adaptive neural network for uncertainty multi-input-multi-output nonlinear systems. Reference [250] proposes the low-complexity prescribed performance feedback control framework. Based on the original prescribed performance function, the framework uses an auxiliary function based on the logarithmic function, which simplifies the error transformation in the original PPC framework. Further, it simplifies the conversion process of the PPC method. Based on this, the approximation-free performance constraint control is designed. Reference [251] considers PPC with input saturation situation and designs PPC based on the adaptive neural network. In [252], the time delay problem in PPC is further considered, and the solution based on adaptive fuzzy output feedback control is proposed. Reference [253] considers the PPC problem of the stochastic nonlinear systems with input saturation and proposes the observer-based adaptive fuzzy performance-constrained control method. Reference [254] proposes the simplified adaptive neural network PPC. Reference [255] proposes the prescribed performance output feedback adaptive neural network tracking control for time-delay stochastic nonlinear systems. Reference [256] proposes the adaptive robust fault-tolerant performance-constrained control based on the PPC method. Reference [257] additionally discusses the situation where the system state or error is equal to the performance constraint boundary. In addition, in theory, the PPC method is applied in various situations, and many corresponding results are derived, such as, model-free or model-indeterminate PPC [258–260], low-complexity prescribed performance feedback control [261], PPC with unknown control direction [262], PPC for multi-agent system [163, 263–265], PPC for switch systems [266–268], prescribed performance fuzzy control [257, 269–272], prescribed performance fault tolerance control [273, 274], PPC for discrete-time systems [275], event-driven PPC [276, 277], etc.

In applying PPC methods, fruitful results have been achieved in recent years. Reference [278] designs the force and position performance constraint controller based on an adaptive neural network for robot contact control and applies performance constraint control to the occasion of robot control. Reference [279] solves the performance-constrained joint-position tracking control problem of the model-free robot using the PPC method. Reference [280] designs the prescribed performance tracking control for the flexible joint robot with unknown dynamics and variable elasticity. Reference [281] designs the adaptive prescribed performance motion control for a servomechanism with friction compensation. In addition, PPC is also widely used in hypersonic vehicles [282, 283], near space vehicles [284], aviation vehicles [285, 286], surface vehicles [287–289], unmanned ships [290, 291], underwater vehicles [292, 293], robot control [294, 295], robotic arm control [296–298], nonlinear servo systems [299], suspension systems [300, 301], network system [302], etc.

1.3.5 Finite-Time Prescribed Performance Control

In the traditional prescribed performance control (PPC), the convergence rate of the closed-loop system is regarded as the primary performance index to be guaranteed. This index partially describes the convergence performance of the closed-loop system but does not explicitly include the settling time parameters. With the continuous increase of transient performance requirements for control tasks, many scholars have begun to pay attention to the time parameters in prescribed performance control problems. Considering that the settling time parameter is involved in the finite-time control theory, some scholars have used the finite-time stability theory to solve the time parameter explicit problem in the prescribed performance control. Therefore, the finite-time prescribed performance control (FTPPC) problem is derived. Unlike the pure finite-time stability problem, the FTPPC focuses on performance constraints/guaranteed conditions and the finite settling time of the closed-loop system. Simultaneously, finite-time stability can also improve the transient performance of the closed-loop system.

Here we briefly analyze the main ideas of the FTPPC method. According to the traditional PPC method, the system model and constraints can be transformed into an auxiliary system with the independent variable $\varepsilon(t) = S^{-1}(e(t)/\rho(t))$ related to the performance constraints as the system state, where $S(\cdot)$ is a function with the same value range as the constraint condition of $e(t)/\rho(t)$, and $e(t)$ and $\rho(t)$ are the state or the error to be constrained and performance constraint functions, respectively. Secondly, for the auxiliary system, the finite/fixed-time controller for the state $\varepsilon(t)$ is designed to achieve the finite/fixed-time convergence of the closed-loop system. The original system state or error always satisfies the constraints and converges. Considering that the auxiliary variable $\varepsilon(t)$ has a deterministic relationship with the original system state or error $e(t)$, therefore, when $\varepsilon(t)$ can converge within the finite time T, it means that $e(t)$ can also converge within T. In this way, constrained-performance control objectives and finite-time stability of the closed-loop system are simultaneously achieved.

Reference [303] designs the finite-time formation control with prescribed performance constraints for the multiple quadrotors. Among them, through the error transformation method in the traditional PPC method, the auxiliary variables about the system error and constraints are obtained. Through the Lyapunov method, the finite-time formation control is designed, which realizes the closed-loop system error finite-time convergence. Meanwhile, the error always meets predefined performance constraints. For the multi-agent systems, [304] first designs the prescribed performance error transformation and then designs the finite-time control with performance constraints through finite-time stability theory. Reference [305] designs the event-driven prescribed performance finite-time control with considering saturated nonlinearities. Reference [306] designs the finite-time prescribed performance fault-tolerant control based on the adaptive fuzzy method. Reference [307] designs the performance constraint function that converges within finite time. At the same time, based on the prescribed performance error transformation, the finite-time stable closed-loop system is designed, which realizes the finite-time stability of the system, and the system state or error always meets the constraints. Reference [308] designs the finite-time prescribed performance control for higher-order stochastic nonlinear systems. Reference [309] designs the finite-time adaptive fuzzy prescribed performance control for stochastic high-order nonlinear systems, which achieves finite-time stability, and the performance constraints are satisfied at the same time. Reference [310] designs the finite-time adaptive fuzzy control based on observers and prescribed performance constraints. Reference [311] designs the finite-time prescribed performance control for nonlinear switching systems based on the finite-time performance constraint functions. In addition, based on the finite-time prescribed performance control method, researchers have carried out much practical exploration and applied it in many practical situations, such as unmanned aerial vehicles [312], horizontal platform systems [313], power system turbine generator [314], robotic arm [315], gyroscope [316], etc. In general, the above achievements are obtained based on the following two main steps. The steps are as follows: obtain the constrained-performance auxiliary system through error transformation as in the PPC method, and then design the controller through finite-time stability theory.

On the other hand, depending on the performance response, the constrained controller that makes the system state or error converge within the predetermined range within a predetermined time also begins to appear. It mainly relies on defining a particular prescribed performance function that converges to a fixed value range within the desired finite time. Under the constraint algorithm of this function combination, the system state or error can be achieved to converge to a predetermined value range within the desired finite time. Reference [317] designs a piecewise constraint function. The first segment of this piecewise function is similar to the traditional prescribed performance function, the second segment is a constant value constraint, which is considered the upper bound on the desired steady-state error, and the segmented point is at the desired settling time T_f. This method aims to distinguish between transient and steady-state intervals by time points and different segments in the piecewise function. When $t < T_f$, the system state or error is constrained by an exponentially decreasing function, and this stage can be considered a transient process. When $t \geq T_f$, it can be considered that the system enters the steady-state stage, and the closed-loop system state or error is less than the constant constraints. In this way, the system is constrained to converge to a limited steady-state error interval within the finite time. Although finite-time similar effects can be achieved, the stability in [317] is asymptotic stability, not finite-time stability. Reference [318] designs the performance constraint function that converges in finite time through piecewise functions. For example, when $t \in [0, T_0]$, $\rho(t) = (\rho_0^p - p\lambda t)^{\frac{1}{p}} + \rho_{T_0}$. When $t > T_0$, $\rho(t) = \rho_{T_0}$, where $\rho_0 > 0$, $\rho_{T_0} > 0$, $p > 0$, $\lambda > 0$ is the parameter to be defined. This means that if the system satisfies the performance constraints, then when $t > T_0$, the system state or error is less than ρ_{T_0}, Although T_0 is considered as finite settling time in [318], the stability of its closed-loop system is also asymptotically stable. Reference [319] also uses the piecewise finite-time performance constraint function similar to the previous references to constrain the transient and steady-state performance. Although the closed-loop system under the controller is asymptotically stable, an effect similar to finite-time convergence can be achieved through the constraint function. Reference [320] considers the nonlinear problem of actuators and designs the adaptive prescribed performance control for nonlinear switching systems based on the finite-time performance constraint function. Reference [321] designs the time-driven performance-constrained control based on the adaptive neural network and the finite-time performance-constrained function. Reference [322] designs the adaptive prescribed performance fault-tolerant control for nonlinear switching systems based on neural networks and finite-time performance constraint functions. Reference [323] designs the distributed adaptive prescribed performance control based on the finite-time performance constraint function and also considers the problem of the unknown control direction. Reference [324] designs the adaptive fuzzy prescribed performance control based on the finite-time performance constraint function, considering the nonlinear uncertainties and actuator faults. Reference [325] designs the adaptive fuzzy prescribed performance control considering input uncertainty based on the finite-time performance constraint function. Based on this technology, [326] shows the application results of the finite-time performance constraint functions in unmanned aerial vehicle control.

Compared with the finite-time control, with this method, the designer can select the performance parameters first and then design the controller. Then, the performance response of the system is theoretically strictly constrained. The closed-loop system can not only achieve finite-time stability but also meet the constraints of prescribed performance at all times. Considering that the predefined constraint time and the finite convergence time can be designed separately, such a performance-constrained control method has more time-parameter design capabilities.

1.3.6 Performance Guaranteed Control for Spacecraft

The performance guaranteed control method, such as prescribed performance, has a strong constraint on the time domain response of the system state or error, including the predefined convergence rate, the overshoot, the steady-state error range, etc. Therefore, it is widely used in control tasks with higher performance requirements. With the continuous improvement of spacecraft on-orbit service missions, the performance requirements of spacecraft relative motion control are getting higher and higher. Therefore, performance guaranteed control has received attention in the field of space missions. Thus, there have been many research results in recent years which take the performance guaranteed method as the primary method, among which the application of the prescribed performance control method is the most representative. This section will briefly introduce these research results.

There have been relatively fruitful research results for the attitude prescribed performance control of rigid-body spacecraft in recent years. Based on the traditional performance constraint control, Ref. [327] designs the prescribed performance robust controller for the attitude tracking problem of rigid-body spacecraft. The basic idea is briefly described as follows: Through the attitude motion model of the rigid body spacecraft, the corresponding attitude tracking error model is obtained. Define the performance parameters of the desired attitude tracking error time domain response, including convergence rate, overshoot, steady-state error, etc. The constraints are constructed according to the performance constraint function and the desired performance parameters, and the error transformation is performed on the constraints to obtain the auxiliary system. According to the relationship between the auxiliary system and the tracking error, the relationship between the state of the auxiliary system and the control input is constructed. The controller is designed based on the auxiliary system, and the auxiliary closed-loop system is stabilized. Considering that the spacecraft receives external disturbances, robustness issues must be considered. The control algorithm obtained in this way can make the spacecraft attitude tracking error converge and satisfy the predefined performance constraints.

Reference [328] designs the spacecraft attitude fault tolerance prescribed performance control. For the attitude tracking control of rigid-body spacecraft with saturation, Ref. [329] designs the prescribed performance control based on adaptive neural networks. For the attitude stabilization and tracking problems of the rigid-body spacecraft, Ref. [330] designs the performance-constrained control based on the low-complexity prescribed performance method, which also achieves tracking error convergence and performance constraints. For the attitude tracking problem of rigid-body spacecraft with input saturation, Ref. [331] designs a fault-tolerant controller with prescribed performance. Reference [328] designs the adaptive fault-tolerant spacecraft attitude tracking control with performance constraints. For the attitude consistency control problem of spacecraft formation flight, Ref. [332] designs an adaptive prescribed performance control based on learning so that the attitude errors of the formation spacecraft meet the performance constraints. Reference [333] applies a piecewise performance constraint function to design a prescribed performance control for spacecraft attitude tracking, where the piecewise performance constraint function sets the expected settling time. For the spacecraft attitude tracking problem, Ref. [334] designs the performance-constrained fault-tolerant control. For the attitude control problem of rigid-body spacecraft, Ref. [335] designs an adaptive fixed-time convergence prescribed performance fault-tolerant control while considering the model uncertainty and other issues. Reference [336] designs the robust fault-tolerant attitude tracking control based on prescribed performance. Reference [337] considers the control-constrained adaptive prescribed performance attitude tracking control problem. Reference [338] examines in detail the performance-constrained attitude control for spacecraft after capture. Reference [339] studies the event-triggered adaptive prescribed performance attitude tracking control for spacecraft with unknown actuator failures. Reference [340] designs the model-free prescribed performance spacecraft attitude tracking control. Reference [341] designs the robust control with prescribed performance based on an adaptive neural network for spacecraft attitude maneuver. Reference [342] designs the adaptive robust finite-time prescribed performance attitude tracking control. For the attitude control problem of the spacecraft assembly and the takeover control problem of the combined spacecraft with unknown rotational inertia, Ref. [343] designs the robust control with prescribed performance, which uses constraints to constrain the transient and steady-state performance of the assembly's attitude motion. For the combined spacecraft takeover control problem with unknown rotational inertia, Ref. [344] designs a learning-based adaptive prescribed performance control. For the same application scenario, Ref. [46] designs an event-driven adaptive prescribed performance control. Reference [51] designs a fault-tolerant prescribed performance control for combined spacecraft attitude with flywheel. Reference [52] applies a piecewise performance constraint function to design a finite-time prescribed performance attitude control for combined spacecraft considering uncertain nonlinear problems.

For the flexible spacecraft attitude control problem, Ref. [345] considers the input saturation problem and designs an observer-based prescribed performance controller. This reference proposes a piecewise performance constraint function of parameter design. When the time exceeds the predefined time, the performance constraint is a constant value. Reference [346] designs an adaptive prescribed performance controller for flexible spacecraft with active vibration suppression. Reference [347] designs the performance constraint function with time parameters and the flexible spacecraft adaptive performance constraint attitude tracking control with considering actuator saturation. For the high-precision attitude stabilization of flexible spacecraft, Ref. [348] designs the fixed-time prescribed performance control.

Regarding relative position control of spacecraft, Ref. [349] studies the problem of adaptive prescribed performance attitude control for close-range spacecraft operation with space motion constraints. Reference [350] designs the performance constraint function according to a predetermined settling time and designs the prescribed performance adaptive control for the relative motion of non-cooperative spacecraft.

In other space applications, Ref. [351] considers the adaptive prescribed performance control problem of uncertain free-floating space manipulators. Reference [352] designs the spacecraft vibration boundary constraint problem considering dual-flexible attachments. In [353], an adaptive sliding-mode performance-constrained control based on disturbance observation is designed to capture a target by space manipulators. Among them, performance constraints provide transient and steady-state performance guarantees for target capture, and finite-time stability also improves the system's transient performance.

1.4 Main Contents of This Book

In this chapter, the research background of this book is presented, and the relative motion control problem of the on-orbit service spacecraft is focused on. Among them, the practical engineering requirements and research statuses of aerospace control design, such as precise motion control, state or error constraints, input constraints, energy optimization, spacecraft game and equilibrium, and cooperation, are introduced. Combined with the design requirements of precise control and constraint control, the research history and status of the performance guaranteed control algorithm are investigated and analyzed in detail. In the background, as mentioned earlier, this book addresses the challenge of the relative motion precise control of the on-orbit service spacecraft. This book considers performance constraints to force spacecraft closed-loop systems to achieve accurate transient and steady-state performance, discusses performance guaranteed conditions that consider settling time, and conducts research to optimize control performance continuously.

Specifically, for the relative motion control and general precise control problems in the spacecraft on-orbit service missions, considering the valuable performance indicators such as settling time, overshoot, and steady-state errors in the attitude or position time-domain performance response, this book is devoted to solving the precise control problem of the spacecraft maneuvering by performance guaranteed control methods. This book is devoted to giving the closed-loop system the ability to guarantee performance. That is, the practical time-domain response of the closed-loop system always meets the desired performance index parameters. Based on this, designers can design the transient and steady-state performance indicators, such as the expected maximum settling time, maximum overshoot range, and maximum steady-state error range, according to the actual task requirements and real physical constraints. The closed-loop system states or tracking errors can be achieved to converge within the predefined steady-state error range within the predefined desired settling time.

In Chap. 2, based on the performance constraint function, by setting the expected error coordinate points in the time domain response curve, this book proposes the continuous time-varying performance guaranteed function that explicitly includes the expected settling time parameter in the time domain analysis, and further designs the simplified system transformation method. In addition, the coordinate systems, spacecraft models, related algorithms, and theories involved in this book are introduced in this chapter.

Under the motivation of the aerospace control design objectives, as mentioned earlier, combined with the control, game, and cooperation issues concerned by aerospace missions, the rest of the book is divided into three parts. The algorithm design and analysis are systematically carried out, and several methods and algorithms are proposed.

Part I, Control, mainly introduces the performance guaranteed control algorithm proposed in this study. Based on the proposed performance function, this book presents the performance guaranteed attitude-asymptotic control methods and the performance guaranteed finite-time control methods for the on-orbit service spacecraft. Considering the settling time parameters, the control methods improve the transient performance. These methods are presented in Chaps. 3, 4, and 5. In the case where the spacecraft tracking error is constrained by the desired performance, considering the actual actuator capability of the spacecraft, this study endows the spacecraft performance guaranteed control with the ability to actively adjust the control input signal through the command filtering technique in Chaps. 6 and 7. This means adding desired constraints to the input signal is possible. Considering the performance constraints and fuel consumption optimization problems of the space missions, this study proposes the optimal or sub-optimal control methods for the on-orbit service spacecraft attitude control system that satisfies performance constraints based on the adaptive dynamic programming technique in Chap. 8.

Part II, Games, mainly introduces the performance guaranteed control algorithm proposed in this study considering cooperative games. Based on the non-zero-sum game algorithm, the performance guaranteed game control strategies are proposed in Chap. 9 for the cooperative control problem of two spacecraft relative motions. Considering the cooperative game control problem of the multi-spacecraft combination, the optimal control methods that satisfy performance constraints are proposed in Chap. 10.

Part III, Cooperation, mainly introduces the multi-spacecraft performance guaranteed control algorithms in this study, considering cooperation and consistency. Considering the network system state constraint problem, the coordinated tracking control with the full-state constraints is presented in Chap. 11. Furthermore, considering the multi-spacecraft attitude and multi-constraint problem, the distributed control algorithm based on state and output feedback is designed in Chap. 12. For multi-spacecraft formation flight, some of the spacecraft can be used to build a flight configuration, and the rest of the spacecraft is contained in that configuration. Such a design can guarantee the formation flight configuration and the motion performance of all spacecraft. By changing the configuration, other spacecraft can be kept within the constraints of the configuration, which also constitutes the performance guaranteed of the spacecraft formation. Control of this behavior can be referred to as *multi-spacecraft formation-containment control*. Considering the multi-spacecraft formation-containment control problem, the two-layer distributed spacecraft formation-containment control based on output feedback is introduced in Chap. 13. Considering other aerospace obstacles that may be encountered during spacecraft formation behavior, the formation-containment control for networked spacecraft with collision avoidance is fully designed in Chap. 14. Further, considering the more complex space mission scenarios with borders and obstacles caused by other close-range spacecraft or on-orbit assembly components, the layered formation-containment control for networked spacecraft in constrained space is designed in Chap. 15.

In summary, the research in this book provides a series of high-precision constrained control methods based on performance guaranteed and explicit settling time for the more complex control problems of spacecraft on-orbit or space-station service missions. These include the general performance guaranteed control method, command-filter-based control method, optimization control method, cooperative game control method, and formation control method. Applying these control methods can make the relative motion of the on-orbit service spacecraft have higher control accuracy and better performance during missions. The research in this book is expected to provide the theoretical basis and the advanced design approaches for the relative motion and general motion control of the on-orbit service spacecraft under high-precision and multi-constraint conditions. These methods and algorithms are expected to have significant theoretical value and application prospects in aerospace engineering.

References

1. Waltz, D.M.: On-orbit servicing of space systems. Krieger Publishing Company (1993)
2. Li, W., Cheng, D., Liu, X., Wang, Y., Shi, W., Tang, Z., Gao, F., Zeng, F., Chai, H., Luo, W., et al.: On-orbit service (oos) of spacecraft: A review of engineering developments. Progress in Aerospace Sciences **108**, 32–120 (2019)
3. Flores-Abad, A., Ma, O., Pham, K., Ulrich, S.: A review of space robotics technologies for on-orbit servicing. Progress in aerospace sciences **68**, 1–26 (2014)
4. Franklin, G.F., Powell, J.D., Emami-Naeini, A., Powell, J.D.: Feedback control of dynamic systems, vol. 4. Prentice hall Upper Saddle River (2002)
5. Khalil, H.K., Grizzle, J.W.: Nonlinear systems. Prentice hall Upper Saddle River, NJ (2002)
6. Bechlioulis, C.P., Rovithakis, G.A.: Robust adaptive control of feedback linearizable mimo nonlinear systems with prescribed performance. IEEE Transactions on Automatic Control **53**(9), 2090–2099 (2008)
7. Kristiansen, R., Nicklasson, P.J., Gravdahl, J.T.: Spacecraft coordination control in 6dof: Integrator backstepping vs passivity-based control. Automatica **44**(11), 2896–2901 (2008)
8. Zhao, L., Jia, Y.: Decentralized adaptive attitude synchronization control for spacecraft formation using nonsingular fast terminal sliding mode. Nonlinear Dynamics **78**(4), 2779–2794 (2014)
9. On-orbit satellite servicing study. Project Report of NASA Goddard Space Flight Center (2010)
10. Shayler, D.J., Harland, D.M.: The Hubble Space Telescope: From Concept to Success. Springer (2015)
11. Flores-Abad, A., Ma, O., Pham, K., Ulrich, S.: A review of space robotics technologies for on-orbit servicing. Progress in aerospace sciences **68**, 1–26 (2014)
12. Stamm, S., Motaghedi, P.: Orbital express capture system: concept to reality. In: Spacecraft Platforms and Infrastructure, vol. 5419, pp. 78–91. International Society for Optics and Photonics (2004)

13. Friend, R.B.: Orbital express program summary and mission overview. In: Sensors and Systems for space applications II, vol. 6958. International Society for Optics and Photonics (2008)
14. Roesler, G.: Robotic servicing of geosynchronous satellites (rsgs). Defense Advanced Research Projects Agency (2017)
15. Barnhart, D.: The phoenix project. The Phoenix Project, Report for the United Nations Committee on the Peaceful Uses of Outer Space
16. Harbaugh, J.: In-space robotic manufacturing and assembly (irma) (2017)
17. Reintsema, D., Sommer, B., Wolf, T., Theater, J., Radthke, A., Naumann, W., Rank, P., Sommer, J.: Deos-the in-flight technology demonstration of german's robotics approach to dispose malfunctioned satellites. In: ESA 11th Symposium on Advanced Space Technologies in Robotics and Automation, ESTEC, Netherlands (2011)
18. Benedict, B.L.: Rationale for need of in-orbit servicing capabilities for geo spacecraft. In: AIAA SPACE 2013 Conference and Exposition, p. 5444 (2013)
19. Gaddis, S.W.: Space technology mission directorate: Game changing development (2015)
20. Yan, R., Wu, Z.: Finite-time attitude stabilization of flexible spacecrafts via reduced-order smdo and ntsmc. Journal of Aerospace Engineering **31**(4), 04018023 (2018)
21. Hu, Q., Tan, X., Akella, M.R.: Finite-time fault-tolerant spacecraft attitude control with torque saturation. Journal of Guidance, Control, and Dynamics **40**(10), 2524–2537 (2017)
22. Sari, N.N., Jahanshahi, H., Fakoor, M.: Adaptive fuzzy pid control strategy for spacecraft attitude control. International Journal of Fuzzy Systems **21**(3), 769–781 (2019)
23. Li, B., Hu, Q., Yu, Y., Ma, G.: Observer-based fault-tolerant attitude control for rigid spacecraft. IEEE Transactions on Aerospace and Electronic Systems **53**(5), 2572–2582 (2017)
24. Zhu, X., Chen, J., Zhu, Z.H.: Adaptive learning observer for spacecraft attitude control with actuator fault. Aerospace Science and Technology **108**, 106389 (2021)
25. Xin, W., Jiayi, W., Xin, L.: A feed-forward wavelet neural network adaptive observer-based fault detection technique for spacecraft attitude control systems. Chinese Journal of Electronics **27**(1), 102–108 (2018)
26. Bang, H., Kim, J., Jung, Y.: Spacecraft attitude control compensating internal payload motion using disturbance observer technique. International Journal of Aeronautical and Space Sciences **20**(2), 459–466 (2019)
27. Wu, B., Shen, Q., Cao, X.: Event-triggered attitude control of spacecraft. Advances in Space Research **61**(3), 927–934 (2018)
28. Xing, L., Wen, C., Liu, Z., Su, H., Cai, J.: An event-triggered design scheme for spacecraft attitude control. In: 2017 12th IEEE Conference on Industrial Electronics and Applications (ICIEA), pp. 1552–1557. IEEE (2017)
29. Gui, H., Vukovich, G.: Adaptive fault-tolerant spacecraft attitude control using a novel integral terminal sliding mode. International Journal of Robust and Nonlinear Control **27**(16), 3174–3196 (2017)
30. Khamis, A., Zydek, D.: Finite horizon optimal nonlinear spacecraft attitude control. The Journal of the Astronautical Sciences **67**(3), 1002–1020 (2020)
31. Sharifi, E., Damaren, C.J.: Nonlinear optimal approach to spacecraft attitude control using magnetic and impulsive actuations. Journal of Guidance, Control, and Dynamics **43**(6), 1154–1163 (2020)
32. Chen, X., Sun, R., Wang, F., Song, D., Jiang, W.: Two-stage unscented kalman filter algorithm for fault estimation in spacecraft attitude control system. IET Control Theory & Applications **12**(13), 1781–1791 (2018)
33. Saki, S., Bolandi, H., et al.: Optimal direct adaptive soft switching multi-model predictive control using the gap metric for spacecraft attitude control in a wide range of operating points. Aerospace Science and Technology **77**, 235–243 (2018)
34. Guo, X.G., Tian, M.E., Li, Q., Ahn, C.K., Yang, Y.H.: Multiple-fault diagnosis for spacecraft attitude control systems using rbfnn-based observers. Aerospace Science and Technology **106**, 106195 (2020)

35. Cai, H., Huang, J.: The leader-following attitude control of multiple rigid spacecraft systems. Automatica **50**(4), 1109–1115 (2014)
36. Massari, M., Zamaro, M.: Application of sdre technique to orbital and attitude control of spacecraft formation flying. Acta Astronautica **94**(1), 409–420 (2014)
37. Juan Song, X., Wei Wang, H., Feng Lu, S.: Adaptive backstepping attitude control for liquid-filled spacecraft without angular velocity measurement. Journal of Aerospace Engineering **34**(3), 04021021 (2021)
38. Weiss, A., Kolmanovsky, I., Bernstein, D.S., Sanyal, A.: Inertia-free spacecraft attitude control using reaction wheels. Journal of Guidance, Control, and Dynamics **36**(5), 1425–1439 (2013)
39. Shen, Q., Yue, C., Goh, C.H., Wang, D.: Active fault-tolerant control system design for spacecraft attitude maneuvers with actuator saturation and faults. IEEE Transactions on Industrial Electronics **66**(5), 3763–3772 (2018)
40. Huang, X., Yan, Y.: Fully actuated spacecraft attitude control via the hybrid magnetocoulombic and magnetic torques. Journal of Guidance, Control, and Dynamics **40**(12), 3358–3360 (2017)
41. Wang, Z., Yuan, J., Shi, Y., Che, D.: Robust adaptive fault tolerant attitude control for post-capture non-cooperative targets with actuator nonlinearities. Transactions of the Institute of Measurement and Control **40**(7), 2116–2128 (2018)
42. Jiang, H., Zhou, B., Li, D., Duan, G.: Data-driven–based attitude control of combined spacecraft with noncooperative target. International Journal of Robust and Nonlinear Control **29**(16), 5801–5819 (2019)
43. Chang, H., Huang, P., Zhang, Y., Meng, Z., Liu, Z.: Distributed control allocation for spacecraft attitude takeover control via cellular space robot. Journal of Guidance, Control, and Dynamics **41**(11), 2499–2506 (2018)
44. Huang, P., Lu, Y., Wang, M., Meng, Z., Zhang, Y., Zhang, F.: Postcapture attitude takeover control of a partially failed spacecraft with parametric uncertainties. IEEE Transactions on Automation Science and Engineering **16**(2), 919–930 (2018)
45. Wu, Y., Han, F., Zheng, M., He, M., Chen, Z., Hua, B., Wang, F.: Attitude control for on-orbit servicing spacecraft using hybrid actuator. Advances in Space Research **61**(6), 1600–1616 (2018)
46. Wei, C., Luo, J., Ma, C., Dai, H., Yuan, J.: Event-triggered neuroadaptive control for post-capture spacecraft with ultralow-frequency actuator updates. Neurocomputing **315**, 310–321 (2018)
47. Wei, C., Luo, J., Dai, H., Yuan, J.: Adaptive model-free constrained control of postcapture flexible spacecraft: a euler–lagrange approach. Journal of Vibration and Control **24**(20), 4885–4903 (2018)
48. She, Y., Li, S.: An extra degree-of-freedom model for combined spacecraft attitude control with unilateral contact constraint. Acta Astronautica **165**, 54–67 (2019)
49. Gao, H., Ma, G., Lv, Y., Guo, Y.: Forecasting-based data-driven model-free adaptive sliding mode attitude control of combined spacecraft. Aerospace Science and Technology **86**, 364–374 (2019)
50. Han, G., Guangfu, M., Yueyong, L., Yanning, G.: Data-driven model-free adaptive attitude control of partially constrained combined spacecraft with external disturbances and input saturation. Chinese Journal of Aeronautics **32**(5), 1281–1293 (2019)
51. Huang, X., Duan, G.: Fault-tolerant attitude tracking control of combined spacecraft with reaction wheels under prescribed performance. ISA transactions **98**, 161–172 (2020)
52. Zhang, X., Xu, T., Wei, C.: Novel finite-time attitude control of postcapture spacecraft with input faults and quantization. Advances in Space Research **65**(1), 297–311 (2020)
53. Chai, Y., Luo, J.: Finite-time fuzzy game-based attitude control for on-orbit cooperative transporting. Journal of the Franklin Institute (2021)
54. Chai, Y., Luo, J., Han, N., Sun, J.: Robust event-triggered game-based attitude control for on-orbit assembly. Aerospace Science and Technology **103**, 105894 (2020)
55. Yang, H., Hu, Q.: Research and experiment on dynamic weight pseudo-inverse control allocation for spacecraft attitude control system. In: 2019 Chinese Control Conference (CCC), pp. 8200–8205. IEEE (2019)

56. Lang, X., de Ruiter, A.: A control allocation scheme for spacecraft attitude stabilization based on distributed average consensus. Aerospace Science and Technology **106**, 106173 (2020)
57. Lee, T.: Relative attitude control of two spacecraft on so (3) using line-of-sight observations. In: 2012 American Control Conference (ACC), pp. 167–172. IEEE (2012)
58. Wu, T.H., Flewelling, B., Leve, F., Lee, T.: Spacecraft relative attitude formation tracking on so (3) based on line-of-sight measurements. In: 2013 American Control Conference, pp. 4820–4825. IEEE (2013)
59. Sun, L., Huo, W., Jiao, Z.: Robust adaptive relative position and attitude control for spacecraft autonomous proximity. ISA transactions **63**, 11–19 (2016)
60. Wang, Y., Ji, H.: Integrated relative position and attitude control for spacecraft rendezvous with iss and finite-time convergence. Aerospace Science and Technology **85**, 234–245 (2019)
61. Liu, W., Geng, Y., Wu, B., Wang, D.: Neural-network-based adaptive event-triggered control for spacecraft attitude tracking. IEEE transactions on neural networks and learning systems **31**(10), 4015–4024 (2019)
62. Li, C., Teo, K.L., Li, B., Ma, G.: A constrained optimal pid-like controller design for spacecraft attitude stabilization. Acta Astronautica **74**, 131–140 (2012)
63. Kalabić, U., Gupta, R., Di Cairano, S., Bloch, A., Kolmanovsky, I.: Constrained spacecraft attitude control on so (3) using reference governors and nonlinear model predictive control. In: 2014 American Control Conference, pp. 5586–5593. IEEE (2014)
64. Sun, C., Dai, R.: Spacecraft attitude control under constrained zones via quadratically constrained quadratic programming. In: AIAA guidance, navigation, and control conference, p. 2010 (2015)
65. Guiggiani, A., Kolmanovsky, I., Patrinos, P., Bemporad, A.: Fixed-point constrained model predictive control of spacecraft attitude. In: 2015 American Control Conference (ACC), pp. 2317–2322. IEEE (2015)
66. Gupta, R., Kalabić, U.V., Di Cairano, S., Bloch, A.M., Kolmanovsky, I.V.: Constrained spacecraft attitude control on so (3) using fast nonlinear model predictive control. In: 2015 American Control Conference (ACC), pp. 2980–2986. IEEE (2015)
67. Guiggiani, A., Kolmanovsky, I., Patrinos, P., Bemporad, A.: Constrained model predictive control of spacecraft attitude with reaction wheels desaturation. In: 2015 European Control Conference (ECC), pp. 1382–1387. IEEE (2015)
68. Wu, C., Xu, R., Zhu, S., Cui, P.: Time-optimal spacecraft attitude maneuver path planning under boundary and pointing constraints. Acta Astronautica **137**, 128–137 (2017)
69. Wu, C., Han, X.: Energy-optimal spacecraft attitude maneuver path-planning under complex constraints. Acta Astronautica **157**, 415–424 (2019)
70. Wu, Y.H., Hu, Y.B., Hua, B., Chen, Z.M., Ge, L.L.: A time suboptimal method for real-time spacecraft attitude agile maneuvering with angular velocity constraint. Proceedings of the Institution of Mechanical Engineers, Part G: Journal of Aerospace Engineering **233**(2), 561–572 (2019)
71. Burlion, L., Biannic, J.M., Ahmed-Ali, T.: Attitude tracking control of a flexible spacecraft under angular velocity constraints. International Journal of Control **92**(7), 1524–1540 (2019)
72. Bayat, F.: Model predictive sliding control for finite-time three-axis spacecraft attitude tracking. IEEE Transactions on Industrial Electronics **66**(10), 7986–7996 (2018)
73. Nicotra, M.M., Liao-McPherson, D., Burlion, L., Kolmanovsky, I.V.: Spacecraft attitude control with nonconvex constraints: An explicit reference governor approach. IEEE Transactions on Automatic Control **65**(8), 3677–3684 (2019)
74. Boïskovic, J.D., Li, S.M., Mehra, R.K.: Robust adaptive variable structure control of spacecraft under control input saturation. Journal of Guidance, Control, and Dynamics **24**(1), 14–22 (2001)
75. Boskovic, J.D., Li, S.M., Mehra, R.K.: Robust tracking control design for spacecraft under control input saturation. Journal of Guidance, Control, and Dynamics **27**(4), 627–633 (2004)
76. Bang, H., Tahk, M.J., Choi, H.D.: Large angle attitude control of spacecraft with actuator saturation. Control engineering practice **11**(9), 989–997 (2003)

77. Wallsgrove, R.J., Akella, M.R.: Globally stabilizing saturated attitude control in the presence of bounded unknown disturbances. Journal of Guidance, Control, and Dynamics **28**(5), 957–963 (2005)
78. Anton, H.: Adaptive spacecraft attitude tracking control with actuator saturation. J. Guid. Control Dyn **33**(5), 1692–1695 (2010)
79. Lin, T., Shijie, X.: Attitude control considering variable input saturation limit for a spacecraft equipped with flywheels. Chinese Journal of Aeronautics **25**(3), 437–445 (2012)
80. Hu, Q., Xiao, B.: Intelligent proportional-derivative control for flexible spacecraft attitude stabilization with unknown input saturation. Aerospace Science and Technology **23**(1), 63–74 (2012)
81. Bustan, D., Pariz, N., Sani, S.K.H.: Robust fault-tolerant tracking control design for spacecraft under control input saturation. ISA transactions **53**(4), 1073–1080 (2014)
82. Bustan, D., Sani, S.H., Pariz, N.: Adaptive fault-tolerant spacecraft attitude control design with transient response control. IEEE/ASME Transactions on Mechatronics **19**(4), 1404–1411 (2013)
83. Li, B., Hu, Q., Ma, G.: Extended state observer based robust attitude control of spacecraft with input saturation. Aerospace Science and Technology **50**, 173–182 (2016)
84. Sun, L., Zheng, Z.: Disturbance-observer-based robust backstepping attitude stabilization of spacecraft under input saturation and measurement uncertainty. IEEE Transactions on Industrial Electronics **64**(10), 7994–8002 (2017)
85. Cao, S., Zhao, Y.: Anti-disturbance fault-tolerant attitude control for satellites subject to multiple disturbances and actuator saturation. Nonlinear Dynamics **89**(4), 2657–2667 (2017)
86. Zou, A.M., de Ruiter, A.H., Kumar, K.D.: Disturbance observer-based attitude control for spacecraft with input mrs. IEEE Transactions on Aerospace and Electronic Systems **55**(1), 384–396 (2018)
87. Liu, Z., Liu, J., Wang, L.: Disturbance observer based attitude control for flexible spacecraft with input magnitude and rate constraints. Aerospace Science and Technology **72**, 486–492 (2018)
88. Hu, Q., Shi, Y., Shao, X.: Adaptive fault-tolerant attitude control for satellite reorientation under input saturation. Aerospace Science and Technology **78**, 171–182 (2018)
89. Zou, A.M., Kumar, K.D.: Finite-time attitude control for rigid spacecraft subject to actuator saturation. Nonlinear Dynamics **96**(2), 1017–1035 (2019)
90. Hu, Q., Liu, Y., Dong, H., Zhang, Y.: Saturated attitude control for rigid spacecraft under attitude constraints. Journal of Guidance, Control, and Dynamics **43**(4), 790–805 (2020)
91. Li, M., Hou, M., Yin, C.: Constrained dual-loop attitude control design for spacecraft. Aerospace Science and Technology **55**, 170–176 (2016)
92. Liu, Q., Wie, B.: Robust time-optimal control of uncertain flexible spacecraft. Journal of guidance, control, and dynamics **15**(3), 597–604 (1992)
93. Shen, H., Tsiotras, P.: Time-optimal control of axisymmetric rigid spacecraft using two controls. Journal of Guidance, Control, and Dynamics **22**(5), 682–694 (1999)
94. Vadali, S., Kraige, L., Junkins, J.: New results on the optimal spacecraft attitude maneuver problem. Journal of Guidance, Control, and Dynamics **7**(3), 378–380 (1984)
95. Carrington, C.K., Junkins, J.: Optimal nonlinear feedback control for spacecraft attitude maneuvers. Journal of Guidance, Control, and Dynamics **9**(1), 99–107 (1986)
96. Tsiotras, P.: Stabilization and optimality results for the attitude control problem. Journal of guidance, control, and dynamics **19**(4), 772–779 (1996)
97. Schaub, H., Junkins, J.L., Robinett, R.D.: New penalty functions and optimal control formulation for spacecraft attitude control problems. Journal of Guidance, Control, and Dynamics **20**(3), 428–434 (1997)
98. Spindler, K.: Optimal control on lie groups with applications to attitude control. Mathematics of Control, Signals and Systems **11**(3), 197–219 (1998)
99. Wu, C.S., Chen, B.S.: Attitude control of spacecraft: mixed h/h approach. Journal of Guidance, Control, and Dynamics **24**(4), 755–766 (2001)

100. Sharma, R., Tewari, A.: Optimal nonlinear tracking of spacecraft attitude maneuvers. IEEE Transactions on Control Systems Technology **12**(5), 677–682 (2004)
101. Hegrenæs, Ø., Gravdahl, J.T., Tøndel, P.: Spacecraft attitude control using explicit model predictive control. Automatica **41**(12), 2107–2114 (2005)
102. Pukdeboon, C., Zinober, A.S.I.: Optimal sliding mode controllers for attitude tracking of spacecraft. In: 2009 IEEE Control Applications, (CCA) & Intelligent Control, (ISIC), pp. 1708–1713. IEEE (2009)
103. Xin, M., Pan, H.: Integrated nonlinear optimal control of spacecraft in proximity operations. International Journal of Control **83**(2), 347–363 (2010)
104. Li, S., Ding, S., Li, Q.: Global set stabilization of the spacecraft attitude control problem based on quaternion. International Journal of Robust and Nonlinear Control: IFAC-Affiliated Journal **20**(1), 84–105 (2010)
105. Horri, N.M., Palmer, P.L., Roberts, M.: Energy optimal spacecraft attitude control subject to convergence rate constraints. Control Engineering Practice **19**(11), 1297–1314 (2011)
106. Xin, M., Pan, H.: Nonlinear optimal control of spacecraft approaching a tumbling target. Aerospace Science and Technology **15**(2), 79–89 (2011)
107. Pukdeboon, C., Zinober, A.S.I.: Control lyapunov function optimal sliding mode controllers for attitude tracking of spacecraft. Journal of the Franklin Institute **349**(2), 456–475 (2012)
108. Pukdeboon, C.: Optimal output feedback controllers for spacecraft attitude tracking. Asian Journal of Control **15**(5), 1284–1294 (2013)
109. Li, B., Hu, Q., Qi, J.: Null-space-based optimal control reallocation for spacecraft stabilization under input saturation. International Journal of Adaptive Control and Signal Processing **29**(6), 705–724 (2015)
110. Gong, L.G., Wang, Q., Dong, C.Y.: Spacecraft output feedback attitude control based on extended state observer and adaptive dynamic programming. Journal of the Franklin Institute **356**(10), 4971–5000 (2019)
111. Cheng, Y.H., Jiang, B., Li, H., Han, X.d.: On-orbit reconfiguration using adaptive dynamic programming for multi-mission-constrained spacecraft attitude control system. International Journal of Control, Automation and Systems **17**(4), 822–835 (2019)
112. Von Neumann, J., Morgenstern, O.: Theory of games and economic behavior, 2nd rev (1947)
113. Isaacs, R.: Differential games: a mathematical theory with applications to warfare and pursuit, control and optimization. Courier Corporation (1999)
114. Vamvoudakis, K.G., Lewis, F.L.: Online actor–critic algorithm to solve the continuous-time infinite horizon optimal control problem. Automatica **46**(5), 878–888 (2010)
115. Kreps, D.M.: Nash equilibrium. In: Game Theory, pp. 167–177. Springer (1989)
116. Vrabie, D., Vamvoudakis, K.G., Lewis, F.L.: Optimal Adaptive Control and Differential Games by Reinforcement Learning Principles. Control, Robotics, Sensors. Institution of Engineering and Technology (2012)
117. Bertsekas, D.: Reinforcement learning and optimal control. Athena Scientific (2019)
118. Kamalapurkar, R., Walters, P., Rosenfeld, J., Dixon, W.: Reinforcement learning for optimal feedback control. Springer (2018)
119. Vamvoudakis, K.G., Lewis, F.L.: Multi-player non-zero-sum games: Online adaptive learning solution of coupled hamilton–jacobi equations. Automatica **47**(8), 1556–1569 (2011)
120. Vamvoudakis, K.G., Wan, Y., Lewis, F.L., Cansever, D.: Handbook of Reinforcement Learning and Control. Springer (2021)
121. Vrabie, D., Lewis, F.: Integral reinforcement learning for online computation of feedback nash strategies of nonzero-sum differential games. In: 49th IEEE Conference on Decision and Control (CDC), pp. 3066–3071. IEEE (2010)
122. Zhang, H., Cui, L., Luo, Y.: Near-optimal control for nonzero-sum differential games of continuous-time nonlinear systems using single-network adp. IEEE transactions on cybernetics **43**(1), 206–216 (2012)
123. Johnson, M., Kamalapurkar, R., Bhasin, S., Dixon, W.E.: Approximate n-player nonzero-sum game solution for an uncertain continuous nonlinear system. IEEE Transactions on Neural Networks and Learning Systems **26**(8), 1645–1658 (2014)

124. Ovalle, A., Hably, A., Bacha, S.: Optimal management and integration of electric vehicles to the grid: Dynamic programming and game theory approach. In: 2015 IEEE International Conference on Industrial Technology (ICIT), pp. 2673–2679. IEEE (2015)
125. Li, H., Liu, D., Wang, D.: Adaptive dynamic programming for solving non-zero-sum differential games. IFAC Proceedings Volumes **46**(20), 586–591 (2013)
126. Song, R., Wei, Q., Zhang, H., Lewis, F.L.: Discrete-time non-zero-sum games with completely unknown dynamics. IEEE Transactions on Cybernetics **51**(6), 2929–2943 (2019)
127. Liu, D., Li, H., Wang, D.: Online synchronous approximate optimal learning algorithm for multi-player non-zero-sum games with unknown dynamics. IEEE Transactions on Systems, Man, and Cybernetics: Systems **44**(8), 1015–1027 (2014)
128. Pontani, M.: Numerical solution of orbital combat games involving missiles and spacecraft. Dynamic Games and Applications **1**(4), 534–557 (2011)
129. Blasch, E.P., Pham, K., Shen, D.: Orbital satellite pursuit-evasion game-theoretical control. In: 2012 11th International Conference on Information Science, Signal Processing and their Applications (ISSPA), pp. 1007–1012. IEEE (2012)
130. Jagat, A., Sinclair, A.J.: Optimization of spacecraft pursuit-evasion game trajectories in the euler-hill reference frame. In: AIAA/AAS astrodynamics specialist conference, p. 4131 (2014)
131. Hafer, W.T., Reed, H.L.: Orbital pursuit-evasion hybrid spacecraft controllers. In: AIAA Guidance, Navigation, and Control Conference, p. 2000 (2015)
132. Jagat, A.: Spacecraft relative motion applications to pursuit-evasion games and control using angles-only navigation. Ph.D. thesis (2015)
133. Sun, S., Zhang, Q., Loxton, R., Li, B.: Numerical solution of a pursuit-evasion differential game involving two spacecraft in low earth orbit. Journal of Industrial and Management Optimization (JIMO) **11**(4), 1127–1147 (2015)
134. Jagat, A., Sinclair, A.J.: Nonlinear control for spacecraft pursuit-evasion game using the state-dependent riccati equation method. IEEE Transactions on Aerospace and Electronic Systems **53**(6), 3032–3042 (2017)
135. Venigalla, C., Scheeres, D.: Spacecraft rendezvous and pursuit/evasion analysis using reachable sets. In: 2018 Space Flight Mechanics Meeting, p. 0219 (2018)
136. Li, Z.y., Zhu, H., Yang, Z., Luo, Y.z.: A dimension-reduction solution of free-time differential games for spacecraft pursuit-evasion. Acta Astronautica **163**, 201–210 (2019)
137. Li, Z.y., Zhu, H., Yang, Z., Luo, Y.z.: Saddle point of orbital pursuit-evasion game under j 2-perturbed dynamics. Journal of Guidance, Control, and Dynamics **43**(9), 1733–1739 (2020)
138. Ye, D., Shi, M., Sun, Z.: Satellite proximate pursuit-evasion game with different thrust configurations. Aerospace Science and Technology **99**, 105715 (2020)
139. Tang, X., Ye, D., Huang, L., Sun, Z., Sun, J.: Pursuit-evasion game switching strategies for spacecraft with incomplete-information. Aerospace Science and Technology **119**, 107112 (2021)
140. Li, Z., Zhu, H., Luo, Y.: An escape strategy in orbital pursuit-evasion games with incomplete information. Science China Technological Sciences **64**(3), 559–570 (2021)
141. Franzini, G., Pollini, L., Innocenti, M.: H-infinity controller design for spacecraft terminal rendezvous on elliptic orbits using differential game theory. In: 2016 American Control Conference (ACC), pp. 7438–7443. IEEE (2016)
142. Sun, G., Zhou, M., Jiang, X.: Non-cooperative spacecraft proximity control considering target behavior uncertainty. Astrodynamics pp. 1–13 (2022)
143. Lang, X., de Ruiter, A.: Non-cooperative differential game based output feedback control for spacecraft attitude regulation. Acta Astronautica **193**, 370–380 (2022)
144. Ren, H., Jiang, B., Ma, Y.: Zero-sum differential game-based fault-tolerant control for a class of affine nonlinear systems. IEEE Transactions on Cybernetics (2022)
145. Meng, Q., Ma, Y., Jiang, B., Mao, Z., Ren, H., Wang, J.: Fault-tolerant control for spacecraft under actuator fault via zero-sum differential game theory. In: 2021 CAA Symposium on Fault Detection, Supervision, and Safety for Technical Processes (SAFEPROCESS), pp. 1–6. IEEE (2021)

146. Venigalla, C., Scheeres, D.J.: Delta-v-based analysis of spacecraft pursuit–evasion games. Journal of Guidance, Control, and Dynamics **44**(11), 1961–1971 (2021)
147. Innocenti, M., Tartaglia, V.: Game theoretic strategies for spacecraft rendezvous and motion synchronization. In: AIAA guidance, navigation, and control conference, p. 0873 (2016)
148. Chai, Y., Luo, J., Han, N., Xie, J.: Linear quadratic differential game approach for attitude takeover control of failed spacecraft. Acta Astronautica **175**, 142–154 (2020)
149. Han, N., Luo, J., Zheng, Z., Sun, J.: Distributed cooperative game method for attitude takeover of failed satellites using nanosatellites. Aerospace Science and Technology **106**, 106151 (2020)
150. Han, N., Luo, J., Zong, L.: Cooperative game method for on-orbit substructure transportation using modular robots. IEEE Transactions on Aerospace and Electronic Systems **58**(2), 1161–1175 (2021)
151. Wu, Y., Zheng, M., He, M., Zhang, D., He, W., Hua, B., Chen, Z., Wang, F.: Cooperative game theory-based optimal angular momentum management of hybrid attitude control actuator. IEEE Access **7**, 6853–6865 (2018)
152. Olfati-Saber, R., Fax, J.A., Murray, R.M.: Consensus and cooperation in networked multi-agent systems. Proceedings of the IEEE **95**(1), 215–233 (2007)
153. Bullo, F., Cortés, J., Martinez, S.: Distributed control of robotic networks: a mathematical approach to motion coordination algorithms, vol. 27. Princeton University Press (2009)
154. Hu, J., Zheng, W.X.: Emergent collective behaviors on coopetition networks. Physics Letters A **378**(26-27), 1787–1796 (2014)
155. Ren, W., Beard, R.W., Atkins, E.M.: Information consensus in multivehicle cooperative control. IEEE Control systems magazine **27**(2), 71–82 (2007)
156. Wu, C.W.: Synchronization in Coupled Chaotic Circuits & Systems, vol. 41. World Scientific (2002)
157. Wu, C.W.: Synchronization in networks of nonlinear dynamical systems coupled via a directed graph. Nonlinearity **18**(3), 1057 (2005)
158. Tanner, H.G., Jadbabaie, A., Pappas, G.J.: Flocking in fixed and switching networks. IEEE Transactions on Automatic control **52**(5), 863–868 (2007)
159. Su, H., Wang, X., Lin, Z.: Flocking of multi-agents with a virtual leader. IEEE transactions on automatic control **54**(2), 293–307 (2009)
160. Zhang, H.T., Cheng, Z., Chen, G., Li, C.: Model predictive flocking control for second-order multi-agent systems with input constraints. IEEE Transactions on Circuits and Systems I: Regular Papers **62**(6), 1599–1606 (2015)
161. Dimarogonas, D.V., Kyriakopoulos, K.J.: On the rendezvous problem for multiple nonholonomic agents. IEEE Transactions on automatic control **52**(5), 916–922 (2007)
162. Su, H., Wang, X., Chen, G.: Rendezvous of multiple mobile agents with preserved network connectivity. Systems & Control Letters **59**(5), 313–322 (2010)
163. Hashim, H.A., El-Ferik, S., Lewis, F.L.: Neuro-adaptive cooperative tracking control with prescribed performance of unknown higher-order nonlinear multi-agent systems. International Journal of Control **92**(2), 445–460 (2019)
164. Tang, Y., Xing, X., Karimi, H.R., Kocarev, L., Kurths, J.: Tracking control of networked multi-agent systems under new characterizations of impulses and its applications in robotic systems. IEEE Transactions on Industrial Electronics **63**(2), 1299–1307 (2015)
165. Zuo, Z.: Nonsingular fixed-time consensus tracking for second-order multi-agent networks. Automatica **54**, 305–309 (2015)
166. Liu, X., Ge, S.S., Goh, C.H.: Formation potential field for trajectory tracking control of multi-agents in constrained space. International Journal of Control **90**(10), 2137–2151 (2017)
167. Sun, X., Sam Ge, S., Zhang, J., Cao, X.: Region tracking control for high-order multi-agent systems in restricted space. IET Control Theory & Applications **10**(4), 396–406 (2016)
168. Han, Z., Guo, K., Xie, L., Lin, Z.: Integrated relative localization and leader–follower formation control. IEEE Transactions on Automatic Control **64**(1), 20–34 (2018)
169. Cao, K., Li, X., Xie, L.: Preview-based formation control. In: 2018 37th Chinese Control Conference (CCC), pp. 7031–7036. IEEE (2018)

170. Mei, J., Ren, W., Ma, G.: Distributed containment control for lagrangian networks with parametric uncertainties under a directed graph. Automatica **48**(4), 653–659 (2012)
171. Yoo, S.J.: Distributed adaptive containment control of uncertain nonlinear multi-agent systems in strict-feedback form. Automatica **49**(7), 2145–2153 (2013)
172. Miao, Z., Thakur, D., Erwin, R.S., Pierre, J., Wang, Y., Fierro, R.: Orthogonal vector field-based control for a multi-robot system circumnavigating a moving target in 3d. In: 2016 IEEE 55th Conference on Decision and Control (CDC), pp. 6004–6009. IEEE (2016)
173. Petersen, C., Fierro, R.: Network-lyapunov technique for spacecraft formation control. In: 2018 AIAA Guidance, Navigation, and Control Conference, p. 1110 (2018)
174. Li, D., Ma, G., He, W., Ge, S.S., Lee, T.H.: Cooperative circumnavigation control of networked microsatellites. IEEE Transactions on Cybernetics **50**(10), 4550–4555 (2019)
175. Li, T., Fu, M., Xie, L., Zhang, J.F.: Distributed consensus with limited communication data rate. IEEE Transactions on Automatic Control **56**(2), 279–292 (2010)
176. Shi, G., Johansson, K.H., Hong, Y.: Reaching an optimal consensus: Dynamical systems that compute intersections of convex sets. IEEE Transactions on Automatic Control **58**(3), 610–622 (2012)
177. Choi, J., Oh, S., Horowitz, R.: Distributed learning and cooperative control for multi-agent systems. Automatica **45**(12), 2802–2814 (2009)
178. Castellano, G., Cimino, M.G., Fanelli, A.M., Lazzerini, B., Marcelloni, F., Torsello, M.A.: A multi-agent system for enabling collaborative situation awareness via position-based stigmergy and neuro-fuzzy learning. Neurocomputing **135**, 86–97 (2014)
179. Zhao, S.: Affine formation maneuver control of multiagent systems. IEEE Transactions on Automatic Control **63**(12), 4140–4155 (2018)
180. Do, K.D.: Bounded controllers for formation stabilization of mobile agents with limited sensing ranges. IEEE Transactions on Automatic Control **52**(3), 569–576 (2007)
181. Ihle, I.A.F., Jouffroy, J., Fossen, T.I.: Formation control of marine surface craft: A lagrangian approach. IEEE Journal of Oceanic Engineering **31**(4), 922–934 (2006)
182. Fidan, B., Gazi, V., Zhai, S., Cen, N., Karataş, E.: Single-view distance-estimation-based formation control of robotic swarms. IEEE Transactions on Industrial Electronics **60**(12), 5781–5791 (2012)
183. Kang, S.M., Park, M.C., Ahn, H.S.: Distance-based cycle-free persistent formation: Global convergence and experimental test with a group of quadcopters. IEEE Transactions on Industrial Electronics **64**(1), 380–389 (2016)
184. Zhao, S., Zelazo, D.: Translational and scaling formation maneuver control via a bearing-based approach. IEEE Transactions on Control of Network Systems **4**(3), 429–438 (2015)
185. Cao, K., Han, Z., Li, X., Xie, L.: Ratio-of-distance rigidity theory with application to similar formation control. IEEE Transactions on Automatic Control **65**(6), 2598–2611 (2019)
186. Han, T., Lin, Z., Zheng, R., Fu, M.: A barycentric coordinate-based approach to formation control under directed and switching sensing graphs. IEEE Transactions on cybernetics **48**(4), 1202–1215 (2017)
187. Lin, Z., Wang, L., Han, Z., Fu, M.: Distributed formation control of multi-agent systems using complex laplacian. IEEE Transactions on Automatic Control **59**(7), 1765–1777 (2014)
188. Lin, Z., Wang, L., Chen, Z., Fu, M., Han, Z.: Necessary and sufficient graphical conditions for affine formation control. IEEE Transactions on Automatic Control **61**(10), 2877–2891 (2015)
189. Ferrari-Trecate, G., Egerstedt, M., Buffa, A., Ji, M.: Laplacian sheep: A hybrid, stop-go policy for leader-based containment control. In: International workshop on hybrid systems: Computation and control, pp. 212–226. Springer (2006)
190. Ji, M., Ferrari-Trecate, G., Egerstedt, M., Buffa, A.: Containment control in mobile networks. IEEE Transactions on Automatic Control **53**(8), 1972–1975 (2008)
191. Cheah, C.C., Hou, S.P., Slotine, J.J.E.: Region-based shape control for a swarm of robots. Automatica **45**(10), 2406–2411 (2009)
192. Notarstefano, G., Egerstedt, M., Haque, M.: Containment in leader–follower networks with switching communication topologies. Automatica **47**(5), 1035–1040 (2011)

193. Li, J., Ren, W., Xu, S.: Distributed containment control with multiple dynamic leaders for double-integrator dynamics using only position measurements. IEEE Transactions on Automatic Control **57**(6), 1553–1559 (2011)
194. Dong, X., Shi, Z., Lu, G., Zhong, Y.: Formation-containment analysis and design for high-order linear time-invariant swarm systems. International Journal of Robust and Nonlinear Control **25**(17), 3439–3456 (2015)
195. Dong, X., Li, Q., Ren, Z., Zhong, Y.: Output formation-containment analysis and design for general linear time-invariant multi-agent systems. Journal of the Franklin Institute **353**(2), 322–344 (2016)
196. Han, L., Dong, X., Li, Q., Ren, Z.: Formation-containment control for second-order multi-agent systems with time-varying delays. Neurocomputing **218**, 439–447 (2016)
197. Zheng, B., Mu, X.: Formation-containment control of second-order multi-agent systems with only sampled position data. International Journal of Systems Science **47**(15), 3609–3618 (2016). 10.1080/00207721.2015.1107148
198. Wang, Y.W., Liu, X.K., Xiao, J.W., Lin, X.: Output formation-containment of coupled heterogeneous linear systems under intermittent communication. Journal of the Franklin Institute **354**(1), 392–414 (2017)
199. Han, T., Chi, M., Guan, Z.H., Hu, B., Xiao, J.W., Huang, Y.: Distributed three-dimensional formation containment control of multiple unmanned aerial vehicle systems. Asian Journal of Control **19**(3), 1103–1113 (2017)
200. Cao, Y., Yu, W., Ren, W., Chen, G.: An overview of recent progress in the study of distributed multi-agent coordination. IEEE Transactions on Industrial informatics **9**(1), 427–438 (2012)
201. Oh, K.K., Park, M.C., Ahn, H.S.: A survey of multi-agent formation control. Automatica **53**, 424–440 (2015)
202. Miller, D.E., Davison, E.J.: An adaptive controller which provides an arbitrarily good transient and steady-state response. IEEE Transactions on Automatic Control **36**(1), 68–81 (1991)
203. Xu, H., Ioannou, P.A.: Robust adaptive control for a class of mimo nonlinear systems with guaranteed error bounds. IEEE Transactions on Automatic Control **48**(5), 728–742 (2003)
204. Xu, H., Ioannou, P.A.: Robust adaptive control of linearizable nonlinear single input systems with guaranteed error bounds. Automatica **40**(11), 1905–1911 (2004)
205. Li, P., Ma, J., Geng, L., Zheng, Z.: Robust adaptive sliding mode control for uncertain nonlinear mimo system with guaranteed error bounds. In: Proceedings of the 33rd Chinese Control Conference, pp. 109–114. IEEE (2014)
206. Li, P., Ma, J., Zheng, Z.: Robust adaptive sliding mode control for uncertain nonlinear mimo system with guaranteed steady state tracking error bounds. Journal of the Franklin Institute **353**(2), 303–321 (2016)
207. Di Cairano, S., Borrelli, F.: Constrained tracking with guaranteed error bounds. In: 52nd IEEE Conference on Decision and Control, pp. 3800–3805. IEEE (2013)
208. Di Cairano, S., Kalabić, U., Berntorp, K.: Vehicle tracking control on piecewise-clothoidal trajectories by mpc with guaranteed error bounds. In: 2016 IEEE 55th Conference on Decision and Control (CDC), pp. 709–714. IEEE (2016)
209. Yan, L., Hsu, L., Xiuxia, S.: A variable structure mrac with expected transient and steady-state performance. Automatica **42**(5), 805–813 (2006)
210. Tee, K.P., Ge, S.S., Tay, E.H.: Barrier lyapunov functions for the control of output-constrained nonlinear systems. Automatica **45**(4), 918–927 (2009)
211. Ngo, K.B., Mahony, R., Jiang, Z.P.: Integrator backstepping using barrier functions for systems with multiple state constraints. In: Proceedings of the 44th IEEE Conference on Decision and Control, pp. 8306–8312. IEEE (2005)
212. Tee, K.P., Ge, S.S.: Control of nonlinear systems with full state constraint using a barrier lyapunov function. In: Proceedings of the 48h IEEE Conference on Decision and Control (CDC) held jointly with 2009 28th Chinese Control Conference, pp. 8618–8623. IEEE (2009)
213. Ren, B., Ge, S.S., Tee, K.P., Lee, T.H.: Adaptive neural control for output feedback nonlinear systems using a barrier lyapunov function. IEEE Transactions on Neural Networks **21**(8), 1339–1345 (2010)

214. Tee, K.P., Ren, B., Ge, S.S.: Control of nonlinear systems with time-varying output constraints. Automatica **47**(11), 2511–2516 (2011)
215. Niu, B., Zhao, J.: Tracking control for output-constrained nonlinear switched systems with a barrier lyapunov function. International Journal of Systems Science **44**(5), 978–985 (2013)
216. Liu, L., Liu, Y.J., Chen, A., Tong, S., Chen, C.P.: Integral barrier lyapunov function-based adaptive control for switched nonlinear systems. Science China Information Sciences **63**(3), 1–14 (2020)
217. Li, Y., Li, T., Jing, X.: Indirect adaptive fuzzy control for input and output constrained nonlinear systems using a barrier lyapunov function. International Journal of Adaptive Control and Signal Processing **28**(2), 184–199 (2014)
218. He, W., Chen, Y., Yin, Z.: Adaptive neural network control of an uncertain robot with full-state constraints. IEEE Transactions on Cybernetics **46**(3), 620–629 (2016)
219. An, H., Xia, H., Wang, C.: Barrier lyapunov function-based adaptive control for hypersonic flight vehicles. Nonlinear Dynamics **88**(3), 1833–1853 (2017)
220. Xu, B., Shi, Z., Sun, F., He, W.: Barrier lyapunov function based learning control of hypersonic flight vehicle with aoa constraint and actuator faults. IEEE transactions on cybernetics **49**(3), 1047–1057 (2018)
221. Liu, L., Liu, Y.J., Li, D., Tong, S., Wang, Z.: Barrier lyapunov function-based adaptive fuzzy ftc for switched systems and its applications to resistance–inductance–capacitance circuit system. IEEE transactions on cybernetics **50**(8), 3491–3502 (2019)
222. Ilchmann, A., Trenn, S.: Input constrained funnel control with applications to chemical reactor models. Systems & control letters **53**(5), 361–375 (2004)
223. Schuster, H., Hackl, C., Westermaier, C., Schroder, D.: Funnel-control for electrical drives with uncertain parameters. In: 2005 International Power Engineering Conference, pp. 1–522. IEEE (2005)
224. Hackl, C.M., Schroder, D.: Funnel-control for nonlinear multi-mass flexible systems. In: IECON 2006-32nd Annual Conference on IEEE Industrial Electronics, pp. 4707–4712. IEEE (2006)
225. Hackl, C.M., Ji, Y., Schroder, D.: Enhanced funnel-control with improved performance. In: 2007 Mediterranean Conference on Control & Automation, pp. 1–6. IEEE (2007)
226. Ilchmann, A., Schuster, H.: Pi-funnel control for two mass systems. IEEE Transactions on Automatic Control **54**(4), 918–923 (2009)
227. Hopfe, N., Ilchmann, A., Ryan, E.P.: Funnel control with saturation: linear mimo systems. IEEE Transactions on Automatic Control **55**(2), 532–538 (2010)
228. Hopfe, N., Ilchmann, A., Ryan, E.P.: Funnel control with saturation: nonlinear siso systems. IEEE Transactions on Automatic Control **55**(9), 2177–2182 (2010)
229. Berger, T., Ilchmann, A., Reis, T.: Zero dynamics and funnel control of linear differential-algebraic systems. Mathematics of Control, Signals, and Systems **24**(3), 219–263 (2012)
230. Hackl, C.M., Hopfe, N., Ilchmann, A., Mueller, M., Trenn, S.: Funnel control for systems with relative degree two. SIAM Journal on Control and Optimization **51**(2), 965–995 (2013)
231. Berger, T., Reis, T.: Funnel control via funnel precompensator for minimum phase systems with relative degree two. IEEE Transactions on Automatic Control **63**(7), 2264–2271 (2017)
232. Chowdhury, D., Khalil, H.K.: Funnel control of higher relative degree systems. In: 2017 American Control Conference (ACC), pp. 598–603. IEEE (2017)
233. Berger, T., Lê, H.H., Reis, T.: Funnel control for nonlinear systems with known strict relative degree. Automatica **87**, 345–357 (2018)
234. Chowdhury, D., Khalil, H.K.: Funnel control for nonlinear systems with arbitrary relative degree using high-gain observers. Automatica **105**, 107–116 (2019)
235. Berger, T., Ilchmann, A., Ryan, E.P.: Funnel control of nonlinear systems. Mathematics of Control, Signals, and Systems **33**(1), 151–194 (2021)
236. Liberzon, D., Trenn, S.: The bang-bang funnel controller for uncertain nonlinear systems with arbitrary relative degree. IEEE Transactions on Automatic Control **58**(12), 3126–3141 (2013)
237. Liu, X., Wang, H., Gao, C., Chen, M.: Adaptive fuzzy funnel control for a class of strict feedback nonlinear systems. Neurocomputing **241**, 71–80 (2017)

238. Wang, H., Zou, Y., Liu, P.X., Liu, X.: Robust fuzzy adaptive funnel control of nonlinear systems with dynamic uncertainties. Neurocomputing **314**, 299–309 (2018)
239. Liu, C., Wang, H., Liu, X., Zhou, Y.: Adaptive fuzzy funnel control for nonlinear systems with input deadzone and saturation. International Journal of Systems Science **51**(9), 1542–1555 (2020)
240. Liu, C., Liu, X., Wang, H., Zhou, Y., Lu, S., Xu, B.: Event-triggered adaptive tracking control for uncertain nonlinear systems based on a new funnel function. ISA transactions **99**, 130–138 (2020)
241. Hackl, C., Kennel, R.: Position funnel control for rigid revolute joint robotic manipulators with known inertia matrix. In: 2012 20th Mediterranean Conference on Control and Automation (MED), pp. 615–620. IEEE (2012)
242. Chen, Q., Tang, X., Nan, Y., Ren, X.: Finite-time neural funnel control for motor servo systems with unknown input constraint. Journal of Systems Science and Complexity **30**(3), 579–594 (2017)
243. Berger, T., Otto, S., Reis, T., Seifried, R.: Combined open-loop and funnel control for under-actuated multibody systems. Nonlinear Dynamics **95**(3), 1977–1998 (2019)
244. Bechlioulis, C.P., Rovithakis, G.A.: Prescribed performance adaptive control of siso feedback linearizable systems with disturbances. In: 2008 16th Mediterranean Conference on Control and Automation, pp. 1035–1040. IEEE (2008)
245. Bechlioulis, C.P., Rovithakis, G.A.: Adaptive control with guaranteed transient and steady state tracking error bounds for strict feedback systems. Automatica **45**(2), 532–538 (2009)
246. Bechlioulis, C.P., Rovithakis, G.A.: Prescribed performance adaptive control for multi-input multi-output affine in the control nonlinear systems. IEEE Transactions on Automatic Control **55**(5), 1220–1226 (2010)
247. Bechlioulis, C.P., Rovithakis, G.A.: Robust partial-state feedback prescribed performance control of cascade systems with unknown nonlinearities. IEEE Transactions on Automatic Control **56**(9), 2224–2230 (2011)
248. Kostarigka, A.K., Rovithakis, G.A.: Prescribed performance output feedback/ observer-free robust adaptive control of uncertain systems using neural networks. IEEE Transactions on Systems, Man, and Cybernetics, Part B (Cybernetics) **41**(6), 1483–1494 (2011)
249. Kostarigka, A.K., Rovithakis, G.A.: Adaptive dynamic output feedback neural network control of uncertain mimo nonlinear systems with prescribed performance. IEEE transactions on neural networks and learning systems **23**(1), 138–149 (2011)
250. Bechlioulis, C.P., Rovithakis, G.A.: A low-complexity global approximation-free control scheme with prescribed performance for unknown pure feedback systems. Automatica **50**(4), 1217–1226 (2014)
251. Yang, Y., Ge, C., Wang, H., Li, X., Hua, C.: Adaptive neural network based prescribed performance control for teleoperation system under input saturation. Journal of the Franklin Institute **352**(5), 1850–1866 (2015)
252. Li, Y., Tong, S.: Prescribed performance adaptive fuzzy output-feedback dynamic surface control for nonlinear large-scale systems with time delays. Information Sciences **292**, 125–142 (2015)
253. Sui, S., Tong, S., Li, Y.: Observer-based fuzzy adaptive prescribed performance tracking control for nonlinear stochastic systems with input saturation. Neurocomputing **158**, 100–108 (2015)
254. Theodorakopoulos, A., Rovithakis, G.A.: A simplified adaptive neural network prescribed performance controller for uncertain mimo feedback linearizable systems. IEEE transactions on neural networks and learning systems **26**(3), 589–600 (2014)
255. Hua, C., Zhang, L., Guan, X.: Decentralized output feedback adaptive nn tracking control for time-delay stochastic nonlinear systems with prescribed performance. IEEE transactions on neural networks and learning systems **26**(11), 2749–2759 (2015)
256. Chen, M., Ge, S.S.: Adaptive neural output feedback control of uncertain nonlinear systems with unknown hysteresis using disturbance observer. IEEE Transactions on Industrial Electronics **62**(12), 7706–7716 (2015)

257. Zhang, J.X., Yang, G.H.: Fuzzy adaptive output feedback control of uncertain nonlinear systems with prescribed performance. IEEE transactions on cybernetics **48**(5), 1342–1354 (2018)
258. Xia, X., Zhang, T., Yi, Y., Shen, Q.: Adaptive prescribed performance control of output feedback systems including input unmodeled dynamics. Neurocomputing **190**, 226–236 (2016)
259. Zhao, K., Song, Y., Ma, T., He, L.: Prescribed performance control of uncertain euler–lagrange systems subject to full-state constraints. IEEE transactions on neural networks and learning systems **29**(8), 3478–3489 (2017)
260. Dimanidis, I.S., Bechlioulis, C.P., Rovithakis, G.A.: Output feedback approximation-free prescribed performance tracking control for uncertain mimo nonlinear systems. IEEE Transactions on Automatic Control **65**(12), 5058–5069 (2020)
261. Theodorakopoulos, A., Rovithakis, G.A.: Low-complexity prescribed performance control of uncertain mimo feedback linearizable systems. IEEE Transactions on Automatic Control **61**(7), 1946–1952 (2015)
262. Wang, W., Wang, D., Peng, Z., Li, T.: Prescribed performance consensus of uncertain nonlinear strict-feedback systems with unknown control directions. IEEE Transactions on Systems, Man, and Cybernetics: Systems **46**(9), 1279–1286 (2015)
263. Hashim, H.A., El-Ferik, S., Lewis, F.L.: Adaptive synchronisation of unknown nonlinear networked systems with prescribed performance. International Journal of Systems Science **48**(4), 885–898 (2017)
264. Liang, H., Zhang, Y., Huang, T., Ma, H.: Prescribed performance cooperative control for multiagent systems with input quantization. IEEE Transactions on cybernetics **50**(5), 1810–1819 (2020)
265. Wang, W., Liang, H., Pan, Y., Li, T.: Prescribed performance adaptive fuzzy containment control for nonlinear multiagent systems using disturbance observer. IEEE transactions on cybernetics **50**(9), 3879–3891 (2020)
266. Li, Y., Tong, S., Liu, L., Feng, G.: Adaptive output-feedback control design with prescribed performance for switched nonlinear systems. Automatica **80**, 225–231 (2017)
267. Zhai, D., Xi, C., An, L., Dong, J., Zhang, Q.: Prescribed performance switched adaptive dynamic surface control of switched nonlinear systems with average dwell time. IEEE Transactions on Systems, Man, and Cybernetics: Systems **47**(7), 1257–1269 (2016)
268. Hua, C., Liu, G., Li, L., Guan, X.: Adaptive fuzzy prescribed performance control for nonlinear switched time-delay systems with unmodeled dynamics. IEEE Transactions on Fuzzy Systems **26**(4), 1934–1945 (2017)
269. Liu, H., Li, S., Cao, J., Li, G., Alsaedi, A., Alsaadi, F.E.: Adaptive fuzzy prescribed performance controller design for a class of uncertain fractional-order nonlinear systems with external disturbances. Neurocomputing **219**, 422–430 (2017)
270. Li, Y., Tong, S.: Adaptive fuzzy control with prescribed performance for block-triangular-structured nonlinear systems. IEEE Transactions on Fuzzy Systems **26**(3), 1153–1163 (2018)
271. Li, Y., Shao, X., Tong, S.: Adaptive fuzzy prescribed performance control of nontriangular structure nonlinear systems. IEEE Transactions on Fuzzy Systems **28**(10), 2416–2426 (2020)
272. Sun, W., Wu, Y.Q., Sun, Z.Y.: Command filter-based finite-time adaptive fuzzy control for uncertain nonlinear systems with prescribed performance. IEEE Transactions on Fuzzy Systems **28**(12), 3161–3170 (2020)
273. Zhang, J.X., Yang, G.H.: Prescribed performance fault-tolerant control of uncertain nonlinear systems with unknown control directions. IEEE Transactions on Automatic Control **62**(12), 6529–6535 (2017)
274. Huang, H., He, W., Li, J., Xu, B., Yang, C., Zhang, W.: Disturbance observer-based fault-tolerant control for robotic systems with guaranteed prescribed performance. IEEE Transactions on Cybernetics **52**(2), 772–783 (2020)
275. Liu, D., Yang, G.H.: Data-driven adaptive sliding mode control of nonlinear discrete-time systems with prescribed performance. IEEE Transactions on Systems, Man, and Cybernetics: Systems **49**(12), 2598–2604 (2019)

276. Qiu, J., Sun, K., Wang, T., Gao, H.: Observer-based fuzzy adaptive event-triggered control for pure-feedback nonlinear systems with prescribed performance. IEEE Transactions on Fuzzy Systems **27**(11), 2152–2162 (2019)
277. Sun, K., Jianbin, Q., Karimi, H.R., Fu, Y.: Event-triggered robust fuzzy adaptive finite-time control of nonlinear systems with prescribed performance. IEEE Transactions on Fuzzy Systems (2021)
278. Bechlioulis, C.P., Doulgeri, Z., Rovithakis, G.A.: Neuro-adaptive force/position control with prescribed performance and guaranteed contact maintenance. IEEE Transactions on Neural Networks **21**(12), 1857–1868 (2010)
279. Karayiannidis, Y., Doulgeri, Z.: Model-free robot joint position regulation and tracking with prescribed performance guarantees. Robotics and Autonomous Systems **60**(2), 214–226 (2012)
280. Kostarigka, A.K., Doulgeri, Z., Rovithakis, G.A.: Prescribed performance tracking for flexible joint robots with unknown dynamics and variable elasticity. Automatica **49**(5), 1137–1147 (2013)
281. Na, J., Chen, Q., Ren, X., Guo, Y.: Adaptive prescribed performance motion control of servo mechanisms with friction compensation. IEEE Transactions on Industrial Electronics **61**(1), 486–494 (2013)
282. Bu, X., Wu, X., Zhu, F., Huang, J., Ma, Z., Zhang, R.: Novel prescribed performance neural control of a flexible air-breathing hypersonic vehicle with unknown initial errors. ISA transactions **59**, 149–159 (2015)
283. Wang, Y., Hu, J.: Improved prescribed performance control for air-breathing hypersonic vehicles with unknown deadzone input nonlinearity. ISA transactions **79**, 95–107 (2018)
284. Yang, Q., Chen, M.: Adaptive neural prescribed performance tracking control for near space vehicles with input nonlinearity. Neurocomputing **174**, 780–789 (2016)
285. Guan, Z., Ma, Y., Zheng, Z., Guo, N.: Prescribed performance control for automatic carrier landing with disturbance. Nonlinear Dynamics **94**(2), 1335–1349 (2018)
286. Xingling, S., Biao, T., Wei, Y., Wendong, Z.: Estimator-based mlp neuroadaptive dynamic surface containment control with prescribed performance for multiple quadrotors. Aerospace Science and Technology **97**, 105620 (2020)
287. Zheng, Z., Feroskhan, M.: Path following of a surface vessel with prescribed performance in the presence of input saturation and external disturbances. IEEE/ASME Transactions on Mechatronics **22**(6), 2564–2575 (2017)
288. Jia, Z., Hu, Z., Zhang, W.: Adaptive output-feedback control with prescribed performance for trajectory tracking of underactuated surface vessels. ISA transactions **95**, 18–26 (2019)
289. Dai, S.L., He, S., Wang, M., Yuan, C.: Adaptive neural control of underactuated surface vessels with prescribed performance guarantees. IEEE transactions on neural networks and learning systems **30**(12), 3686–3698 (2019)
290. Dai, S.L., He, S., Lin, H., Wang, C.: Platoon formation control with prescribed performance guarantees for usvs. IEEE Transactions on Industrial Electronics **65**(5), 4237–4246 (2017)
291. He, S., Wang, M., Dai, S.L., Luo, F.: Leader–follower formation control of usvs with prescribed performance and collision avoidance. IEEE Transactions on Industrial Informatics **15**(1), 572–581 (2019)
292. Bechlioulis, C.P., Karras, G.C., Heshmati-Alamdari, S., Kyriakopoulos, K.J.: Trajectory tracking with prescribed performance for underactuated underwater vehicles under model uncertainties and external disturbances. IEEE Transactions on Control Systems Technology **25**(2), 429–440 (2016)
293. Li, J., Du, J., Sun, Y., Lewis, F.L.: Robust adaptive trajectory tracking control of underactuated autonomous underwater vehicles with prescribed performance. International Journal of Robust and Nonlinear Control **29**(14), 4629–4643 (2019)
294. Psomopoulou, E., Theodorakopoulos, A., Doulgeri, Z., Rovithakis, G.A.: Prescribed performance tracking of a variable stiffness actuated robot. IEEE Transactions on Control Systems Technology **23**(5), 1914–1926 (2015)

295. Karayiannidis, Y., Papageorgiou, D., Doulgeri, Z.: A model-free controller for guaranteed prescribed performance tracking of both robot joint positions and velocities. IEEE Robotics and Automation Letters **1**(1), 267–273 (2016)
296. Wang, M., Yang, A.: Dynamic learning from adaptive neural control of robot manipulators with prescribed performance. IEEE Transactions on Systems, Man, and Cybernetics: Systems **47**(8), 2244–2255 (2017)
297. Jing, C., Xu, H., Niu, X.: Adaptive sliding mode disturbance rejection control with prescribed performance for robotic manipulators. ISA transactions **91**, 41–51 (2019)
298. Guo, Q., Zhang, Y., Celler, B.G., Su, S.W.: Neural adaptive backstepping control of a robotic manipulator with prescribed performance constraint. IEEE transactions on neural networks and learning systems **30**(12), 3572–3583 (2019)
299. Wang, S., Na, J., Ren, X.: Rise-based asymptotic prescribed performance tracking control of nonlinear servo mechanisms. IEEE Transactions on Systems, Man, and Cybernetics: Systems **48**(12), 2359–2370 (2018)
300. Huang, Y., Na, J., Wu, X., Liu, X., Guo, Y.: Adaptive control of nonlinear uncertain active suspension systems with prescribed performance. ISA transactions **54**, 145–155 (2015)
301. Liu, Y.J., Zeng, Q., Tong, S., Chen, C.P., Liu, L.: Actuator failure compensation-based adaptive control of active suspension systems with prescribed performance. IEEE Transactions on Industrial Electronics **67**(8), 7044–7053 (2019)
302. Liu, Y., Liu, X., Jing, Y., Zhou, S.: Adaptive backstepping h_{∞} tracking control with prescribed performance for internet congestion. ISA transactions **72**, 92–99 (2018)
303. Hua, C., Chen, J., Li, Y.: Leader-follower finite-time formation control of multiple quadrotors with prescribed performance. International Journal of Systems Science **48**(12), 2499–2508 (2017)
304. Li, P., Xu, S., Chu, Y., Chen, W., Zhang, Z.: Finite-time leader-following rendezvous for Euler–Lagrange multi-agent systems with an uncertain leader. Transactions of the Institute of Measurement and Control **40**(6), 1766–1775 (2018)
305. Zheng, Z., Lau, G.K., Xie, L.: Event-triggered control for a saturated nonlinear system with prescribed performance and finite-time convergence. International Journal of Robust and Nonlinear Control **28**(17), 5312–5325 (2018)
306. Su, H., Zhang, W.: Finite-time prescribed performance adaptive fuzzy fault-tolerant control for nonstrict-feedback nonlinear systems. International Journal of Adaptive Control and Signal Processing **33**(9), 1407–1424 (2019)
307. Yin, Z., Luo, J., Wei, C.: Robust prescribed performance control for euler–lagrange systems with practically finite-time stability. European Journal of Control **52**, 1–10 (2020)
308. Sui, S., Chen, C.P., Tong, S.: Finite-time adaptive fuzzy prescribed performance control for high-order stochastic nonlinear systems. IEEE Transactions on Fuzzy Systems (2021)
309. Fu, Z., Wang, N., Song, S., Wang, T.: Adaptive fuzzy finite-time tracking control of stochastic high-order nonlinear systems with a class of prescribed performance. IEEE Transactions on Fuzzy Systems (2020)
310. Cui, G., Yu, J., Shi, P.: Observer-based finite-time adaptive fuzzy control with prescribed performance for nonstrict-feedback nonlinear systems. IEEE Transactions on Fuzzy Systems (2020)
311. Huang, S., Yan, Z., Zeng, G., Zhang, Z., Zhu, Z.: Finite-time prescribed performance control of switched nonlinear systems with input quantisation. International Journal of Systems Science pp. 1–17 (2020)
312. Yu, Z., Zhang, Y., Liu, Z., Qu, Y., Su, C.Y., Jiang, B.: Decentralized finite-time adaptive fault-tolerant synchronization tracking control for multiple uavs with prescribed performance. Journal of the Franklin Institute **357**(16), 11830–11862 (2020)
313. Tran, X.T., Oh, H.: Prescribed performance adaptive finite-time control for uncertain horizontal platform systems. ISA transactions **103**, 122–130 (2020)
314. Liang, Q., Yang, Q., Meng, W., Li, Y.: Adaptive finite-time control for turbo-generator of power systems with prescribed performance. Asian Journal of Control (2021)

315. An, S.Y., Chen, M., Wang, H.Q., Wu, L.B.: Fast finite-time dynamic surface tracking control of a single-joint manipulator system with prescribed performance. International Journal of Systems Science pp. 1–13 (2021)
316. Zhang, R., Xu, B., Zhao, W.: Finite-time prescribed performance control of mems gyroscopes. Nonlinear Dynamics pp. 1–12 (2020)
317. Liu, Y., Liu, X., Jing, Y.: Adaptive neural networks finite-time tracking control for non-strict feedback systems via prescribed performance. Information Sciences **468**, 29–46 (2018)
318. Jing, Y., Liu, Y., Zhou, S.: Prescribed performance finite-time tracking control for uncertain nonlinear systems. Journal of Systems Science and Complexity **32**(3), 803–817 (2019)
319. Liu, Y., Liu, X., Jing, Y.: Adaptive fuzzy finite-time stability of uncertain nonlinear systems based on prescribed performance. Fuzzy Sets and Systems **374**, 23–39 (2019)
320. Wang, X., Wu, Q., Yin, X.: Adaptive finite-time prescribed performance control of switched nonlinear systems with unknown actuator dead-zone. International Journal of Systems Science **51**(1), 133–145 (2020)
321. Gao, C., Zhang, C., Liu, X., Wang, H., Wang, L.: Event-triggering based adaptive neural tracking control for a class of pure-feedback systems with finite-time prescribed performance. Neurocomputing **382**, 221–232 (2020)
322. Wang, X., Niu, B., Zhao, P., Song, X.: Neural networks-based adaptive finite-time prescribed performance fault-tolerant control of switched nonlinear systems. International Journal of Adaptive Control and Signal Processing **35**(4), 532–548 (2021)
323. He, S., Li, X.: Decentralised adaptive prescribed performance finite-time tracking control for a class of nonlinear interconnected systems with unknown control directions. International Journal of Control pp. 1–20 (2020)
324. Wang, H., Bai, W., Zhao, X., Liu, P.X.: Finite-time-prescribed performance-based adaptive fuzzy control for strict-feedback nonlinear systems with dynamic uncertainty and actuator faults. IEEE Transactions on Cybernetics (2021)
325. Zhou, X., Gao, C., Li, Z.g., Ouyang, X.y., Wu, L.b.: Observer-based adaptive fuzzy finite-time prescribed performance tracking control for strict-feedback systems with input dead-zone and saturation. Nonlinear Dynamics **103**(2), 1645–1661 (2021)
326. Jiang, T., Huang, J., Li, B.: Composite adaptive finite-time control for quadrotors via prescribed performance. Journal of the Franklin Institute **357**(10), 5878–5901 (2020)
327. Zhou, Z.G., Zhang, Y.A., Shi, X.N., Zhou, D.: Robust attitude tracking for rigid spacecraft with prescribed transient performance. International Journal of Control **90**(11), 2471–2479 (2017)
328. Hu, Q., Shao, X., Guo, L.: Adaptive fault-tolerant attitude tracking control of spacecraft with prescribed performance. IEEE/ASME Transactions on Mechatronics **23**(1), 331–341 (2017)
329. Yin, Z., Luo, J., Wei, C.: Novel adaptive saturated attitude tracking control of rigid spacecraft with guaranteed transient and steady-state performance. Journal of Aerospace Engineering **31**(5), 04018062 (2018)
330. Luo, J., Yin, Z., Wei, C., Yuan, J.: Low-complexity prescribed performance control for spacecraft attitude stabilization and tracking. Aerospace Science and Technology **74**, 173–183 (2018)
331. Shao, X., Hu, Q., Shi, Y., Jiang, B.: Fault-tolerant prescribed performance attitude tracking control for spacecraft under input saturation. IEEE Transactions on Control Systems Technology **28**(2), 574–582 (2018)
332. Wei, C., Luo, J., Dai, H., Duan, G.: Learning-based adaptive attitude control of spacecraft formation with guaranteed prescribed performance. IEEE transactions on cybernetics **49**(11), 4004–4016 (2018)
333. Yin, Z., Suleman, A., Luo, J., Wei, C.: Appointed-time prescribed performance attitude tracking control via double performance functions. Aerospace Science and Technology **93**, 105337 (2019)
334. Liu, M., Shao, X., Ma, G.: Appointed-time fault-tolerant attitude tracking control of spacecraft with double-level guaranteed performance bounds. Aerospace Science and Technology **92**, 337–346 (2019)

335. Tao, M., Chen, Q., He, X., Sun, M.: Adaptive fixed-time fault-tolerant control for rigid spacecraft using a double power reaching law. International Journal of Robust and Nonlinear Control **29**(12), 4022–4040 (2019)
336. Yue, C., Wang, F., Cao, X., Shen, Q., Chen, X.: Robust fault-tolerant attitude tracking with guaranteed prescribed performance. Journal of the Franklin Institute **357**(1), 229–253 (2020)
337. Xia, K., Son, H.: Guaranteed performance based adaptive attitude tracking of spacecraft with control constraints. Advances in Space Research **65**(3), 1095–1104 (2020)
338. Huang, X., Biggs, J.D., Duan, G.: Post-capture attitude control with prescribed performance. Aerospace Science and Technology **96**, 105572 (2020)
339. Wang, C., Guo, L., Wen, C., Hu, Q., Qiao, J.: Event-triggered adaptive attitude tracking control for spacecraft with unknown actuator faults. IEEE Transactions on Industrial Electronics **67**(3), 2241–2250 (2019)
340. Hu, Y., Geng, Y., Wu, B., Wang, D.: Model-free prescribed performance control for spacecraft attitude tracking. IEEE Transactions on Control Systems Technology **29**(1), 165–179 (2020)
341. Fu, J., Liu, M., Cao, X., Li, A.: Robust neural-network-based quasi-sliding-mode control for spacecraft-attitude maneuvering with prescribed performance. Aerospace Science and Technology **112** (2021)
342. Yao, Q.: Robust adaptive finite-time prescribed performance attitude tracking control of spacecraft. International Journal of Aeronautical and Space Sciences pp. 1–11 (2021)
343. Luo, J., Wei, C., Dai, H., Yin, Z., Wei, X., Yuan, J.: Robust inertia-free attitude takeover control of postcapture combined spacecraft with guaranteed prescribed performance. ISA transactions **74**, 28–44 (2018)
344. Wei, C., Luo, J., Dai, H., Bian, Z., Yuan, J.: Learning-based adaptive prescribed performance control of postcapture space robot-target combination without inertia identifications. Acta Astronautica **146**, 228–242 (2018)
345. Zhang, C., Ma, G., Sun, Y., Li, C.: Observer-based prescribed performance attitude control for flexible spacecraft with actuator saturation. ISA transactions **89**, 84–95 (2019)
346. Zhang, C., Ma, G., Sun, Y., Li, C.: Prescribed performance adaptive attitude tracking control for flexible spacecraft with active vibration suppression. Nonlinear Dynamics **96**(3), 1909–1926 (2019)
347. Tao, J., Zhang, T., Liu, Q.: Novel finite-time adaptive neural control of flexible spacecraft with actuator constraints and prescribed attitude tracking performance. Acta Astronautica **179**, 646–658 (2021)
348. Golestani, M., Esmaeilzadeh, S.M., Mobayen, S.: Fixed-time control for high-precision attitude stabilization of flexible spacecraft. European Journal of Control **57**, 222–231 (2021)
349. Shao, X., Hu, Q., Shi, Y.: Adaptive pose control for spacecraft proximity operations with prescribed performance under spatial motion constraints. IEEE Transactions on Control Systems Technology **29**(4), 1405–1419 (2020)
350. Xia, K., Zou, Y.: Neuroadaptive saturated control for relative motion based noncooperative spacecraft proximity with prescribed performance. Acta Astronautica **180**, 361–369 (2021)
351. Lu, X., Jia, Y.: Adaptive coordinated control of uncertain free-floating space manipulators with prescribed control performance. Nonlinear Dynamics **97**(2), 1541–1566 (2019)
352. Ma, J., Wen, H., Jin, D.: Pde model-based boundary control of a spacecraft with double flexible appendages under prescribed performance. Advances in Space Research **65**(1), 586–597 (2020)
353. Zhu, Y., Qiao, J., Guo, L.: Adaptive sliding mode disturbance observer-based composite control with prescribed performance of space manipulators for target capturing. IEEE Transactions on Industrial Electronics **66**(3), 1973–1983 (2018)

Chapter 2
Spacecraft Model and Preliminaries

Abstract The *performance guaranteed control problem for spacecraft* is solved by the designed control algorithms in this book. For the single spacecraft system and the proposed control algorithms, its related attitude model, performance guaranteed technique, and control theory are common, whether it is about control, filtering, or optimization. For the multi-spacecraft system and the proposed control algorithms, its associated differential games and graph theory are common. This chapter will serve as the preliminaries and focus on the spacecraft model, stability theory, performance guaranteed technique, command-filter technique, optimization methods, differential games, and graph theory involved in this book.

2.1 Introduction

The performance guaranteed control problem for the single spacecraft is thoroughly studied in each chapter in Part I, the performance guaranteed differential game methods for multi-spacecraft systems are fully studied in Part II, and the performance guaranteed cooperative approach for multi-spacecraft systems is fully studied in Part III. The various controllers are designed with similar models and different fundamental theories. To keep the book concise, this chapter presents the relevant common issues involved in the performance guaranteed control of spacecraft.

In this chapter, the space coordinate system is first given, used to describe the attitude motion of the spacecraft based on quaternion and the relative orbital motion of spacecraft formation. Combined with the attitude dynamics of the rigid-body spacecraft, the corresponding tracking error model of the spacecraft attitude is derived. Considering the formation behavior of multiple spacecraft, the networked multi-spacecraft relative motion systems are constructed. As the research basis, this chapter introduces the stability theories and control techniques relevant to the results of these studies.

Y. Gao and D. Li, *Spacecraft Maneuver with Performance Guaranteed*,
https://doi.org/10.1007/978-981-99-4653-2_2

2.2 Coordinate Systems

The description of the spacecraft attitude motion requires more than two reference coordinate systems. Considering the spacecraft's on-orbit service mission, more coordinate systems are required in the engineering. To describe the attitude tracking motion of the on-orbit service spacecraft relative to the target spacecraft in space during the on-orbit service mission, we first define the coordinate systems as shown in Fig. 2.1.

In order to form the basis of the attitude and position of the on-orbit service spacecraft with respect to the target spacecraft, we use the standard definition of the Earth-Centered Inertial (ECI) frame $\mathfrak{F}_I$ as the essential reference coordinate frame [1]. It is attached to the center of the Earth O_I with x_I axis towards spring equinox γ, z_I axis towards celestial north, y_I axis being mutually perpendicular to the x_I and z_I axis such that forms a right-hand coordinate frame.

The attitude movement of the target spacecraft and the on-orbit service spacecraft can be described by the rotation of their body-fixed Cartesian frames $\mathfrak{F}_{bt}$ and $\mathfrak{F}_{bs}$ relative to the $\mathfrak{F}_I$. The defined frames $\mathfrak{F}_{bt}$ and $\mathfrak{F}_{bs}$ are located at centroid on the target spacecraft and the on-orbit service spacecraft, respectively. The coordinate axes are the main inertia axes of the corresponding spacecraft. Moreover, the relative attitude motion between them can be expressed by the relative rotation of $\mathfrak{F}_{bs}$ and $\mathfrak{F}_{bt}$.

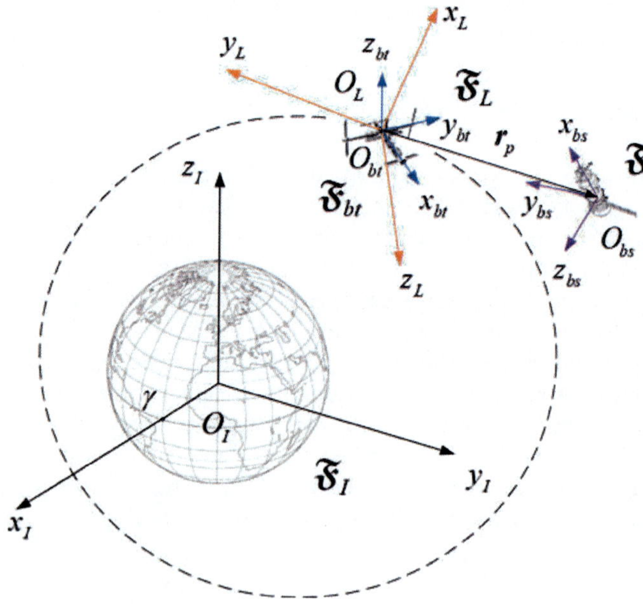

Fig. 2.1 The space coordinate frames

Since some space missions are related to the attitude relative to the orbital coordinate system, the definition of the relevant coordinate system is also given in this section. Considering the relative orbital motion of the spacecraft, the Local-Vertical/Local-Horizontal (LVLH) reference frame $\mathfrak{F}_L$ is defined as shown in Fig. 2.1. The origin of $\mathfrak{F}_L$ is located at the centroid on the target spacecraft, the x_L axis points to the radial direction from the Earth centroid, the z_L axis points to the orbital angular momentum direction, and y_L axis being mutually perpendicular to the x_L and z_L axis such that forms a right-hand coordinate frame. Simultaneously, we define a series of rotation matrices to describe the rotational motion between two coordinate systems, R_{ts} represents the rotation from $\mathfrak{F}_{bs}$ to $\mathfrak{F}_{bt}$, The rotation matrices $R_{li}(\mathfrak{F}_I \to \mathfrak{F}_L)$, $R_{si}(\mathfrak{F}_I \to \mathfrak{F}_{bs})$, $R_{ti}(\mathfrak{F}_I \to \mathfrak{F}_{bt})$, are similarly defined.

2.3 Spacecraft Systems

2.3.1 *Quaternion*

The spacecraft attitude motion can be described by the direction cosine matrix, Euler angle, Euler parameter (quaternion), Rodrigues parameter, modified Rodrigues parameter, etc. [1–4]. Quaternions q defined in $\mathbb{R}^4$ are described as follows [2, 5–7],

$$q = q_0 + q_1 i + q_2 j + q_3 k \tag{2.1}$$

where $q_0, q_1, q_2, q_3 \in \mathbb{R}$. Complex units i, j, k satisfy $i^2 = j^2 = k^2 = -1, i = jk$, $j = ki$, and $k = ij$. For the convenience of expression, the quaternion is written in a compact form as $q = [q_0, q_v^T]^T$, where q_0 is the scalar part of the quaternion, $q_v = [q_1, q_2, q_3]^T$ is the vector part. For any quaternion q, q_1 and q_2, there are the following basic operations and properties:

(1) Addition and subtraction:

$$q_1 \pm q_2 = [q_{10} \pm q_{20}, q_{1v}^T \pm q_{2v}^T]^T \tag{2.2}$$

(2) Proportion calculation:

$$\lambda q = [\lambda q_0, \lambda q_v^T]^T, \lambda \in \mathbb{R} \tag{2.3}$$

(3) Euclidean norm:

$$||q||^2 = q \cdot q = q_0^2 + q_v^T q_v \tag{2.4}$$

(4) Quaternion multiplication:

$$q_1 \circ q_2 = [q_{10}q_{20} - q_{1v}^T q_{2v}, (q_{10}q_{2v} + q_{20}q_{1v} + q_{1v} \times q_{2v})^T]^T \triangleq Q(q_1)q_2 \tag{2.5}$$

where $Q(q_1)$ represents the parameter matrix about q_1.

$$Q(q_1) \triangleq \begin{bmatrix} q_{10} & -q_{1v}^T \\ q_{1v} & q_{10} I_{3\times3} + q_{1v}^\times \end{bmatrix} \tag{2.6}$$

where $I_{3\times3}$ represents the 3×3 unit matrix, and $a^\times$ represents the skew-symmetric matrix of the vector $\forall a = [a_1, a_2, a_3]^T$, which is defined as follows

$$a^\times \triangleq \begin{bmatrix} 0 & -a_3 & a_2 \\ a_3 & 0 & -a_1 \\ -a_2 & a_1 & 0 \end{bmatrix} \tag{2.7}$$

When $||q||^2 = 1$, q is called a unit quaternion. According to Euler's rotation theorem, in three-dimensional Euler space, the rotation of an object can be described by the unit direction vector (Eulerian rotation axis, fixed axis through fixed point) $e \in \mathbb{R}^3$ and the rotation angle (Eulerian angle) $\theta \in \mathbb{R}$ [1]. This rotation process can be further described by the unit quaternion. Considering the fixed point as the coordinate origin, the unit quaternion related to rotation is defined as $q = [q_0, q_v^T]^T$, where $q_0 = \cos(\frac{\theta}{2})$, $q_v = e \sin(\frac{\theta}{2})$.

Unit quaternions are used to describe spacecraft attitude motions in Part I. Considering the inherent constraints of the unit quaternion, that is, $||q||^2 = 1$. Thus, in the system, we only need to reflect the vector part of the quaternion, and the scalar part is determined by the constraints. Under the condition of the unit quaternion constraint, the spacecraft attitude motion described by quaternion can still be a three-dimensional system. After this section, the quaternion-based kinematics and dynamics model of spacecraft attitude is derived.

2.3.2 *Spacecraft Attitude Systems*

Considering the tracking problem of a rigid spacecraft with the body-fixed frame located at the center of mass, the attitude motion of the on-orbit service spacecraft, as well as the orientation of the body-fixed frame $\mathfrak{F}_{bs}$ with respect to the inertial frame $\mathfrak{F}_I$, can be described by the unit quaternion $q_a = [q_0, q_v^T]^T \in \mathbb{R}^4$, which is subject to the constraint [8].

$$q_v^T q_v + q_0^2 = 1 \tag{2.8}$$

The dynamic model for the described rigid spacecraft can be expressed as follows [9]:

$$J_s\dot{\omega}_a = -\omega_a^\times J_s\omega_a + \tau_a + d_a \tag{2.9}$$

$$\dot{q}_v = \frac{1}{2}(q_v^\times \omega_a + q_0\omega_a) \tag{2.10}$$

$$\dot{q}_0 = -\frac{1}{2}q_v^T \omega_a \tag{2.11}$$

where $J_s \in \mathbb{R}^{3\times 3}$ is the constant, positive-definite, symmetric inertia matrix of the on-orbit service spacecraft, $\omega_a \in \mathbb{R}^3$ is the angular velocity of the body-fixed frame $\mathfrak{F}_{bs}$ with respect to the inertial reference frame $\mathfrak{F}_I$ expressed in $\mathfrak{F}_{bs}$, $\tau_a \in \mathbb{R}^3$ is a vector of control torques in $\mathfrak{F}_{bs}$ with unit N·m, $d_a \in \mathbb{R}^3$ expressed in $\mathfrak{F}_{bs}$ is a vector of disturbance torques with unit N·m, and the notation $a^\times, \forall a = [a_1, a_2, a_3]^T$, denotes the following skew-symmetric matrix:

$$a^\times \triangleq \begin{bmatrix} 0 & -a_3 & a_2 \\ a_3 & 0 & -a_1 \\ -a_2 & a_1 & 0 \end{bmatrix} \tag{2.12}$$

For on-orbit spacecraft, there are many external disturbances. The external disturbances considered in this book are all environmental disturbance torques when the spacecraft is in orbit, such as gravity gradient torque, solar pressure torque, aerodynamic torque, geomagnetic torque, space radiation torque, and so on. In practical engineering, the magnitude of these disturbances is very small, and attitude control is hardly affected in a short time. In order to be more rigorous in theory, the relevant controllers in Part I will consider these disturbances and express them with disturbance terms, and the assumption of the external disturbance torques received by the spacecraft is given. Such an assumption not only meets the needs of theoretical derivation but also conforms to the practical situation of engineering.

Assumption 2.1 The external disturbance torques received by the spacecraft, that is, the external disturbances, are bounded.

The rotation matrix that brings $\mathfrak{F}_I$ onto $\mathfrak{F}_{bs}$, denoted by $R_a(q_0, q_v) \in SO(3)$, is defined as follows:

$$R_a = (q_0^2 - q_v^T q_v)I_3 + 2q_v q_v^T - 2q_0 q_v^\times \tag{2.13}$$

where I_3 is the 3×3 identity matrix, and the angular velocity ω_a can be computed as follows:

$$\omega_a = 2(q_0\dot{q}_v - q_v\dot{q}_0) - 2q_v^\times \dot{q}_v \tag{2.14}$$

In order to obtain the attitude error model, we assume that the desired attitude of the on-orbit service spacecraft can be described by the desired unit quaternion $q_{ad}(t) = [q_{d0}, q_{dv}^T]^T \in \mathbb{R}^4$ and define the desired angular velocity $\omega_d \in \mathbb{R}^3$ through the following kinematic equations:

$$\dot{q}_{dv} = \frac{1}{2}\left(q_{dv}^{\times}\omega_d + q_{d0}\omega_d\right) \tag{2.15}$$

$$\dot{q}_{d0} = -\frac{1}{2}q_{dv}^{T}\omega_d \tag{2.16}$$

$$\omega_d = 2(q_{d0}\dot{q}_{dv} - q_{dv}\dot{q}_{d0}) - 2q_{dv}^{\times}\dot{q}_{dv} \tag{2.17}$$

The corresponding rotation matrix $R_d(q_{d0}, q_{dv}) \in SO(3)$ is defined as follows:

$$R_d = (q_{d0}^2 - q_{dv}^T q_{dv})I_3 + 2q_{dv}q_{dv}^T - 2q_{d0}q_{dv}^{\times} \tag{2.18}$$

To quantify the error between the actual and desired spacecraft attitudes, we define the rotation matrix $R_e(q_{ev}, q_{e0}) \in SO(3)$ and quaternion tracking error $q_e(t) = [q_{e0}, q_{ev}^T]^T \in \mathbb{R}^4$ as follows:

$$R_e = RR_d^T = (q_{e0}^2 - q_{ev}^T q_{ev})I_3 + 2q_{ev}q_{ev}^T - 2q_{e0}q_{ev}^{\times} \tag{2.19}$$

$$q_{ev} = q_{d0}q_v - q_0 q_{dv} + q_v^{\times} q_{dv} \tag{2.20}$$

$$q_{e0} = q_0 q_{d0} + q_v^T q_{dv} \tag{2.21}$$

Based on the preceding formulation, the tracking error of the angular velocity $\omega_e \in \mathbb{R}^3$ is defined as follows:

$$\omega_e = \omega_a - R_e\omega_d \tag{2.22}$$

Computing (2.9)–(2.11), (2.15), (2.16), (2.20)–(2.22), we obtain the open-loop tracking error dynamics as follows

$$\begin{aligned} J_s\dot{\omega}_e = &-(\omega_e + R_e\omega_d)^{\times} J_s(\omega_e + R_e\omega_d) \\ &+ J_s(\omega_e^{\times} R_e\omega_d - R_e\dot{\omega}_d) + \tau_a + d_a \end{aligned} \tag{2.23}$$

$$\dot{q}_{ev} = \frac{1}{2}(q_{ev}^{\times} + q_{e0}I_3)\omega_e \tag{2.24}$$

$$\dot{q}_{e0} = -\frac{1}{2}q_{ev}^T\omega_e \tag{2.25}$$

where $\dot{R}_e = -\omega_e^{\times} R_e$.

Considering the systematization of controller design, simplified expression, and scalability of the model, this study converts the above-mentioned spacecraft attitude tracking kinematics and dynamics model into the Euler-Lagrange (EL) system form.

To express the open-loop tracking error dynamics, we rewrite (2.24) as follows:

$$\dot{q}_{ev} = \frac{1}{2}T\omega_e \tag{2.26}$$

where $T = q_{ev}^{\times} + q_{e0}I_3$ is a Jacobian-type matrix.

After taking the time derivative of (2.26) and pre multiplying both sides of the resulting expression by $T^{-T} J_s T^{-1}$, we obtain,

$$M_a \ddot{q}_{ev} = \frac{1}{2} M_a \dot{T} \omega_e + \frac{1}{2} P^T J_s \dot{\omega}_e \tag{2.27}$$

where $M_a = P^T J_s P$ and $P = T^{-1}$.

Considering that $q_e = [q_{e0}, q_{ev}^T]^T$ is a unit quaternion, $T = q_{ev}^{\times} + q_{e0} I_3$, and J_s are known constant-valued matrices, then M_a is bounded.

After substituting (2.23) into the right-hand side of (2.27), we can obtain the following EL expression for the open-loop tracking error dynamics of the relative attitude motion

$$M_a \ddot{q}_{ev} + C_a \dot{q}_{ev} + G_a + d = \tau \tag{2.28}$$

where

$$\tau = \frac{1}{2} P^T \tau_a \tag{2.29}$$

$$d = -\frac{1}{2} P^T d_a \tag{2.30}$$

$$C_a = -M_a \dot{P}^{-1} P - 2P^T (J_s P \dot{q}_{ev})^{\times} P \tag{2.31}$$

$$\begin{aligned} G_a =& P^T [(P\dot{q}_{ev})^{\times} J_s R_e \omega_d] + P^T [(R_e \omega_d)^{\times} J_s P \dot{q}_{ev}] + \frac{1}{2} P^T [(R_e \omega_d)^{\times} J_s R_e \omega_d] \\ & - \frac{1}{2} P^T J_s [(2P\dot{q}_{ev})^{\times} R_e \omega_d - R_e \dot{\omega}_d]. \end{aligned} \tag{2.32}$$

2.3.3 Spacecraft Formation Systems

In the Local-Vertical-Local-Horizontal (LVLH) rotating frame $\mathfrak{F}_L$, the relative orbital motion of the ith spacecraft can be modeled as [3]

$$\begin{aligned} &\ddot{x}_i - 2\omega_0 \dot{y}_i - \omega_0^2 x_i + \frac{\mu_e (R_0 + x_i)}{R_i^3} - \frac{\mu_e}{R_0^2} = \frac{\tau_{ix}}{m_i}, \\ &\ddot{y}_i - 2\omega_0 \dot{x}_i - \omega_0^2 y_i + \frac{\mu_e y_i}{R_i^3} = \frac{\tau_{iy}}{m_i}, \\ &\ddot{z}_i + \frac{\mu_e z_i}{R_i^3} = \frac{\tau_{iz}}{m_i}, \end{aligned} \tag{2.33}$$

where m_i denotes the mass of the ith spacecraft, μ_e is the earth gravitational constant, R_0 is the orbit radius, R_i is the distance between the geocenter and the ith spacecraft,

$\omega_0 = \sqrt{\mu_e / R_0^3}$ is the orbit angular velocity, $q_i \triangleq \left[x_{i,} y_i, z_i\right]^T$ is the relative coordinate of the ith spacecraft in the LVLH frame, and $\tau_i = \left[\tau_{ix}, \tau_{iy}, \tau_{iz}\right]^T$ represents the control force.

$$C_i = m_i \begin{bmatrix} 0 & -2\omega_0 & 0 \\ 2\omega_0 & 0 & 0 \\ 0 & 0 & 0 \end{bmatrix},$$

$$g_i = m_i \begin{bmatrix} -\omega_0^2 x_i + \frac{\mu_e (R_0 + x_i)}{R_i^3} - \frac{\mu_e}{R_0^2} \\ -\omega_0^2 y_i + \frac{\mu_e y_i}{R_i^3} \\ \frac{\mu_e z_i}{R_i^3} \end{bmatrix}.$$

Let $M_i = m_i I_3$, and then the relative orbit motion equations can be rewritten in the form of the Euler-Lagrange (EL) system. The spacecraft attitude and relative orbit models in this book can be expressed as EL systems. Therefore, the above models use letters and symbols similar to attitude EL expressions.

When the number of spacecraft is small, traditional control algorithm design can solve the problem of relative motion between spacecraft. When the number of spacecraft increases significantly, the networked spacecraft system can be applied to describe all spacecraft. This description method can describe the dynamics and kinematics behavior of all spacecraft in a general formula and has the advantages of simple structure, intuitive algorithm, and unified design. Moreover, when increasing individual spacecraft in the systems, only the system dimension needs to be changed, and the algorithm is universal.

We consider the multi-spacecraft formation systems comprising N leader spacecraft labeled as $1, 2, \ldots, N$ and M follower spacecraft labeled as $N+1, N+2, \ldots, N+M$. Similar to the EL expression of spacecraft attitude motion in the previous subsection, considering that the spacecraft relative orbital dynamics and kinematics models Eq. (2.33) can be abstractly expressed as multiple Euler-Lagrange systems (MELSs), the multi-spacecraft formation systems are modeled by the following MELSs equation

$$M_i\left(q_i\right) \ddot{q}_i + C_i\left(q_i, \dot{q}_i\right) \dot{q}_i + g_i\left(q_i\right) = \tau_i\left(t\right), \tag{2.34}$$

where $q_i \in \mathbb{R}^p$, $i = 1, 2, \ldots, N+M$ denotes the position vector of spacecraft i, $M_i\left(q_i\right) \in \mathbb{R}^{p \times p}$ is the symmetric positive definite inertia matrix, $C_i\left(q_i, \dot{q}_i\right) \in \mathbb{R}^{p \times p}$ is the vector of Coriolis-like and centrifugal anti-symmetric matrix, $g_i\left(q_i\right) \in \mathbb{R}^p$ denotes the nonlinear term, $\tau_i \in \mathbb{R}^p$ represents the control input, and $p \in \mathbb{R}$ denotes system dimension. If environmental disturbances are considered, then model (2.34) is rewritten as follows,

$$M_i\left(q_i\right) \ddot{q}_i + C_i\left(q_i, \dot{q}_i\right) \dot{q}_i + g_i\left(q_i\right) = \tau_i\left(t\right) - f_{di}\left(t\right), \tag{2.35}$$

where $f_{di} \in \mathbb{R}^p$ denotes the external disturbances.

Of course, the formation behaviors described by the above MELSs model include but are not limited to the relative orbital motion of the spacecraft. For formation behavior, when multi-spacecraft attitude formation motion is considered, it is only necessary to replace the above model expressed by EL with the attitude model given in the previous subsection expressed by EL. Similarly, when the dynamic behavior or process of objects can be described by the EL system, the formation system composed of these objects can also be expressed by MELSs (2.34). The properties of the EL equation are given as follows:

Property 2.1 [10] *For any i, $M_i(q_i)$ is bounded, which is bounded as*

$$0 < k_{\underline{m}}\|q\|^2 \leq q^T M_i(q_i)\, q \leq k_{\overline{m}}\|q\|^2,\ \forall q \in \mathbb{R}^p, \tag{2.36}$$

where $k_{\underline{m}}$ and $k_{\overline{m}}$ are positive constants.

Property 2.2 [11] *For any i, $M_i(q_i)$ is symmetric and positive definite.*

Property 2.3 [12, 13] *The matrix $\dot{M}_i(q_i) - 2C_i(q_i, \dot{q}_i)$ is skew-symmetric.*

2.4 Relevant Knowledge of Nonlinear System Control

2.4.1 Relevant Stability

The stability theory is crucial for dynamic systems in the control theory and engineering design. The main research findings in this book are on spacecraft dynamics systems. Based on Lyapunov stability theory, controller design and closed-loop system stability analysis are carried out. This subsection introduces the relevant theories one by one.

Consider the nonlinear autonomous systems [14]

$$\dot{x} = f(x) \tag{2.37}$$

where $f : D \to \mathbb{R}^n$ is a locally Lipschitz map from a domain $D \subset \mathbb{R}^n$ into $\mathbb{R}^n$, and the initial value of the system state is $x_0 = x(0)$. Assume that $x_{eq} \in D$ is an equilibrium point of the equation (2.37); that is, $f(x_{eq}) = 0$. Without loss of generality, for a system whose equilibrium is not at the origin, it is always possible to shift the equilibrium to the origin through variable substitution. Therefore, the equilibrium points studied are all at the origin, that is, $x_{eq} = 0$. Simultaneously, considering that the control input $u = f_u(x)$, $f_u : D \to \mathbb{R}^n$ is the function of the system state x, therefore, the control input u is not explicitly included in (2.37).

Definition 2.1 [14] (*The equilibrium point*) Consider a certain point $x_{eq} \in D$ in the state space of the system (2.37), if the system moves to this point, that is, $x(t) = x_{eq}$, the system state x will always remain at this point, then $x(t) = x_{eq}$ is called an equilibrium point of the system (2.37).

Definition 2.2 [14] (*Lyapunov Stable*) The equilibrium point $x = x_{eq} = 0$ of (2.37) is
(1) stable if, for each $\varepsilon > 0$, there is $\delta = \delta(\varepsilon) > 0$, and $\forall t \geq 0$ such that

$$||x(0)|| < \delta \Rightarrow ||x(t)|| < \varepsilon. \tag{2.38}$$

(2) unstable if it is not stable.
(3) asymptotically stable if it is stable and δ can be chosen such that

$$||x(0)|| < \delta \Rightarrow \lim_{t\to\infty} x(t) = 0, \tag{2.39}$$

where $||\cdot||$ represents the Euclidean norm.

Definition 2.3 [14] (*Uniformly Ultimately Bounded, UUB*) The equilibrium point $x = x_{eq} = 0$ of (2.37) is
(1) uniformly bounded/stable,if there exists a positive constant c independent of t_0, $t_0 \geq 0$, for each $a \in (0, c)$, there exists $\beta = \beta(a) > 0$ independent of t_0, for $\forall t \geq t_0$, such that

$$||x(t_0)|| \leq a \Rightarrow ||x(t)|| \leq \beta; \tag{2.40}$$

(2) globally uniformly bounded/stable, if (2.40) holds for any large a;
(3) uniformly ultimately bounded/stable, if there are positive constants b and c independent of t_0, $t_0 \geq 0$, for any $a \in (0, c)$, there is $T = T(a, b) \geq 0$ independent of t_0, satisfies

$$||x(t_0)|| \leq a \Rightarrow ||x(t)|| \leq b, \forall t \geq t_0 + T, \tag{2.41}$$

and the final boundary is b;
(4) globally uniformly ultimately bounded/stable, if (2.41) holds for any large a.

Definition 2.4 [15, 16] (*Semi-Globally Uniformly Ultimately Bounded, SGUUB*) If for any closed set $\Omega \subseteq \mathbb{R}^n$ and all $x(t_0) = x_0 \in \Omega$, there exist constants $\mu > 0$ and $T(\mu, x_0)$, so that $||x(t)|| < \mu$ for $\forall t > t_0 + T$, then the equilibrium point of the system (2.37) is semi-globally uniformly ultimately bounded/stable.

Definition 2.5 [17, 18] (*Finite-time Stable, FTS*) The origin is said to be a (locally) finite-time-stable equilibrium of (2.37) if there exists an open neighborhood $N \subseteq D$ of the origin and a function $T : N\backslash\{0\} \to (0, \infty)$, called the settling-time function, such that the following statements hold:

(1) finite-time convergence: for every $x \in N\backslash\{0\}$, $g_t(x)$ is defined on $t \in [0, T(x))$, $g_t(x) \in N\backslash\{0\}$ for all $t \in [0, T(x))$, and $\lim_{t\to T(x)} g_t(x) = 0$;

(2) Lyapunov stability: for every open neighborhood $N_\varepsilon : 0 \in N_\varepsilon \subseteq N$, there exists an open subset $N_\delta : 0 \in N_\delta \subseteq N$ containing 0 such that, for every $x \in N_\delta \backslash \{0\}$, $g_t(x) \in N_\varepsilon, t \in [0, T(x))$.
The origin is said to be a globally finite-time-stable equilibrium if it is a finite-time-stable equilibrium with $N = D = \mathbb{R}^n$.

Definition 2.6 [19] (*Finite-time Uniformly Ultimately Bounded, FTUUB*) The equilibrium point $x = 0$ of the system (2.37) is finite-time uniformly ultimately bounded, if there is a finite settling time T, such that $\forall t > T$, the system is uniformly ultimately bounded.

Lemma 2.1 [14] (Lyapunov Stability) *Let $x = x_{eq} = 0$ be an equilibrium point for (2.37) and $D \subset \mathbb{R}^n$ be a domain containing $x = 0$. Let $V : D \to \mathbb{R}$ be a positive definite continuously differentiable function such that*
(1) the equilibrium point $x = x_{eq} = 0$ is stable, if $V(0) = 0$, there is $V(x) > 0$ in $D \backslash \{0\}$, and $\dot{V}(x) \leq 0$ in D;
(2) the equilibrium point $x = x_{eq} = 0$ is asymptotically stable, if (1) holds, and $\dot{V}(x) < 0$ in $D \backslash \{0\}$;
(3) the equilibrium point $x = x_{eq} = 0$ is globally asymptotically stable, if (2) holds and $||x|| \to \infty \Rightarrow V(x) \to \infty$.

We call $V(x) > 0$ a Lyapunov function of the system (2.37).

Lemma 2.2 [16] (Boundedness) *Considering the equilibrium point $x = x_{eq} = 0$ of the system (2.37), $D \subset \mathbb{R}^n$ is the domain containing the origin. Consider the positive definite continuously differentiable function $V : D \to \mathbb{R}$, and $V(x_0)$ is bounded, the real-valued function $\varepsilon(t) \in \mathfrak{L}_\infty$. $V(x)$ and system state x are bounded if $\dot{V}(x) \leq -\kappa_1 V(x) + \kappa_2 \varepsilon(t)$, where $\kappa_1 > 0$ and κ_2 are constants. Furthermore, $V(x)$ and system state x are bounded if $\dot{V}(x) \leq -\kappa_3 V(x) + \kappa_4$, where $\kappa_3 > 0$ and κ_4 are constants.*

Lemma 2.3 [17, 18] (Finite-time Stability) *Considering the equilibrium point $x = x_{eq} = 0$ of the system (2.37), $D \subset \mathbb{R}^n$ is the domain containing the origin. Consider the positive definite continuously differentiable function $V : D \to \mathbb{R}$, real numbers $\kappa > 0$ and $\alpha \in (0, 1)$, the neighborhood of the equilibrium point $U \subset D$, and V is positive definite on U, $\dot{V} + \kappa V^\alpha$ is semi-negative definite on U, where $\dot{V}(x) = \frac{\partial V}{\partial x} f(x)$, then the equilibrium point of the system is finite-time stable, and for all x in the open neighborhood of the equilibrium point, the finite settling time of the system is $T(x) \leq \frac{1}{\kappa(1-\alpha)} V^{1-\alpha}(x)$. If $U = \mathbb{R}^n$ and V is radially unbounded, then the equilibrium point of the system is globally finite-time stable.*

Lemma 2.4 [20, 21] (Practical Finite-time Stability, PFS) *Considering the equilibrium point $x = x_{eq} = 0$ of the system (2.37), $D \subset \mathbb{R}^n$ is the domain of definition containing the origin. Suppose that there exist positive definite continuous function $V(x)$ in D satisfying*

$$\dot{V}(x) \leq -\alpha V^p(x) + \epsilon \tag{2.42}$$

where α, p, and ϵ are positive constants satisfying $p \in (0, 1)$, then the trajectory of system is practical finite-time stable. The system state can converge to the neighborhood near the equilibrium point within the finite time T, and the boundary of its neighborhood is

$$\left\{ \lim_{t \to T} (x) | V(x) < [\epsilon(1-\theta)^{-1}\alpha^{-1}]^{1/p} \right\}. \tag{2.43}$$

The upper bound of the settling time T satisfies

$$T \le \frac{V^{1-p}(x(0))}{\alpha\theta(1-p)} \tag{2.44}$$

where θ is arbitrary constant satisfying $\theta \in (0, 1)$.

Lemma 2.5 [19] (Finite-time Uniformly Ultimately Boundedness, FTUUB) *Considering the equilibrium point $x = x_{eq} = 0$ of the system (2.37), $D \subset \mathbb{R}^n$ is the domain of definition containing the origin. Consider positive definite continuously differentiable Lyapunov functions $V(x) : D \to \mathbb{R}$, positive constants $p_1 \in (0, 1)$, $p_2 < p_1$, $\kappa_1 > 0$, $\kappa_2 > 0$, and $\kappa_3 > 0$. The equilibrium point of the system is finite-time uniformly ultimately bounded, if $\dot{V}(x) \le -\kappa_1 V^{p_1}(x) - \kappa_2 V(x) + \kappa_3 V^{p_2}(x)$. The system state converges to the compact set Ω of the neighborhood of the equilibrium point within the finite time T, where,*

$$T \le \frac{1}{(\kappa_2 - \kappa_5)(1-p_1)} \ln\left[1 + \frac{\kappa_2 - \kappa_5}{\kappa_1 - \kappa_4} V^{1-p_1}(x_0)\right] \tag{2.45}$$

$$D = \{x : \kappa_4 V^{p_1 - p_2}(x) + \kappa_5 V^{1-p_2}(x) < \kappa_3\} \tag{2.46}$$

κ_4 and κ_5 are arbitrary positive constants satisfying $\kappa_4 \in (0, \kappa_1)$ and $\kappa_5 \in (0, \kappa_2)$.

Lemma 2.6 [22] (Multivariable Super-Twisting Sliding Mode Observer) *For the system*

$$\dot{\sigma} = -k_1 \frac{\sigma}{||\sigma||^{1/2}} + z - k_2\sigma + \gamma(t, \sigma) \tag{2.47}$$

$$\dot{z} = -k_3 \frac{\sigma}{||\sigma||} - k_4\sigma + \phi(t) \tag{2.48}$$

where $\sigma \in \mathbb{R}^n$ and $z \in \mathbb{R}^n$ denote the state vector, and the terms $\gamma(t, \sigma) \in \mathbb{R}^n$ $\phi(t) \in \mathbb{R}^n$ are assumed to satisfy $||\gamma(t, \sigma)|| \le \delta_1 ||\sigma||$ and $||\phi(t)|| \le \delta_2$ for known scalar bounds $\delta_1, \delta_2 > 0$, there exist a range of values for the gains k_1, k_2, k_3, k_4, such that the variables σ and $\dot{\sigma}$ are forced to zero in finite time and remain zero for all subsequent time.

Lemma 2.7 [23] (Rayleigh-Ritz Theorem) *Let $A \in \mathbb{R}^{n \times n}$ be a positive definite real symmetric matrix, then all eigenvalues of A are positive real numbers. Let $\lambda_{\min}$ and*

$\lambda_{\max}$ *represent the minimum and maximum eigenvalues of A respectively, then for* $\forall x \in \mathbb{R}^n$, *there is*

$$\lambda_{\min}||x||^2 \le x^T A x \le \lambda_{\max}||x||^2, \tag{2.49}$$

where $||\cdot||$ *represents the Euclidean norm.*

Lemma 2.8 [24] *If* $0 < p = p_1/p_2 \le 1$, *where* p_1 *and* p_2 *are positive integer odd numbers, then the following inequality for* $\forall x, y \in \mathbb{R}$ *holds,*

$$|x^p - y^p| \le 2^{1-p}|x - y|^p. \tag{2.50}$$

Lemma 2.9 [24, 25] *For any* $x, y \in \mathbb{R}$, $c > 0$, $d > 0$, *the following inequality holds,*

$$|x|^c|y|^d \le c/(c+d)\ |x|^{c+d} + d/(c+d)\ |y|^{c+d}. \tag{2.51}$$

Lemma 2.10 [25, 26] *For any* $x_i \in \mathbb{R}$, $i = 1, 2, \ldots, n$, *and the real number* $p \in (0, 1]$, *the following inequality holds,*

$$(|x_1| + \cdots + |x_n|)^p \le |x_1|^p + \cdots + |x_n|^p \le n^{1-p}(|x_1| + \cdots + |x_n|)^p. \tag{2.52}$$

Lemma 2.11 [27] (Young's Inequality) *For* $\forall x, y \in \mathbb{R}^n$, *the following inequality holds,*

$$x^T y \le \frac{a^p}{p}\|x\|^p + \frac{1}{qa^q}\|y\|^q \tag{2.53}$$

where $a > 0$, $p > 1$, $q > 1$, $(p-1)(q-1) = 1$.

2.4.2 Performance Guaranteed Technique with Pre-defined Maximum Settling Time Function

As one of the core techniques of this book, the performance guaranteed technique covered in Part I and Part II is given in this section.

The concept of the performance guaranteed constraints appeared more clearly in [28]. A model reference adaptive control problem was proposed in [28]: forcing the system error to be less than an (arbitrarily small) prespecified constant after an (arbitrarily short) prespecified period of time, with an (arbitrarily small) prespecified upper bound on the amount of overshoot. This idea provides a method to constrain the transient and steady-state performance of the system, and it can also constrain the settling time indirectly. This design method only depends on the desired control indicators and is not related to the system and controller parameters. Compared with the finite-time control, utilizing this method, the designer can first select the performance parameters, and then design the controller, where the performance of the system is theoretically strictly constrained. Based on this scheme, some research results that use constant value segmented constraints to control errors have begun to appear [29, 30]. Since then, there have been results that use time-varying functions to constrain errors, such as funnel control [31, 32].

After 2008, [33, 34] proposed prescribed performance control (PPC) methods, including performance functions and error transformation techniques. These developed methods make it possible for designers to set the convergence rate, overshoot, and steady-state errors in advance. The latest PPC research results so far mainly include theoretical applications [35], engineering applications [36, 37], and theoretical extensions [38], all of which use PPC to provide a performance guarantee.

Inspired by PPC design, in fact, by using the valid points on the performance function curves and attaching some reasonable definitions, the maximum settling time can be intuitively expressed. Considering the spacecraft attitude tracking error q_{ev} as described in the previous section, to achieve the complete guaranteed of the transient and steady-state performance of the tracking error, (the constraints on the transient performance parameters of the closed-loop system, such as the settling time, convergence rate, overshoot, steady-state error, etc.), the constraint functions need to be designed first. To constrain the settling time of the closed-loop system, we first define the *pre-defined maximum settling time* (PMST) as follows.

Definition 2.7 The *pre-defined maximum settling time* $t_{ps} > 0$ is the upper bound of the expected closed-loop system settling time that is specified in advance according to the practical task requirements.

Suppose we can propose some control algorithms so that the appropriate expected settling time is set theoretically to make the closed-loop system always converge to any small neighborhood of the equilibrium point. In that case, this time can be called the PMST. This control algorithm can be named the *PMST-Control* (PMSTC) method.

Remark 2.1 PMST can also be considered the settling time's upper bound guarantee. This definition implies that we can artificially set the upper bound of the settling time of any closed-loop system. If the algorithm can be designed to satisfy PMST, the actual settling time is always less than the upper bound of the preset settling time. The upper bound of the settling time under finite time theory is related to the initial state of the system and the parameters of the controller. To design different settling times, it is necessary to change the relevant parameters of the controller, and this parameter adjustment may affect the performance of the controller itself. Different from finite-time control, the preset settling time proposed here refers to the time that we can arbitrarily set, which is not affected by the parameters of the system and control algorithm. Although the practical settling time of the closed-loop system is still related to the system state and controller parameters, this method can isolate the design and selection of the maximum settling time from the controller parameters adjustment so that the more reasonable pre-designed settling time and control inputs can be obtained simultaneously.

Based on the defined PMST, combined with the PPC method, we can design a performance constraint-like control function named the *pre-defined maximum settling time constraint function* (PMSTF) as follows, that can set the settling time upper bound, the steady-state error upper bound, and the overshoot upper bound in advance.

Definition 2.8 A smooth function $\rho(t) : \mathbb{R}^+ \to \mathbb{R}^+ \backslash 0$ will be called a PMSTF if: (1) $\rho(t)$ is positive and decreasing; (2) $\rho(t_{ps}) = \rho_t$, where $\rho_t > 0$ is the upper bound of the expected steady-state error specified in advance; (3) $\lim_{t\to\infty} \rho(t) = \rho_t - \varepsilon_\rho$, where $0 < \varepsilon_\rho < \rho_t$ is a small constant to provide an asymptote.

Example 2.1 According to Definition 2.8, we construct a conditional PMSTF as follows and apply it in this Part. Other functions with such properties can also be used as time constraint functions.

$$\rho\left(t_{ps}, t\right) = \left(\rho_0 - \rho_t + \varepsilon_\rho\right) \exp\left(\frac{t}{t_{ps}} \ln \frac{\varepsilon_\rho}{\rho_0 - \rho_t + \varepsilon_\rho}\right) + \rho_t - \varepsilon_\rho \tag{2.54}$$

where $\rho_0 = \rho(0)$.

Remark 2.2 The performance function in the PPC [39] can constrain the convergence rate; however, it does not directly indicate the settling time. The main reason is that the constraint function is asymptotic, and the constraint curve converges to the steady-state error value at infinity. Therefore, the settling time cannot be explicitly given. Inspired by the performance function in [39], the new performance constraint function proposed in this book plans the expected settling time through a point (t_{ps}, ρ_t) so that the expected settling time can be manifested and designed in advance. Satisfying the constraints of this constraint function means that the corresponding variables must converge to the neighborhood of the equilibrium point with ρ_t as the maximum radius within t_{ps}. Due to the existence of the parameter ε_ρ, the practical steady-state neighborhood will be smaller, and this method has a certain degree of conservation.

Remark 2.3 Different from the performance function in [38], the performance function proposed in this book is a continuous exponential function, which is more explicit and contains pre-defined time parameters. The performance function in [38] is a piecewise low-complexity function.

After the definition, we can use this function to constrain the variables that need to be controlled, shown as in Fig. 2.2. For the tracking control of full-state feedback, we have the following formula

$$\begin{aligned} -\delta_i \rho_i(t_{ps}, t) < q_{evi}(t) < \rho_i(t_{ps}, t), \quad q_{evi}(0) \geq 0 \\ -\rho_i(t_{ps}, t) < q_{evi}(t) < \delta_i \rho_i(t_{ps}, t), \quad q_{evi}(0) < 0 \end{aligned} \tag{2.55}$$

where $q_{evi}(t)$ is the ith dimension element of the spacecraft attitude tracking error vector, $i = 1, 2, 3$. The subscript i in the above inequalities indicates the constraint

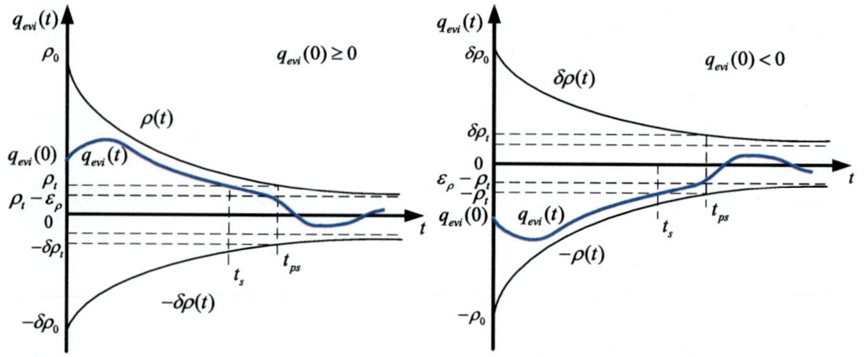

Fig. 2.2 PMSTF (2.54) and the constraint condition (2.55). (For the generality of the constraint conditions, the parameters of the constraint function are not marked with the subscript i in this figure. t_s is the time when the closed-loop system error/state converges within the range of the maximum steady-state error ρ_t and no longer escapes)

functions and parameters as defined in Eq. (2.54) for each element. $\delta_i \in [0, 1]$ is the overshoot limit parameter, when $\delta_i = 0$, the closed-loop system is not allowed to overshoot. Other related parameters are defined as Definition 2.8. This condition is the constraint on each element of the spacecraft tracking error vector. It can be seen that under the framework of performance constraints, the maximum settling time, overshoot, and steady-state error of each element of the tracking error vector can be constrained respectively by the performance constraint function (2.54) and the overshoot limit parameter δ_i. For spacecraft closed-loop systems, if the constraint condition (2.55) holds, the closed-loop system control error will converge within a specific error range within a pre-defined particular time.

To introduce inequality constraints (2.55) into the control design, a mapping is required to convert inequality into an auxiliary system. The main idea is that inequality conditions constrain the signal within a certain range, which is similar to stability and boundedness. Therefore, if the range $(-\delta_i \rho_i(t), \rho_i(t))$ or $(-\rho_i(t), \delta_i \rho_i(t))$ constrained by the inequality is mapped to $(-\infty, +\infty)$. Thus, when the variables of the mapping are bounded, stable, or convergent, the corresponding variables always satisfy the constraints. Simultaneously, such a mapping can simplify the control design, that is, convert the inequality constraint conditions into stability conditions, and only need to design the control algorithm through the traditional control design method [33, 34].

To transform the inequality constraints (2.55) into equations and apply them to the control design, some transformations are required. The new error transformation is designed here. By (2.55), we are easy to find when $q_{evi}(0) \geq 0$, one has

$$-1 < \frac{2q_{evi}(t)}{(1+\delta_i)\,\rho_i(t)} + \frac{\delta_i - 1}{1+\delta_i} < 1, \tag{2.56}$$

when $q_{evi}(0) < 0$, one has

$$-1 < \frac{2q_{evi}(t)}{(\delta_i + 1)\,\rho_i(t)} + \frac{1-\delta_i}{\delta_i + 1} < 1. \tag{2.57}$$

We also define variables as follows,

$$\vartheta_i = \begin{cases} \frac{2q_{evi}(t)}{(1+\delta_i)\rho_i(t)} + \frac{\delta_i - 1}{1+\delta_i},\ q_{evi}(0) \geq 0 \\ \frac{2q_{evi}(t)}{(1+\delta_i)\rho_i(t)} + \frac{1-\delta_i}{1+\delta_i},\ q_{evi}(0) < 0 \end{cases} \tag{2.58}$$

This indicates that if we can guarantee that $-1 < \vartheta_i < 1$ holds, then (2.56) and (2.57) can be achieved, and (2.55) holds. The control objectives are achieved. Therefore, we transform the constraint (2.55) into the constraint $-1 < \vartheta_i < 1$. According to the previous description, To constrain ϑ_i, we can use functions with similar ranges. An equivalent mapping and a smooth function $\varphi : \mathbb{R} \to \mathbb{R}$ need to be defined, which satisfies the following conditions,

$$\begin{aligned} \lim_{\sigma \to +\infty} \varphi(\sigma) &= 1 \\ \lim_{\sigma \to -\infty} \varphi(\sigma) &= -1 \end{aligned} \tag{2.59}$$

where σ is considered to be the transform variable after mapping in the PMSTC method, which can also be called an *error-transform auxiliary variable* or a *performance-constrained auxiliary variable*. With this transform mapping $\vartheta_i = \varphi(\sigma)$, the constraint domain of the inequality will be mapped into a new domain. As long as σ is bounded stable or convergent, then the constraint $-1 < \vartheta_i < 1$ can be satisfied. That is to say, the constraint curve can sufficiently constrain the corresponding variable to the expected transient and steady-state error range, and the practical settling time is less than the PMST. Therefore, we transform the control problem under the inequality constraint into the stability problem that can be solved by the Lyapunov theory.

For multi-input multi-output systems, the subscript i is used to distinguish the elements of the vector. Take the function that satisfies the definition of the smooth function φ as the mapping function, for example, $\varphi(\sigma_i) = \tanh(\sigma_i)$, where $\tanh(\cdot)$ is the hyperbolic tangent function. Taking $\vartheta_i = \varphi(\sigma_i)$ and doing the inverse transformation, one has,

$$\sigma_i = \varphi^{-1}(\vartheta_i). \tag{2.60}$$

To satisfy the constraint $-1 < \vartheta_i < 1$, just make σ_i bounded. This means that the corresponding control needs to be designed to make σ_i bounded, stable, or convergent. Then the control objective changes from making the spacecraft tracking error meet the constraints to designing the control algorithm to make the mapping variable σ_i (performance-constrained auxiliary variable) bounded, stable, or convergent. For convenience, some preparatory derivations are given here. Taking the derivative with respect to time of σ_i, one has,

$$\dot{\sigma}_i = \frac{\partial \varphi^{-1}(\vartheta_i)}{\partial \vartheta_i}\dot{\vartheta}_i = r_i\dot{\vartheta}_i \tag{2.61}$$

where $r_i = \partial\varphi^{-1}(\vartheta_i)/\partial\vartheta_i$.

Remark 2.4 In order to limit overshoot, $q_{evi} = (1-\delta_i)\rho_i(t)/2$ will be the tracking path of the error. This is considered a steady-state error due to limiting the overshoot. When $q_{evi}(0) = 0$, δ_i can not be 0.

Remark 2.5 Considering the design $\varphi(\sigma_i) = \tanh(\sigma_i)$, therefore, according to $r_i = \partial\varphi^{-1}(\vartheta_i)/\partial\vartheta_i$, r_i^{-1} exists.

The set $\Omega_\rho = \{q_{ev}|||q_{evi}|_{t\geq t_{ps}}|| \leq \rho_t\}$ is defined as the *predefined maximum steady-state error region*. If the control algorithm is designed by $\sigma_i = \varphi^{-1}(\vartheta_i)$ such that the performance constraints (2.55) can be satisfied, then it can be achieved, when $t \geq t_{ps}$, $||q_{evi}|_{t\geq t_{ps}}|| \leq \rho_t$. Then, the closed-loop system tracking error q_{ev} converges to the equilibrium point neighborhood Ω_ρ within the predefined time t_{ps}.

The above analysis is based on the performance constraints of the spacecraft attitude tracking error q_{ev}. In order to be more general and consider the performance guaranteed by general systems, the following theorems are proposed.

Theorem 2.1 (Predefined Maximum Settling Time Boundedness, PMSTB) *Considering the system (2.37) with the state $x \in \mathbb{R}^n$, define $\sigma_i = \varphi^{-1}(\vartheta_i(x_i, \rho_i))$ as a performance-constrained auxiliary variable with respect to the system state x, if σ_i is bounded and converges to the equilibrium point neighborhood within a predefined maximum settling time t_{ps}, that is, x satisfies*

$$\begin{aligned} -\delta_i\rho_i(t_{ps},t) < x_i(t) < \rho_i(t_{ps},t),\ x_i(0) \geq 0 \\ -\rho_i(t_{ps},t) < x_i(t) < \delta_i\rho_i(t_{ps},t),\ x_i(0) < 0 \end{aligned} \tag{2.62}$$

where σ_i, $\varphi^{-1}(\cdot)$, ϑ_i, $\rho_i(t_{ps},t)$, δ_i are as defined above.

Proof If σ_i is bounded, then $||\vartheta_i(x_i,\rho_i)|| < 1$, then the condition (2.62) holds. Considering Definition 2.8, then when $t = t_{ps}$, there is $\rho\left(t_{ps}\right) = \rho_t$, and when $t > t_{ps}$, there is $\rho(t) < \rho_t$. Considering the condition (2.62) holds, then there must be when $t \geq t_{ps}$, $||x_i(t)|| \leq \rho_i(t) \leq \rho_t$, then $x(t)$ converges to the equilibrium point neighborhood Ω_ρ within t_{ps}. □

Theorem 2.2 (Bounds of PMSTB) *Considering the variable definition (2.58) and the error transformation (2.60), if the theorem 2.1 is satisfied, that is, the performance-constrained auxiliary variable σ_i is UUB, and its boundary is $||\sigma_i|| \leq \epsilon_\sigma$, ϵ_σ is a non-zero positive constant. Its bounded set with respect to σ_i is Ω_σ, and suppose that $t_\sigma \in (0,+\infty]$ is the time that σ_i converges to Ω_σ, then, when $t_\sigma \leq t_{ps}$, $x_i(t)$ is bounded by*

$$||x_i|| \leq \frac{1}{2}(1+\delta_i)\rho_t\epsilon_\vartheta + \frac{1}{2}\rho_t(1-\delta_i) \tag{2.63}$$

when $t_\sigma > t_{ps}$ or $t_\sigma \to +\infty$, $x_i(t)$ is bounded by

$$||x_i|| < \rho_t. \tag{2.64}$$

Proof Considering that the error mapping auxiliary variable σ_i is bounded by $||\sigma_i|| \leq \epsilon_\sigma$, according to $\vartheta_i = \varphi(\sigma_i)$, where $\varphi(\cdot)$ is a monotonically increasing function, it can be seen that $||\vartheta_i|| \leq \varphi(\epsilon_\sigma) \leq \epsilon_\vartheta$, ϵ_ϑ is a non-zero positive constant. Considering

$$\vartheta_i = \begin{cases} \frac{2x_i(t)}{(1+\delta_i)\rho_i(t)} + \frac{\delta_i - 1}{1+\delta_i}, x_i(0) \geq 0 \\ \frac{2x_i(t)}{(1+\delta_i)\rho_i(t)} + \frac{1-\delta_i}{1+\delta_i}, x_i(0) < 0 \end{cases}, \tag{2.65}$$

then,

$$x_i(t) = \begin{cases} \frac{1}{2}(1+\delta_i)\rho_i(t)\left(\vartheta_i - \frac{\delta_i - 1}{1+\delta_i}\right), x_i(0) \geq 0 \\ \frac{1}{2}(1+\delta_i)\rho_i(t)\left(\vartheta_i - \frac{1-\delta_i}{1+\delta_i}\right), x_i(0) < 0 \end{cases} \tag{2.66}$$

when $t > t_{ps}$, one has

$$x_i(t) \leq \begin{cases} \frac{1}{2}(1+\delta_i)\rho_t\vartheta_i + \frac{1}{2}\rho_t(1-\delta_i), x_i(0) \geq 0 \\ \frac{1}{2}(1+\delta_i)\rho_t\vartheta_i + \frac{1}{2}\rho_t(\delta_i - 1), x_i(0) < 0 \end{cases} \tag{2.67}$$

If σ_i converges to the set Ω_σ within t_{ps}, considering $||\vartheta_i|| \leq \epsilon_\vartheta$, then

$$x_i(t) \leq \begin{cases} \frac{1}{2}(1+\delta_i)\rho_t\epsilon_\vartheta + \frac{1}{2}\rho_t(1-\delta_i), x_i(0) \geq 0 \\ \frac{1}{2}(1+\delta_i)\rho_t\epsilon_\vartheta + \frac{1}{2}\rho_t(\delta_i - 1), x_i(0) < 0 \end{cases} \tag{2.68}$$

Combining the above cases, the boundary is obtained

$$||x_i|| \leq \frac{1}{2}(1+\delta_i)\rho_t\epsilon_\vartheta + \frac{1}{2}\rho_t(1-\delta_i) \tag{2.69}$$

If σ_i converges to the set Ω_σ at time $t > t_{ps}$, because $||\vartheta_i|| < 1$, so there is

$$||x_i|| < \frac{1}{2}(1+\delta_i)\rho_t + \frac{1}{2}\rho_t(1-\delta_i) = \rho_t. \tag{2.70}$$

This theorem mainly gives the boundary relationship between σ_i and the system state x_i. According to Theorem 2.1, as long as σ_i is bounded, x_i can be guaranteed to converge to the set Ω_ρ. However, depending on the class of stability that σ_i can achieve, a more precise steady-state error bound for x_i can be obtained.

Theorem 2.2 is about estimating the more accurate steady-state error bound of x_i under performance constraints by the bound of σ_i. In Theorem 2.2, when $t_\sigma < +\infty$, σ_i is finite-time stable, then x_i is also finite-time stable. According to the relationship between t_σ and t_{ps}, a more precise boundary of x_i can be obtained. When $t_\sigma \to +\infty$, σ_i and x_i are asymptotically stable or UUB, the bounds of x_i can be obtained as Theorem 2.1. It should be noted that the boundary proposed by this theorem is highly conservative, and the practical boundary will be smaller. Through this theorem, the comprehensive constraints of the performance constraint function on transient and steady-state processes can be fully verified. Simultaneously, it can be noticed that when designing a finite-time stable σ_i, the boundary of the system will be smaller, and the performance will be better.

Remark 2.6 For the multi-input multi-output system, the performance constraints selected are the performance constraints for each dimension element of the closed-loop system. To make each dimension satisfy the corresponding performance constraints, there should be a feasible region between the selected constraints. That is, there can be no contradictory relationship between the constraints of any element and other constraints. Considering that the selected constraints are described by exponentially decreasing constraint functions, such constraints have corresponding feasible regions for attitude control described by quaternions.

In summary, this section provides the basic transformation principles of the performance guaranteed algorithm. The auxiliary variable σ_i can be used to provide a scheme for the design of the performance guaranteed control algorithm in the subsequent chapters. The specific control algorithm design will be given in the subsequent chapters.

2.4.3 Command Filter

Command filtering technique refers to the method of improving the command signal (control input signal or related signals generated during the control process) into the modified command signal with the desired amplitude, rate, and bandwidth constraints through the command filter (CF) [40–42]. Under the framework of the traditional command filtering backstepping method, there are usually several command filters and corresponding compensators, which filter and compensate the command signals in multiple links, respectively. The relationship between the single command filter and the filtered signal is shown in Fig. 2.3. The signal x_d is the signal to be filtered, and it is input to the command filter to obtain a new command signal x_c with the same dimension as x_d being improved in amplitude, rate, and bandwidth.

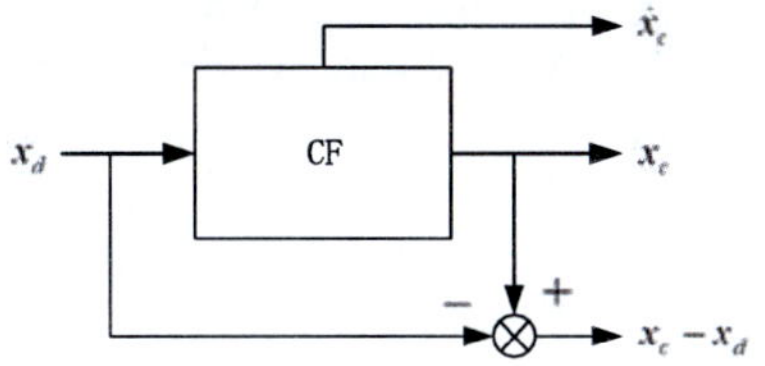

Fig. 2.3 The command filter (CF) structure. ($\dot{x}_c$ exists if and only if the system is above second order)

For the second-order nonlinear system of spacecraft attitude, the second-order nonlinear command filter is adopted as follows [42–44],

$$\begin{aligned} \dot{\varsigma}_1 &= \varsigma_2 \\ \dot{\varsigma}_2 &= 2\eta\omega_n \left[\Upsilon_R \left(\frac{\omega_n^2}{2\eta\omega_n} \left(\Upsilon_M \left(x_d \right) - \varsigma_1 \right) \right) - \varsigma_2 \right] \end{aligned} \tag{2.71}$$

where x_d is the signal to be filtered, the filter state variables ς_1 and ς_2 satisfy $\varsigma_1(0) = \varsigma_2(0) = 0$, $\eta \in (0, 1]$ represents the damping ratio, $\omega_n > 0$ represents the natural frequency. The output signals x_c and $\dot{x}_c$ with amplitude, rate, and bandwidth constraints are defined as,

$$\begin{aligned} x_c &= \varsigma_1 \\ \dot{x}_c &= \varsigma_2 \end{aligned} \tag{2.72}$$

The $\Upsilon_R(\cdot)$ and $\Upsilon_M(\cdot)$ in (2.71) are determined by the Eq. (2.73) definition, representing the constraint function, where $\chi_{\max}$ and $\chi_{\min}$ are constants, representing the expected amplitude range of the signal χ [42].

$$\Upsilon_{R/M}(\chi) = \begin{cases} \chi_{\max}, & \chi \geq \chi_{\max} \\ \chi, & \chi_{\min} < \chi < \chi_{\max} \\ \chi_{\min}, & \chi \leq \chi_{\min} \end{cases} \tag{2.73}$$

For the second-order nonlinear command filter (2.71), we only need to input the bounded signal x_d to be filtered, the continuous bounded signal x_c and its first derivative $\dot{x}_c$ can be obtained. According to the nature of the command filter, x_c is the signal about the input x_d adjusted for amplitude, rate, and bandwidth. In general spacecraft control design, the theoretically designed controller can achieve closed-loop system stability or convergence objectives. However, the control command signals (inputs) themselves may have problems such as larger peak value, larger rate, stronger jitter, signal value greater than the actuator capability (input saturation), and signal delay. Such signals are not suitable for direct input into spacecraft systems. Considering the many advantages of the command filter application, we can use the command filter to actively adjust the unreasonable command signal and make it suitable for the actuator of the spacecraft. The signal filtered by the command will have the desired amplitude, rate, bandwidth, etc., and work within a reasonable expected range. Moreover, the difference between the filter output signal and the input signal is bounded. For the analysis of the properties of the command filter, see [40, 42].

As shown in Fig. 2.3, due to the existence of the command filter, there is a difference $x_c - x_d$ between the filter output signal x_c and the input signal x_d. Therefore, it is necessary to design corresponding compensators in the control algorithm to compensate for this signal loss to ensure the stability of the closed-loop system. Regarding the design of the compensator, it will be explained and designed in the following chapters.

2.4.4 Optimal Control Based on Adaptive Dynamic Programming

Consider the following nonlinear time-varying system with $x \in \mathbb{R}^n$, $f(x) \in \mathbb{R}^n$, $g(x) \in \mathbb{R}^{n\times m}$, and control input $u(t) \in \mathbb{R}^m$.

$$\dot{x} = f(x) + g(x)u \tag{2.74}$$

where $x \in \mathbb{R}^n$ indicates system state variables, $f(x) \in \mathbb{R}^n$ and $g(x) \in \mathbb{R}^{n\times m}$ are smooth continuous functions, $u(t) \in \mathbb{R}^m$ represents control input.

Assume that $f(x) + g(x)u$ is Lipschitz continuous and the closed-loop system is stabilizable on a set $\Omega_x \subseteq \mathbb{R}^n$ that contains the origin. And the state x, $f(x)$ and $g(x)$ are known.

The infinite horizon integral cost function is defined as [45]

$$V(x_0) \triangleq \int_0^\infty \left(Q(x) + u^T R u\right) \mathrm{d}\tau \tag{2.75}$$

where $Q(x)$ is the positive definite parameter of x, and $R \in \mathbb{R}^{m\times m}$ is a known symmetric positive definite matrix.

Define the Hamiltonian as follows [45, 46]

$$H(x, u, \nabla_x V) = Q(x) + u^T R u + \left(\nabla_x V\right)^T \cdot (f(x) + g(x)u) \tag{2.76}$$

where $\nabla_x V$ represents the partial derivative of V with respect to x. And if we give an admissible control policy $\mu \in U(\Omega_x)$, $U(\Omega_x)$ represents the set of admissible control policies [47], the nonlinear Lyapunov equation can be obtained as follows [45]

$$Q(x) + \mu^T R \mu + \left(\nabla_x V\right)^T \cdot (f(x) + g(x)\mu) = 0 \tag{2.77}$$

If we can find an admissible control policy μ^* such that the cost index (2.75) is minimized, that is to say, we can get the optimal cost function or the value function as follows,

$$V^*(x_0) = \min_{\mu \in U} \left\{ \int_0^{\infty} \left(Q(x) + u^T R u \right) \mathrm{d}\tau \right\} \tag{2.78}$$

then such a control μ^* is the optimal control policy. V^* needs to satisfy the Hamilton-Jacobi-Bellman (HJB) equation as follows.

$$\min_{\mu \in U} \left\{ H\left(x, \mu, \nabla_x V^*\right) \right\} = 0 \tag{2.79}$$

So, assuming that (2.79) exists and is unique, then we can get the optimal control policy

$$\mu^*(x) = -\frac{1}{2} R^{-1} g^T(x) \nabla_x V^* \tag{2.80}$$

Substituting (2.80) into (2.77), we can get the Hamilton-Jacobi-Bellman (HJB) equation in terms of $\nabla_x V^*$

$$Q(x) + \left(\nabla_x V^*\right)^T f(x) - \frac{1}{4}\left(\nabla_x V^*\right)^T g R^{-1} g^T \nabla_x V^* = 0 \tag{2.81}$$

In order to obtain the optimal solution of the control policy, we need to solve the above HJB equation to obtain $\nabla_x V^*$, and then obtain the optimal control policy μ^*.

For the complex nonlinear system and strong nonlinear controller design, the HJB equation has strong nonlinearity, and it is very difficult to solve. Therefore, the solution of the HJB equation can be approximated by an adaptive neural network to obtain the corresponding optimal control. Based on the above optimal control theory, the *adaptive dynamic programming* (ADP) method is designed and applied to solve the HJB equation through an adaptive algorithm and obtain the optimal control solution [45, 48–50].

2.4.5 *Differential Nonzero-Sum Games*

We consider the Nonzero-sum (NZS) games problem of the following form [51–53]

$$\dot{x} = f(x) + \sum_{j=1}^{N} g_j(x) u_j(t) \tag{2.82}$$

with the state vector $x \in \mathbb{R}^n$, N control input $u_j(t) \in \mathbb{R}^{m_j}$, and the smooth nonlinear dynamics $f(x) \in \mathbb{R}^n$ and $g_j(x) \in \mathbb{R}^{n \times m_j}$, which are also assumed to satisfy the Lipschitz condition on a compact set $\Omega \subseteq \mathbb{R}^n$.

For each spacecraft, we define the following value functions as

$$J_i(x_0, u_1, \ldots, u_N) = \int_0^{\infty} \left(Q_i(x) + \sum_{j=1}^{N} u_j^T R_{ij} u_j \right) \mathrm{d}t \tag{2.83}$$

where Q_i is positive definite, R_{ii} and R_{ij} are positive definite symmetric matrices, x_0 is the initial state, $i, j = 1, \ldots, N$.

If we can find a feedback control policy pair $\mu = \{\mu_1, \ldots, \mu_N\}$, μ can stabilize (2.82) on the set Ω, (2.83) is finite, and μ_i is continuous on the set Ω, then we defined it as admissible on the set Ω, and denote it by $\mu_i \in \Psi(\Omega)$ [51]. Then we define the value function for the admissible control policies as

$$V_i(x, \mu_1, \ldots, \mu_N) = \int_t^{\infty} \left(Q_i(x(\tau)) + \sum_{j=1}^{N} \mu_j^T(\tau) R_{ij} \mu_j(\tau) \right) \mathrm{d}\tau \tag{2.84}$$

The optimal admissible control policy pair $\mu^* = \{\mu_1^*, \ldots, \mu_N^*\}$ means that such a control strategy can minimize the value function (2.84) for each spacecraft. If this optimal strategy μ^* can satisfy the following definition, then such a strategy is a Nash equilibrium solution.

Definition 2.9 [54] An N-tuple of policies $\{\mu_1^*, \ldots, \mu_N^*\}$ with $\mu_i^* \in \Psi(\Omega)$ is said to constitute a Nash equilibrium for an N-player games, if the following N inequalities are satisfied for all $\mu_i \in \Psi(\Omega)$:

$$J_i^*(\mu_1^*, \ldots, \mu_i^*, \ldots, \mu_N^*) \le J_i(\mu_1^*, \ldots, \mu_i, \ldots, \mu_N^*), \, i \in \mathbf{N} \tag{2.85}$$

Taking the derivative of (2.84), we can get the derivative equation of the value function

$$0 = r_i(x, \mu_1, \ldots, \mu_N) + (\nabla V_i)^T \left(f + \sum_{j=1}^{N} g_j \mu_j \right) \tag{2.86}$$

where $r_i(x, u_1, \ldots, u_N) = x^T Q_i x + \sum_{j=1}^{N} u_j^T R_{ij} u_j$, $\nabla V_i = \partial V_i / \partial x$, and $V_i(0) = 0$. So the Hamiltonian functions and the optimal value functions can be expressed as follows

$$H_i(x, \nabla V_i, \mu_1, \ldots, \mu_N) = r_i(x, \mu_1, \ldots, \mu_N) + (\nabla V_i)^T \left(f + \sum_{j=1}^{N} g_j \mu_j \right) \tag{2.87}$$

$$V_i^*(x) = \min_{\mu_i} \int_t^{\infty} \left(x^T(\tau) Q_i x(\tau) + \sum_{j=1}^{N} \mu_j^T(\tau) R_{ij} \mu_j(\tau) \right) \mathrm{d}\tau \tag{2.88}$$

By utilizing the stationarity conditions $\partial H_i / \partial \mu_i = 0$, we can get the optimal control strategy for each spacecraft as

$$\mu_i^* = -\frac{1}{2} R_{ii}^{-1} g_i^T \nabla V_i^* \tag{2.89}$$

And the coupled HJ equation can be described as

$$
\begin{aligned}
0 =& r_i(x, \mu_1^*, ..., \mu_N^*) + (\nabla V_i^*)^T \left(f + \sum\nolimits_{j=1}^{N} g_j \mu_j^*\right) \\
=& x^T Q_i x + \frac{1}{4}\sum\nolimits_{j=1}^{N} (\nabla V_j^*)^T g_j R_{jj}^{-T} R_{ij} R_{jj}^{-1} g_j^T \nabla V_j^* \\
&+ (\nabla V_i^*)^T f - \frac{1}{2}(\nabla V_i^*)^T \sum\nolimits_{j=1}^{N} g_j R_{jj}^{-1} g_j^T \nabla V_j^*
\end{aligned}
\tag{2.90}
$$

where $V_i^*(0) = 0$.

By solving the above HJ equation, we can obtain V_i^*, so we can obtain the optimal control policies (2.89). At the same time, this pair of optimal controls also reach a Nash equilibrium solution. However, it is often difficult for nonlinear systems to obtain the HJ equation's solution. Therefore, we can use the neural network approximation method to obtain the corresponding solution.

2.4.6 Graph Theory

We consider the communication topology as a directed graph. We utilize $\mathcal{G} = (\mathcal{V}, \mathcal{E})$ to describe the communication graph, where $\mathcal{V} = \mathcal{V}_l \cup \mathcal{V}_f$ and $\mathcal{E} \subseteq \mathcal{V} \times \mathcal{V}$ denote the vertex set and the arc set, respectively. Denote $\mathcal{N}_i$ as the in-neighbour set of node i, where $\mathcal{N}_i = \{j : (j, i) \in \mathcal{E}\}$. The adjacency matrix $\mathcal{A} = [a_{ij}] \in \mathbb{R}^{(N+M)\times(N+M)}$ is defined as $a_{ij} > 0$ if $(j, i) \in \mathcal{E}$, and $a_{ij} = 0$ otherwise. We define $a_{ii} = 0$, and the Laplacian matrix is defined as $\mathcal{L} = [l_{ij}] \in \mathbb{R}^{(N+M)\times(N+M)}$, where $l_{ii} = \sum_{j=1, j\neq i}^{k} a_{ij}$ and $l_{ij} = -a_{ij}, i \neq j$. For convenience, we use a virtual agent 0 to denote the virtual center of desired formation, and only parts of the leaders can access information agent 0.

Thus the corresponding Laplacian matrix in this chapter can be written as

$$
\mathcal{L} = \begin{bmatrix} 0 & 0_{1\times N} & 0_{1\times M} \\ \mathcal{L}_{L2} & \mathcal{L}_{L1} & 0_{N\times M} \\ 0_{M\times 1} & \mathcal{L}_{F2} & \mathcal{L}_{F1} \end{bmatrix}, \tag{2.91}
$$

where $\mathcal{L}_{L1}$ represents the interaction graph among the leaders, $\mathcal{L}_{L2}$ the information flow from the virtual agent to the leaders, $\mathcal{L}_{F1}$ the interaction graph among the followers, and $\mathcal{L}_{F2}$ the information interaction between the leaders and followers. Let $\mathcal{G}_L$ denote the directed graph associated with $N + 1$ agents labeled as 0 to N. Then, an assumption on the graph is made.

Assumption 2.2 There exists a directed spanning tree in $\mathcal{G}_L$, where the root of $\mathcal{G}_L$ is the agent 0. There exists at least one leader has a directed path to each follower.

Based on Assumption 2.2, an useful lemma in a directed graph is recalled.

Lemma 2.12 *Under Assumption 2.2, the following statements hold:*

(1) $\mathcal{L}_{L1}$ *and* $\mathcal{L}_{F1}$ *are non-singular;*
(2) All eigenvalues of $\mathcal{L}_{L1}$ *have positive real parts;*
(3) Each entry of $-\mathcal{L}_{L1}^{-1}\mathcal{L}_{L2}$ *is nonnegative and each row sum of* $-\mathcal{L}_{L1}^{-1}\mathcal{L}_{L2}$ *is equal to one. Similarly, each entry of* $-\mathcal{L}_{F1}^{-1}\mathcal{L}_{F2}$ *is nonnegative and each row sum of* $-\mathcal{L}_{F1}^{-1}\mathcal{L}_{F2}$ *is equal to one.*

Proof The proof is similar to that in [55–57]. □

2.5 Brief Summary

This chapter introduces the mathematical theoretical basis required for subsequent chapters in Part I, II, and III, and proposes a predefined maximum settling time constraint function and its error mapping transformation method. First, according to the object and task of this research, the space coordinate system related to the commonly used spacecraft system is established. Based on the quaternion's description of object rotation, the attitude-tracking error model of the spacecraft system is deduced. Considering the networked system description, the spacecraft formation systems based on relative orbital motion are modeled. This chapter presents the basic stability theory needed for this study. Then, the primary control theory required in the following chapters is introduced, including the performance guaranteed technique with predefined maximum settling time function and its error mapping transformation method proposed in this study, the command filtering technique, the optimal control theory based on adaptive dynamic programming, the differential games, and the graph theory.

References

1. Junkins, J.L., Schaub, H.: Analytical mechanics of space systems. American Institute of Aeronautics and Astronautics (2009)
2. Markley, F.L., Crassidis, J.L.: Fundamentals of spacecraft attitude determination and control (2014)
3. Schaub, H., Junkins, J.L.: Analytical mechanics of space systems. Aiaa (2003)
4. Mazzini, L.: Flexible Spacecraft Dynamics, Control and Guidance: Technologies by Giovanni Campolo. Springer (2015)
5. Dong, H., Hu, Q., Akella, M.R., Mazenc, F.: Partial lyapunov strictification: Dual-quaternion-based observer for 6-dof tracking control. IEEE Transactions on Control Systems Technology **27**(6), 2453–2469 (2019). DOI https://doi.org/10.1109/TCST.2018.2864723
6. Dong, H., Hu, Q., Friswell, M.I., Ma, G.: Dual-quaternion-based fault-tolerant control for spacecraft tracking with finite-time convergence. IEEE Transactions on Control Systems Technology **25**(4), 1231–1242 (2016)

7. Dong, H.: Spacecraft integrated attitude-position control using dual quaternions. Ph.D. thesis, Harbin Institute of Technology (2017)
8. Hughes, P.C.: Spacecraft attitude dynamics. Courier Corporation (2012)
9. Costic, B., Dawson, D., De Queiroz, M., Kapila, V.: Quaternion-based adaptive attitude tracking controller without velocity measurements. Journal of Guidance, Control, and Dynamics **24**(6), 1214–1222 (2001)
10. He, W., Chen, Y., Yin, Z.: Adaptive neural network control of an uncertain robot with full-state constraints. IEEE Transactions on Cybernetics **46**(3), 620–629 (2016)
11. He, W., Dong, Y.: Adaptive fuzzy neural network control for a constrained robot using impedance learning. IEEE transactions on neural networks and learning systems **29**(4), 1174–1186 (2018)
12. Jung, S.: Improvement of tracking control of a sliding mode controller for robot manipulators by a neural network. International Journal of Control, Automation and Systems **16**(2), 937–943 (2018)
13. He, W., Dong, Y.: Adaptive Fuzzy Neural Network Control for a Constrained Robot Using Impedance Learning. IEEE Transactions on Neural Networks and Learning Systems in press (2017). DOI https://doi.org/10.1109/TNNLS.2017.2665581
14. Khalil, H.K., Grizzle, J.W.: Nonlinear systems, vol. 3. Prentice hall Upper Saddle River, NJ (2002)
15. Ge, S.S., Hang, C.C., Lee, T.H., Zhang, T.: Stable adaptive neural network control, vol. 13. Springer Science & Business Media (2013)
16. Ge, S.S., Wang, C.: Adaptive neural control of uncertain mimo nonlinear systems. IEEE Transactions on Neural Networks **15**(3), 674–692 (2004)
17. Bhat, S.P., Bernstein, D.S.: Continuous finite-time stabilization of the translational and rotational double integrators. IEEE Transactions on automatic control **43**(5), 678–682 (1998)
18. Bhat, S.P., Bernstein, D.S.: Finite-time stability of continuous autonomous systems. SIAM Journal on Control and Optimization **38**(3), 751–766 (2000)
19. Hu, Q., Jiang, B.: Continuous finite-time attitude control for rigid spacecraft based on angular velocity observer. IEEE Transactions on Aerospace and Electronic Systems **54**(3), 1082–1092 (2017)
20. Zhu, Z., Xia, Y., Fu, M.: Attitude stabilization of rigid spacecraft with finite-time convergence. International Journal of Robust and Nonlinear Control **21**(6), 686–702 (2011)
21. Jiang, B.: Research on finite-time control problem for second order system. Ph.D. thesis, Harbin Institute of Technology (2018)
22. Nagesh, I., Edwards, C.: A multivariable super-twisting sliding mode approach. Automatica **50**(3), 984–988 (2014)
23. Horn, R.A., Johnson, C.R.: Matrix analysis. Cambridge university press (2012)
24. Qian, C., Lin, W.: A continuous feedback approach to global strong stabilization of nonlinear systems. IEEE Transactions on Automatic Control **46**(7), 1061–1079 (2001)
25. Beckenbach, E.F., Bellman, R.: Inequalities. Springer Science & Business Media (2012)
26. Hardy, G.H., Littlewood, J.E.: An inequality. Mathematische Zeitschrift **40**(1), 1–40 (1936)
27. Ji, H.B., Xi, H.S.: Adaptive output-feedback tracking of stochastic nonlinear systems. IEEE Transactions on Automatic Control **51**(2), 355–360 (2006)
28. Miller, D.E., Davison, E.J.: An adaptive controller which provides an arbitrarily good transient and steady-state response. IEEE Transactions on Automatic Control **36**(1), 68–81 (1991)
29. Xu, H., Ioannou, P.A.: Robust adaptive control for a class of mimo nonlinear systems with guaranteed error bounds. IEEE Transactions on Automatic Control **48**(5), 728–742 (2003)
30. Yan, L., Hsu, L., Xiuxia, S.: A variable structure mrac with expected transient and steady-state performance. Automatica **42**(5), 805–813 (2006)
31. Ilchmann, A., Trenn, S.: Input constrained funnel control with applications to chemical reactor models. Systems & control letters **53**(5), 361–375 (2004)
32. Liberzon, D., Trenn, S.: The bang-bang funnel controller for uncertain nonlinear systems with arbitrary relative degree. IEEE Transactions on Automatic Control **58**(12), 3126–3141 (2013)

33. Bechlioulis, C.P., Rovithakis, G.A.: Robust adaptive control of feedback linearizable mimo nonlinear systems with prescribed performance. IEEE Transactions on Automatic Control **53**(9), 2090–2099 (2008)
34. Bechlioulis, C.P., Rovithakis, G.A.: Adaptive control with guaranteed transient and steady state tracking error bounds for strict feedback systems. Automatica **45**(2), 532–538 (2009)
35. Zhang, J.X., Yang, G.H.: Fuzzy adaptive output feedback control of uncertain nonlinear systems with prescribed performance. IEEE transactions on cybernetics **48**(5), 1342–1354 (2018)
36. Luo, J., Yin, Z., Wei, C., Yuan, J.: Low-complexity prescribed performance control for spacecraft attitude stabilization and tracking. Aerospace Science and Technology **74**, 173–183 (2018)
37. Kanakis, G.S., Dimeas, F., Rovithakis, G.A., Doulgeri, Z.: Prescribed contact establishment of a robot with a planar surface under position and stiffness uncertainties. Robotics and Autonomous Systems **104**, 99–108 (2018)
38. Yin, Z., Luo, J., Wei, C.: Robust prescribed performance control for euler–lagrange systems with practically finite-time stability. European Journal of Control **52**, 1–10 (2020)
39. Bechlioulis, C.P., Rovithakis, G.A.: Prescribed performance adaptive control for multi-input multi-output affine in the control nonlinear systems. IEEE Transactions on Automatic Control **55**(5), 1220–1226 (2010)
40. Farrell, J.A., Polycarpou, M., Sharma, M., Dong, W.: Command filtered backstepping. IEEE Transactions on Automatic Control **54**(6), 1391–1395 (2009)
41. Farrell, J.A., Polycarpou, M.M.: Adaptive approximation based control: unifying neural, fuzzy and traditional adaptive approximation approaches. John Wiley & Sons (2006)
42. Sonneveldt, L., Chu, Q., Mulder, J.: Constrained adaptive backstepping flight control: application to a nonlinear f-16/matv model. In: AIAA guidance, navigation, and control conference and exhibit, p. 6413 (2006)
43. Wang, Y., Cao, L., Zhang, S., Hu, X., Yu, F.: Command filtered adaptive fuzzy backstepping control method of uncertain non-linear systems. IET Control Theory & Applications **10**(10), 1134–1141 (2016)
44. Wang, Y., Chen, W., Zhang, S., Zhu, J., Cao, L.: Command-filtered incremental backstepping controller for small unmanned aerial vehicles. Journal of Guidance, Control, and Dynamics **41**(4), 954–967 (2018)
45. Vamvoudakis, K.G., Lewis, F.L.: Online actor–critic algorithm to solve the continuous-time infinite horizon optimal control problem. Automatica **46**(5), 878–888 (2010)
46. Jiang, Y., Jiang, Z.P.: Global adaptive dynamic programming for continuous-time nonlinear systems. IEEE Transactions on Automatic Control **60**(11), 2917–2929 (2015)
47. Abu-Khalaf, M., Lewis, F.L.: Nearly optimal control laws for nonlinear systems with saturating actuators using a neural network hjb approach. Automatica **41**(5), 779–791 (2005)
48. Bertsekas, D.: Reinforcement learning and optimal control. Athena Scientific (2019)
49. Kamalapurkar, R., Walters, P., Rosenfeld, J., Dixon, W.: Reinforcement learning for optimal feedback control. Springer (2018)
50. Vrabie, D., Vamvoudakis, K.G., Lewis, F.L.: Optimal Adaptive Control and Differential Games by Reinforcement Learning Principles. Control, Robotics, Sensors. Institution of Engineering and Technology (2012)
51. Vamvoudakis, K.G., Lewis, F.L.: Multi-player non-zero-sum games: Online adaptive learning solution of coupled hamilton–jacobi equations. Automatica **47**(8), 1556–1569 (2011)
52. Vamvoudakis, K.G., Lewis, F.L.: Non-zero sum games: Online learning solution of coupled hamilton-jacobi and coupled riccati equations. In: 2011 IEEE International Symposium on Intelligent Control, pp. 171–178 (2011). DOI https://doi.org/10.1109/ISIC.2011.6045401
53. Zhao, D., Zhang, Q., Wang, D., Zhu, Y.: Experience replay for optimal control of nonzero-sum game systems with unknown dynamics. IEEE transactions on cybernetics **46**(3), 854–865 (2015)
54. Başar, T., Olsder, G.J.: Dynamic noncooperative game theory. SIAM (1998)
55. Mei, J., Ren, W., Ma, G.: Distributed containment control for Lagrangian networks with parametric uncertainties under a directed graph. Automatica **48**(4), 653–659 (2012). DOI https://doi.org/10.1016/j.automatica.2012.01.020

56. Meng, Z., Lin, Z., Ren, W.: Robust cooperative tracking for multiple non-identical second-order nonlinear systems. Automatica **49**(8), 2363–2372 (2013)
57. Meng, Z., Ren, W., You, Z.: Distributed finite-time attitude containment control for multiple rigid bodies. Automatica **46**(12), 2092–2099 (2010). DOI https://doi.org/10.1016/j.automatica.2010.09.005

Part I
Performance Guaranteed Control for Single Spacecraft

In this book, the performance guaranteed control and maneuvering issues of spacecraft are mainly considered in Part I. The control and maneuvering of the spacecraft can be either the control problem related to attitude, orbit maintenance, or relative orbital motion. Considering the current technological development and demand, spacecraft on-orbit service missions' control design and technology research are representative and widely applicable. In Part I, taking the attitude control of the single on-orbit service spacecraft as an example, the general performance guaranteed control algorithm for spacecraft is designed. Similar problems, such as relative orbit control, can also be transformed into similar system models to be solved.

Chapter 3
Spacecraft Performance Guaranteed Asymptotic Control with Settling Time Constraints

Abstract In spacecraft on-orbit service missions, strict state constraints and more precise control are essential guarantees for mission success. Effective constraints on the time-domain response parameters of the closed-loop system state, that is, to ensure the performance of the closed-loop system, can meet the state constraints and improve the task objective of steady-state control accuracy. Effectively constraining the time-domain response parameters of the closed-loop system state, that is, guaranteeing the performance of the closed-loop system, can promote the realization of the mission objectives of the strict state constraints and control accuracy improvement. This chapter proposes the asymptotically stable performance guaranteed control algorithm based on the common objectives of achieving performance guaranteed control for aerospace systems. The algorithm proposed in this chapter is the basic form and framework of the performance guaranteed control algorithm studied in this book, including the performance guaranteed constraint function, fundamental error transformation, performance constraint auxiliary variable, virtual auxiliary system, disturbance suppression, control algorithm design, and rigorous theoretical proof. Under the proposed control algorithm, the state of the spacecraft attitude tracking closed-loop system converges to the expected steady-state error interval within the expected maximum settling time. The transient trajectory of the closed-loop system state satisfies the strict constraints described by the performance constraint function. Finally, the numerical simulation analysis verifies the rationality and effectiveness of the proposed control algorithm.

3.1 Introduction

According to the research purpose of performance guaranteed control considered in Chap. 1, based on the spacecraft model, stability theorem introduced in Chap. 2, and the proposed performance guaranteed technique with pre-defined maximum settling time function in Chap. 2, this chapter proposes the corresponding spacecraft performance guaranteed asymptotic control with time parameters.

Specifically, it is hoped that the controller proposed in this chapter can make the spacecraft attitude tracking error converge to a predefined interval and the actual

Y. Gao and D. Li, *Spacecraft Maneuver with Performance Guaranteed*,
https://doi.org/10.1007/978-981-99-4653-2_3

settling time is less than the predefined maximum settling time t_{ps}. Simultaneously, the transient and steady-state performance of the closed-loop system can satisfy the performance constraints (2.55). Combined with the specific performance requirement parameters of the spacecraft, this section describes the problems or control objectives studied in this chapter as follows:

Problem 3.1 Consider the spacecraft attitude tracking error model (2.28), and select predefined related parameters according to practical mission requirements, including predefined maximum settling time t_{ps}, the maximum steady-state error ρ_t, the maximum overshoot χ_t, design the performance guaranteed control such that:
(1) $t_s \leq t_{ps}$, where t_s is the time when the closed-loop system state/error converges within the range of the maximum steady-state error ρ_t and no longer escapes;
(2) the transient performance of the closed-loop system meets the performance constraints (2.55);
(3) $-\delta\rho_t < q_{evi}|_{t\geq t_{ps}} < \rho_t$ when $q_{evi}(0) > 0$, or $-\rho_t < q_{evi}|_{t\geq t_{ps}} < \delta\rho_t$ when $q_{evi}(0) < 0$, where $q_{evi}|_{t\geq t_{ps}}$ represents the closed-loop system state/error at $t \geq t_{ps}$, $\delta \in [0, 1]$ is an adjustable parameter;
(4) the maximum overshoot of the closed-loop system is less than the predefined maximum overshoot χ_t, that is, the practical overshoot of the system is limited by adjusting δ;
(5) the closed-loop system has the property of disturbance suppression; that is, when there are strong external disturbances, the closed-loop system still satisfies the above performance constraints.

Then, according to the control problem presented in this chapter, design the performance guaranteed controller $\tau \in \mathbb{R}^n$ or $\tau_a \in \mathbb{R}^n$ for the spacecraft (τ and τ_a are defined in (2.28)). So that when $t \to +\infty$, the spacecraft closed-loop system state/error $q_{ev} \to 0$, that is, converge to the equilibrium point. Alternatively, $q_{ev} \to \Theta$, where $\Theta = \{|q_{evi}| \leq \iota_i\}$ denotes a small neighborhood containing the equilibrium point, $\iota_i \in \mathbb{R}^+$ and $\iota_i \leq \rho_t$ are the sufficiently small positive constant related to the controller parameters and the initial state of the closed-loop system. Alternatively, achieve the stability when $t \to T$, where $T < +\infty$ denotes the finite time. When the system is asymptotically stable, $\iota_i = \rho_t$ and $T = t_{ps}$. When the system is finite-time stable, $\iota_i \leq \rho_t$ and $T \leq \min\{t_{ps}, t_{finite}\}$, where t_{finite} represents the finite settling time of the closed-loop system.

Remark 3.1 These control objectives imply that, for the controller proposed in this chapter, the time response of the closed-loop system should be fully described by humans. The closed-loop system can converge to the specified steady-state error range within the specified time. At the same time, transient behavior, such as overshoot, can be limited.

The selection of performance indicators in the traditional performance-constrained control is theoretically arbitrary, which also means that the control input in traditional performance-constrained control is theoretically arbitrary. However, in practical applications, the selection of predefined performance constraints often needs to consider the actual situation. Therefore, for real physical systems, such as the spacecraft systems described in this book, we will describe expected performance index values as 'appropriate' rather than 'arbitrary'. For example, considering the fact that the input of the spacecraft attitude actuator is bounded, the performance constraints often do not strictly select an ultra-small constraint interval. Instead, according to the actual requirements of the task, the corresponding expected transient and steady-state constraints that are more accurate than the traditional expected value are given.

For a controllable spacecraft attitude tracking system, this chapter aims to achieve the control task objectives described in Problem 3.1 through the performance guaranteed control design rather than directly improving the capability of the spacecraft actuator itself. Compared with the traditional controller, the controller proposed can realize the transient and steady-state performance constraints of the closed-loop system and further improve the accuracy, safety, and reliability of the spacecraft attitude control.

Remark 3.2 The differences and connections between the predefined maximum settling time, finite-time stability, fixed-time stability, asymptotic stability, and uniformly ultimately bounded (UUB) need to be explained. The algorithms proposed in this book are mainly performance guaranteed control algorithms with the predefined maximum settling time parameter. Its primary connotation is that the closed-loop system state/error converges to the predefined maximum steady-state error with the boundary ρ_t within the predefined maximum settling time t_{ps}. This only requires the auxiliary variable σ_i defined in Sect. 2.4.2 to be stable. In order to make σ_i stable, the controller can be designed so that the closed-loop system satisfies asymptotic stability, UUB, finite-time stability, or fixed-time stability. Regardless of which controller design method is used, as long as the closed-loop system is stable, the aforementioned performance constraints can be satisfied. Different stability will bring different properties, which will be analyzed in detail here. If the closed-loop system with respect to σ_i is asymptotically stable or UUB, then the settling time that σ_i converges to 0 or within the neighborhood of the equilibrium point (boundary described by UUB) is infinite in theory. That is, the spacecraft attitude can converge to the range described by ρ_t within the predefined settling time t_{ps}, but the settling time that the state/error converges to the equilibrium point or its smaller neighborhood is infinite. If the closed-loop system with respect to σ_i is finite-time stable, then σ_i converges to the equilibrium point or its neighborhood within the finite time. This means that not only the spacecraft attitude tracking error converges to the range described by ρ_t within the predefined time t_{ps}, but also the error can converge to the equilibrium point or its smaller neighborhood (the neighborhood boundary is less than ρ_t) within the finite time t_{finite} (this finite time is related to the initial value of the system). Then the maximum time for the spacecraft attitude tracking error to converge to the range described by ρ_t is $t \leq \min\{t_{ps}, t_{finite}\}$, which enhances the transient and steady-state performance of the closed-loop system. If the closed-loop system with respect to σ_i is fixed-time stable, then unlike finite-time stability, the fixed-time t_{fix} parameter is independent of the initial value of the system. The algorithm proposed later in this book still behaves similarly to finite-time stability when the closed-loop system is asymptotically stable or bounded. That is, it converges to a certain range within a certain time. However, its theory is not the finite-time or fixed-time stability of the closed-loop system. In this book, the control is called a finite-time control algorithm only if the closed-loop system satisfies the finite-time stability theory. Otherwise, the performance guaranteed condition provides only a performance constraint on the upper bound of the settling time.

Remark 3.3 A clarification is needed about the asymptotically stable control in this book. The asymptotic stability of the system is fully stated in Definition 2.2, which requires the system state to converge asymptotically to the equilibrium point. For the systems considering the external disturbances, it is often only concluded that the closed-loop system is UUB, which means that the closed-loop system state asymptotically converges to the neighborhood of the equilibrium point and cannot converge to the equilibrium point. In order to name the controller explicitly. Suppose the designed controller can make the closed-loop system converge asymptotically without considering the disturbances or make the closed-loop system asymptotically converge to the neighborhood of the equilibrium point when considering the disturbances. In that case, such controllers are considered asymptotically stable control algorithms in this book.

This chapter is arranged as follows: In Sect. 3.2, based on the spacecraft tracking model and the performance constraints with the predefined settling time, an intuitive performance guaranteed asymptotically stable control is designed, which has the advantages of simple structure and fewer parameters. Section 3.3 presents the numerical simulation analysis of the proposed control algorithm. Section 3.4 gives the conclusions of this chapter.

3.2 Spacecraft Attitude Performance Guaranteed Asymptotic Control

Considering the spacecraft attitude tracking system (2.28), we design an auxiliary system that satisfies the performance constraints (2.55) with predefined maximum settling time t_{ps} as follows, where λ_i is a positive constant coefficient, $\sigma_i(t, t_{ps})$ is the *performance-constrained auxiliary variable*, see (2.60) for the definition, t_{ps} is the predefined maximum settling time. The relationship between t_{ps} and $\sigma_i(t, t_{ps})$ is shown in Sect. 2.4.2.

$$\zeta_i(t, t_{ps}) = \dot{\sigma}_i(t, t_{ps}) + \lambda_i \sigma_i(t, t_{ps}) \tag{3.1}$$

Here the new state variable $\zeta_i(t, t_{ps})$ of the auxiliary system is the auxiliary variable of the sliding mode surface. The main idea is to use the quasi-sliding-mode-surface to reorganize the spacecraft attitude tracking system with constraints into a first-order nonlinear system. Then the corresponding controller can be designed through a simple control method. It should be noted that both $\sigma_i(t, t_{ps})$ and $\zeta_i(t, t_{ps})$ are related to t_{ps}, but do not explicitly contain t_{ps}. In the following text, $\sigma_i(t, t_{ps})$ and $\zeta_i(t, t_{ps})$ will be abbreviated as σ_i and ζ_i.

Taking the derivative of the auxiliary system (3.1) with respect to time results in

$$\dot{\zeta}_i = \alpha_i + \beta_i \ddot{q}_{evi}(t) \tag{3.2}$$

where

$$\begin{aligned} \alpha_i = & -\frac{2r_i \dot{q}_{evi}(t)\dot{\rho}_i(t)}{(1+\delta_i)\rho_i{}^2(t)} - \frac{2r_i \ddot{\rho}_i(t) q_{evi}(t) + 2r_i \dot{\rho}_i(t)\dot{q}_{evi}(t)}{(1+\delta_i)\rho_i{}^2(t)} \\ & + \frac{4r_i q_{evi}(t)\dot{\rho}_i^2(t)}{(1+\delta_i)\rho_i{}^3(t)} + \dot{r}_i\dot{\vartheta}_i + \lambda_i r_i \dot{\vartheta}_i \end{aligned} \tag{3.3}$$

$$\beta_i = \frac{2r_i}{(1+\delta_i)\rho_i(t)} \tag{3.4}$$

Express (3.2) into a compact form, we have

$$\dot{\zeta} = \ddot{\sigma} + \Gamma\dot{\sigma} = A + B\ddot{q}_{ev}(t) \tag{3.5}$$

where $\zeta = [\zeta_1, \ldots, \zeta_n]^T$, $\sigma = [\sigma_1, \ldots, \sigma_n]^T$, $q_{ev} = [q_{ev1}, \ldots, q_{evn}]^T$, $A = [\alpha_1, \ldots, \alpha_n]^T$, $B = \text{diag}(\beta_1, \ldots, \beta_n)$, $\Gamma = \text{diag}(\lambda_1, \ldots, \lambda_n)$, $n = 1, 2, 3$.

Remark 3.4 According to Remark 2.5, B^{-1} exists. Simultaneously, considering that the matrix A and B is known, and thus can be utilized in controller design.

Substituting (2.28) into (3.5), we get

$$\dot{\zeta} = \ddot{\sigma} + \Gamma\dot{\sigma} = D + E\tau + F \tag{3.6}$$

where

$$D = A + BM_a{}^{-1}[-C_a(q_{ev}, \dot{q}_{ev})\dot{q}_{ev} - G_a(q_{ev})] \tag{3.7}$$

$$E = BM_a{}^{-1} \tag{3.8}$$

$$F = -BM_a{}^{-1}d \tag{3.9}$$

The above system can be regarded as a new virtual auxiliary system. By designing the controller that can make ζ bounded or asymptotically stable for the above system, the spacecraft attitude tracking closed-loop system is bounded or asymptotically stable. Also, the tracking error satisfies associated transient and steady-state performance constraints with the predefined maximum settling time.

It can be seen from the above that the function term $M_a{}^{-1}d$ is bounded. In the control design of this section, there is a function term B. In order to enhance the robustness of the closed-loop system, the Multivariable Super-Twisting Sliding

Mode Observer proposed in [1] is used to compensate for the disturbance term in this section. According to [1], the observer can thoroughly observe the external disturbances when the system model is known. Considering the practical situation, we still consider the observation error of the disturbance observer, and the observation error is bounded.

According to Lemma 2.6, for the system $\dot{x} = f + g\nu + d$ in $\mathbb{R}^n$, where $x \in \mathbb{R}^n$ represents the state vector, $f \in \mathbb{R}^n$ is the known nonlinear term, $g \in \mathbb{R}^{n\times n}$ is the known non-singular nonlinear term, $\nu \in \mathbb{R}^n$ is the control input, $d \in \mathbb{R}^n$ is the unknown bounded disturbance and uncertainty, and $\dot{d}$ is bounded. Define $\tilde{y}_0 = y_0 - x$ and $\tilde{y}_1 = y_1 - d$. From Lemma 2.6, the finite-time disturbance observer can be designed to observe d as follows,

$$\dot{\tilde{y}}_0 = -k_1 \frac{\tilde{y}_0}{||\tilde{y}_0||^{1/2}} + \tilde{y}_1 - k_2 \tilde{y}_0 \tag{3.10}$$

$$\dot{\tilde{y}}_1 = -k_3 \frac{\tilde{y}_0}{||\tilde{y}_0||} - k_4 \tilde{y}_0 - \dot{d} \tag{3.11}$$

Substituting $\tilde{y}_0$ and $\tilde{y}_1$ into the above equation and simplifying, the general form of the observer for system $\dot{x} = f + g\nu + d$ is obtained as follows,

$$\dot{y}_0 = -k_1 \frac{y_0 - x}{||y_0 - x||^{1/2}} + y_1 - k_2 (y_0 - x) + f + g\nu \tag{3.12}$$

$$\dot{y}_1 = -k_3 \frac{y_0 - x}{||y_0 - x||} - k_4 (y_0 - x) \tag{3.13}$$

where k_1, k_2, k_3, and k_4 represent positive constant gains, y_0 and y_1 are observations for state x and unknown disturbance d.

So far, we design the following observer to observe the uncertainty F in (3.6),

$$\dot{y}_0 = -\kappa_{y1} \frac{y_0 - \zeta}{||y_0 - \zeta||^{1/2}} + y_1 - \kappa_{y2} (y_0 - \zeta) + D + E\tau \tag{3.14}$$

$$\dot{y}_1 = -\kappa_{y3} \frac{y_0 - \zeta}{||y_0 - \zeta||} - \kappa_{y4} (y_0 - \zeta) \tag{3.15}$$

where κ_{y1}, κ_{y2}, κ_{y3}, and κ_{y4} represent positive constant gains, y_0 and y_1 are observations for state ζ and unknown disturbance F, i.e. $y_0 \to \zeta$, $y_1 \to F$. Considering that in this controller, the model is precisely known, so only the y_1 item is needed. Based on the previous analysis and derivation, we propose the following theorem.

Theorem 3.1 *For the spacecraft attitude tracking system (2.28) with considering external disturbances, if Assumption 2.1 is satisfied and the controller τ is chosen as*

$$\tau = -\kappa E^{-1} \zeta(t, t_{ps}) - E^{-1} D - E^{-1} y_1 \tag{3.16}$$

where $\kappa \geq \varrho^2/2$, $\varrho$is a positive constant, then, the variable $\zeta(t, t_{ps})$ of the auxiliary system is uniformly ultimately bounded (UUB). The performance-constrained auxiliary variable $\sigma(t, t_{ps})$ is UUB. The tracking error q_{ev} of the closed-loop system is bounded within the predefined time (predefined maximum settling time boundedness). q_{ev} converges to the neighborhood of equilibrium point Ω_ρ within the predefined time t_{ps}, and satisfies the transient and steady-state performance constraints (2.55).

Proof The candidate Lyapunov functions for ζ are selected as follows,

$$V = \frac{1}{2}\zeta^T\zeta \tag{3.17}$$

Find the derivative of (3.17) with respect to time and substitute the asymptotically stable controller. We get

$$\begin{aligned}
\dot{V} &= \zeta^T\dot{\zeta} = \zeta^T(D + E\tau + F) \\
&= \zeta^T(D + E(-\kappa E^{-1}\zeta - E^{-1}D - E^{-1}y_1) + F) \\
&= \zeta^T(-\kappa\zeta + F_y) = -\kappa\zeta^T\zeta + \zeta^T F_y \\
&\leq -\kappa\zeta^T\zeta + \frac{\varrho^2}{2}\zeta^T\zeta + \frac{1}{2\varrho^2}||F_y||^2 \\
&= -(\kappa - \frac{\varrho^2}{2})\zeta^T\zeta + \frac{1}{2\varrho^2}||F_y||^2
\end{aligned} \tag{3.18}$$

where $F_y = -y_1 + F$ is the observation error of the multivariable super-twisting sliding mode observer. According to the properties of the super-twisting observer, the observation error is bounded. So, F_y is bounded, that is, there is F_M makes $||F_y|| \leq F_M$, then the above formula can be written as:

$$\dot{V} \leq -(\kappa - \frac{\varrho^2}{2})\zeta^T\zeta + \frac{1}{2\varrho^2}F_M^2 = -2(\kappa - \frac{\varrho^2}{2})V + \frac{1}{2\varrho^2}F_M^2 \tag{3.19}$$

Therefore we have

$$V \leq V(0)e^{-2(\kappa-\varrho^2/2)t} + \frac{F_M^2}{4\varrho^2(\kappa - \varrho^2/2)}(1 - e^{-2(\kappa-\varrho^2/2)t}) \tag{3.20}$$

then

$$||\zeta|| \leq \sqrt{2V} \leq \zeta_m = \sqrt{2V(0)e^{-2(\kappa-\varrho^2/2)t} + \frac{F_M^2}{2\varrho^2(\kappa - \varrho^2/2)}(1 - e^{-2(\kappa-\varrho^2/2)t})} \tag{3.21}$$

Therefore, ζ is uniformly ultimately bounded. Next, we prove the boundedness of σ.

Consider the compact form of (3.1),

$$\zeta = \dot{\sigma} + \lambda\sigma \tag{3.22}$$

where $\lambda = \min\{\lambda_1, ..., \lambda_n\}$, and $\lambda \geq 1/2$.

The candidate Lyapunov functions for σ is selected as follows,

$$V_\sigma = \frac{1}{2}\sigma^T\sigma \tag{3.23}$$

Find the derivative of (3.23) with respect to time, and substitute (3.22) to get

$$\begin{aligned} \dot{V}_\sigma &= \sigma^T\dot{\sigma} = -\lambda\sigma^T\sigma + \sigma^T\zeta \\ &\leq -(\lambda - \frac{1}{2})\sigma^T\sigma + \frac{1}{2}\zeta_m^2. \end{aligned} \tag{3.24}$$

Therefore we have

$$V_\sigma \leq V_\sigma(0)e^{-2(\lambda-1/2)t} + \frac{\zeta_m^2}{4(\lambda - 1/2)}(1 - e^{-2(\lambda-1/2)t}) \tag{3.25}$$

then

$$||\sigma|| \leq \sqrt{2V_\sigma} \leq \sqrt{2V_\sigma(0)e^{-2(\lambda-1/2)t} + \frac{\zeta_m^2}{2(\lambda - 1/2)}(1 - e^{-2(\lambda-1/2)t})} \tag{3.26}$$

Therefore, σ is UUB. According to the performance constraints, $\vartheta_i = \varphi(\sigma_i)$ satisfies the constraint $-1 < \vartheta_i < 1$ when σ is UUB. Then the spacecraft attitude tracking error q_{ev} satisfies the constraint condition (2.55), and the closed-loop system is UUB. According to Theorem 2.1, the practical settling time is less than the predefined maximum settling time t_{ps}, transient performance such as overshoot and settling rate satisfies the constraints, and the steady-state error satisfies the constraints. □

3.3 Simulation Analysis

In this section, we perform simulation analysis on the controller (3.16) in Sect. 3.2. This section first presents the initial conditions and parameters for the simulation. Considering the background of the on-orbit service spacecraft mission, although this study takes attitude motion as the research object, this section still gives relevant information about the spacecraft's orbit. The orbital parameters of the target spacecraft are as follows.

Considering that the spacecraft is very close to the target spacecraft in the short-range on-orbit service mission. Therefore, the orbital parameters of the spacecraft

Table 3.1 The orbital parameters of the target spacecraft

Parameters	Values
Semimajor axis	6.85×10^6 m
Eccentricity	0.0073
Inclination	60°
Argument of periapsis	0°
Longitude of the ascending node	0°
Mean anomaly	0°

Table 3.2 The initial attitude parameters of the spacecraft

Parameters	Values
q_{v0}	$[0.6, 0.3, -0.3]^T$ rad
ω_{a0}	$[0, 0, 0]^T$ rad/s
J_s	$\begin{bmatrix} 100 & -10 & 5 \\ -10 & 120 & -3 \\ 5 & -3 & 100 \end{bmatrix}$ $\text{kg} \cdot \text{m}^2$

to be controlled are also shown in Table 3.1. Next, consider the initial parameters of the spacecraft as shown in Table 3.2. Without loss of generality, the parameters of the Tables 3.1 and 3.2 can be changed according to specific tasks.

Next, consider the desired attitude of the on-orbit service spacecraft. In the numerical simulation of this study, it is considered that the spacecraft is tracking the attitude of the target spacecraft. The target spacecraft is inertial oriented, that is, $\mathfrak{F}_{bt}$ and $\mathfrak{F}_I$ coincide. Therefore, q_{dv} can be obtained via the relative motion between $\mathfrak{F}_{bs}$ and $\mathfrak{F}_I$. Define R_{li} as the rotation matrix of the target spacecraft from the inertial system to LVLH coordinate system, $r_{li-j,k}$ is the matrix element in row j, column k, then $q_{d0} = 1/2\ (1 + r_{li-11} + r_{li-22} + r_{li-33})^{1/2}$.

$$q_{dv} = \begin{bmatrix} (r_{li-23} - r_{li-32})/(4q_{d0}) \\ (r_{li-31} - r_{li-13})/(4q_{d0}) \\ (r_{li-12} - r_{li-21})/(4q_{d0}) \end{bmatrix} \tag{3.27}$$

According to the spacecraft attitude tracking model in Sect. 2.3, the tracking error quaternion and the angular velocity tracking error can be obtained. To demonstrate the performance of the proposed controller and to facilitate comparison with similar controllers, the error-index is defined as $I_a = ||q_{ev}|| + ||\omega_e||$. To illustrate the energy consumption of the proposed controller, the energy consumption index $E_a = \sum_{i=1}^{n} \int_0^t |\tau_{ai}\ (t)|^2 \mathrm{d}t$ is defined as in [2]. This index describes the electrical or chemical energy consumed by actuators such as thrusters or reaction flywheels during the stabilization of an attitude closed-loop system.

Table 3.3 The initial parameters of the proposed controller (3.16)

Parameters	Values	Parameters	Values
t_{ps}	100s	δ_i	1
ρ_0	1	λ_i	1
ρ_t	0.02	κ	0.08
ε_ρ	0.01		

Then, the parameters selected in the simulation are given. These performance indicators are approximately selected from the control engineering requirements of the practical situation and simulation scenarios, as shown in the table below.

In Table 3.3, the parameter t_{ps} is the predefined maximum settling time. ρ_0 is the maximum initial value of the predefined performance boundary, which needs to be larger than the initial error quaternion. Considering that all quaternion elements are less than or equal to 1, thus, ρ_0 is taken as 1. ρ_t and ε_ρ provide steady-state error bounds. δ_i is about whether to limit the overshoot, λ_i is the sliding mode surface scale coefficient in the controller, and κ is the controller gain parameter. The observer parameters are $\kappa_{y1} = 1$, $\kappa_{y2} = 1$, $\kappa_{y3} = 0.004$, $\kappa_{y4} = 0.004$. These values are based on adjusting the parameters to make the observed effects and control input values reasonable.

Considering the external disturbances suffered by the spacecraft, we take the disturbance variable as follows [3]

$$d_a = \begin{bmatrix} 0.06\sin(0.5t) + 2\text{rand}_a(t) \\ 0.04\sin(0.5t + \pi/2\,) + 2\text{rand}_a(t) \\ 0.04\sin(0.5t + \pi/3) + 2\text{rand}_a(t) \end{bmatrix} \text{N}\cdot\text{m} \tag{3.28}$$

where $\text{rand}_a(t)$ represents a zero-mean Gaussian random signal, and the variance is set to 0.01. It should be pointed out that, in order to demonstrate the disturbance rejection ability of the controller proposed fully, the disturbance selected in the simulation is larger than the practical situation, and the selected value is still within the reasonable range in orbit.

Comparative simulation results are presented in this section to compare the controller proposed in this chapter with existing state-of-the-art controllers. For the convenience of representation, the controller (3.16) proposed in Sect. 3.2 is marked as the controller 'A'. The proposed controller is compared with the traditional prescribed performance controller in [4] and the finite-time controller in [5], and label the two contrasting controllers as controllers 'B' and 'C', respectively. In the simulation results and analysis, the symbols 'A', 'B', and 'C' represent the above three controllers. It should be noted that there is a system model adaptation in controller 'B'. In order to visually compare the difference between performance constraints and controllers, the adaptive model is omitted in controller 'B', and the accurate model is directly used.

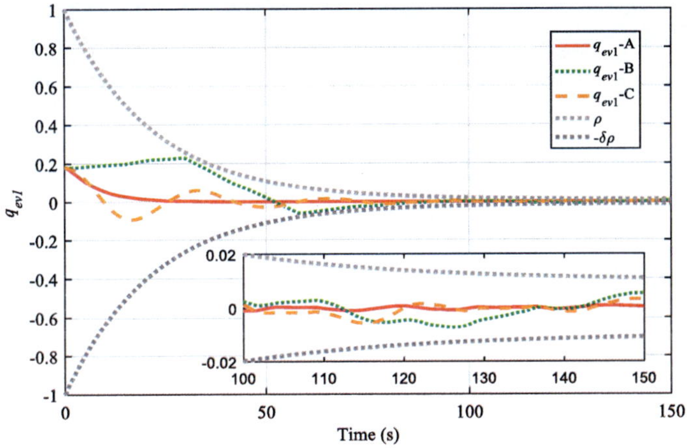

Fig. 3.1 The spacecraft attitude tracking error q_{ev1}

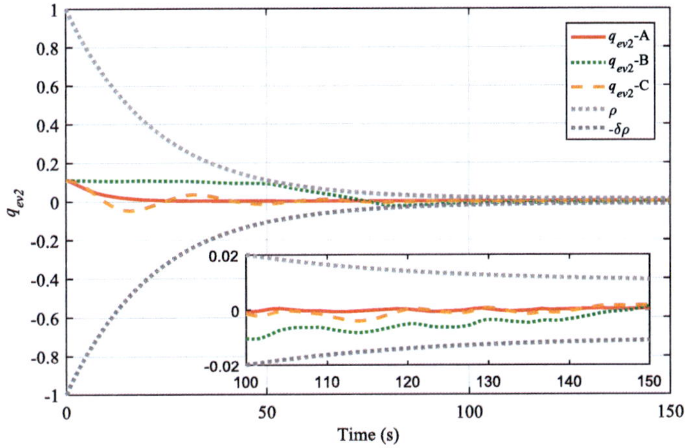

Fig. 3.2 The spacecraft attitude tracking error q_{ev2}

Figures 3.1, 3.2, 3.3, 3.5, 3.6, 3.7, 3.8, 3.9 and 3.10 are the comparative simulation results of the proposed performance guaranteed controller. For clarity, the simulation analysis given in this section considers bounded external disturbances.

Figures 3.1, 3.2 and 3.3 show the tracking errors of the closed-loop system under the proposed controller (3.16) and the comparison controllers 'B' and 'C'. To facilitate the comparison of different controllers' performance responses, each dimension's tracking error is plotted separately. Through these three comparative simulation graphs, it can be seen that under the action of the proposed controller, the tracking error of the spacecraft attitude closed-loop system always meets the performance constraints, ensuring that it can converge to the predefined steady-state error interval

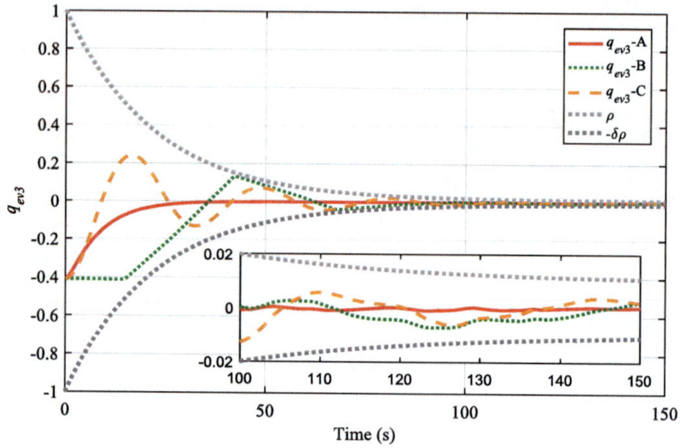

Fig. 3.3 The spacecraft attitude tracking error q_{ev3}

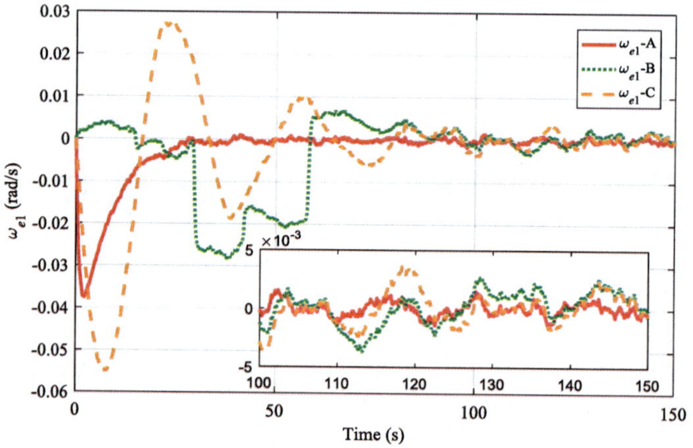

Fig. 3.4 The angular velocity tracking error ω_{e1}

within the predefined time. Furthermore, from the numerical simulation curves, the transient process is smoother. In contrast, controller 'B' is a traditional prescribed performance control, and its control effect appears when the tracking error is close to the boundary. Therefore, the proposed controller can give the closed-loop system faster transient performance. For controller 'C', although this controller can enable the closed-loop system to converge within the finite time, it lacks the performance constraint guarantees.

Figures 3.4, 3.5 and 3.6 present the spacecraft attitude angular velocity tracking error. It can be seen from the numerical simulation curve that, compared with the controllers 'B' and 'C', under the proposed controller, during the entire control process, the attitude angular velocity tracking error of the closed-loop system is smoother, the

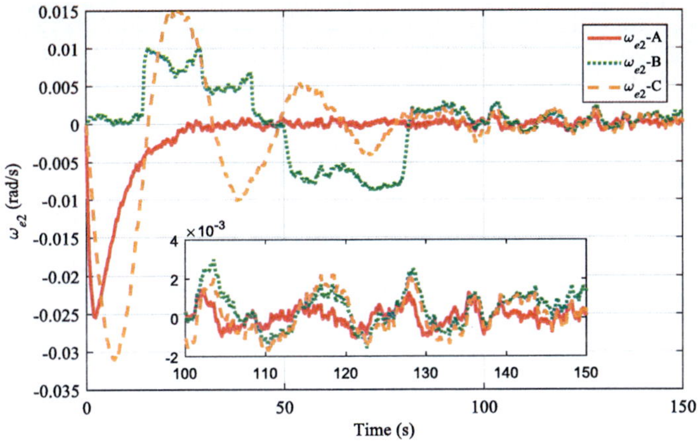

Fig. 3.5 The angular velocity tracking error ω_{e2}

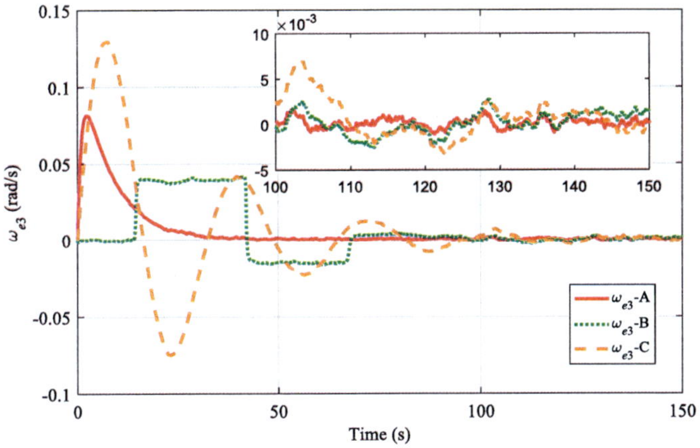

Fig. 3.6 The angular velocity tracking error ω_{e3}

variation range is smaller, and the transient and steady-state performance is better. It is worth noting that in this simulation situation, the initial angular velocity error is about $[3.87 \times 10^{-4}, -2.60 \times 10^{-4}, -0.0010]^T$ rad/s.

Figure 3.7 shows the comparison of the error index $I_a = ||q_{ev}|| + ||\omega_e||$ of the three controllers. It can be seen from the comparison curves that under the proposed controller, the error-index converges faster, the steady-state error is smaller, and the disturbance suppression ability is stronger. From Fig. 3.8, it can be seen that under the proposed controller, when the transient and steady-state performance of the closed-loop system is better, the controller proposed in this section consumes similar or less energy than the comparative controllers.

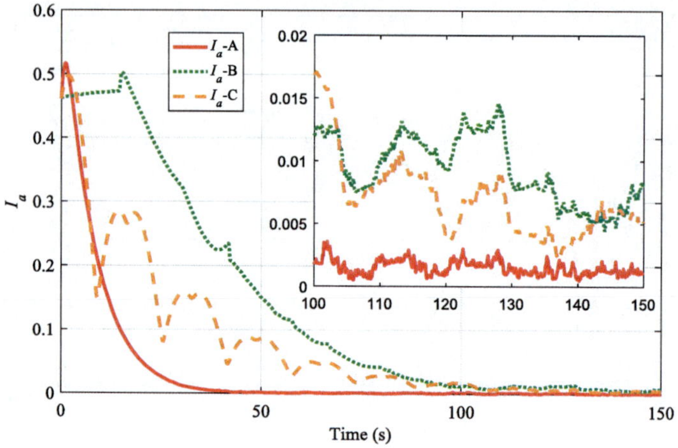

Fig. 3.7 The error index I_a

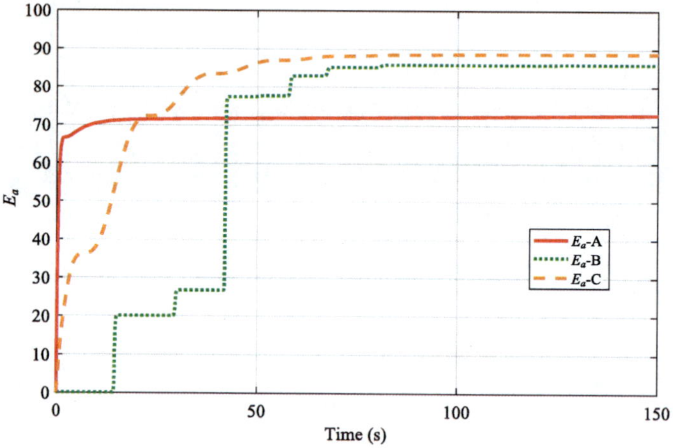

Fig. 3.8 The energy consumption

Figure 3.9 is the control input curve of the proposed controller. It can be seen from Fig. 3.9 that the proposed control input mainly acts at the beginning of the control, so the peak torque in the initial control stage is large, and when the state converges, the control input decreases. Figure 3.10 is the numerical reference for the external disturbances suffered by the spacecraft in the numerical simulation. Considering that the control input proposed has a large peak value at the beginning of the control, the controller is further improved in subsequent chapters.

Through numerical simulation, it is fully demonstrated that under the action of the proposed controller, the closed-loop system can obtain better transient and steady-state performance by consuming similar or less energy, and the performance is guar-

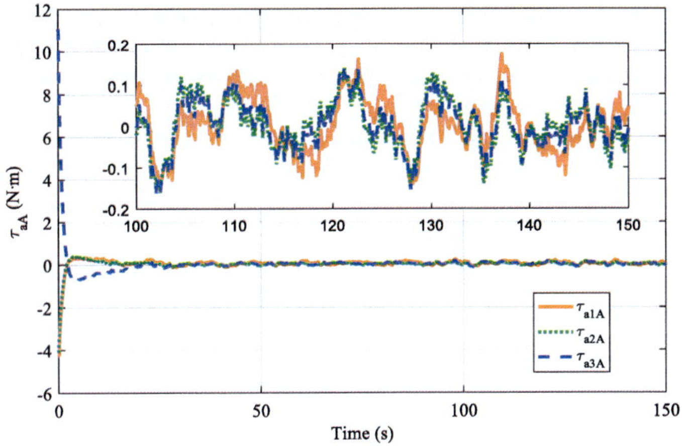

Fig. 3.9 The control input of 'A'

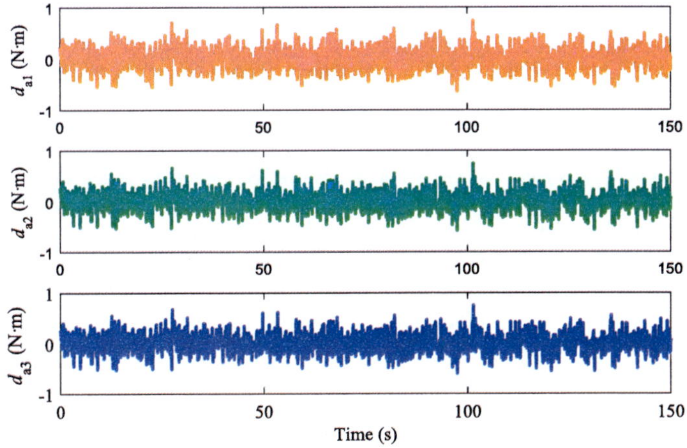

Fig. 3.10 The external disturbances

anteed. That is, it is ensured that the maximum settling time of the tracking error is less than or equal to the predefined maximum settling time, the practical steady-state error is less than the predefined values, and the overshoot satisfies the constraints.

3.4 Conclusion

Based on the predefined maximum settling time constraint function proposed in Chap. 2, this chapter designs the auxiliary system with the predefined maximum settling time and predefined transient performance constraints. This chapter proposes the spacecraft asymptotic performance guaranteed control with considering the time parameters (predefined maximum settling time parameters). Specifically, the controller proposed in this chapter is a sliding-mode-like controller based on performance constraints and auxiliary transformation systems. The asymptotically stable performance guaranteed control algorithm is proposed by defining performance-constrained auxiliary variables, designing sliding mode surfaces, and candidate Lyapunov functions. The stability of its closed-loop system is proven. Under the proposed controller's action, the closed-loop system's tracking error can converge to the predefined maximum steady-state error interval under the constraint within the maximum settling time parameter. Finally, the rationality and effectiveness of the controller proposed in this chapter are proved by the comparisons and numerical simulation experiments with advanced controllers in similar works of literature.

References

1. Nagesh, I., Edwards, C.: A multivariable super-twisting sliding mode approach. Automatica **50**(3), 984–988 (2014)
2. Jiang, B., Li, C., Ma, G.: Finite-time output feedback attitude control for spacecraft using "adding a power integrator" technique. Aerospace Science and Technology **66**, 342–354 (2017)
3. Zou, A.M.: Finite-time output feedback attitude tracking control for rigid spacecraft. IEEE Transactions on Control Systems Technology **22**(1), 338–345 (2013)
4. Bechlioulis, C.P., Rovithakis, G.A.: Robust adaptive control of feedback linearizable mimo nonlinear systems with prescribed performance. IEEE Transactions on Automatic Control **53**(9), 2090–2099 (2008)
5. Du, H., Li, S., Qian, C.: Finite-time attitude tracking control of spacecraft with application to attitude synchronization. IEEE Transactions on Automatic Control **56**(11), 2711–2717 (2011)

Chapter 4
Spacecraft Finite-Time Performance Guaranteed Control

Abstract For the attitude control system of the on-orbit service spacecraft with strong motion constraints and high control precision requirements, this chapter proposes the finite-time performance guaranteed control algorithms for the spacecraft. The performance guaranteed control algorithm provides the closed-loop system with the determined transient and steady-state motion intervals by directly constraining the state time-domain response of the closed-loop system. This enables the state of the closed-loop system to be strictly constrained by the desired performance parameters, such as the desired maximum settling time, minimum convergence rate, maximum overshoot, and maximum steady-state error. On this basis, the finite-time performance guaranteed algorithm proposed in this chapter further improves the accuracy of the settling time description. The conservatism of the closed-loop system is relaxed through the more precise settling time design. Simultaneously, the finite-time algorithm also improves the transient performance of the system, such as the smoother state trajectory, etc. Under the proposed control algorithms, the transient and steady-state behaviors of the spacecraft attitude closed-loop system meet the predefined performance constraints. Furthermore, the closed-loop system can converge to the neighborhood of the equilibrium point within the finite time. Synchronously, the closed-loop system has a strong disturbance suppression ability. Finally, the rigorous theoretical proof and numerical simulation analysis verify the system's stability and the control algorithms' rationality and effectiveness.

4.1 Introduction

Based on the spacecraft model, stability theorems described in Chap. 2, and the performance constraints with the predefined maximum settling time proposed in this book, this chapter proposes the finite-time performance guaranteed control algorithms for spacecraft attitude with the predefined settling time. Specifically, it is hoped that the control algorithms proposed in this chapter can make the actual settling time of the spacecraft attitude tracking closed-loop system state converge to the predefined interval within the predefined maximum settling time t_{ps}. The transient

Y. Gao and D. Li, *Spacecraft Maneuver with Performance Guaranteed*,
https://doi.org/10.1007/978-981-99-4653-2_4

and steady-state performance of the closed-loop system can satisfy the performance constraints (2.55). Simultaneously, the closed-loop system is finite-time stable.

In the previous chapters, the basic techniques for performance guaranteed control have been more comprehensively described. Here, considering the finite-time stability expectations studied in this chapter, finite-time stability is briefly introduced. In the 1960s, the possibility of finite time was proposed [1], making up the theoretical blank of control theory for designing the settling time in advance. In the late 1990s, due to the proposal of the homogeneous theory [2], and the finite-time Lyapunov stability theorem [3], the research on the finite-time theory achieved breakthrough and development. After that, many research results on finite-time control were obtained. The theoretical methods mainly include the terminal sliding mode (TSM) method [4], the homogeneous system (HS) method [5], and the adding a power integrator (API) technique [6]. Compared with the HS method, the API technique can perform stability analysis on disturbed systems [7]. Compared with TSM, the API does not have singularity problems and can give a complete expression of the upper bound of the settling time [8]. Therefore, the API becomes a practical design method for finite-time controllers. Based on the finite-time theory, the fixed-time method [9] was proposed. This method can calculate the upper bound of the settling time without initial state values. The finite-time and fixed-time methods can indirectly design the upper bound of the settling time by changing the controller parameters. However, adjusting the parameters to set the upper bound of the settling time will sacrifice the controller performance, and the practical application of this method is thus limited. Moreover, these methods also do not have the ability to strictly restrict the transient and steady-state behavior of closed-loop systems.

As analyzed by Remark 3.2 in Chap. 3, although the performance guaranteed of the predefined time makes it possible to design the closed-loop system settling time in advance, the control algorithm proposed in Chap. 3 cannot make the closed-loop system stable within the finite time theoretically. That is, the closed-loop system is asymptotically stable under the performance guaranteed control algorithm. Moreover, the performance constraints and the predefined maximum settling time are only the boundaries of the performance parameters, so the stability of the closed-loop system is relatively conservative. For more precise performance requirements or narrower expected trajectory motion ranges, if the performance index parameters are blindly reduced, a higher price may be paid for the practical system, which is not expected by the system design. Under this requirement, finite-time stability can be considered to further enhance the performance guaranteed algorithm because of its stronger descriptiveness of the convergence boundary and the settling time. The finite-time performance guaranteed control design, on the one hand, improves the accuracy, time descriptiveness, low-cost enhancement and multi-parameter designability of the closed-loop system. On the other hand, performance boundary guarantees are also provided. It is worth noting that the finite-time performance guaranteed algorithm has two time parameters, namely the predefined maximum settling time parameter t_{ps} in the performance guaranteed algorithm and the settling time parameter t_{finite} in the finite-time stability. Furthermore, t_{ps} is actively designed, t_{finite} is passively adjusted according to the system and parameters, etc. Introducing a predefined settling time can also enhance the initiative of finite-time stability design. At the same time, the two parameters are designed separately to form the time parameter design isolation or decoupling. t_{ps} constrains the upper bound of the settling time, and t_{finite} finds a smaller upper bound on the actual settling time of the closed-loop system. Although t_{finite} will still be affected by t_{ps} to a certain extent, to the greatest extent, it gives the control designer the ability to actively intervene in the actual settling time of the closed-loop system, without considering the system parameters at all. (The design of t_{finite} is constrained by the initial state of the system, controller parameters, and system parameters.)

Combined with the specific situation of spacecraft attitude tracking, the problems or control objectives studied in this chapter are described as follows.

Problem 4.1 Consider the spacecraft attitude tracking error model (2.28), and select predefined related parameters according to practical mission requirements, including predefined maximum settling time t_{ps}, the maximum steady-state error ρ_t, the maximum overshoot χ_t, design the performance guaranteed control such that:
(1) the conditions (performance guaranteed) in Problem 3.1 are satisfied;
(2) the tracking error converges within the finite time;
(3) the closed-loop system has the property of disturbance suppression. That is, when strong external disturbances exist, the closed-loop system still satisfies the above performance constraints.

According to the control problem presented in this chapter, design the performance guaranteed control algorithms $\tau \in \mathbb{R}^n$ or $\tau_a \in \mathbb{R}^n$ for the spacecraft (τ and τ_a are defined in (2.28)). So that when $t \to T$, where $T < +\infty$ denotes the finite time, the spacecraft closed-loop system state/error $q_{ev} \to 0$, that is, converge to the equilibrium point. Alternatively, $q_{ev} \to \Theta$, where $\Theta = \{|q_{evi}| \leq \iota_i\}$ denotes a small neighborhood containing the equilibrium point, $\iota_i \in \mathbb{R}^+$ and $\iota_i\ le\rho_t$ are the sufficiently small positive constant related to the controller parameters and the initial state of the closed-loop system. When the system is finite-time stable, $\iota_i \leq \rho_t$ and $T \leq \min\{t_{ps}, t_{finite}\}$, where t_{finite} represents the finite settling time of the closed-loop system.

In addition, for practical spacecraft systems, the selection of predefined performance constraints often needs to consider practical situations. Therefore, the performance constraints in practical applications are often not chosen to be very narrow.

This chapter is arranged as follows. Based on the performance guaranteed control techniques described in the previous chapters, this chapter directly gives the finite-time performance guaranteed control algorithm design when external disturbances are not considered in Sect. 4.2 and gives complete proof. The proposed control algorithm is based on adding a power integrator finite-time method. Further, the finite-time performance guaranteed algorithm with strong robustness is proposed when external disturbances are considered in Sect. 4.3. Numerical simulations and analyses of these two algorithms are presented in Sect. 4.4. Section 4.5 gives the conclusions of this chapter.

4.2 Control Design and Stability Analysis Without Disturbances

The same as the definition of the auxiliary system (3.6) in the previous chapter, consider the spacecraft attitude tracking error model (2.28), where τ_a is the control input of the spacecraft attitude, $\tau = \frac{1}{2}P^T\tau_a$ is the control input for the equivalent Euler–Lagrange system. In this section, the spacecraft finite-time performance guaranteed control with predefined performance constraints is designed. The relationship between the predefined time and the finite time will be given later.

Theorem 4.1 *Under no disturbances, consider the spacecraft attitude error system (2.28) with a known smooth reference trajectory. If the controller* τ *is chosen as*

$$\tau = -\kappa_1\kappa_E\left(\dot{\sigma}^p + \kappa_2^p\sigma\right)^{2/p-1} - E^{-1}D + E^{-1}\Gamma\dot{\sigma} \tag{4.1}$$

where

$$\begin{aligned} \kappa_1 &\geq \frac{1}{E_{\min}}(2-1/p)\,2^{1-1/p}\,\kappa_2^{1+p}\left(\frac{2^{1-1/p}+np}{1+p}+\frac{2^{1-1/p}\,n}{\kappa_2}+\kappa_3\right)\\ \kappa_2 &\geq \frac{2^{1-1/p}\,p+n}{1+p}+\kappa_3\\ \kappa_3 &> 0 \end{aligned} \tag{4.2}$$

where $1 < p = p_1/p_2 < 2$, p_1, p_2 *are positive odd integers,* κ_E *is the judgment function, when E is negative* $\kappa_E = -1$, *otherwise* $\kappa_E = 1$, *then* σ *can converge in finite time, and the finite settling time of* σ *satisfies* $T_1 \leq V(0)^{1-d/2}/\lambda_V(1-d/2) < +\infty$. *The closed-loop tracking error* q_{ev} *for all states is bounded concerning the inequality (2.55) all the time. When* $q_{evi} \leq \rho_t$, *the settling time of* q_{evi} *satisfies* $t|_{q_{evi}\leq\rho_t} \leq t_{ps}$, *where* t_{ps} *is the expected settling time pre-defined in advance. In addition, the transient and steady-state performance of the closed-loop system can also be described.*

Proof The API method is used to design the controller and prove the stability of the closed-loop system. We first design the virtual input $\dot{\sigma}^*$ so that σ converges in finite time. After that, $\dot{\sigma}^*$ is proved to be tracked in finite time. Finally, the proof that the closed-loop system converges in finite time is completed.

The candidate Lyapunov function for the auxiliary system can be selected as

$$V = V_0 + \sum\nolimits_{i=1}^{n} V_i,\ n = 3 \tag{4.3}$$

where

$$V_0 = \frac{1}{2}\sigma^T\sigma \tag{4.4}$$

$$V_i = \frac{1}{(2-1/p)\,2^{1-1/p}\,\kappa_2^{1+p}} \int_{\dot{\sigma}_i^*}^{\dot{\sigma}_i} \left(s^p - \dot{\sigma}_i^{*p}\right)^{2-1/p} \mathrm{d}s \tag{4.5}$$
$$i = 1, 2, \ldots, n$$

and $\dot{\sigma}^* = -\kappa_2 \sigma^{1/p}$, which the virtual control can stabilize V_0. By Propositions B1 and B2 in [10], $\int_{\dot{\sigma}_i^*}^{\dot{\sigma}_i} (s^p - \dot{\sigma}_i^{*p})^{2-1/p}\,\mathrm{d}s$ are differentiable and positive definite. To facilitate the derivation, the auxiliary variable is defined as

$$\xi_i = \dot{\sigma}_i^p - \dot{\sigma}_i^{*p} \tag{4.6}$$

Next, we will calculate $\dot{V}_0$, $\sum_{i=1}^n \dot{V}_i$ and $\dot{V}$ in the following three steps.
Step 1: Calculate $\dot{V}_0$.
Taking the derivative of (4.4), we have

$$\dot{V}_0 = \sigma^T \dot{\sigma} = \sum\nolimits_{i=1}^n \sigma_i \dot{\sigma}_i^* + \sum\nolimits_{i=1}^n \sigma_i (\dot{\sigma}_i - \dot{\sigma}_i^*) \tag{4.7}$$

By Lemma 2.8, we obtain

$$\sigma_i (\dot{\sigma}_i - \dot{\sigma}_i^*) \le 2^{1-1/p} |\sigma_i| |\dot{\sigma}_i^p - \dot{\sigma}_i^{*p}|^{1/p} \tag{4.8}$$

By Lemma 2.9, we have

$$\sigma_i (\dot{\sigma}_i - \dot{\sigma}_i^*) \le 2^{1-1/p} \left(\frac{p}{1+p} |\sigma_i|^d + \frac{1}{1+p} |\xi_i|^d \right) \tag{4.9}$$

where $d = 1 + 1/p$.
Substituting (4.9) into (4.7) results in

$$\begin{aligned} \dot{V}_0 &= -\kappa_2 \sum\nolimits_{i=1}^n \sigma_i \sigma_i^{1/p} + \sum\nolimits_{i=1}^n \sigma_i \left(\dot{\sigma}_i - \dot{\sigma}_i^*\right) \\ &\le \sum\nolimits_{i=1}^n 2^{1-1/p} \left(\frac{p}{1+p} |\sigma_i|^d + \frac{1}{1+p} |\xi_i|^d \right) - \kappa_2 \sum\nolimits_{i=1}^n \sigma_i^{1+1/p} \\ &= -\left(\kappa_2 - 2^{1-1/p} \frac{p}{1+p} \right) \sum\nolimits_{i=1}^n |\sigma_i|^d + 2^{1-1/p} \frac{1}{1+p} \sum\nolimits_{i=1}^n |\xi_i|^d \end{aligned} \tag{4.10}$$

Step 2: Calculate $\sum_{i=1}^n \dot{V}_i$.
To design the system control input τ based on the virtual control input $\dot{\sigma}^*$, taking the derivative of V_i, we have

$$\begin{aligned} \dot{V}_i = &-\frac{1}{2^{1-1/p}\,\kappa_2^{1+p}} \frac{\partial \dot{\sigma}_i^{*p}}{\partial t} \int_{\dot{\sigma}_i^*}^{\dot{\sigma}_i} \left(s^p - \dot{\sigma}_i^{*p}\right)^{1-1/p} \mathrm{d}s \\ &+ \frac{1}{(2-1/p)\,2^{1-1/p}\,\kappa_2^{1+p}} \xi_i^{2-1/p}\, \bar{\tau}_i,\ i = 1, 2, \ldots, n \end{aligned} \tag{4.11}$$

where $\bar{\tau}_i = \partial\dot{\sigma}_i/\partial t$. According to $\dot{\sigma}^* = -\kappa_2 \sigma^{1/p}$ and (3.6), we have

$$\frac{\partial \dot{\sigma}^{*p}}{\partial t} = \left[\frac{\partial \dot{\sigma}_1^{*p}}{\partial t}, \ldots, \frac{\partial \dot{\sigma}_i^{*p}}{\partial t}, \ldots, \frac{\partial \dot{\sigma}_n^{*p}}{\partial t}\right]^T = -\kappa_2^p \dot{\sigma} \tag{4.12}$$

Noticing that

$$\begin{aligned} \left|\frac{\partial}{\partial t}\dot{\sigma}_i^{*p}\right| &\le \kappa_2^p (|\dot{\sigma}_1| + \cdots + |\dot{\sigma}_i| + \cdots \\ &+ |\dot{\sigma}_n|) = \kappa_2^p \sum\nolimits_{i=1}^n |\dot{\sigma}_i| \end{aligned} \tag{4.13}$$

By imposing the result in (4.13), (4.11) is written as follows

$$\begin{aligned} \dot{V}_i \le & \frac{\left(\sum_{i=1}^n |\dot{\sigma}_i|\right) \left|\dot{\sigma}_i - \dot{\sigma}_i^*\right|}{2^{1-1/p}\,\kappa_2} |\xi_i|^{1-1/p} \\ &+ \frac{\xi_i^{2-1/p}\,\bar{\tau}_i}{(2 - 1/p\,)\,2^{1-1/p}\,\kappa_2^{1+p}} \end{aligned} \tag{4.14}$$

By Lemmas 2.8 and 2.9, for any $i, m = 1, 2, \ldots, n$, we have

$$\begin{aligned} |\dot{\sigma}_m| \left|\dot{\sigma}_i - \dot{\sigma}_i^*\right| |\xi_i|^{1-1/p} &\le 2^{1-1/p}\, |\dot{\sigma}_m|\, |\xi_i|^{1/p}\, |\xi_i|^{1-1/p} \\ &\le 2^{1-1/p} \left(\frac{(2^{1-1/p} + \kappa_2)\,p}{1+p} |\xi_i|^d \right. \\ &\left. + \frac{2^{1-1/p}}{1+p} |\xi_m|^d + \frac{\kappa_2}{1+p} |\sigma_m|^d\right) \end{aligned} \tag{4.15}$$

Substituting (4.15) into (4.14) results in

$$\begin{aligned} \dot{V}_i \le & \left(\frac{2^{1-1/p} + \kappa_2}{\kappa_2} \frac{np}{1+p} |\xi_i|^d \right. \\ &\left. + \frac{1}{\kappa_2} \frac{2^{1-1/p}}{1+p} \sum\nolimits_{m=1}^n |\xi_m|^d + \frac{1}{1+p} \sum\nolimits_{m=1}^n |\sigma_m|^d \right) \\ &+ \frac{1}{(2 - 1/p\,)\,2^{1-1/p}\,\kappa_2^{1+p}} \xi_i^{2-1/p}\,\bar{\tau}_i \end{aligned} \tag{4.16}$$

Step 3: Calculate $\dot{V}$.
Taking the derivative of (4.3), we have

$$\dot{V} = \dot{V}_0 + \sum\nolimits_{i=1}^n \dot{V}_i \tag{4.17}$$

Substituting (4.10) and (4.16) into (4.17) and considering $|\sigma_i|^d = |\sigma_i|^{(1+p)/p} = \sigma_i^d$, $|\xi_i|^d = \xi_i^d$, we have

$$\begin{aligned}\dot{V} \le & -\left(\kappa_2 - \frac{2^{1-1/p}\, p + n}{1+p}\right)\sum_{i=1}^{n} |\sigma_i|^d \\ & + \left(\frac{2^{1-1/p} + np}{1+p} + \frac{2^{1-1/p}\, n}{\kappa_2}\right)\sum_{i=1}^{n} |\xi_i|^d \\ & + \frac{\sum_{i=1}^{n} \xi_i^{2-1/p}\, \bar{\tau}_i}{(2-1/p\,)\, 2^{1-1/p}\, \kappa_2^{1+p}}\end{aligned} \tag{4.18}$$

According to the Theorem 4.1 proposed, if we take the controller as (4.1), then

$$\ddot{\sigma} = -\kappa_1 E_*\left(\dot{\sigma}^p + \kappa_2^p \sigma\right)^{2/p\,-1} \tag{4.19}$$

where $E_* = \mathrm{diag}(|E_1|, \ldots, |E_i|, \ldots, |E_n|)$, $i = 1, \ldots, n$. Furthermore, it always exists $|E_i| \ge E_{\min} > 0$, where $E_{\min}$ is a positive constant.

Then we have

$$\bar{\tau}_i = \partial\dot{\sigma}_i/\partial t = -\kappa_1 |E_i|(\dot{\sigma}_i^p + \kappa_2^p \sigma_i)^{2/p\,-1} = -\kappa_1 |E_i| \xi_i^{2/p\,-1} \tag{4.20}$$

Substituting (4.20) into (4.18), we can obtain

$$\begin{aligned}\dot{V} \le & -\left(\kappa_2 - \frac{2^{1-1/p}\, p + n}{1+p}\right)\sum_{i=1}^{n} |\sigma_i|^d \\ & + \left(\frac{2^{1-1/p} + np}{1+p} + \frac{2^{1-1/p}\, n}{\kappa_2}\right)\sum_{i=1}^{n} |\xi_i|^d \\ & - \frac{\kappa_1 \sum_{i=1}^{n} |E_i| \xi_i^d}{(2-1/p\,)\, 2^{1-1/p}\, \kappa_2^{1+p}}\end{aligned} \tag{4.21}$$

Considering the value of $|E_i|$, we have

$$|E_i| \xi_i^{1+1/p} \ge E_{\min} \xi_i^{1+1/p} \tag{4.22}$$

$$-\sum_{i=1}^{n} |E_i| \xi_i^{1+1/p} \le -E_{\min} \sum_{i=1}^{n} \xi_i^{1+1/p} \tag{4.23}$$

So that (4.21) can be rewritten as

$$\begin{aligned}\dot{V} \le & -\left(\kappa_2 - \frac{2^{1-1/p}\, p + n}{1+p}\right)\sum\nolimits_{i=1}^{n} |\sigma_i|^d \\ & + \left(\frac{2^{1-1/p} + np}{1+p} + \frac{2^{1-1/p}\, n}{\kappa_2}\right)\sum\nolimits_{i=1}^{n} |\xi_i|^d \\ & - \frac{\kappa_1 E_{\min} \sum_{i=1}^{n} \xi_i^d}{(2-1/p)\, 2^{1-1/p} \kappa_2^{1+p}}\end{aligned} \tag{4.24}$$

To simplify the above inequality, we have

$$\begin{aligned}\dot{V} \le & -\left(\kappa_2 - \frac{2^{1-1/p}\, p + n}{1+p}\right)\sum\nolimits_{i=1}^{n} \sigma_i^d \\ & - \left(\frac{\kappa_1 E_{\min}}{(2-1/p)\, 2^{1-1/p} \kappa_2^{1+p}} - \frac{2^{1-1/p} + np}{1+p} - \frac{2^{1-1/p}\, n}{\kappa_2}\right)\sum\nolimits_{i=1}^{n} \xi_i^d\end{aligned} \tag{4.25}$$

Take $\kappa_3 > 0$, the above inequality becomes

$$\dot{V} \le -\kappa_3 \sum\nolimits_{i=1}^{n} \sigma_i^d - \kappa_3 \sum\nolimits_{i=1}^{n} \xi_i^d \tag{4.26}$$

where κ_1, κ_2 and κ_3 need to satisfy the following conditions

$$\begin{aligned} & \kappa_2 - \frac{2^{1-1/p}\, p + n}{1+p} > \kappa_3 \\ & \kappa_1 \frac{E_{\min}}{(2-1/p)\, 2^{1-1/p} \kappa_2^{1+p}} - \frac{2^{1-1/p} + np}{1+p} - \frac{2^{1-1/p}\, n}{\kappa_2} > \kappa_3\end{aligned} \tag{4.27}$$

The ranges of values of κ_1, κ_2 are obtained

$$\begin{aligned} & \kappa_2 > \frac{2^{1-1/p}\, p + n}{1+p} + \kappa_3 \\ & \kappa_1 > \frac{(2-1/p)\, 2^{1-1/p} \kappa_2^{1+p}}{E_{\min}}\left(\frac{2^{1-1/p} + np}{1+p} + \frac{2^{1-1/p}\, n}{\kappa_2} + \kappa_3\right)\end{aligned} \tag{4.28}$$

From the above derivation, we know that the closed-loop system is stable. Next, we prove that $V(t)$ can converge to the origin in finite time.

Through the calculation of (4.11) and Lemma 2.8, we can get

$$V_i \le \frac{|\dot{\sigma}_i - \dot{\sigma}_i^*||\dot{\sigma}_i^p - \dot{\sigma}_i^{*p}|^{2-1/p}}{(2-1/p)2^{1-1/p} \kappa_2^{1+p}} \le \frac{\xi_i^2}{(2-1/p)\kappa_2^{1+p}} \tag{4.29}$$

Hence,

$$V = V_0 + \sum\nolimits_{i=1}^{n} V_i \le \kappa_4 \left(\sum\nolimits_{i=1}^{n} \sigma_i^2 + \sum\nolimits_{i=1}^{n} \xi_i^2 \right) \tag{4.30}$$

where $\kappa_4 = \max\{1/2, 1/(2 - 1/p)\kappa_2^{1+p}\}$.

By Lemma 2.10, we have

$$V^{d/2} \le \kappa_4^{d/2} \left(\sum\nolimits_{i=1}^{n} \sigma_i^d + \sum\nolimits_{i=1}^{n} \xi_i^d \right) \tag{4.31}$$

Noticing

$$\begin{aligned} \dot{V} + \lambda_V V^{d/2} \le & -\kappa_3 \sum\nolimits_{i=1}^{n} \sigma_i^d - \kappa_3 \sum\nolimits_{i=1}^{n} \xi_i^d \\ & + \lambda_V \kappa_4^{d/2} \sum\nolimits_{i=1}^{n} \sigma_i^d + \lambda_V \kappa_4^{d/2} \sum\nolimits_{i=1}^{n} \xi_i^d \end{aligned} \tag{4.32}$$

simplify the above inequality, and we get

$$\begin{aligned} \dot{V} + \lambda_V V^{d/2} \le & -\left(\kappa_3 - \lambda_V \kappa_4^{d/2} \right) \sum\nolimits_{i=1}^{n} \sigma_i^d \\ & -\left(\kappa_3 - \lambda_V \kappa_4^{d/2} \right) \sum\nolimits_{i=1}^{n} \xi_i^d \end{aligned} \tag{4.33}$$

Because of $\kappa_3 - \lambda_V \kappa_4^{d/2} \ge 0, \lambda_V > 0, \lambda_V \le \kappa_3/\kappa_4^{d/2}$, we can take $\lambda_V = \kappa_3/(2\kappa_4^{d/2})$, then

$$\dot{V} + \lambda_V V^{d/2} \le 0 \tag{4.34}$$

Note that $0 < d/2 < 1$, by Lemma 2.3, the closed-loop system is finite-time stable with the finite settling time

$$T_1 \le \frac{V(0)^{1-d/2}}{\lambda_V (1 - d/2)} < +\infty \tag{4.35}$$

This implies $\sigma \to 0$, $\dot{\sigma} \to 0$ in finite time and $\dot{\sigma} = \sigma = 0$ for $\forall t \ge T_1$. $\sigma = 0$ means that not only the system converges in finite time, but the transient and steady-state performance of the closed-loop system satisfies the performance constraints (2.55). □

The above proof process shows that the controller force $\dot{\sigma}$ and σ to reach zero in finite time. Thus, $-1 < \vartheta_i = \varphi(\sigma_i) < 1$ holds. Furthermore, (2.55) is achieved. In turn, the control objectives are finally achieved.

Remark 4.1 If the power exponent $p = 1$ in the controller (4.1), then the finite-time controller will become as follows,

$$\tau = -\kappa_1 \kappa_E (\dot{\sigma} + \kappa_2 \sigma) - E^{-1} D + E^{-1} \Gamma \dot{\sigma} \tag{4.36}$$

where $\kappa_1 \geq (\kappa_2^2/E_{\min})((1+n)/2 + n/\kappa_2 + \kappa_3)$, $\kappa_2 \geq (1+n)/2 + \kappa_3$, $\kappa_3 > 0$. The controller (4.36) can be considered a simple conventional asymptotical full-state feedback controller with the proposed PMSTC method. Under this controller, σ of the closed-loop system will asymptotically converge, and the convergence rate is exponential. So, the closed-loop tracking error for all states is bounded with respect to the inequality (2.55) all the time.

4.3 Control Design and Stability Analysis with Disturbances

Considering the finite-time controller based on performance constraints designed in the previous section, if the external disturbance is considered, the nonlinear term $F = -BM_a{}^{-1}d$ about the disturbance in the model (2.28) needs to be considered. Considering the robustness enhancement of the system, based on the previous section, the robust finite-time controller based on the performance guaranteed is further designed in this section.

According to Lemma 2.6, and the analysis in the previous section, the general form of the observer for the system $\dot{x} = f + gv + d$ is as follows,

$$\dot{y}_0 = -k_1 \frac{y_0 - x}{||y_0 - x||^{1/2}} + y_1 - k_2 (y_0 - x) + f + gv \tag{4.37}$$

$$\dot{y}_1 = -k_3 \frac{y_0 - x}{||y_0 - x||} - k_4 (y_0 - x) \tag{4.38}$$

where k_1, k_2, k_3, and k_4 represent positive constant gains, y_0 and y_1 are observations for state x and unknown disturbance d.

So far, we design the following observer to observe the uncertainty F in $\ddot{\sigma} = D - \Gamma\dot{\sigma} + E\tau + F$,

$$\dot{y}_0 = -\kappa_{y1} \frac{y_0 - \dot{\sigma}}{||y_0 - \dot{\sigma}||^{1/2}} + y_1 - \kappa_{y2} (y_0 - \dot{\sigma}) + D - \Gamma\dot{\sigma} + E\tau \tag{4.39}$$

$$\dot{y}_1 = -\kappa_{y3} \frac{y_0 - \dot{\sigma}}{||y_0 - \dot{\sigma}||} - \kappa_{y4} (y_0 - \dot{\sigma}) \tag{4.40}$$

where κ_{y1}, κ_{y2}, κ_{y3}, and κ_{y4} represent positive constant gains, y_0 and y_1 are observations for state $\dot{\sigma}$ and unknown disturbance F, i.e. $y_0 \to \dot{\sigma}$, $y_1 \to F$. Considering that in this controller, the model is precisely known, so only the y_1 item is needed. And the tracking error of y_1 for $-BM_a^{-1}d$ converges within the finite time T_y.

According to the robust term y_1 designed by the above observer, the modified controller (4.1) is as follows,

$$\tau = -\kappa_1 \kappa_E \left(\dot{\sigma}^p + \kappa_2^p \sigma\right)^{2/p-1} - E^{-1}D - E^{-1}y_1 + E^{-1}\Gamma\dot{\sigma} \tag{4.41}$$

Considering the observation error of the observer, define $F_y = -y_1 + F$. According to the properties of the observer, the observation error is bounded. The boundary is $||F_y|| \le f_d$, where f_d is a positive number. The following Theorem is proposed.

Theorem 4.2 *Considering the spacecraft error system (2.28) with a known smooth reference trajectory under the finite-time controller (4.41), then σ and $\dot{\sigma}$ will converge to the region Ω in finite time T_2, where*

$$\Omega = \left\{ \|\xi\| \le \left[\frac{n^{1/(2p)}\,\kappa_5}{\kappa_7\,(1-\kappa_8)}\right]^{p/(2-p)} \right\} \tag{4.42}$$

$$T_2 \le \frac{2pV^{(p-1)/(2p)}\ (0)}{\kappa_9\,(p-1)} + T_y < +\infty \tag{4.43}$$

and

$$\kappa_5 = \frac{f_d}{(2-1/p\,)\,2^{1-1/p}\,\kappa_2^{1+p}} \tag{4.44}$$

$$\kappa_6 = \kappa_2 - \frac{2^{1-1/p}\ p+n}{1+p} \tag{4.45}$$

$$\kappa_7 = \frac{\kappa_1 E_{\min}}{(2-1/p\,)\,2^{1-1/p}\,\kappa_2^{1+p}} - \frac{2^{1-1/p}\ +np}{1+p} - \frac{2^{1-1/p}\,n}{\kappa_2} \tag{4.46}$$

$$1 > \kappa_8 > 0 \tag{4.47}$$

$$\kappa_9 = \min\left\{2^{\frac{1+p}{2p}}\kappa_6,\ \kappa_7\kappa_8(2-1/p\,)^{\frac{1+p}{2p}}\kappa_2^{\frac{(1+p)^2}{2p}}\right\} \tag{4.48}$$

Proof Considering the disturbance d in (2.28) based on Assumption 2.1 and previous analysis of $F = -BM_a^{-1}d$, we substitute the controller (4.41) into the definition of $\bar{\tau}_i$, then

$$\bar{\tau}_i = -\kappa_1 |E_i| \xi_i^{2/p-1} + F_{yi} \tag{4.49}$$

We rewrite (4.18) as

$$\begin{aligned}\dot{V} \leq & -\left(\kappa_2 - \frac{2^{1-1/p}\ p+n}{1+p}\right)\sum\nolimits_{i=1}^{n} |\sigma_i|^d \\ & + \left(\frac{2^{1-1/p}\ +np}{1+p} + \frac{2^{1-1/p}\ n}{\kappa_2}\right)\sum\nolimits_{i=1}^{n} |\xi_i|^d \\ & + \frac{1}{(2-1/p\ )\, 2^{1-1/p}\, \kappa_2^{1+p}} \cdot \\ & \sum\nolimits_{i=1}^{n} \xi_i^{2-1/p} \left(-\kappa_1 |E_i| \xi_i^{2/p\ -1} + F_{yi}\right)\end{aligned} \tag{4.50}$$

Then, we can obtain

$$\begin{aligned}\dot{V} \leq & -\left(\kappa_2 - \frac{2^{1-1/p}\ p+n}{1+p}\right)\sum\nolimits_{i=1}^{n} \sigma_i^d \\ & -\left(\frac{\kappa_1 E_{\min}}{(2-1/p\ )\, 2^{1-1/p}\, \kappa_2^{1+p}} - \frac{2^{1-1/p}\ +np}{1+p}\right. \\ & \left. -\frac{2^{1-1/p}\ n}{\kappa_2}\right)\sum\nolimits_{i=1}^{n} \xi_i^d + \kappa_5 \sum\nolimits_{i=1}^{n} |\xi_i|^{2-1/p}\end{aligned} \tag{4.51}$$

Simplify (4.51), we can get

$$\begin{aligned}\dot{V} \leq & -\kappa_3 \sum\nolimits_{i=1}^{n} \sigma_i^d - \kappa_3 \sum\nolimits_{i=1}^{n} \xi_i^d \\ & + \kappa_5 \sum\nolimits_{i=1}^{n} |\xi_i|^{2-1/p}\end{aligned} \tag{4.52}$$

where $\kappa_5 = \frac{f_d}{(2-1/p\ )2^{1-1/p}\, \kappa_2^{1+p}}$.

If we take κ_6 and κ_7 as in (4.45) and (4.46), inequality (4.52) can be written as

$$\begin{aligned}\dot{V} \leq & -\kappa_6 \sum\nolimits_{i=1}^{n} \sigma_i^d - \kappa_7 \sum\nolimits_{i=1}^{n} \xi_i^d \\ & + \kappa_5 \sum\nolimits_{i=1}^{n} |\xi_i|^{2-1/p}\end{aligned} \tag{4.53}$$

Considering the last item of the above inequality, by Lemma 2.10, we have

$$\kappa_5 \sum\nolimits_{i=1}^{n} |\xi_i|^{2-1/p} \leq n^{1/(2p)}\, \kappa_5 \left(\xi^T \xi\right)^{1-1/(2p)} \tag{4.54}$$

where $\xi = [\xi_1, \ldots, \xi_n]^T$.

Similarly, we can get the following inequalities

$$-\kappa_6 \sum\nolimits_{i=1}^{n} \sigma_i^d \leq -\kappa_6 \left(\sigma^T \sigma\right)^{(1+p)/(2p)} \tag{4.55}$$

$$-\kappa_7 \sum\nolimits_{i=1}^{n} \xi_i^d \leq -\kappa_7 \left(\xi^T \xi\right)^{(1+p)/(2p)} \tag{4.56}$$

Substituting (4.54), (4.55) and (4.56) into (4.53), we have

$$\begin{aligned}\dot{V} \leq & -\kappa_6\left(\sigma^T\sigma\right)^{(1+p)/(2p)} - \kappa_7\kappa_8\left(\xi^T\xi\right)^{(1+p)/(2p)} \\ & - n^{1/(2p)}\,\kappa_5\left(\xi^T\xi\right)^{1-1/(2p)} \cdot \\ & \left[\frac{\kappa_7\,(1-\kappa_8)}{n^{1/(2p)}\,\kappa_5}\left(\xi^T\xi\right)^{(2-p)/(2p)} - 1\right]\end{aligned} \tag{4.57}$$

where $0 < \kappa_8 < 1$.

If ξ is inside of the set Ω in (4.42), ξ is bounded. When ξ is outside of the set Ω, which means ξ satisfying

$$\xi^T\xi \geq \left[\frac{n^{1/(2p)}\,\kappa_5}{\kappa_7\,(1-\kappa_8)}\right]^{2p/(2-p)} \tag{4.58}$$

then we can obtain

$$\dot{V} \leq -\kappa_6(\sigma^T\sigma)^{(1+p)/(2p)} - \kappa_7\kappa_8(\xi^T\xi)^{(1+p)/(2p)}. \tag{4.59}$$

By (4.29),

$$\begin{aligned}\dot{V} \leq & -\kappa_6(2V_0)^{(1+p)/(2p)} - \kappa_7\kappa_8(\xi^T\xi)^{(1+p)/(2p)} \\ \leq & -2^{(1+p)/(2p)}\,\kappa_6 V_0^{(1+p)/(2p)} \\ & -\kappa_7\kappa_8((2-1/p\,)\kappa_2^{1+p}\sum\nolimits_{i=1}^{n} V_i)^{(1+p)/(2p)} \\ \leq & -\kappa_9 V^{(1+p)/(2p)}\end{aligned} \tag{4.60}$$

where

$$\begin{aligned}\kappa_9 = \min\{ & 2^{(1+p)/(2p)}\,\kappa_6, \\ & \kappa_7\kappa_8(2-1/p\,)^{(1+p)/(2p)}\,\kappa_2^{(1+p)^2/(2p)}\,\}\end{aligned} \tag{4.61}$$

By Lemma 2.3, the closed-loop system is finite-time stable, considering the observation error finite settling time T_y, σ will converge to the region Ω in finite time $T_2 \leq \frac{2pV^{(p-1)/(2p)}(0)}{\kappa_9(p-1)} + T_y < +\infty$. This means that the closed-loop system can still be stable, with disturbance rejection and strong robustness under the consideration of disturbances. σ converges in the finite time, which means that the tracking error of the closed-loop system still satisfies the performance constraints when considering the disturbances. The proof is completed. □

The above proof process shows that the controller can cause σ and $\dot{\sigma}$ to reach set Ω in finite time in the presence of bounded disturbances. So that $-1 < \vartheta_i = \varphi(\sigma_i) < 1$ holds. Furthermore, the control objectives (2.55) are finally achieved.

Remark 4.2 (*The meaning of time parameters T_1 and T_2*) In the above process, it is proved that σ can converge to equilibrium point or neighborhood of the equilibrium point within finite time $t_{fs} > 0$, which is defined as $t_{fs} = T_1$ (without disturbance) and $t_{fs} = T_2$ (with disturbance). t_{fs} theoretically describes the maximum time for σ to converge into the theoretical boundary, and it is bounded, i.e., $t_{fs} < +\infty$. At the same time, this means that q_{evi} will also converge in finite time when $\delta = 1$. Besides, under the controller proposed in this chapter, when $\delta \neq 1$, the tracking error converges to $(1 - \delta_i)\rho_i/2$ instead of 0 in finite time t_{fs}, which can also be considered as the cost paid for adjusting the overshoot. From a definition perspective, t_{fs} is different from t_s, where t_s is the time when the closed-loop system state/error converges within the range of the maximum steady-state error ρ_t and no longer escapes.

Remark 4.3 (*Relationship between three time parameters*) This chapter directly or indirectly relates to three time parameters, which are: (1) the actual settling time t_s; (2) PMST t_{ps} defined in this book; (3) settling time upper bound t_{fs} derived by finite-time theory. t_{ps} is the maximum value of the actual settling time t_s, and t_s is generally smaller than t_{ps}. That is, under the proposed controller, there is always $t_s \leq t_{ps}$. Therefore, by setting the value of t_{ps} in advance, the actual settling time is constrained within the preset range described by t_{ps}. For the relationship between t_{ps} and t_{fs}, they are different. In this chapter, the finite-time stability theory is applied to stabilize σ in finite time. Such a signal reaches $\sigma = 0$ or the neighborhood of $\sigma = 0$ in finite time, and the tracking error satisfies the constraints of the pre-defined upper bound of settling time t_{ps} proposed in this chapter. In this method, as long as it can be proved that σ is stable or finite-time stable, the constraint of the actual settling time can be realized. There is no significant relationship between t_{ps} and t_{fs}. That is, the former can be larger or smaller than the latter. When $t_{fs} < t_{ps}$, the tracking error will converge more quickly. When $t_{fs} \geq t_{ps}$, the tracking error convergence rate will decrease, but the constraint of the actual settling time will still be satisfied. t_{fs} is related to controller parameter selection.

Remark 4.4 One of the innovations of this chapter is that the controller parameter design is separated from the maximum settling time setting. t_s is the actual settling time under the controller and t_{ps} is artificially specified in advance. t_{fs} is determined by the controller parameters selection and the system's initial conditions. In a conventional finite-time controller, the controller parameters need to be adjusted in order to adjust t_{fs}, which may degrade the performance of the control input and closed-loop system. On the other hand, in order to obtain a better control input, it may affect the selection of t_{fs}. The controller proposed in this chapter separates the design of t_{fs} and the maximum settling time. It can be considered that the maximum settling time represented by t_{fs} is replaced by t_{ps}. So that the design of the controller parameters does not affect the selection of the maximum settling time. Therefore, both the maximum settling time constraint and the good controller performance are realized. It should be noted that the final control performance is still affected by various parameters. The term 'separation', as used herein, refers to separating the selection of parameters from the setting of maximum settling time when designing the controller.

Remark 4.5 The proposed algorithm requires fewer parameters to adjust. The complicated process of adjusting parameters is avoided. Better control performance and stronger robustness are obtained.

4.4 Simulation Analysis

4.4.1 Simulation Parameters

This subsection first presents the initial conditions and parameters for the simulation. Considering the background of the on-orbit service spacecraft mission, although this study takes attitude motion as the research object, this section still gives relevant information about the spacecraft's orbit. The orbital parameters of the target spacecraft are as follows.

Considering that the spacecraft is very close to the target spacecraft in the short-range on-orbit service mission. Therefore, the orbital parameters of the spacecraft to be controlled are also shown in Table 4.1. Next, consider the initial parameters of the spacecraft as shown in Table 4.2. Without loss of generality, the parameters of the Tables 4.1 and 4.2 can be changed according to specific tasks.

Next, consider the desired attitude of the on-orbit service spacecraft. In the numerical simulation of this study, it is considered that the spacecraft is tracking the attitude of the target spacecraft. The target spacecraft is inertial oriented, that is, $\mathfrak{F}_{bt}$ and $\mathfrak{F}_I$ coincide. Therefore, q_{dv} can be obtained via the relative motion between $\mathfrak{F}_{bs}$ and

Table 4.1 The orbital parameters of the target spacecraft

Parameters	Values
Semimajor axis	6.85×10^6 m
Eccentricity	0.0073
Inclination	60 deg
Argument of periapsis	0 deg
Longitude of the ascending node	0 deg
Mean anomaly	0 deg

Table 4.2 The initial attitude parameters of the spacecraft

Parameters	Values
q_{v0}	$[0.6, 0.3, -0.3]^T$ rad
ω_{a0}	$[0, 0, 0]^T$ rad/s
J_s	$\begin{bmatrix} 100 & -10 & 5 \\ -10 & 120 & -3 \\ 5 & -3 & 100 \end{bmatrix}$ $\text{kg} \cdot \text{m}^2$

$\mathfrak{F}_I$. Define R_{li} as the rotation matrix of the target spacecraft from the inertial system to LVLH coordinate system, $r_{li-j,k}$ is the matrix element in row j, column k, then $q_{d0} = 1/2\left(1 + r_{li-11} + r_{li-22} + r_{li-33}\right)^{1/2}$.

$$q_{dv} = \begin{bmatrix} (r_{li-23} - r_{li-32})/(4q_{d0}) \\ (r_{li-31} - r_{li-13})/(4q_{d0}) \\ (r_{li-12} - r_{li-21})/(4q_{d0}) \end{bmatrix} \tag{4.62}$$

According to the spacecraft attitude tracking model in Sect. 2.3, the tracking error quaternion and the angular velocity tracking error can be obtained. To demonstrate the performance of the proposed controller and to facilitate comparison with similar controllers, the error-index is defined as $I_a = ||q_{ev}|| + ||\omega_e||$. To illustrate the energy consumption of the proposed controller, the energy consumption index $E_a = \sum_{i=1}^{n} \int_0^t |\tau_{ai}(t)|^2 dt$ is defined as in [11]. This index describes the electrical or chemical energy consumed by actuators such as thrusters or reaction flywheels during the stabilization of an attitude closed-loop system.

For the controller proposed in this study, the selection of parameters needs to be considered. In this study, the basis for parameter selection is mainly summarized as follows: (1) Adjust the controller parameters to make the control effect better and the energy consumption smaller; (2) In the comparison simulation, the controller proposed in this study and the controller to be compared respectively adjust their parameters to make the energy consumption similar, compare the control effect of the closed-loop system, or adjust the parameters to make the control effect similar and compare the energy consumption; (3) If the principles and structures of the controllers to be compared are similar, the same or similar parameters will be used to compare the differences between the controllers themselves directly.

At the same time, this section presents the parameters selected in the simulation of the finite-time performance guaranteed controller (4.41) in Sect. 4.2, as shown in the following table.

In Table 4.3, the parameter t_{ps} is the predefined maximum settling time. ρ_0 is the maximum initial value of the predefined performance boundary, which needs to be larger than the initial error quaternion. Considering that all quaternion elements are less than or equal to 1, thus, ρ_0 is taken as 1. ρ_t and ε_ρ provide steady-state error bounds. δ_i is about whether to limit the overshoot, λ_i is the sliding mode surface scale coefficient in the controller, p is the exponent in the controller, and κ_1 and κ_2 are the controller gain parameters. The observer parameters are $\kappa_{y1} = 1$, $\kappa_{y2} = 1$,

Table 4.3 The initial parameters of the proposed controller (4.41)

Param	Value	Param	Value	Param	Value
ρ_0	1	t_{ps}	20s	p	11/9
ρ_t	0.02	δ_i	1	κ_1	2
ε_ρ	0.01	λ_i	1	κ_2	0.5

$\kappa_{y3} = 0.002$, and $\kappa_{y4} = 0.002$. These values are based on adjusting the parameters to make the observed effects and control input values reasonable. It should be noted that the controller (4.41) adds a function term that considers robustness. When the disturbances are not considered, its form is (4.1). The simulation analysis part directly gives the results by considering the disturbances.

Considering the external disturbances to the spacecraft, we take the disturbance variable as follows [12],

$$d_a = \begin{bmatrix} 0.3\sin(0.5t) + 3\,\text{rand}_a(t) \\ 0.06\sin(0.5t + \pi/2) + 0.6\,\text{rand}_a(t) \\ 0.2\sin(0.5t + \pi/3) + 2\,\text{rand}_a(t) \end{bmatrix} \text{N} \cdot \text{m} \tag{4.63}$$

where $\text{rand}_a(t)$ represents a zero-mean Gaussian random signal, and the variance is set to 0.01. It should be pointed out that, in order to fully demonstrate the disturbance rejection ability of the controller proposed, the disturbance selected in the simulation is larger than the practical situation, and the selected value is still within the reasonable range in orbit.

4.4.2 *Performance and Comparison*

To compare the controller proposed in this study with existing state-of-the-art controllers, this section labels the spacecraft attitude finite-time performance guaranteed controller (4.41) proposed as controller 'A'. After setting the exponent parameter p of the controller (4.41) to 1, the resulting asymptotic controller of a similar structure is marked as controller 'B'. The traditional prescribed performance controller in [13] is marked as controller 'C'. The finite-time controller in [14] is marked as controller 'D'. In the simulation diagram and analysis, the symbols 'A', 'B', 'C', and 'D' represent the above four controllers. Like the previous chapter, controller 'C' uses an exact model. It should be noted that the controllers designed by Theorems 4.1 and 4.2 are expressions about τ, and the control input of the spacecraft is τ_a. This section gives the simulation curve of τ_a directly.

Figures 4.1, 4.2, 4.3, 4.4, 4.5, 4.6, 4.7, 4.8, 4.9, 4.10, 4.11, 4.12, 4.13 and 4.14 are the comparative simulation results of the finite-time performance guaranteed controller proposed in Sect. 4.2. The simulation analysis presented in this section considers bounded external disturbances.

Figures 4.1, 4.2 and 4.3 shows the attitude tracking error curve of the spacecraft under the action of four controllers. It can be seen intuitively from the figures that the attitude tracking error can be converged under the action of the four controllers. Moreover, the tracking errors under the action of controllers 'A', 'B', and 'C' all meet the predefined performance constraints. However, under the action of the traditional finite-time controller 'D', the tracking error does not necessarily meet the predefined performance constraints. That is, the tracking error curve will cross the constraint region. Compared with the traditional prescribed performance controller 'C', the

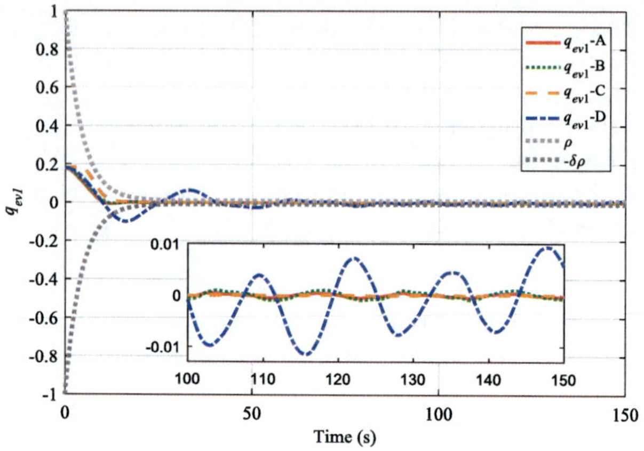

Fig. 4.1 The spacecraft attitude tracking error q_{ev1}

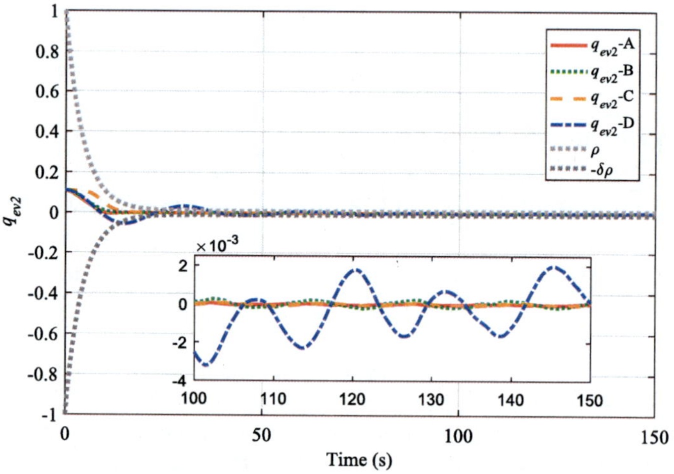

Fig. 4.2 The spacecraft attitude tracking error q_{ev2}

proposed controller has faster convergence speed and better transient performance, and the steady-state error curve fluctuation is smaller. Compared with controller 'B' whose power exponent is set to 1, the proposed controller has the property of finite-time convergence. Figures 4.4, 4.5 and 4.6 show the spacecraft attitude angular velocity tracking error under the action of four controllers. By comparison, it can be seen that under the action of the controller proposed, the attitude angular velocity tracking error change is smoother, the floating range is smaller, and the transient performance is better.

Figures 4.7 and 4.8 respectively show the spacecraft attitude error index $I_a = ||q_{ev}|| + ||\omega_e||$ and the energy consumption E_a. Through the controller comparison,

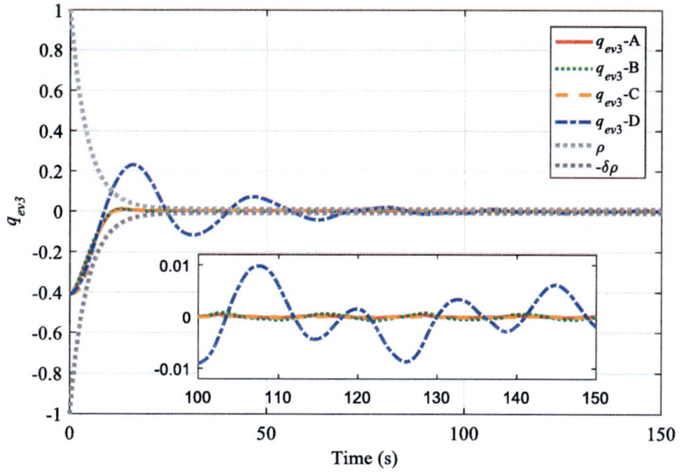

Fig. 4.3 The spacecraft attitude tracking error q_{ev3}

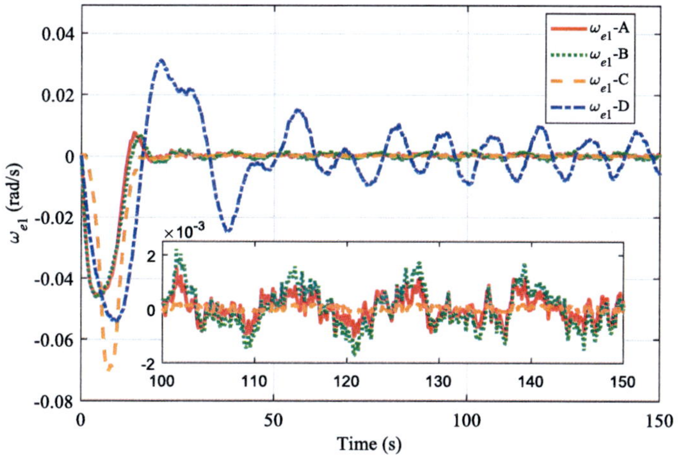

Fig. 4.4 The angular velocity tracking error ω_{e1}

it can be seen that the controller in Sect. 4.2 has a faster convergence rate and smaller steady-state error when the energy consumption of the four controllers is similar. Meanwhile, Figs. 4.9 and 4.10 give the control input curves of the proposed controller.

In order to further illustrate the role of the performance constraint function with the predefined maximum settling time parameters on the error response of the closed-loop system, this section adds the performance constraint parameter comparison simulation, as shown in Figs. 4.11, 4.12 and 4.13. Among them, Fig. 4.11 considers the constraint effect of different maximum settling time parameters t_{ps} on the track-

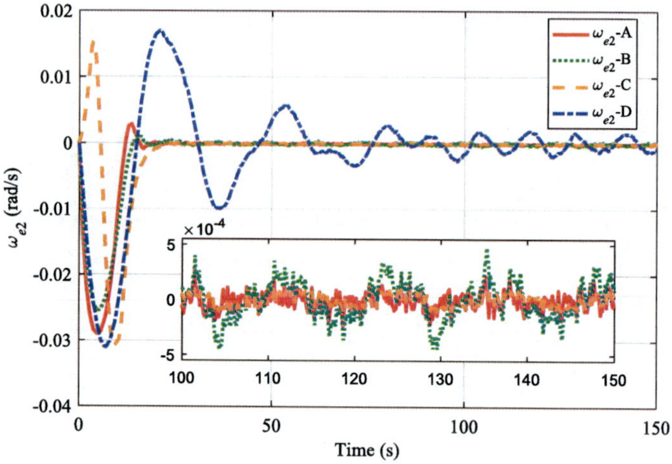

Fig. 4.5 The angular velocity tracking error ω_{e2}

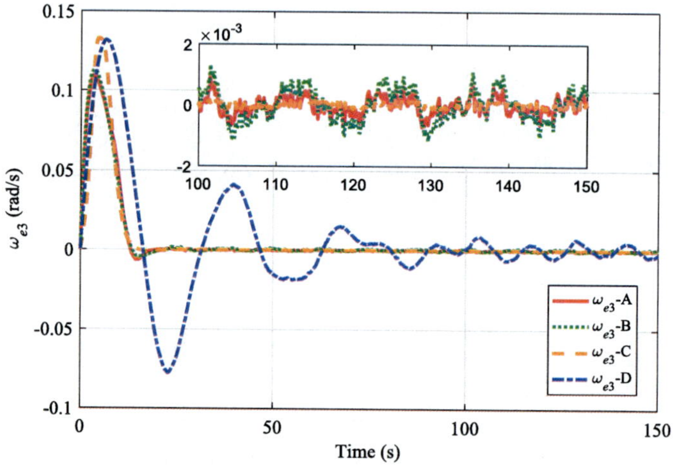

Fig. 4.6 The angular velocity tracking error ω_{e3}

ing error. It can be seen from the simulation that when the maximum settling time parameter is smaller, the error convergence rate is larger, and the error will converge to the predefined steady-state interval in a shorter time. Figure 4.12 considers the constraint effect of different steady-state error parameters ρ_t on the tracking error. It can be seen from the simulation that even if external disturbances are considered when the predefined steady-state error parameter is smaller, the steady-state error of the closed-loop system is smaller. Figure 4.13 considers the constraining effect of different overshoot parameters δ on the tracking error. It can be seen from the simulation that the overshoot parameter has obvious constraints on the overshoot amount.

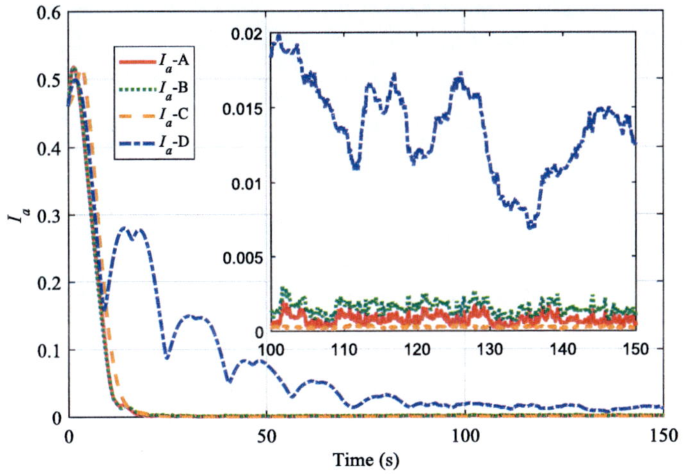

Fig. 4.7 The error index I_a

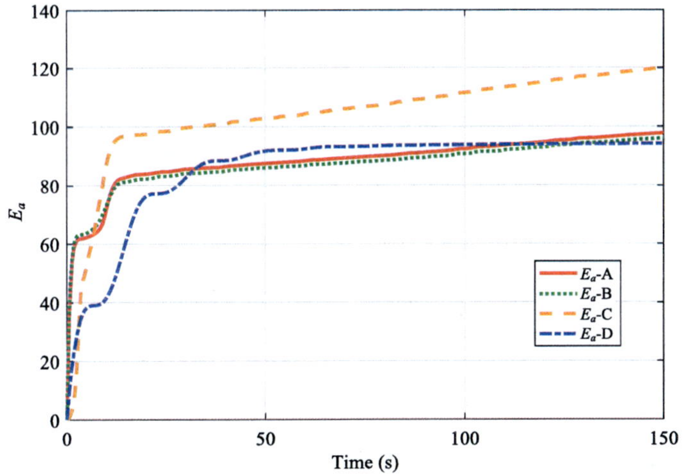

Fig. 4.8 The energy consumption

When $\delta = 0$, there is no overshoot in the closed-loop system, but the steady-state error will increase. Finally, Fig. 4.14 presents the external disturbances considered in the simulation.

Through numerical simulation, it is fully demonstrated that under the action of the controller proposed in Sect. 4.2, the closed-loop systems can achieve better transient and steady-state performance by consuming less energy. Furthermore, the performance is guaranteed. That is, the maximum settling time of the tracking error is guaranteed to be less than or equal to the predefined maximum settling time, the

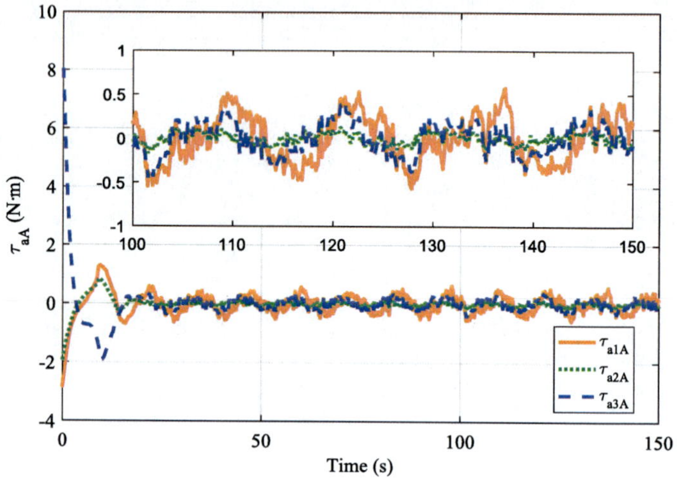

Fig. 4.9 The control input of 'A'

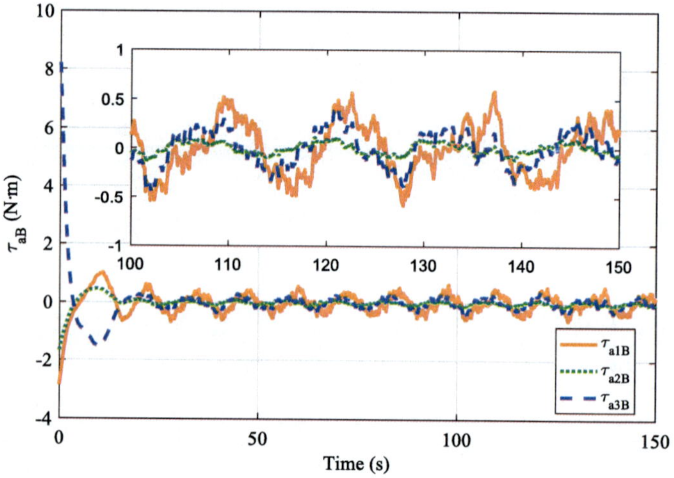

Fig. 4.10 The control input of 'B'

steady-state error is less than the predefined steady-state error, and the overshoot satisfies the constraints. Considering that the controller proposed has the ability to converge in the finite time, the transient performance can be further improved by further tuning the controller parameters. Considering the strong performance constraints of the controller proposed, it is suitable for practical spacecraft attitude control applications.

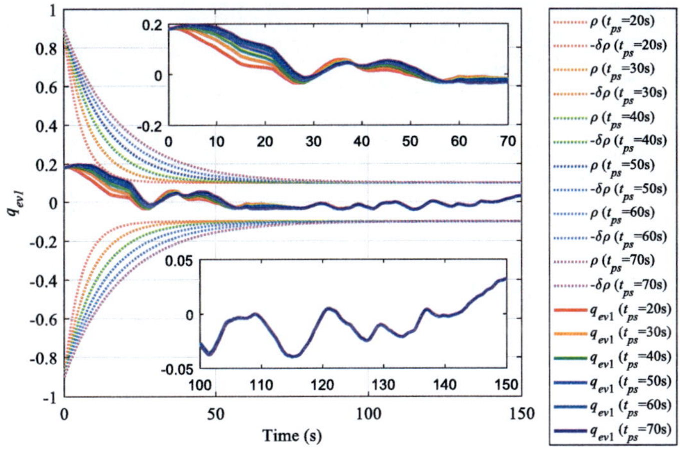

Fig. 4.11 The effect of different maximum settling time parameters t_{ps} on tracking error

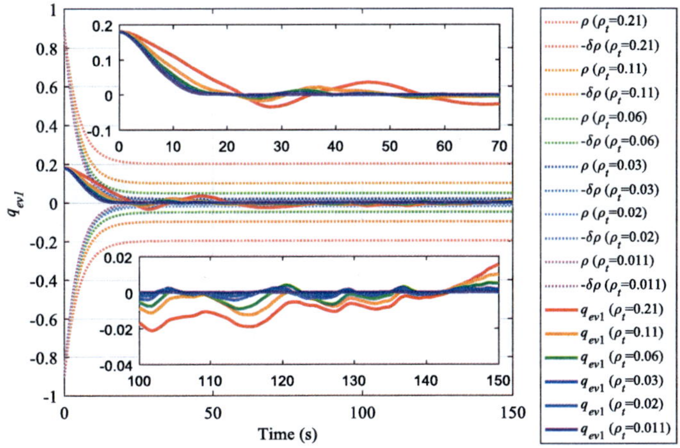

Fig. 4.12 The effect of different steady-state error parameters ρ_t on tracking error

4.5 Conclusion

Considering the effect of the asymptotically stable controller proposed in Chap. 3, the tracking error of the closed-loop system can converge to the predefined maximum steady-state error interval within the maximum settling time parameter. However, the constrained asymptotic stability is highly conservative, which means that the settling time constraints and the steady-state error interval for performance guaranteed are the largest. Therefore, to further accurately design the closed-loop system settling time and steady-state error range and improve the error convergence rate under per-

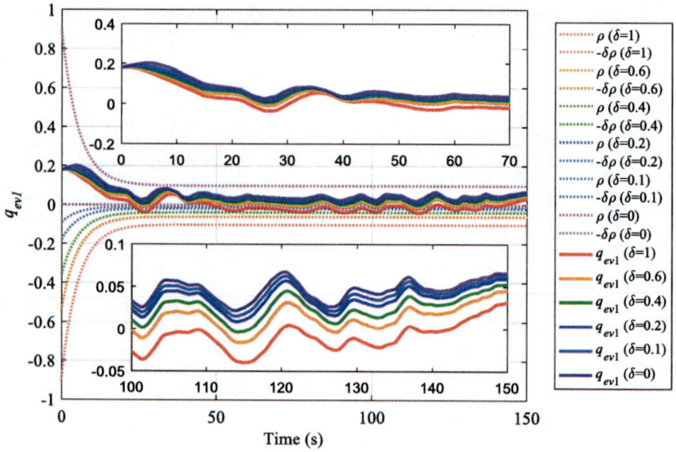

Fig. 4.13 The effect of different overshoot parameters δ on tracking error

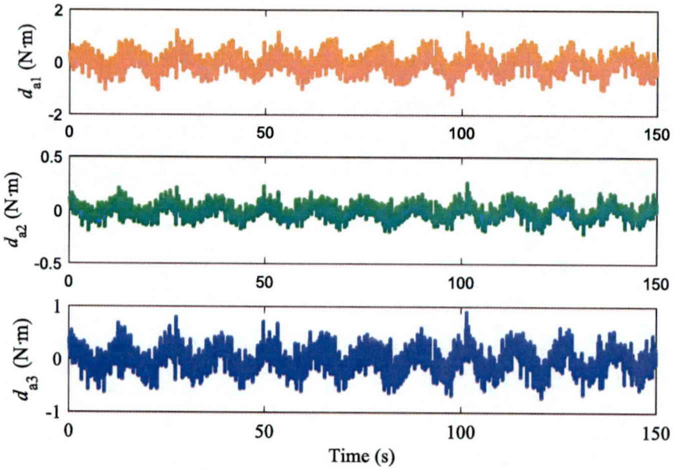

Fig. 4.14 The external disturbances

formance constraints, this chapter proposes the finite-time performance guaranteed control for spacecraft attitude based on adding a power integrator method. This not only enables the closed-loop system to converge to a predefined steady-state error interval within a predefined settling time but also allows the closed-loop system to converge to a more accurate region faster based on finite-time stability. The spacecraft finite-time performance guaranteed control algorithm is first proposed in this chapter. The external disturbance is further considered based on this algorithm, and an improved algorithm is proposed. The rationality of the algorithms is demonstrated through rigorous theoretical analysis and proof. Finally, the effectiveness of the controllers proposed in this chapter is demonstrated by the comparisons and numerical simulation analysis with advanced controllers in similar literature.

References

1. Rang, E.: Isochrone families for second-order systems. IEEE Transactions on Automatic Control **8**(1), 64–65 (1963)
2. Bhat, S.P., Bernstein, D.S.: Finite-time stability of homogeneous systems. In: Proceedings of the 1997 American control conference (Cat. No. 97CH36041), vol. 4, pp. 2513–2514. IEEE (1997)
3. Bhat, S.P., Bernstein, D.S.: Lyapunov analysis of finite-time differential equations. In: Proceedings of 1995 American Control Conference-ACC'95, vol. 3, pp. 1831–1832. IEEE (1995)
4. Yu, X., Man, Z.: Model reference adaptive control systems with terminal sliding modes. International Journal of Control **64**(6), 1165–1176 (1996)
5. Bhat, S.P., Bernstein, D.S.: Geometric homogeneity with applications to finite-time stability. Mathematics of Control, Signals and Systems **17**(2), 101–127 (2005)
6. Huang, X., Lin, W., Yang, B.: Global finite-time stabilization of a class of uncertain nonlinear systems. Automatica **41**(5), 881–888 (2005)
7. Ding, S., Li, S., Li, Q.: Stability analysis for a second-order continuous finite-time control system subject to a disturbance. Journal of Control Theory and Applications **7**(3), 271–276 (2009)
8. Hu, Q., Jiang, B., Friswell, M.I.: Robust saturated finite time output feedback attitude stabilization for rigid spacecraft. Journal of Guidance, Control, and dynamics **37**(6), 1914–1929 (2014)
9. Polyakov, A.: Nonlinear feedback design for fixed-time stabilization of linear control systems. IEEE Transactions on Automatic Control **57**(8), 2106–2110 (2012). DOI https://doi.org/10.1109/TAC.2011.2179869
10. Qian, C., Lin, W.: A continuous feedback approach to global strong stabilization of nonlinear systems. IEEE Transactions on Automatic Control **46**(7), 1061–1079 (2001)
11. Jiang, B., Li, C., Ma, G.: Finite-time output feedback attitude control for spacecraft using "adding a power integrator" technique. Aerospace Science and Technology **66**, 342–354 (2017)
12. Zou, A.M.: Finite-time output feedback attitude tracking control for rigid spacecraft. IEEE Transactions on Control Systems Technology **22**(1), 338–345 (2013)
13. Bechlioulis, C.P., Rovithakis, G.A.: Robust adaptive control of feedback linearizable mimo nonlinear systems with prescribed performance. IEEE Transactions on Automatic Control **53**(9), 2090–2099 (2008)
14. Du, H., Li, S., Qian, C.: Finite-time attitude tracking control of spacecraft with application to attitude synchronization. IEEE Transactions on Automatic Control **56**(11), 2711–2717 (2011)

Chapter 5
Spacecraft Performance Guaranteed Control with Improved Transient Performance

Abstract The attitude tracking control algorithm design for the on-orbit service spacecraft needs to consider strict performance constraints and control accuracy for mission precision control and safety requirements. Simultaneously, it is more desirable to design the trajectory of the attitude motion directly. The performance guaranteed control algorithm can clearly describe the feasible region or trajectory range of the closed-loop system state. For the closed-loop system trajectories, this range is only the boundary and envelope of motion, and the desired precise motion trajectory cannot be described. Adding more stringent motion constraints or reducing the feasible region will reduce the system tolerance, which is not conducive to the transient process of the closed-loop system and may increase the control input excessively. To enhance the trajectory description ability of the closed-loop system under the performance constraints, the performance guaranteed control algorithm with improved transient performance is proposed in this chapter. Using the ratio between multiple state elements of the closed-loop system and the Ratio Persistence property, the performance guaranteed algorithm that can describe the trajectory of the closed-loop system is designed. With the improvement of the algorithm, the state of the spacecraft attitude tracking closed-loop system not only satisfies the predefined performance guaranteed constraints but also the trajectory of the closed-loop system approaches a straight line in the three-dimensional state space. The closed-loop system is finite-time stable and has a strong disturbance suppression ability. Finally, the rigorous theoretical proof verifies the stability of the closed-loop system. The sufficient numerical simulation analysis and comparative analysis verify the rationality, effectiveness, and advantages of the proposed control algorithms.

Y. Gao and D. Li, *Spacecraft Maneuver with Performance Guaranteed*,
https://doi.org/10.1007/978-981-99-4653-2_5

5.1 Introduction

Rendezvous and docking [1, 2] is an essential basic technology in the on-orbit services, enabling the service spacecraft to approach the target spacecraft. The final stage of the rendezvous and docking process requires the strict and precise operation and control of the relative attitude and relative orbit or position. This chapter will focus on the control process of the spacecraft's attitude relative motion in the near-distance rendezvous and docking phase during the on-orbit service.

For the close-range control of the on-orbit service spacecraft [3], we usually design controllers for multi-input-multi-output (MIMO) systems, such as PID control [4], adaptive control [5], robust control [6], sliding mode control [7, 8], finite-time control [9–11], prescribed performance control [12, 13], etc. However, we all face the rarely discussed problems, the describable closed-loop system trajectory, the relationship between the direction of the time-varying error and the direction of the output trajectory, and the relationship between the state elements of the closed-loop system.

Based on the spacecraft model and stability theorems described in Chap. 2, and the corresponding controllers proposed in Chaps. 3 and 4, the spacecraft closed-loop system can achieve stable and performance guaranteed. The performance constraints provide a better ability to describe closed-loop systems' state transient and steady-state performance. These bring higher stability and accuracy guarantees in spacecraft attitude relative motion control for rendezvous and docking missions. However, performance constraints can only limit state motion within the feasible region. It cannot further describe the deterministic transient behavior of the system state in the feasible region. Therefore, we still face the challenge that the state trajectories of closed-loop systems cannot be described deterministically.

The closed-loop system trajectory can be analyzed by the relationship between the state/error vector, the system's initial point, and the equilibrium point. Suppose the direction of the time-varying error is always on the line between the initial error and the origin. If the error can converge to the origin, the error trajectory always remains on this line. It implies that the trajectory of the closed-loop system movement during the error convergence process is the shortest path or the straight line. This provides an accurate description of the transient trajectory of the closed-loop system, which we describe as the describable straight-line convergence trajectory. It should be noted that the shortest path is only for the closed-loop system state (error). This is the shortest path in the direction of state convergence, not necessarily the shortest path of the actual physical system. If the closed-loop system state is the actual physical parameter, then the shortest path of this actual physical parameter can be obtained. However, most current controllers do not have such properties, so real-time error often deviates from this straight line in the control process. In such a process, excessive motion components orthogonal to the direction from the initial point to the origin will only increase the uncertainty of the transient trajectory, the unnecessary energy consumption, and the wear of spacecraft structural components. This is catastrophic for the spacecraft, and we hope to overcome such control or motion losses through advanced control methods. If we want to make the trajectory of the closed-loop system descriptive and tend to be straight, we can force any pair of the nonzero elements of the error to maintain a fixed ratio. When one of the nonzero elements converges to 0, the other nonzero elements will also converge to 0 simultaneously, which means that the tracking error elements will converge synchronously.

In the application scenario where the service spacecraft is close to the target spacecraft, considering the spacecraft's closed-loop system, we hope that the tracking errors will always remain on the line from the initial point to the equilibrium point. In this way, each element of the errors can be synchronized to converge, and any pair of error elements are in the fixed ratio. The trajectory of the closed-loop system tends to be a straight line, and the convergence process is smoother and softer to protect the spacecraft's structure. Especially for the latest intelligent and fully automated on-orbit service tasks, including on-orbit assembly, on-orbit operations, etc., synchronized convergence is even more critical because of the describable straight-line convergence trajectory. The property named Ratio Persistence (RP) for the system defined in [14] provides us with the technique to solve the above problem. That is, the state of the system that satisfies the RP property can converge synchronously with the describable straight-line trajectory. The RP property implies that the ratio of the state elements is time-invariant in forwarding time. The synchronized controllers for the general first-order system are proposed [14]. Under these controllers, all states of the simple systems can converge synchronously.

On the other hand, we also hope that the performance guaranteed and the actual settling time can be designed theoretically for high-precision control requirements. Simultaneously, the performance guaranteed design can also provide strict constraints on the system to guarantee the motion boundaries of the system.

In the control algorithm design in the previous chapters, the objectives of performance guaranteed control have been achieved. This chapter wishes to further describe the transient behavior under performance constraints via the ability of the system state to converge synchronously. Therefore, based on the norm-normalized sign function and ratio persistence property, this chapter designs the performance guaranteed control with improved or enhanced transient performance. The control algorithm proposed in this chapter is the improvement and enhancement of the performance guaranteed control algorithm. These improvements and enhancements are in: the corresponding closed-loop system has strong performance guaranteed and index constraint ability; the finite-time stability can be achieved; the transient performance meets the ratio persistence property; the synchronous convergence is realized; the motion trajectories tend to be straight lines in Euclidean space.

Specifically, it is hoped that the control algorithms proposed in this chapter can make the practical settling time of the spacecraft attitude tracking closed-loop system state converge to the predefined interval less than the predefined maximum settling time t_{ps}. The transient and steady-state performance of the closed-loop system can satisfy the performance constraints (2.55). Meanwhile, the transient motion of the closed-loop system can be described and tends to be straight in Euclidean space. Combined with the specific situation of spacecraft attitude tracking, the mission objectives of this chapter are highlighted here:

Problem 5.1 Consider the spacecraft attitude tracking error model (2.28), and select predefined related parameters according to practical mission requirements, including predefined maximum settling time t_{ps}, the maximum steady-state error ρ_t, the maximum overshoot χ_t, design the performance guaranteed control such that:
(1) the conditions in Problem 3.1 are satisfied;
(2) the tracking error converges within the finite time;
(3) the elements of the error converge synchronously;
(4) the convergence trajectory of the closed-loop state (tracking error) tends to be the straight line;
(5) the closed-loop system has the property of disturbance suppression. That is, when strong external disturbances exist, the closed-loop system still satisfies the above performance constraints.

Considering the scope of constraints and the selection of indicators, as in the previous chapters, the selection of predefined performance constraints needs to consider the practical situation. According to the practical requirements of the task, correspondingly more accurate expected transient and steady-state constraints than traditional expected values are selected. However, extremely small expected constraint intervals are usually not selected.

For the controllable spacecraft attitude tracking system, this chapter aims to achieve the control task objective as described in Problem 5.1 through the performance guaranteed control design. Compared with the traditional control algorithms, the control algorithms proposed in this chapter can achieve the guaranteed performance of the closed-loop system transient and steady-state. They can also describe the state trajectory of the closed-loop system. This further improves spacecraft attitude control accuracy, safety, and reliability.

This chapter is arranged as follows. In Sect. 5.2, the norm-normalized sign function and ratio persistence property are first introduced. In Sect. 5.3, the performance guaranteed control algorithm is designed based on the norm-normalized sign function, the algorithm considering disturbances is further designed, and rigorous theoretical proofs are given. The transient performance of the closed-loop system is intuitively improved by applying the norm-normalized sign function. It improves transient performance with the advantages of maximum settling time guarantees and performance constraints. Section 5.4 presents the numerical simulation analysis of the control algorithm design. Section 5.5 gives the conclusion of this chapter.

5.2 Norm-Normalized Sign Function and Ratio Persistence

In this section, the definition of the *norm-normalized sign function* is first introduced, and then the *ratio persistence* property is introduced [14–16]. The transient system trajectory of the closed-loop system that satisfies the ratio persistence property is a straight line from the initial point to the equilibrium point [15]. Therefore, we hope to take this property to improve further the transient performance of closed-loop systems under performance guaranteed control. Simultaneously, it is possible to describe the transient behavior of the closed-loop system more accurately.

We introduce the *norm-normalized sign function* [15, 17] as

$$\mathrm{sign_{NN}}(x) \triangleq \begin{cases} \dfrac{x}{\|x\|}, x \neq 0 \\ 0\,, x = 0 \end{cases} \tag{5.1}$$

$$\mathrm{sig}^{\alpha}_{\mathrm{NN}}(x) = ||x||^{\alpha}\mathrm{sign_{NN}}(x) \tag{5.2}$$

where α is a positive constant.

For the MIMO system, we can find that the error or state vectors mapped by the function $\mathrm{sign_{NN}}(\cdot)$ have the same directions as the vectors before mapping. Between any two non-zero elements of the vector, a constant ratio coefficient $1/\|x\|$ will be maintained along the direction of the vector. With a fixed ratio coefficient, when one of the non-zero elements converges, the other non-zero elements converge synchronously. Moreover, the convergence trajectory is along the direction of the initial vector. This means that with norm-normalized mapping, the fixed ratio is maintained between any non-zero elements of the vector. More generally, we describe the above phenomenon in terms of the *ratio persistence* property of the closed-loop system and define as follows.

Definition 5.1 [16] (*Ratio Persistence property, RP*) The closed-loop state x of the system $\dot{x} = f(x)$, which is defined as (2.37) is ratio persistent if for $x \neq 0$, we have

$$\frac{x}{\|x\|} = \varsigma \frac{f(x)}{\|f(x)\|} \tag{5.3}$$

where ς is the direction of ratio persistence, which can be either 1 or -1.

Satisfying the RP property means that all dimensional elements of the state in the MIMO system can converge simultaneously. We explain this below.

Lemma 5.1 [14] *If the system state x is ratio persistent, for $\forall x_i(t), x_j(t) \neq 0, i, j \in \{1, 2, \ldots, n\}, i \neq j$, there are nonzero constant ratio coefficients c_{ij} such that*

$$\frac{x_i(t)}{x_j(t)} = \frac{x_i(0)}{x_j(0)} = c_{ij}. \tag{5.4}$$

Proof If $x = 0$, then the system is stable. If $x \neq 0$, in other words in the set $N_0\backslash\{0\}$, according to Definition 5.1, we can rewrite (5.3) as

$$\dot{x} = \delta(x)x \tag{5.5}$$

where the function $\delta(x) : N_0\backslash\{0\} \to \mathbb{R}$ is defined as

$$\delta(x) \triangleq \varsigma^{-1} \frac{\|f(x)\|}{\|x\|} \tag{5.6}$$

According to (5.3), $\delta(x) \neq 0$ when $x \neq 0$. And for $\forall x_i(t),\ x_j(t) \neq 0,\ i, j \in \{1, 2, \ldots, n\}$, considering $\dot{x}_i = \delta(x)x_i$, $\dot{x}_j = \delta(x)x_j$, we have

$$\frac{\mathrm{d}}{\mathrm{d}t}\left(\frac{x_i(t)}{x_j(t)}\right) = \frac{\dot{x}_i(t)x_j(t) - x_i(t)\dot{x}_j(t)}{x_j^2(t)} = 0 \tag{5.7}$$

So we know that $x_i(t)/x_j(t) = x_i(0)/x_j(0) = c_{ij}$, where c_{ij} is a non-zero constant. □

Theorem 5.1 [17] *Considering the non-zero state variable $x \in \mathbb{R}^n$ of the system $\dot{x} = f(x) + g(x)u$, if different u are designed so that the closed-loop systems $\dot{x} = f_1(x)$ and $\dot{x} = f_2(x)$ are both RP, then the closed-loop system $\dot{x} = \kappa_{f_1} f_1(x) \pm \kappa_{f_2} f_2(x)$ is RP, $\kappa_{f_1}, \kappa_{f_2} \in \mathbb{R}$ are constants.*

Proof Considering the two closed-loop systems $\dot{x} = f_1(x)$ and $\dot{x} = f_2(x)$ satisfying the RP property respectively, we have

$$\frac{x}{\|x\|} = \kappa_{x1} \frac{f_1(x)}{\|f_1(x)\|} \tag{5.8}$$

$$\frac{x}{\|x\|} = \kappa_{x2} \frac{f_2(x)}{\|f_2(x)\|} \tag{5.9}$$

where κ_{x1} and κ_{x2} are the direction of RP as described in Definition 5.1. Transform the above equations, we have

$$f_1(x) = \kappa_{x1}^{-1} \frac{\|f_1(x)\|}{\|x\|} x \tag{5.10}$$

$$f_2(x) = \kappa_{x2}^{-1} \frac{\|f_2(x)\|}{\|x\|} x \tag{5.11}$$

Then,

$$\kappa_{f_1} f_1(x) \pm \kappa_{f_2} f_2(x) = \gamma_x x \tag{5.12}$$

where $\gamma_x = \kappa_{f_1}\kappa_{x1}^{-1} \frac{\|f_1(x)\|}{\|x\|} \pm \kappa_{f_2}\kappa_{x2}^{-1} \frac{\|f_2(x)\|}{\|x\|}$ and $\gamma_x \in \mathbb{R}$. For the closed-loop system $\dot{x} = \kappa_{f_1} f_1(x) \pm \kappa_{f_2} f_2(x)$, we have $\dot{x} = \gamma_x x$. So, for any pair of state elements $\dot{x}_i = \gamma_x x_i$ and $\dot{x}_j = \gamma_x x_j$ with $i \neq j$ and $i, j = 1, 2, \ldots, n$, we have

$$\frac{\mathrm{d}}{\mathrm{d}t}\left(\frac{x_i}{x_j}\right) = \frac{\dot{x}_i x_j - x_i \dot{x}_j}{x_j^2} = 0 \tag{5.13}$$

Therefore, x_i/x_j is a constant. The closed-loop system $\dot{x} = \kappa_{f_1} f_1(x) \pm \kappa_{f_2} f_2(x)$ satisfies the RP property. □

Corollary 5.1 [17] (Linear Combination of RP) *Considering the non-zero state variable $x \in \mathbb{R}^n$ of the systems $\dot{x} = f(x) + g(x)u$ with different u, if there are $f_i(x) \in \mathbb{R}^n, i = 1, 2, \ldots, m$, so that the closed-loop systems $\dot{x} = f_i(x)$ satisfy the RP property, then the closed-loop system $\dot{x} = \sum_{i=1}^{n}(\pm\kappa_i f_i(x))$ is RP, $\kappa_i \in \mathbb{R}$ is a constant.*

Proof According to Theorem 5.1, it is easy to know that the closed-loop system $\dot{x} = f_i(x) \pm \kappa_{i+1} f_{i+1}(x)$ satisfies the RP property. Let $f_{ik}(x) = f_i(x) \pm \kappa_{i+1} f_{i+1}(x)$, then $\dot{x} = f_{ik}(x) \pm \kappa_{i+2} f_{i+2}(x)$ satisfies the RP property. Repeating the above process, Corollary 5.1 is proved.

Combining the norm-normalized sign function and the ratio persistence property above, we give some closed-loop system forms that satisfy the RP property. That is to say, and if the controller is designed so that the closed-loop system can fulfill the following example forms, then the closed-loop system satisfies the RP property. The ratio between the state elements of the closed-loop system is fixed. All elements of the system state converge synchronously, with no overshoot and no steady-state errors. The closed-loop system trajectory is a straight line from the initial point to the equilibrium point.

Example 5.1 [17] The following systems satisfy RP property, where ζ and ξ are vectors

$$\dot{\zeta} = -\alpha \text{sig}_{\text{NN}}^{p}(\zeta) \tag{5.14}$$

$$\dot{\xi} = -\sum_{i=1}^{n} \alpha_i \text{sig}_{\text{NN}}^{p_i}(\xi) \tag{5.15}$$

For (5.14), when $\zeta = 0$, there is $\dot{\zeta} = 0$. When $\zeta \neq 0$, we have

$$\dot{\zeta} = -\alpha \text{sig}_{\text{NN}}^{p}(\zeta) = -\alpha \|\zeta\|^{p} \text{sign}_{\text{NN}}(\zeta) = -\alpha \|\zeta\|^{p} \frac{\zeta}{\|\zeta\|} \tag{5.16}$$

then

$$\frac{\zeta}{\|\zeta\|} = \frac{-\alpha \text{sig}_{\text{NN}}^{p}(\zeta)}{-\alpha \|\zeta\|^{p}} = \frac{-\alpha \text{sig}_{\text{NN}}^{p}(\zeta)}{-\alpha \left\|\text{sig}_{\text{NN}}^{p}(\zeta)\right\|} = -\frac{-\alpha \text{sig}_{\text{NN}}^{p}(\zeta)}{\left\|-\alpha \text{sig}_{\text{NN}}^{p}(\zeta)\right\|} \tag{5.17}$$

where $\alpha > 0$. According to the definition, we know that the system (5.14) satisfies the RP property. The system (5.15) is a linear combination of (5.14), and according to Theorem 5.1, (5.15) also satisfies the RP property.

For a closed-loop system that satisfies the RP property, the ratio between any pair of non-zero state elements is fixed. The state elements of each dimension converge to the equilibrium point synchronously. During the entire convergence process, the convergence trajectory of its closed-loop system is a straight line from the initial point to the equilibrium point. According to such properties, we can design the controller to make the performance guaranteed auxiliary system further satisfy the RP property. In this way, the transient performance of the closed-loop system exhibits the characteristics of ratio persistence and linear trajectory. The latter section further strengthens the deterministic description of the transient behavior of spacecraft attitude tracking errors.

5.3 Control Design with Improved Transient Performance

The performance guaranteed techniques have a good ability to describe the state of closed-loop systems or tracking error transient and steady-state performance. However, performance constraints can only make the state or tracking error move in the feasible region but cannot further describe the deterministic transient behavior of the system state or tracking error in the feasible region. In this section, it is desired to describe further the transient behavior under performance guaranteed. Therefore, the performance guaranteed controllers with improved or enhanced transient performance are designed based on the norm-normalized sign function and ratio persistence property. These proposed controllers are improvements and enhancements of the performance guaranteed controller, which not only has strong performance constraint ability but also has improved transient performance and finite-time stability. Precisely, the closed-loop system trajectory satisfies the performance constraints and is a straight line or tends to a straight line.

5.3.1 *Control Design and Stability Analysis Without Disturbances*

First, without considering external disturbances, the performance guaranteed controller with improved transient performance is designed for system (2.28) with $d = 0$. Considering the relevant links of the performance guaranteed control, we list the function terms and auxiliary systems required for performance constraint transformation as follows. Under the framework of the performance guaranteed control, $\sigma_i(q_{evi})$ is also defined as (2.60), if $\sigma_i(q_{evi})$ converges or is bounded, then the closed-loop system tracking error q_{evi} satisfies the performance constraints (2.55) and maximum settling time constraint T_{ps}. In order to facilitate the derivation, we first find $\dot{\sigma}_i$, where $r_i = \partial\varphi^{-1}(\vartheta_i)/\partial\vartheta_i$, $\dot{\vartheta}_i = \frac{2\dot{q}_{evi}(t)}{(1+\delta)\rho(t)} - \frac{2\dot{\rho}(t)q_{evi}(t)}{(1+\delta)\rho^2(t)}$.

$$\dot{\sigma}_i = \frac{\partial\varphi^{-1}(\vartheta_i)}{\partial\vartheta_i}\dot{\vartheta}_i = r_i\dot{\vartheta}_i \tag{5.18}$$

$$\ddot{\sigma}_i = \dot{r}_i\dot{\vartheta}_i + r_i\ddot{\vartheta}_i = \alpha_i + \beta_i\ddot{q}_{evi}(t) \tag{5.19}$$

where

$$\begin{aligned}\alpha_i = &+\dot{r}_i\dot{\vartheta}_i - \frac{2r_i\dot{q}_{evi}(t)\dot{\rho}_i(t)}{(1+\delta_i)\rho_i{}^2(t)} + \frac{4r_iq_{evi}(t)\dot{\rho}_i^2(t)}{(1+\delta_i)\rho_i{}^3(t)}\\ &-\frac{2r_i\ddot{\rho}_i(t)q_{evi}(t)+2r_i\dot{\rho}_i(t)\dot{q}_{evi}(t)}{(1+\delta_i)\rho_i{}^2(t)}\end{aligned} \tag{5.20}$$

$$\beta_i = \frac{2r_i}{(1+\delta_i)\rho_i(t)} \tag{5.21}$$

By expressing (5.19) into a compact form, the available auxiliary systems for the spacecraft attitude tracking error to meet the performance constraints are as follows

$$\ddot{\sigma} = A + B\ddot{q}_{ev} \tag{5.22}$$

where $\sigma = [\sigma_1, \ldots, \sigma_n]^T$ $A = [\alpha_1, \ldots, \alpha_n]^T$, $B = \text{diag}(\beta_1, \ldots, \beta_n)$.

For the above auxiliary system, the control objective of the performance guaranteed can be achieved by using the control algorithm design in the previous chapter. In this chapter, the intention is to add describable enhanced transient behavior to the controller by applying the norm-normalized sign function and ratio persistence property. Therefore, the next step is to design the controller based on (5.22) such that the closed-loop system with respect to σ satisfies the RP property.

First, the sliding mode surface ζ about σ is designed as follows so that σ satisfies the RP property when the sliding mode surface converges.

$$\zeta = \dot{\sigma} + \kappa_1 \text{sig}_{\text{NN}}^{p_1}(\sigma) \tag{5.23}$$

For convenience, we calculate the derivative of ζ with respect to time as follows

$$\begin{aligned}\dot{\zeta} &= \ddot{\sigma} + \kappa_1\frac{\text{d}}{\text{d}t}\text{sig}_{\text{NN}}^{p_1}(\sigma) = \ddot{\sigma} + \phi\\ &= A - BM_a^{-1}[C_a\dot{q}_{ev} + G_a] + BM_a^{-1}Q\tau_a + \phi\end{aligned} \tag{5.24}$$

where $Q = \frac{1}{2}P^T$,

$$\phi = \kappa_1(p_1 - 1)\|\sigma\|^{p_1-3}\sigma\sigma^T\cdot\dot{\sigma} + \kappa_1\|\sigma\|^{p_1-1}I_n\cdot\dot{\sigma} \tag{5.25}$$

Note that there is a singularity problem in (5.25), that is, when $\sigma = 0$ and $\dot{\sigma} \neq 0$, $||\sigma||^{-3}$ and $||\sigma||^{-1}$ are infinite or unbounded.

For the derivation of traditional symbolic functions, to avoid such singularity problems, the threshold value ϖ is set to judge singularity [18]. Based on this, the non-singular form switching function of $\text{sig}_{\text{NN}}(x)$ is defined as follows

$$\frac{\mathrm{d}}{dt}(\mathrm{sig}_{\mathrm{NN}}^{p_N+1}(x)) = \begin{cases} p_N \|x\|^{p_N-2} x x^T \cdot \dot{x} + \|x\|^{p_N} I_n \cdot \dot{x}, \\ \text{if } \|x\| \geq \varpi, \dot{x} \neq 0 \\ p_N \|\iota\|^{p_N-2} \iota^T \cdot \dot{x} + \|\iota\|^{p_N} I_n \cdot \dot{x}, \\ \text{if } \|x\| < \varpi, \dot{x} \neq 0 \\ 0 \text{, if } \dot{x} = 0 \end{cases} \tag{5.26}$$

where ϖ and ι are both small positive constants, p_N is the power exponent. Therefore, ϕ is rewritten into the switching function form as follows to avoid singularity problems.

$$\phi = \begin{cases} \kappa_1 (p_1 - 1) \|\sigma\|^{p_1-3} \sigma\sigma^T \cdot \dot{\sigma} + \kappa_1 \|\sigma\|^{p_1-1} I_n \cdot \dot{\sigma} \text{ ,} \\ \text{if } \|\sigma\| \geq \varpi, \dot{\sigma} \neq 0 \\ \kappa_1 (p_1 - 1) \|\iota\|^{p_1-3} \iota^T \cdot \dot{\sigma} + \kappa_1 \|\iota\|^{p_1-1} I_n \cdot \dot{\sigma} \text{ ,} \\ \text{if } \|\sigma\| < \varpi, \dot{\sigma} \neq 0 \\ 0, \text{if } \dot{\sigma} = 0 \end{cases} \tag{5.27}$$

The following controller is proposed, the performance guaranteed controller that improves transient performance without considering disturbances,

$$\begin{aligned} \tau_a = & -\kappa_2 Q^{-1} M_a B^{-1} \mathrm{sig}_{\mathrm{NN}}^{p_2}(\zeta) - Q^{-1} M_a B^{-1} A \\ & + Q^{-1}(C_a \dot{q}_{ev} + G_a) - Q^{-1} M_a B^{-1} \phi \end{aligned} \tag{5.28}$$

Moreover, the corresponding theorem is expressed as follows.

Theorem 5.2 *Consider the spacecraft attitude tracking error model (2.28), and the performance constraints (2.55), define the auxiliary system that satisfies the performance constraints (5.22), define the sliding mode surface (5.23). When the external disturbances are not considered, if the controller is selected as (5.28), then the auxiliary system (5.22) is finite-time stable. The sliding mode surface* ζ *and the auxiliary system state* $\sigma\,(\zeta = 0)$ *satisfy the RP property. The finite settling time of* ζ *is* $T_{11} \leq \frac{V_1^{\frac{1-p_2}{2}}(\zeta(0))}{2^{\frac{p_2-1}{2}}\kappa_2(1-p_2)}$. *When* $\zeta = 0$, *the finite settling time of* σ *is* $T_{12} \leq \frac{V_2^{\frac{1-p_1}{2}}(\sigma(0))}{2^{\frac{p_1-1}{2}}\kappa_1(1-p_1)}$. *Then the finite settling time of the auxiliary system state* σ *is* $T_{1m} \leq T_{11} + T_{12}$. *Then the spacecraft attitude tracking error satisfies the constraint condition (2.55). The time for the spacecraft attitude tracking error to converge to the predefined steady-state error range satisfies* $T_{q_{ev}} \leq \min\{T_{ps}, T_{1m}\}$, *where* T_{ps} *is the predefined maximum settling time.*

Proof **Step 1**. Analyze the convergence of the sliding mode variable and the RP condition of ζ.

The candidate Lyapunov function with respect to ζ is selected as

$$V_1 = \frac{1}{2}\zeta^T \zeta \tag{5.29}$$

Taking the derivative of (5.29) with respect to time and substituting controller (5.28), we have

$$\begin{aligned}\dot{V}_1 = \zeta^T\dot{\zeta} &= \zeta^T A - \zeta^T B M_a^{-1}(C_a\dot{q}_{ev} + G_a) + \zeta^T B M_a^{-1} Q\tau_a + \zeta^T\phi \\ &= \zeta^T A - \zeta^T B M_a^{-1}(C_a\dot{q}_{ev} + G_a) + \zeta^T B M_a^{-1} Q(-\kappa_2 Q^{-1} M_a B^{-1}\mathrm{sig}_{\mathrm{NN}}^{p_2}(\zeta) \\ &\quad - Q^{-1} M_a B^{-1} A + Q^{-1}(C_a\dot{q}_{ev} + G_a) - Q^{-1} M_a B^{-1}\phi) + \zeta^T\phi \\ &= -\kappa_2\zeta^T\mathrm{sig}_{\mathrm{NN}}^{p_2}(\zeta) = -\kappa_2||\zeta||^{1+p_2} = -2^{\frac{1+p_2}{2}}\kappa_2 V_1^{\frac{1+p_2}{2}}\end{aligned} \tag{5.30}$$

It can be seen from the above formula that V_1 is finite-time stable, and the finite settling time of ζ is

$$T_{11} \leq \frac{V_1^{\frac{1-p_2}{2}}(\zeta(0))}{2^{\frac{1+p_2}{2}}\kappa_2(\frac{1-p_2}{2})} = \frac{V_1^{\frac{1-p_2}{2}}(\zeta(0))}{2^{\frac{p_2-1}{2}}\kappa_2(1-p_2)} \tag{5.31}$$

Substituting (5.28) into (5.24), we have

$$\begin{aligned}\dot{\zeta} &= A - B M_a^{-1}(C_a\dot{q}_{ev} + G_a) + B M_a^{-1} Q\tau_a + \phi \\ &= A - B M_a^{-1}(C_a\dot{q}_{ev} + G_a) + B M_a^{-1} Q(-\kappa_2 Q^{-1} M_a B^{-1}\mathrm{sig}_{\mathrm{NN}}^{p_2}(\zeta) \\ &\quad - Q^{-1} M_a B^{-1} A + Q^{-1}(C_a\dot{q}_{ev} + G_a) - Q^{-1} M_a B^{-1}\phi) + \phi \\ &= -\kappa_2\mathrm{sig}_{\mathrm{NN}}^{p_2}(\zeta)\end{aligned} \tag{5.32}$$

Based on the definition, we know that ζ_1 is ratio persistent.

Step 2. Analyze the convergence and the RP property of σ when $\zeta = 0$.

The candidate Lyapunov function with respect to σ is selected as

$$V_2 = \frac{1}{2}\sigma^T\sigma \tag{5.33}$$

When $\zeta = 0$, according to (5.23), we have

$$\dot{\sigma} = -\kappa_1\mathrm{sig}_{\mathrm{NN}}^{p_1}(\sigma) \tag{5.34}$$

Similarly, σ is ratio persistent.

Taking the derivative of (5.33) with respect to time and substituting (5.34), we have

$$\dot{V}_2 = \sigma^T\dot{\sigma} = -\kappa_1\sigma^T\mathrm{sig}_{\mathrm{NN}}^{p_1}(\sigma) = -\kappa_1||\sigma||^{1+p_1} = -2^{\frac{1+p_1}{2}}\kappa_1 V_2^{\frac{1+p_1}{2}} \tag{5.35}$$

Then when $\zeta = 0$, the finite settling time of σ is

$$T_{12} \leq \frac{V_2^{\frac{1-p_1}{2}}(\sigma(0))}{2^{\frac{1+p_1}{2}}\kappa_1(\frac{1-p_1}{2})} = \frac{V_2^{\frac{1-p_1}{2}}(\sigma(0))}{2^{\frac{p_1-1}{2}}\kappa_1(1-p_1)} \tag{5.36}$$

Combining the above two steps, in the closed-loop system, the finite settling time of σ is $T_{1m} \leq T_{11} + T_{12}$. Then, the auxiliary system (5.22) satisfying the performance constraints is finite-time stable and has the settling time T_{1m}. Considering the performance constraints (2.55), the tracking error of the spacecraft attitude closed-loop system satisfies the performance constraints, and the maximum time for the error to converge to the expected error range is $T_{q_{ev}} \leq \min\{T_{ps}, T_{1m}\}$. □

5.3.2 Control Design and Stability Analysis with Disturbances

Considering the system (2.28) with external disturbances $d \neq 0$, in this section, the performance guaranteed robust controller with improved transient performance and disturbance rejection capability based on the norm-normalized sign function and ratio persistence property is designed.

With considering the disturbances, (5.22) becomes,

$$\begin{aligned} \ddot{\sigma} &= A + B\ddot{q}_{ev} \\ &= A - BM_a^{-1}(C_a\dot{q}_{ev} + G_a) - BM_a^{-1}d + BM_a^{-1}Q\tau_a \end{aligned} \tag{5.37}$$

Then the nonlinear term about the disturbances in (5.37) is $-BM_a{}^{-1}d$. To enhance the robustness of the closed-loop system, based on the previous section, a robust finite-time controller is further designed.

According to Lemma 2.6, the general form of an observer for the system $\dot{x} = f + gv + d$ is as follows,

$$\dot{y}_0 = -k_1\frac{y_0 - x}{||y_0 - x||^{1/2}} + y_1 - k_2(y_0 - x) + f + gv \tag{5.38}$$

$$\dot{y}_1 = -k_3\frac{y_0 - x}{||y_0 - x||} - k_4(y_0 - x) \tag{5.39}$$

where k_1, k_2, k_3, and k_4 represent the positive constant gains, y_0 and y_1 are the observed values of state x and unknown disturbance d, respectively.

So far, we designed the following observer to observe the uncertain function term $-BM_a^{-1}d$ in $\ddot{\sigma} = A - BM_a^{-1}(C_a\dot{q}_{ev} + G_a) - BM_a^{-1}d + BM_a^{-1}Q\tau_a$,

$$\dot{y}_0 = -\kappa_{y1}\frac{y_0 - \dot{\sigma}}{||y_0 - \dot{\sigma}||^{1/2}} + y_1 - \kappa_{y2}(y_0 - \dot{\sigma}) + \varkappa \tag{5.40}$$

$$\dot{y}_1 = -\kappa_{y3}\frac{y_0 - \dot{\sigma}}{||y_0 - \dot{\sigma}||} - \kappa_{y4}(y_0 - \dot{\sigma}) \tag{5.41}$$

where $\varkappa = A - BM_a^{-1}(C_a\dot{q}_{ev} + G_a) + BM_a^{-1}Q\tau_a$, $\kappa_{y1}, \kappa_{y2}, \kappa_{y3}, \kappa_{y4}$ represent the positive constant gains. y_0 and y_1 are observations of state $\dot{\sigma}$ and unknown disturbance $-BM_a^{-1}d$, i.e. $y_0 \to \dot{\sigma}$ $y_1 \to -BM_a^{-1}d$. Considering that in this controller, the model is precisely known, so only the y_1 item is needed. Define the observation error of the observer as $F_y = y_1 - BM_a^{-1}d$, then there is a finite time T_y in which F_y converges. According to the property of the observer, the observation error is bounded by $||F_y|| \leq \varepsilon_y$, where ε_y is a positive constant.

The modified sliding surface is designed as follows,

$$\zeta_d = \dot{\sigma} + \kappa_1 \mathrm{sig}_{\mathrm{NN}}^{p_1}(\sigma) + \kappa_3\sigma \tag{5.42}$$

where $\kappa_3 \geq \varrho^2/2$, ϱ is a positive constant. Since $\dot{\sigma} = -\kappa_3\sigma$ is ratio persistent, therefore, the above sliding surface is ratio persistent when $\zeta_d = 0$.

Its derivative with respect to time is,

$$\begin{aligned}\dot{\zeta}_d &= \ddot{\sigma} + \kappa_3\dot{\sigma} + \phi = A + B\ddot{q}_{ev} + \kappa_3\dot{\sigma} + \phi \\ &= A - BM_a^{-1}(C_a\dot{q}_{ev} + G_a) - BM_a^{-1}d + BM_a^{-1}Q\tau_a + \kappa_3\dot{\sigma} + \phi\end{aligned} \tag{5.43}$$

Considering that the controller designed in this section is a modification of the controller in the previous section with considering external disturbances, unless otherwise specified, the same symbols as those in the previous section have the same definitions. We propose the controller as follows,

$$\begin{aligned}\tau_{ad} = &-\kappa_2 Q^{-1}M_aB^{-1}\mathrm{sig}_{\mathrm{NN}}^{p_2}(\zeta_d) - \kappa_4 Q^{-1}M_aB^{-1}\zeta_d + Q^{-1}M_aB^{-1}y_1 \\ &- Q^{-1}M_aB^{-1}A + Q^{-1}(C_a\dot{q}_{ev} + G_a) - Q^{-1}M_aB^{-1}(\phi + \kappa_3\dot{\sigma})\end{aligned} \tag{5.44}$$

where $\kappa_4 \geq \varsigma^2/2$, ς is a positive constant.

Theorem 5.3 *Consider the spacecraft attitude tracking error model (2.28), and the performance constraints (2.55), define the auxiliary system that satisfies the performance constraints (5.22), define the sliding mode surface (5.42). When the external disturbances are considered, if Assumption 2.1 is satisfied and the controller is selected as (5.44), then the auxiliary system (5.22) is finite-time uniformly ultimately bounded. The finite settling time of ζ_d is*

$$T_{21} \leq T_y + \frac{1}{(2(\kappa_4 - \frac{\varsigma^2}{2}) - \kappa_6)(\frac{1-p_2}{2})}\ln\left[1 + \frac{2(\kappa_4 - \frac{\varsigma^2}{2}) - \kappa_6}{2^{\frac{1+p_2}{2}}\kappa_2 - \kappa_5}V_3^{\frac{1-p_2}{2}}(\zeta(0))\right].$$

When $||\zeta|| \leq \varepsilon_\zeta$, ε_ζ is the final boundary of $||\zeta||$, and the finite settling time of σ is

$$T_{22} \leq \frac{1}{(2(\kappa_3 - \frac{\varrho^2}{2}) - \kappa_8)(\frac{1-p_1}{2})} \ln\left[1 + \frac{2(\kappa_3 - \frac{\varrho^2}{2}) - \kappa_8}{2^{\frac{1+p_1}{2}}\kappa_1 - \kappa_7} V_4^{\frac{1-p_1}{2}}(\sigma(0))\right].$$

Then the finite settling time of the auxiliary system state σ is $T_{2m} \leq T_{21} + T_{22}$. Then the spacecraft attitude tracking error satisfies the constraint condition (2.55). The time for the spacecraft attitude tracking error to converge to the predefined steady-state error range satisfies $T_{q_{ev}} \leq \min\{T_{ps}, T_{2m}\}$, where T_{ps} is the predefined maximum settling time. The relevant parameters are defined later.

Proof **Step 1**. Analyze the convergence of the sliding mode variable and the RP property of ζ_d.

The candidate Lyapunov functions for ζ_d are selected as follows,

$$V_3 = \frac{1}{2}{\zeta_d}^T \zeta_d \tag{5.45}$$

Taking the derivative with respect to time for (5.45) and substituting the controller into it, we get

$$\begin{aligned}
\dot{V}_3 = \zeta_d^T \dot{\zeta}_d &= \zeta_d^T A - \zeta_d^T B M_a^{-1}(C_a \dot{q}_{ev} + G_a) \\
&\quad - \zeta_d^T B M_a^{-1} d + \zeta_d^T B M_a^{-1} Q \tau_{ad} + \zeta_d^T(\phi + \kappa_3 \dot{\sigma}) \\
&= \zeta_d^T A - \zeta_d^T B M_a^{-1}(C_a \dot{q}_{ev} + G_a) \\
&\quad - \zeta_d^T B M_a^{-1} d + \zeta_d^T B M_a^{-1} Q(-\kappa_2 Q^{-1} M_a B^{-1} \mathrm{sig}_{\mathrm{NN}}^{p_2}(\zeta_d) + Q^{-1} M_a B^{-1} y_1 \\
&\quad - \kappa_4 Q^{-1} M_a B^{-1} \zeta_d - Q^{-1} M_a B^{-1} A + Q^{-1}(C_a \dot{q}_{ev} + G_a) \\
&\quad - Q^{-1} M_a B^{-1}(\phi + \kappa_3 \dot{\sigma})) + \zeta_d^T(\phi + \kappa_3 \dot{\sigma}) \\
&= -\kappa_2 \zeta_d^T \mathrm{sig}_{\mathrm{NN}}^{p_2}(\zeta_d) - \kappa_4 \zeta_d^T \zeta_d - \zeta_d^T \varepsilon_y \\
&\leq -\kappa_2 \zeta_d^T \mathrm{sig}_{\mathrm{NN}}^{p_2}(\zeta_d) - \kappa_4 \zeta_d^T \zeta_d + \frac{\varsigma^2}{2}||\zeta_d||^2 + \frac{1}{2\varsigma^2}||\varepsilon_y||^2 \\
&\leq -\kappa_2 \zeta_d^T \mathrm{sig}_{\mathrm{NN}}^{p_2}(\zeta_d) - \left(\kappa_4 - \frac{\varsigma^2}{2}\right) \zeta_d^T \zeta_d + \frac{1}{2\varsigma^2}||\varepsilon_y||^2
\end{aligned} \tag{5.46}$$

Simplifying the above formula, we get

$$\dot{V}_3 \leq -2^{\frac{1+p_2}{2}} \kappa_2 V_3^{\frac{1+p_2}{2}} - 2\left(\kappa_4 - \frac{\varsigma^2}{2}\right) V_3 + \varepsilon_d \tag{5.47}$$

Considering the boundedness of ε_y, write $\frac{1}{2\varsigma^2}||\varepsilon_y||^2 \leq \varepsilon_d$. According to Lemma 2.5, ζ_d is finite-time uniformly ultimately bounded. Its bounded neighborhood is

$$D_\zeta = \{\zeta | \kappa_5 V_3^{\frac{1+p_2}{2}} + \kappa_6 V_3 < \varepsilon_d\} \tag{5.48}$$

where $\kappa_5 \in (0, 2^{\frac{1+p_2}{2}}\kappa_2)$, $\kappa_6 \in (0, 2(\kappa_4 - \frac{\varsigma^2}{2}))$. Considering the finite settling time T_y of the observation error, then the finite settling time of ζ is,

$$T_{21} \le T_y + \frac{1}{(2(\kappa_4 - \frac{\varsigma^2}{2}) - \kappa_6)(\frac{1-p_2}{2})} \ln\left[1 + \frac{2(\kappa_4 - \frac{\varsigma^2}{2}) - \kappa_6}{2^{\frac{1+p_2}{2}}\kappa_2 - \kappa_5} V_3^{\frac{1-p_2}{2}}(\zeta(0))\right] \tag{5.49}$$

Define $||\zeta|| \le \varepsilon_\zeta$, then when ζ converges to D_ζ, there is

$$\dot{\sigma} \le -\kappa_1 \mathrm{sig}_{\mathrm{NN}}^{p_1}(\sigma) - \kappa_3\sigma + \varepsilon_\zeta \tag{5.50}$$

Step 2. Analyze the convergence and the RP property of σ when ζ converges to D_ζ.

Select the candidate Lyapunov function about σ as follows,

$$V_4 = \frac{1}{2}\sigma^T\sigma \tag{5.51}$$

Find the derivative of (5.51) with respect to time and substitute (5.50), we have

$$\begin{aligned} \dot{V}_4 = \sigma^T\dot{\sigma} &\le -\kappa_1\sigma^T \mathrm{sig}_{\mathrm{NN}}^{p_1}(\sigma) - \kappa_3\sigma^T\sigma + \sigma^T\varepsilon_\zeta \\ &\le -\kappa_1\sigma^T \mathrm{sig}_{\mathrm{NN}}^{p_1}\left(\sigma\right) - \left(\kappa_3 - \frac{\varrho^2}{2}\right)\sigma^T\sigma + \frac{1}{2\varrho^2}\varepsilon_\zeta^2 \\ &\le -2^{\frac{1+p_1}{2}}\kappa_1 V_4^{\frac{1+p_1}{2}} - 2\left(\kappa_3 - \frac{\varrho^2}{2}\right)V_4 + \frac{1}{2\varrho^2}\varepsilon_\zeta^2 \end{aligned} \tag{5.52}$$

According to Lemma 2.5, σ is finite-time uniformly ultimately bounded. Its bounded neighborhood is

$$D_\sigma = \left\{\sigma | \kappa_7 V_4^{\frac{1+p_1}{2}} + \kappa_8 V_4 < \frac{1}{2\varrho^2}\varepsilon_\zeta^2\right\} \tag{5.53}$$

where $\kappa_7 \in (0, 2^{\frac{1+p_1}{2}}\kappa_1)$, $\kappa_8 \in (0, 2(\kappa_3 - \frac{\varrho^2}{2}))$. The finite settling time is,

$$T_{22} \le \frac{1}{(2(\kappa_3 - \frac{\varrho^2}{2}) - \kappa_8)(\frac{1-p_1}{2})} \ln\left[1 + \frac{2(\kappa_3 - \frac{\varrho^2}{2}) - \kappa_8}{2^{\frac{1+p_1}{2}}\kappa_1 - \kappa_7} V_4^{\frac{1-p_1}{2}}(\sigma(0))\right]. \tag{5.54}$$

Then, when ζ converges to D_ζ, σ is finite-time uniformly ultimately bounded. Then σ in the closed-loop system is finite-time uniformly ultimately bounded, with the finite settling time $T_{2m} \le T_{21} + T_{22}$. Considering the performance constraints and the maximum settling time constraints, the time for the spacecraft attitude tracking error to converge to the predefined boundary is $T_{q_{ev}} \le \min\{T_{ps}, T_{2m}\}$. □

Due to the consideration of external disturbances, ζ and σ will no longer satisfy the RP property. However, due to the existence of the $\mathrm{sig}_{\mathrm{NN}}(\cdot)$ function in the controller, the closed-loop system still has a strong ability to improve transient performance by approaching the RP property. When the external disturbance infinitely approaches 0, the system infinitely approximates the RP property, and the closed-loop system trajectory infinitely approximates a straight line.

5.4 Simulation Analysis

5.4.1 Simulation Parameters

This subsection first presents the initial conditions and parameters for the simulation. Considering the background of the on-orbit service spacecraft mission, although this study takes attitude motion as the research object, this section still gives relevant information about the spacecraft's orbit. The orbital parameters of the target spacecraft are as follows.

Considering that the spacecraft is very close to the target spacecraft in the short-range on-orbit service mission. Therefore, the orbital parameters of the spacecraft to be controlled are also shown in Table 5.1. Next, consider the initial parameters of the spacecraft as shown in Table 5.2. Without loss of generality, the parameters of the Tables 5.1 and 5.2 can be changed according to specific tasks.

Next, consider the desired attitude of the on-orbit service spacecraft. In the numerical simulation of this study, it is considered that the spacecraft is tracking the attitude

Table 5.1 The orbital parameters of the target spacecraft

Parameters	Values
Semimajor axis	6.85×10^6 m
Eccentricity	0.0073
Inclination	60 deg
Argument of periapsis	0 deg
Longitude of the ascending node	0 deg
Mean anomaly	0 deg

Table 5.2 The initial attitude parameters of the spacecraft

Parameters	Values
q_{v0}	$[0.6, 0.3, -0.3]^T$ rad
ω_{a0}	$[0, 0, 0]^T$ rad/s
J_s	$\begin{bmatrix} 100 & -10 & 5 \\ -10 & 120 & -3 \\ 5 & -3 & 100 \end{bmatrix}$ $\mathrm{kg} \cdot \mathrm{m}^2$

of the target spacecraft. The target spacecraft is inertial oriented, that is, $\mathfrak{F}_{bt}$ and $\mathfrak{F}_I$ coincide. Therefore, q_{dv} can be obtained via the relative motion between $\mathfrak{F}_{bs}$ and $\mathfrak{F}_I$. Define R_{li} as the rotation matrix of the target spacecraft from the inertial system to LVLH coordinate system, $r_{li-j,k}$ is the matrix element in row j, column k, then $q_{d0} = 1/2\ (1 + r_{li-11} + r_{li-22} + r_{li-33})^{1/2}$.

$$q_{dv} = \begin{bmatrix} (r_{li-23} - r_{li-32})/(4q_{d0}) \\ (r_{li-31} - r_{li-13})/(4q_{d0}) \\ (r_{li-12} - r_{li-21})/(4q_{d0}) \end{bmatrix} \tag{5.55}$$

According to the spacecraft attitude tracking model in Sect. 2.3, the tracking error quaternion and the angular velocity tracking error can be obtained. To demonstrate the performance of the proposed controller and to facilitate comparison with similar controllers, the error-index is defined as $I_a = ||q_{ev}|| + ||\omega_e||$. To illustrate the energy consumption of the proposed controller, the energy consumption index $E_a = \sum_{i=1}^{n} \int_0^t |\tau_{ai}\ (t)|^2 \mathrm{d}t$ is defined as in [19]. This index describes the electrical or chemical energy consumed by actuators such as thrusters or reaction flywheels during the stabilization of an attitude closed-loop system.

For the controller proposed in this study, the selection of parameters needs to be considered. In this study, the basis for parameter selection is mainly summarized as follows: (1) Adjust the controller parameters to make the control effect better and the energy consumption smaller; (2) In the comparison simulation, the controller proposed in this study and the controller to be compared respectively adjust their parameters to make the energy consumption similar, compare the control effect of the closed-loop system, or adjust the parameters to make the control effect similar and compare the energy consumption; (3) If the principles and structures of the controllers to be compared are similar, the same or similar parameters will be used to compare the differences between the controllers themselves directly.

At the same time, this section presents the parameters selected in the simulation of the spacecraft performance guaranteed controller (5.44) with improved transient performance in Sect. 5.3, as shown in the following table.

In Table 5.3, the parameter t_{ps} is the predefined maximum settling time. ρ_0 is the maximum initial value of the predefined performance boundary, which needs to be larger than the initial error quaternion. Considering that all quaternion elements are less than or equal to 1, thus, ρ_0 is taken as 1. ρ_t and ε_ρ provide steady-state error bounds. δ_i is about whether to limit the overshoot, λ_i is the sliding mode surface scale coefficient in the controller, p_1 and p_2 are the exponents in the controller, and κ_1 and κ_2 are the controller gain parameters. The observer parameters are $\kappa_{y1} = 0.01$, $\kappa_{y2} = 0.01$, $\kappa_{y3} = 0.001$, and $\kappa_{y4} = 0.001$. These values are based on adjusting the parameters to make the observed effects and control input values reasonable.

Considering the external disturbances to the spacecraft, we take the disturbance variable as follows [20],

Table 5.3 The initial parameters of the proposed controller (5.44)

Param	Value	Param	Value
t_{ps}	50s	p_1	0.5
ρ_0	1	p_2	0.5
ρ_t	0.02	κ_1	0.1
ε_ρ	0.01	κ_2	0.085
δ_i	1	κ_3	0.01
λ_i	1	κ_4	0.01

$$d_a = \begin{bmatrix} 0.24\sin(0.5t) + 2.4\,\mathrm{rand}_a(t) \\ 0.016\sin(0.5t + \pi/2) + 0.96\,\mathrm{rand}_a(t) \\ 0.16\sin(0.5t + \pi/3) + 0.96\,\mathrm{rand}_a(t) \end{bmatrix} \mathrm{N}\cdot\mathrm{m} \tag{5.56}$$

where $\mathrm{rand}_a(t)$ represents a zero-mean Gaussian random signal, and the variance is set to 0.01. It should be pointed out that, to fully demonstrate the disturbance rejection ability of the controller proposed, the disturbance selected in the simulation is larger than the practical situation, and the selected value is still within the reasonable range in orbit.

5.4.2 *Performance and Comparison*

The controllers proposed in this chapter can be regarded as an improved version of the controllers in the previous two chapters, Chaps. 3 and 4. Therefore, the controllers in this chapter will be compared with the controllers in the previous two chapters. Simultaneously, to illustrate the significance of the norm-normalized sign functions used in this chapter, the controller in this section will also be compared with the controller using the traditional sign function. This section labels the proposed controller (5.44) in Sect. 5.3 as controller 'A'. After all the norm-normalized sign functions in the controller (5.44) are replaced by traditional symbol functions [14, 16, 17], the resulting controller is marked as controller 'B'. Label the asymptotic controller in Sect. 3.2 as controller 'C'. Label the finite-time controller in Sect. 4.3 as controller 'D'. In the simulation diagram and analysis, the symbols 'A', 'B', 'C', and 'D' represent the above four controllers.

Figures 5.1, 5.2, 5.3, 5.4, 5.5, 5.6, 5.7, 5.8, 5.9, 5.10, 5.11, 5.12, 5.13 and 5.14 are the comparative simulation results of the spacecraft performance guaranteed control with improved transient performance proposed in Sect. 5.3. The simulation analysis presented in this section considers bounded external disturbances.

Figures 5.1 and 5.2 show the trajectories of the spacecraft attitude tracking error under the action of the controller proposed in Sect. 5.3 and the controller to be compared. The obvious transient performance improvement can be seen in these figures.

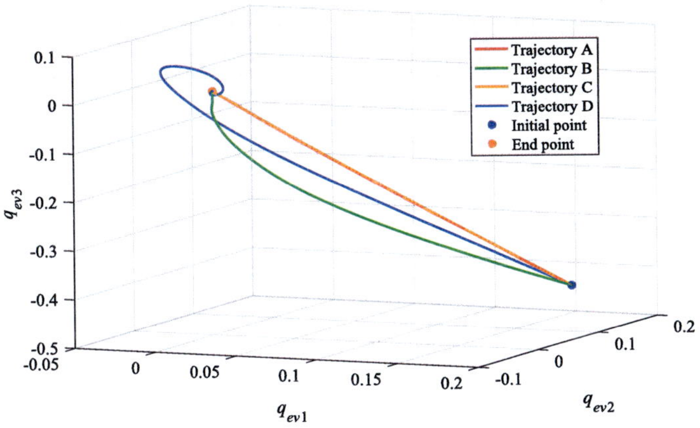

Fig. 5.1 The attitude tracking error trajectory (without disturbances)

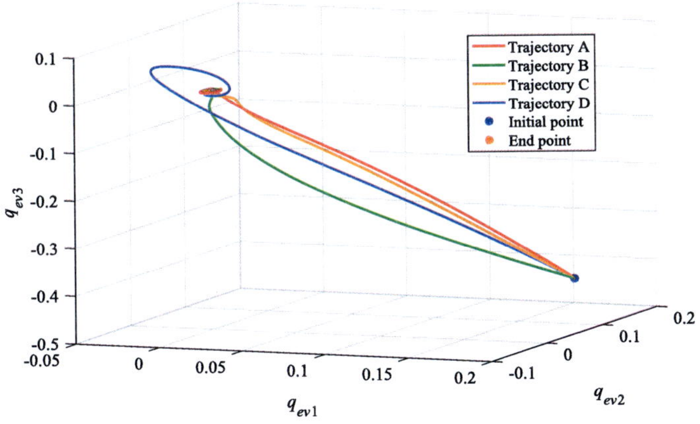

Fig. 5.2 The attitude tracking error trajectory (with disturbances)

The tracking error trajectory is a straight line when the disturbances are not considered and is approximately a straight line when the disturbances are considered. Thus, based on performance constraints, overshoot, and excess motion during convergence are further suppressed.

Figures 5.3, 5.4, 5.5 and 5.6 show the spacecraft attitude tracking error curves under the action of four controllers. All four controllers are performance guaranteed controllers. Therefore, under the action of the four controllers, the tracking errors all satisfy the same performance constraints. Compared with the controllers proposed in Sects. 3.2 and 4.3, the controller proposed in Sect. 5.3 has the faster convergence rate with almost no unnecessary overshoot. Compared with controller 'B', the controller proposed can make each dimension element of the error converge to the equilibrium

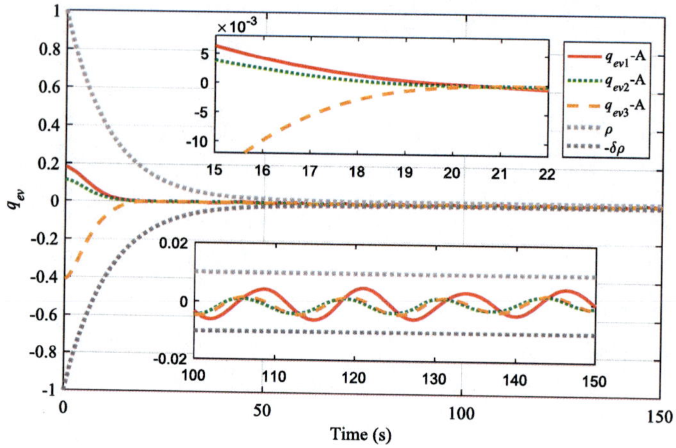

Fig. 5.3 The spacecraft attitude tracking error q_{ev} under controller 'A'

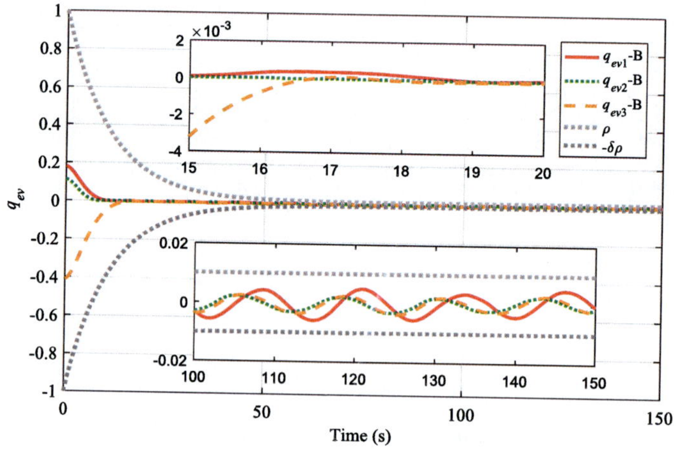

Fig. 5.4 The spacecraft attitude tracking error q_{ev} under controller 'B'

point or its neighborhood simultaneously and synchronously. Figures 5.7, 5.8 and 5.9 are the spacecraft attitude angular velocity tracking error.

Figures 5.10 and 5.11 are the spacecraft attitude tracking error indices $I_a = ||q_{ev}|| + ||\omega_e||$ and the energy consumption E_a. By comparing the two figures, it can be found that under the action of the controller proposed, the closed-loop system needs to consume less energy to achieve similar control effects. Figures 5.12 and 5.13 are the control inputs of the corresponding controllers, respectively. Figure 5.14 is the external disturbances considered in the numerical simulation.

Through the numerical simulation, it is fully demonstrated that under the action of the controller proposed in Sect. 5.3, the closed-loop system can obtain better transient

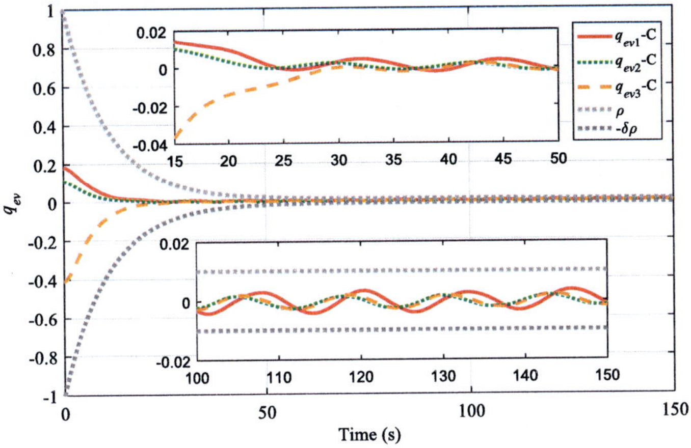

Fig. 5.5 The spacecraft attitude tracking error q_{ev} under controller 'C'

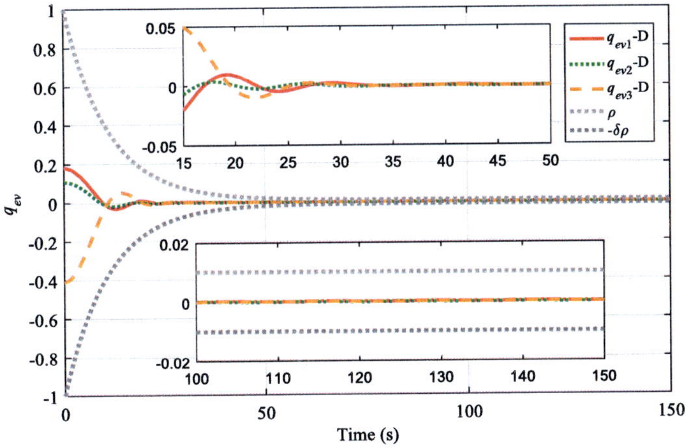

Fig. 5.6 The spacecraft attitude tracking error q_{ev} under controller 'D'

and steady-state performance by consuming less energy. Moreover, the performance is guaranteed. That is, the maximum settling time of the tracking error is guaranteed to be less than or equal to the predefined maximum settling time, the steady-state error is less than the predefined steady-state error interval, and the overshoot satisfies the constraints. Considering that the norm-normalized sign functions are applied to the controller in this chapter, the trajectory of the transient process of the tracking error of the closed-loop system is approximately a straight line. Thus, the transient performance of the closed-loop system is further enhanced. Simultaneously, considering that the controller proposed has the ability to converge in the finite time, the transient performance can be further improved by further adjusting the controller

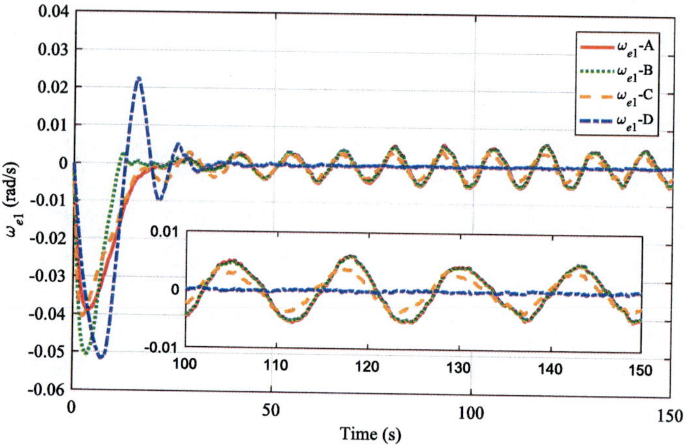

Fig. 5.7 The angular velocity tracking error ω_{e1}

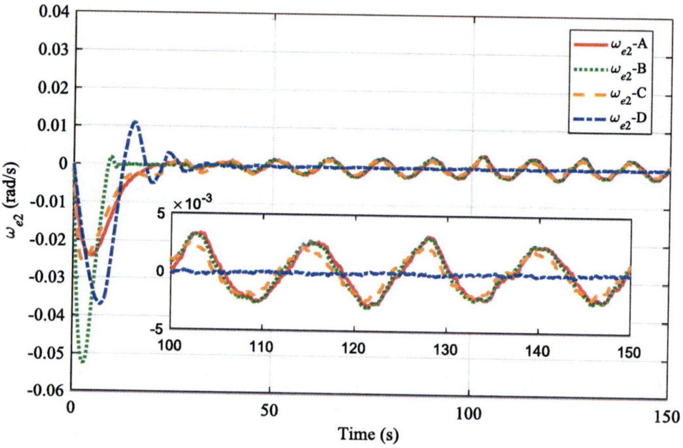

Fig. 5.8 The angular velocity tracking error ω_{e2}

parameters. Considering the strong performance constraints and better transient performance of the controller proposed, it is suitable for practical spacecraft attitude precision control applications.

5.5 Conclusion

The control algorithms proposed in this chapter improve the closed-loop system transient performance of the spacecraft performance guaranteed controllers in Chaps. 3 and 4. The performance guaranteed controllers proposed in the previous chapters can

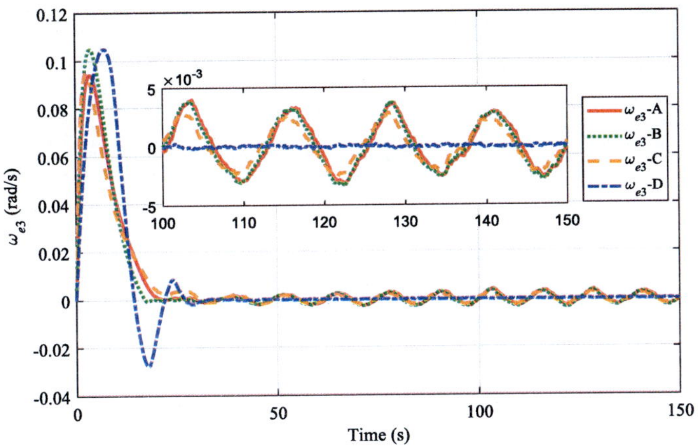

Fig. 5.9 The angular velocity tracking error ω_{e3}

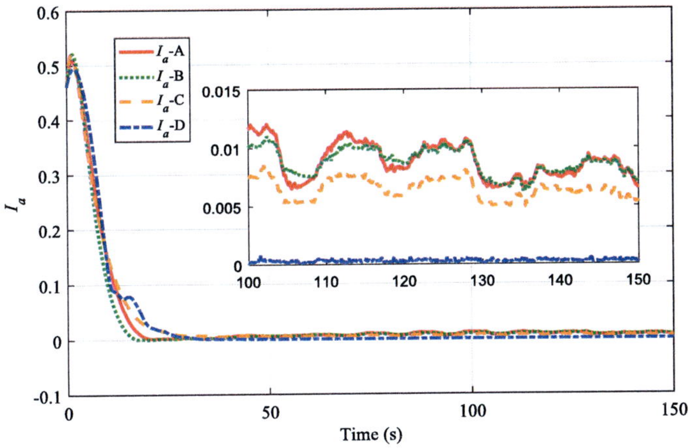

Fig. 5.10 The error index I_a

only provide performance constraints. They cannot directly affect the closed-loop system trajectories, which means that as long as the closed-loop system tracking error trajectories are within the constraints, these trajectories may be unknown and undesired. This performance remains to be further improved. Therefore, using the norm-normalized sign function and ratio persistence property, this chapter proposes the spacecraft performance guaranteed control with improved transient performance. Under the effect of this controller, the tracking errors of the closed-loop system can not only converge to the predefined steady-state error interval within the predefined maximum settling time but also satisfy the performance constraints at all times. Furthermore, the closed-loop system trajectory is approximately a straight line, which can be described more accurately. The tracking trajectory of the actual system further

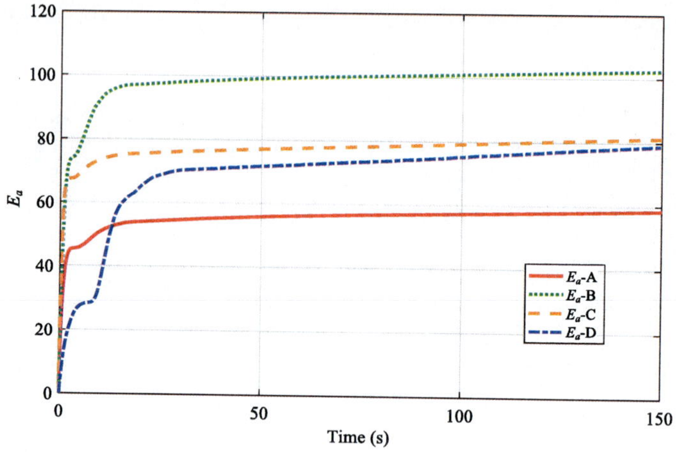

Fig. 5.11 The energy consumption

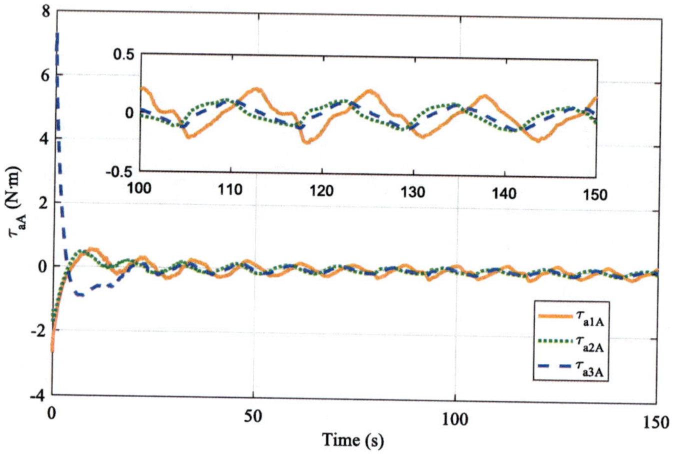

Fig. 5.12 The control input of 'A'

improves the transient performance of the closed-loop system and reduces energy consumption. Considering the external disturbances, the improved control algorithm is proposed based on the proposed controller. The rationality of the controllers is analyzed and proved by rigorous theoretical derivation. Finally, the numerical simulation analysis and comparisons with advanced controllers prove the effectiveness of the controllers proposed in this chapter.

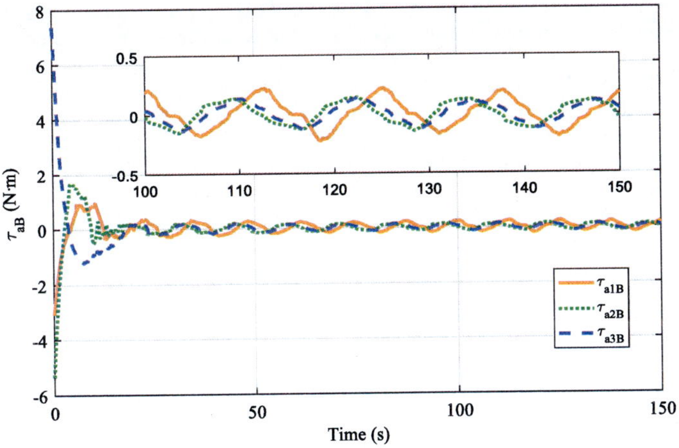

Fig. 5.13 The control input of 'B'

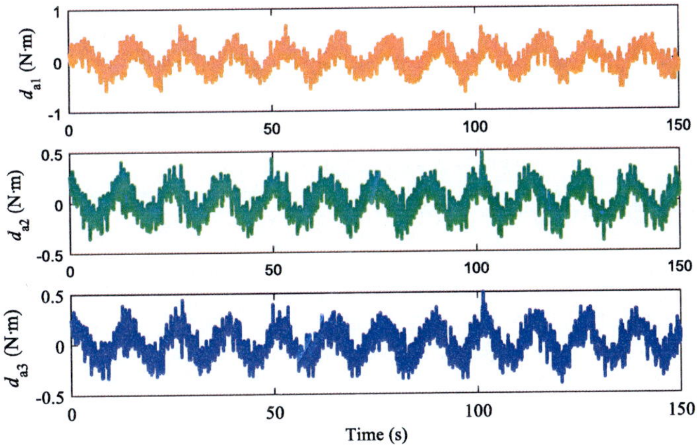

Fig. 5.14 The external disturbances

References

1. Fehse, W.: Automated rendezvous and docking of spacecraft. Cambridge university press (2003)
2. Luo, Y., Sun, Z.: Safe rendezvous scenario design for geostationary satellites with collocation constraints. Astrodynamics **1**(1), 71–83 (2017)
3. Floresabad, A., Ma, O., Pham, K., Ulrich, S.: A review of space robotics technologies for on-orbit servicing. Progress in aerospace sciences **68**, 1–26 (2014)
4. Da Fonseca, I.M., Goes, L.C., Seito, N., da Silva Duarte, M.K., de Oliveira, É.J.: Attitude dynamics and control of a spacecraft like a robotic manipulator when implementing on-orbit servicing. Acta Astronautica **137**, 490–497 (2017)

5. Ahmed, J., Coppola, V.T., Bernstein, D.S.: Adaptive asymptotic tracking of spacecraft attitude motion with inertia matrix identification. Journal of Guidance, Control, and Dynamics **21**(5), 684–691 (1998)
6. Shi, L., Kinkaid, N., Katupitiya, J.: Robust control for satellite attitude regulation during on-orbit assembly. IEEE Transactions on Aerospace and Electronic Systems **52**(1), 49–59 (2016)
7. Li, Q., Yuan, J., Wang, H.: Sliding mode control for autonomous spacecraft rendezvous with collision avoidance. Acta Astronautica **151**, 743–751 (2018)
8. Cao, L., Qiao, D., Xu, J.: Suboptimal artificial potential function sliding mode control for spacecraft rendezvous with obstacle avoidance. Acta Astronautica **143**, 133–146 (2018)
9. Wang, J., Liang, H., Sun, Z., Zhang, S., Liu, M.: Finite-time control for spacecraft formation with dual-number-based description. Journal of Guidance, Control, and Dynamics **35**(3), 950–962 (2012)
10. Wu, S., Radice, G., Gao, Y., Sun, Z.: Quaternion-based finite time control for spacecraft attitude tracking. Acta Astronautica **69**(1-2), 48–58 (2011)
11. Hu, Q., Li, B., Zhang, A.: Robust finite-time control allocation in spacecraft attitude stabilization under actuator misalignment. Nonlinear Dynamics **73**(1), 53–71 (2013)
12. Wei, C., Chen, Q., Liu, J., Yin, Z., Luo, J.: An overview of prescribed performance control and its application to spacecraft attitude system. Proceedings of the Institution of Mechanical Engineers, Part I: Journal of Systems and Control Engineering **235**(4), 435–447 (2021). DOI https://doi.org/10.1177/0959651820952552
13. Luo, J., Wei, C., Dai, H., Yin, Z., Wei, X., Yuan, J.: Robust inertia-free attitude takeover control of postcapture combined spacecraft with guaranteed prescribed performance. ISA transactions **74**, 28–44 (2018)
14. Li, D., Tee, K.P., Xie, L., Yu, H.: Time-synchronized control for disturbed systems. IEEE Transactions on Cybernetics (2021)
15. Li, D., Ge, S.S., Lee, T.H.: Fixed-time-synchronized consensus control of multi-agent systems. IEEE Transactions on Control of Network Systems (2020)
16. Li, D., Yu, H., Tee, K.P., Wu, Y., Ge, S.S., Lee, T.H.: On time-synchronized stability and control. IEEE Transactions on Systems, Man, and Cybernetics: Systems (2021)
17. Gao, Y., Li, D., Ge, S.S.: Time-synchronized tracking control for 6-dof spacecraft in rendezvous and docking. IEEE Transactions on Aerospace and Electronic Systems **58**(3), 1676–1691 (2022). DOI https://doi.org/10.1109/TAES.2021.3124865
18. Liu, H., Zhang, T.: Adaptive neural network finite-time control for uncertain robotic manipulators. Journal of Intelligent & Robotic Systems **75**(3), 363–377 (2014)
19. Jiang, B., Li, C., Ma, G.: Finite-time output feedback attitude control for spacecraft using "adding a power integrator" technique. Aerospace Science and Technology **66**, 342–354 (2017)
20. Zou, A.M.: Finite-time output feedback attitude tracking control for rigid spacecraft. IEEE Transactions on Control Systems Technology **22**(1), 338–345 (2013)

Chapter 6
Spacecraft Performance Guaranteed Control with Command Filter

Abstract In order to ensure the stability and safety of the on-orbit service spacecraft attitude motion, the attitude tracking system often enhances the strict state performance guaranteed design requirements. The performance guaranteed control algorithms provide sufficient state-constrained design methods for closed-loop systems. For spacecraft attitude control systems, actuators usually have limited response capabilities to control command signals, such as maximum output torques, rate, magnitude/amplitude, and bandwidth limitations. In this chapter, considering the control input command signal regulation, the performance guaranteed control algorithm based on the command filter technique is proposed. The command filter is configured between the theoretical control input and the actual control input. Before the theoretical control signal is transmitted to the actuator, it needs to be filtered by the command filter. This filter can constrain the signal's rate, magnitude/amplitude, and bandwidth, smooth the command signals and avoid peak interference. In this way, the response of the actuators is closer to the theoretical design. The spacecraft attitude tracking closed-loop system can not only meet the performance guaranteed constraints but also obtain acceptable control inputs within the reasonable range. The spacecraft attitude tracking closed-loop system is asymptotically stable and has strong disturbance suppression ability. The proposed controller's rationality and effectiveness are further demonstrated through the proof and simulation.

6.1 Introduction

In the previous chapters, the performance guaranteed control algorithms with asymptotic stability, finite-time stability, and transient performance improvement of the spacecraft are designed, and ideal control effects are achieved. In particular, the predefined maximum settling time and the corresponding performance constraints of the closed-loop system can be satisfied. However, actuators often have limited control capabilities for experimental spacecraft, including limited magnitude/amplitude, rate, bandwidth of the control input, etc. The related control algorithms designed in the previous chapters should have considered the capability of the practical spacecraft attitude control actuators. Even though the corresponding parameters can be

Y. Gao and D. Li, *Spacecraft Maneuver with Performance Guaranteed*,
https://doi.org/10.1007/978-981-99-4653-2_6

adjusted to make the control input meet the requirements of the actuators, in different situations, the control input value and the capacity of the actuator may differ. In order to theoretically solve the matching problem between control input and actuator capability, this chapter further studies the control input/command improvement problem in performance guaranteed control algorithms.

Specifically, in many cases, even under the control constraints, the tracking error of the closed-loop system has better transient and steady-state performance. However, the control signal of the controller could be better. For example, to achieve a smaller steady-state error or a faster convergence rate, the designed controller may output a larger control signal, which the existing system may not accept. For spacecraft attitude control, excessively large or higher frequency control signals are not conducive to the actual control and may even cause harm to the spacecraft's actuator.

The proposal of the command filter (CF) algorithm brings the possibility of processing control command signals [1]. They faced the aircraft longitude control problem and designed a controller which directly accommodates magnitude, rate, and bandwidth constraints on the aircraft states and the actuator signals. Since then, command filters have been widely used in various algorithms to adjust control command signals. Reference [2] omitted the derivative calculation of the processed signal by introducing a command filter. Reference [3] used command filters to implement the constraints on the control surfaces and the virtual control states. Reference [4] designed an improved command filter backstepping control algorithm by adding an inner-loop filter structure. For stochastic time-delayed nonstrict feedback systems with unknown input saturation, [5] addressed the problem of adaptive neural tracking control with the command filter technique. References [6, 7] respectively solved the adaptive fuzzy control problem for nonlinear systems with unknown control directions and the adaptive neural networks control problem for MIMO nonlinear systems with full-state constraints. Reference [8] proposed an alternate approach following the command-filtered backstepping principle to control the terminal voltage and stabilize the speed of a diesel generator through the excitation system and did an in-depth numerical analysis on the performance of the controller.

The command filter has the ability to actively adjust the amplitude, rate, and bandwidth of the control command signals. By selecting the desired amplitude, rate, bandwidth, and other parameters of the control signal in the filter, the signal adjusted by the filter can meet the capability of the spacecraft actuators. It only needs to input the theoretically designed control command into the filter. That is, the filtered control command signal can be obtained. Using the filtered control signal as the control input of the closed-loop system, the closed-loop system that satisfies the performance guaranteed and the actuator/control input constraints can be obtained. Combined with the specific situation of spacecraft attitude tracking, the mission objectives of this chapter are highlighted here:

Problem 6.1 Consider the spacecraft attitude tracking error model (2.28), and select predefined related parameters according to practical mission requirements, including predefined maximum settling time t_{ps}, the maximum steady-state error ρ_t, the maximum overshoot χ_t, design the performance guaranteed control such that:

(1) the conditions in Problem 3.1 are satisfied;
(2) the amplitude and rate constraints on the control input/command signals of the closed-loop system are satisfied;
(3) the closed-loop system has the property of disturbance suppression. That is, when strong external disturbances exist, the closed-loop system still satisfies the above performance constraints.

Among them, (1) in Problem 6.1 is the same as in Problem 3.1, that is, the closed-loop system is required to satisfy the predefined maximum settling time constraints and related performance constraints (performance guaranteed). Item (2) in Problem 6.1 highlights the controller's ability to constrain the control input/command signal. Then, according to the control problem proposed in this chapter, the performance guaranteed controller $\tau_d \in \mathbb{R}^n$ is designed for the spacecraft. The control input $\tau_a \in \mathbb{R}^n$ (for definition see (2.28)) obtained after the theoretically designed controller τ_d passes through the command filter is introduced into the spacecraft system, and the closed-loop system can be obtained. So that when $t \to +\infty$, the spacecraft attitude tracking closed-loop system state $q_{ev} \to 0$, that is, it converges to the equilibrium point. Alternatively, $q_{ev} \to \Theta$, where $\Theta = \{|q_{evi}| \le \iota_i, i = 1, 2, 3\}$ represents a small neighborhood containing equilibrium points, $\iota_i \in \mathbb{R}^+$ and $\iota_i \le \rho_t$ is a sufficiently small positive constant relative to the controller parameters and the initial state of the closed-loop system, that is, the closed-loop system converges into the neighborhood of the equilibrium point. When the system is asymptotically stable, there are $\iota_i = \rho_t$ and $T = t_{ps}$. Furthermore, the control command signal satisfies the constraints described by the desired signal amplitude and rate.

It should also be noted that due to the existence of the command filter, there is a difference between the control input τ_a and the theoretically derived control command signal τ_d. Likewise, if the command filter is used, there is a difference between the signal at the input and the signal at the output of the filter. Therefore, it is necessary to design the corresponding compensators to compensate for these signal errors. For an asymptotically stable system, the compensators themselves can be asymptotically stable. For a finite-time stable system, the compensator itself needs to be finite-time stable.

Remark 6.1 If the controller is designed for a closed-loop system first, then command filtering is applied to the control signal. This makes the filtered control variable the control input to the system. Due to the error in the signal before and after filtering, the introduction of the filter may affect the stability of the closed-loop system. The command filtering techniques applied in this chapter have the sufficient analysis of the signals before and after the command filter. The signals before and after the command filter are introduced into the controller design and stability analysis through the compensators. Therefore, the controller design in this chapter fully considers the stability of the closed-loop system with the command filter. Then under the corresponding controller, the closed-loop system containing the command filters can be stable.

This chapter considers the control input boundary constraints and other issues, and the control input boundedness comes from the boundedness of the actual spacecraft actuators. However, the boundedness of the control input does not mean that an arbitrarily small range of the control input amplitude can be selected but needs to be set according to the actual situation. This means that the primary purpose of the controller design in this chapter is to theoretically design the controller through the command filter and adjust the control signal that does not match the actuator rather than deliberately reducing the amplitude range of the control input through constraints.

In summary, based on the basic performance guaranteed algorithm and the regulation effect of the command filter on the control input/command signal, for the on-orbit service spacecraft attitude tracking control, the constraint function and the constraint controller hope to be designed that explicitly contain the pre-defined maximum settling time parameters. This controller can also constrain the transient and steady-state behaviors simultaneously. Moreover, the quality of the control signal hopes to be modified through the command filter so that the control signal can be applied to actual spacecraft attitude control. Therefore, in this chapter, the performance guaranteed control algorithm is designed based on the command filter. Here are the innovations of this chapter as follows.

(1) The pre-defined time constraint function is utilized, which can describe the maximum settling time. The error mapping transformation method is utilized, which can solve the asymmetric constraint problem with a symmetric function. Unlike the currently prescribed performance functions, the proposed function explicitly contains the desired maximum settling time variable and directly restricts the upper bound of the actual settling time.

(2) The control input quality of the performance guaranteed control algorithm is fully considered, and the performance guaranteed control algorithm based on the command filter is designed. This gives the controller proposed in this chapter the ability to actively adjust the command quality and solve the input saturation problem.

(3) The control algorithms proposed in this chapter are applied to the on-orbit service spacecraft attitude tracking control. On the one hand, in space missions, settling time and closed-loop system performance are guaranteed. On the other hand, the spacecraft control signal can be actively adjusted to make the signal received by the actuator smoother and can be improved in rate, magnitude, and bandwidth. In addition, energy consumption is reduced.

This chapter is arranged as follows. In Sect. 6.2, the traditional command filtering backstepping method is used to design the spacecraft attitude tracking command filtering backstepping control with command filters. In this section, the performance guaranteed method has not been considered. The primary purpose is to provide the design framework of the spacecraft command filtering backstepping method and the control algorithm design basis for the subsequent sections. In Sect. 6.3, based on the predefined maximum settling time constraints proposed in this book, an asymptotically stable command filter control algorithm based on performance guaranteed is designed. Section 6.4 gives the numerical simulation analysis of the proposed control algorithms. Section 6.5 gives the conclusion of this chapter.

6.2 Spacecraft Attitude Asymptotic Control Based on Command Filter

6.2.1 *Control Design*

In order to further design a spacecraft attitude tracking performance guaranteed control based on the command filter technique, the conventional command filtering backstepping method is first introduced in this section. It is applied to spacecraft attitude tracking control. In this subsection, the control algorithm is designed first without considering the disturbances, which can be considered a nominal case. Definitions of some related variables are given preferentially.

Considering the spacecraft attitude tracking error model (2.28), for the convenience of unifying symbols, define the system tracking error $\tilde{x}_1 = x_1 = q_{ev}$, and the tracking error of control to virtual control in backstepping method $\tilde{x}_2 = \dot{q}_{ev} - x_{2c}$. Among them, x_{2c} is the signal output by the command filter after x_{2d} has passed through the command filter (2.71). x_{2d} is the expectation $x_2 = \dot{q}_{ev}$ designed by backstepping. When $\dot{q}_{ev} = x_{2d}$ in the system, the system is stable.

The controller designed by theory is defined as τ_d, which passes through the command filter, and the obtained control input τ_a is the actual input of the spacecraft attitude error model (2.28).

On the other hand, due to the existence of the filter, the signals received by the system are the output of the filter, not the expected control and expected virtual control through theoretical design. There are deviations between the output of the filter and the input, so compensators for these deviations need to be designed.

In this study, the command filters are added between the virtual control x_{2d} and x_{2c} and the control signals τ_d and τ_a, respectively. So the compensator ξ_1 with respect to $x_{2c} - x_{2d}$ and the compensator ξ_2 with respect to $\tau_a - \tau_d$ are defined. The composite variable $z_1 = \tilde{x}_1 - \xi_1$ and $z_2 = \tilde{x}_2 - \xi_2$ are designed with respect to $\tilde{x}_1$, ξ_1, $\tilde{x}_2$, and ξ_2. The detailed design steps will be given later.

The design process of the controller will be divided into the following two steps.

Step 1. Design the virtual auxiliary variable x_{2d} and the compensator ξ_1 to stabilize z_1.

First, to stabilize the spacecraft attitude tracking error $\tilde{x}_1$ and the compensator ξ_1, design auxiliary variables z_1 as follows,

$$z_1 = \tilde{x}_1 - \xi_1 \tag{6.1}$$

The candidate Lyapunov functions for z_1 are selected as follows,

$$V_{cf1} = \frac{1}{2} z_1^T z_1 \tag{6.2}$$

It is worth noting that the closed-loop system error comes from two aspects. One is the tracking error; the other is the error of the command signal caused by the filter, so the auxiliary variable is designed as (6.1), if z_1 can be stabilized, then both errors will converge.

The derivative of $\tilde{x}_1$ with respect to time is as follows

$$\begin{aligned}\dot{\tilde{x}}_1 &= \dot{x}_1 = x_2 \\ &= (x_{2d} - x_{2d} + x_{2c} - x_{2c} + x_2) \\ &= x_{2d} + (x_{2c} - x_{2d}) + \tilde{x}_2\end{aligned} \tag{6.3}$$

If x_{2d} is selected as

$$x_{2d} = -c_1 \tilde{x}_1 \tag{6.4}$$

where $c_1 > 0$ is a constant, then

$$\dot{z}_1 = \dot{\tilde{x}}_1 - \dot{\xi}_1 = -c_1 \tilde{x}_1 + (x_{2c} - x_{2d}) + \tilde{x}_2 - \dot{\xi}_1 \tag{6.5}$$

and $\dot{\xi}_1$ can be taken as

$$\dot{\xi}_1 = -c_1 \xi_1 + (x_{2c} - x_{2d}) + \xi_2 \tag{6.6}$$

Then, (6.3) and (6.5) can be rewritten as

$$\dot{\tilde{x}}_1 = -c_1 \tilde{x}_1 + (x_{2c} - x_{2d}) + \tilde{x}_2 \tag{6.7}$$

$$\begin{aligned}\dot{z}_1 =& -c_1 \tilde{x}_1 + (x_{2c} - x_{2d}) + \tilde{x}_2 - (-c_1 \xi_1 + (x_{2c} - x_{2d}) + \xi_2) \\ =& -c_1 \tilde{x}_1 + \tilde{x}_2 + c_1 \xi_1 - \xi_2\end{aligned} \tag{6.8}$$

The derivative of (6.2) with respect to time is obtained

$$\dot{V}_{cf1} = -c_1 z_1^T z_1 + z_1^T (\tilde{x}_2 - \xi_2) = -c_1 z_1^T z_1 + z_1^T z_2 \tag{6.9}$$

Step 2. Design the controller τ_d and the compensator ξ_2 to stabilize z_2. The candidate Lyapunov function for z_2 is selected as follows

$$V_{cf2} = \frac{1}{2} z_2^T z_2 \tag{6.10}$$

Taking the derivative of z_2 with respect to time, one has

$$\begin{aligned}\dot{z}_2 &= \dot{\tilde{x}}_2 - \dot{\xi}_2 \\ &= M_a^{-1}(-C_a \dot{q}_{ev} - G_a) + M_a^{-1} Q \tau_a - M_a^{-1} d - \dot{x}_{2c} - \dot{\xi}_2 \\ &= M_a^{-1}(-C_a \dot{q}_{ev} - G_a) + M_a^{-1} Q \tau_d + M_a^{-1} Q (\tau_a - \tau_d) \\ &\quad - M_a^{-1} d - \dot{x}_{2c} - \dot{\xi}_2\end{aligned} \tag{6.11}$$

where $Q = \frac{1}{2} P^T$, P for definition see (2.28), $\dot{x}_{2c}$ can be obtained by filter.

The controller and compensator can be designed as follows

$$\tau_d = -Q^{-1} M_a \left(c_2 \tilde{x}_2 + M_a^{-1} \left(-C_a \dot{q}_{ev} - G_a \right) - \dot{x}_{2c} + z_1 \right) \tag{6.12}$$

$$\dot{\xi}_2 = -c_2 \xi_2 + M_a^{-1} Q \left(\tau_a - \tau_d \right) \tag{6.13}$$

where $c_2 > 1/2$ is a constant.

Substituting (6.12) and (6.13) into (6.11), one has

$$\begin{aligned}
\dot{z}_2 &= M_a^{-1}(-C_a \dot{q}_{ev} - G_a) + M_a^{-1} Q \tau_d + M_a^{-1} Q \left(\tau_a - \tau_d \right) \\
&\quad - M_a^{-1} d - \dot{x}_{2c} - \dot{\xi}_2 \\
&= M_a^{-1}(-C_a \dot{q}_{ev} - G_a) \\
&\quad - M_a^{-1} Q Q^{-1} M_a \left(c_2 \tilde{x}_2 + M_a^{-1}(-C_a \dot{q}_{ev} - G_a) - \dot{x}_{2c} + z_1 \right) \\
&\quad + M_a^{-1} Q \left(\tau_a - \tau_d \right) - M_a^{-1} d - \dot{x}_{2c} + c_2 \xi_2 \\
&\quad - M_a^{-1} Q \left(\tau_a - \tau_d \right) \\
&= -c_2 z_2 - z_1 - M_a^{-1} d
\end{aligned} \tag{6.14}$$

Therefore, the derivative of (6.10) with respect to time is obtained as

$$\dot{V}_{cf2} = z_2^T \dot{z}_2 = -c_2 z_2^T z_2 - z_2^T z_1 - z_2^T M_a^{-1} d \tag{6.15}$$

So far, the traditional command-filter-based controller and compensator applied to spacecraft attitude tracking control have been designed. The stability analysis is carried out in the next section.

6.2.2 *Stability Analysis*

The corresponding controller and compensators are designed through the above derivation, and some preliminary calculations are made. Then the following theorem is proposed, and the corresponding proof is given.

Theorem 6.1 *Consider the spacecraft attitude tracking error system (2.28) under disturbances with a known smooth reference trajectory. If Assumption 2.1 is satisfied and the controller, compensators, and command filters are chosen as (6.12), (6.6), (6.13) and (2.71), and the differences between the input and output signals of the command filters are bounded. The tracking error of the closed-loop system can be semiglobally ultimately uniformly bounded (SGUUB). Through the command filter, the rate, magnitude/amplitude, and bandwidth of control input can be constrained.*

Proof The candidate Lyapunov function for V_{cf1} and V_{cf2} is selected as follows

$$V_{cf} = V_{cf1} + V_{cf2} \tag{6.16}$$

Taking the derivative of (6.16) and substituting (6.9) and (6.15), one has

$$\dot{V}_{cf} = \dot{V}_{cf1} + \dot{V}_{cf2} = -c_1 z_1^T z_1 - c_2 z_2^T z_2 - z_2^T M_a^{-1} d \tag{6.17}$$

According to Young's inequality (Lemma 2.11) and the matrix norm properties, the following process for $-z_2^T M_a^{-1} d$ in (6.17) can be processed.

$$\begin{aligned} -z_2^T M_a^{-1} d &\le \left\| z_2^T M_a^{-1} d \right\| \le \left\| z_2 \right\| \left\| M_a^{-1} d \right\| \\ &\le \frac{1}{2} \left\| z_2 \right\|^2 + \frac{1}{2} \left\| M_a^{-1} d \right\|^2 \\ &= \frac{1}{2} z_2^T z_2 + \frac{1}{2} \left\| M_a^{-1} d \right\|^2 \end{aligned} \tag{6.18}$$

Substituting (6.18) into (6.17), one has

$$\begin{aligned} \dot{V}_{cf} &= \dot{V}_{cf1} + \dot{V}_{cf2} = -c_1 z_1^T z_1 - c_2 z_2^T z_2 - z_2^T M_a^{-1} d \\ &\le -c_1 z_1^T z_1 - c_2 z_2^T z_2 + \frac{1}{2} z_2^T z_2 + \frac{1}{2} \left\| M_a^{-1} d \right\|^2 \end{aligned} \tag{6.19}$$

Considering the disturbances and M_a are bounded, let d_m be a constant and satisfy the following formula

$$d_m \ge \frac{1}{2} \left\| M_a^{-1} d \right\|^2 \tag{6.20}$$

Then (6.19) is rewritten as follows

$$\dot{V}_{cf} \le -c_1 z_1^T z_1 - \left(c_2 - \frac{1}{2} \right) z_2^T z_2 + d_m \le -c_3 V_{cf} + d_m \tag{6.21}$$

where $0 < c_3 \le \min \left\{ 2c_1, 2\left(c_2 - \frac{1}{2} \right) \right\}$.

According to the above formula, the following formulas can be obtained.

$$V_{cf}(t) \le V_{cf}(0) e^{-c_3 t} + \frac{d_m}{c_3} \left(1 - e^{-c_3 t} \right) \tag{6.22}$$

$$\| z_1 \| \le \sqrt{2 V_{cf}} \le \sqrt{V_{cf}(0) e^{-c_3 t} + \frac{d_m}{c_3} \left(1 - e^{-c_3 t} \right)} \tag{6.23}$$

Equation (6.22) shows that the closed-loop system tracking error (including tracking error and compensator parameters) is semiglobally ultimately uniformly bounded (SGUUB).

Next, the convergence of the compensators is considered, and the candidate Lyapunov function for ξ_1 and ξ_2 is selected as follows

$$V_\xi = \frac{1}{2}\xi_1^T \xi_1 + \frac{1}{2}\xi_2^T \xi_2 \tag{6.24}$$

Taking the time derivative of (6.24), one has

$$\begin{aligned}
\dot{V}_\xi &= \xi_1^T \dot{\xi}_1 + \xi_2^T \dot{\xi}_2 \\
&= -c_1\xi_1^T \xi_1 + \xi_1^T (x_{2c} - x_{2d}) \\
&\quad + \xi_1^T \xi_2 - c_2\xi_2^T \xi_2 + \xi_2^T M_a^{-1} Q (\tau_a - \tau_d) \\
&= -c_1\xi_1^T \xi_1 - c_2\xi_2^T \xi_2 + \xi_1^T (x_{2c} - x_{2d}) \\
&\quad + \xi_1^T \xi_2 + \xi_2^T M_a^{-1} Q (\tau_a - \tau_d)
\end{aligned} \tag{6.25}$$

According to the properties of the CF, assume $\|x_{2c} - x_{2d}\| \le \varepsilon_1$, $\|\tau_a - \tau_d\| \le \varepsilon_2$, and $\left\|M_a^{-1}Q\right\| \le \varepsilon_3$. Each item in (6.25) can be handled separately as follows

$$\begin{aligned}
\xi_1^T (x_{2c} - x_{2d}) &\le \left\|\xi_1^T (x_{2c} - x_{2d})\right\| \\
&\le \left\|\xi_1\right\| \|(x_{2c} - x_{2d})\| \le \varepsilon_1 \left\|\xi_1\right\|
\end{aligned} \tag{6.26}$$

$$\begin{aligned}
\xi_1^T \xi_2 &\le \left\|\xi_1^T \xi_2\right\| \le \left\|\xi_1\right\| \|\xi_2\| \\
&\le \frac{1}{2}\left(\xi_1^T \xi_1 + \xi_2^T \xi_2\right)
\end{aligned} \tag{6.27}$$

$$\xi_2^T M_a^{-1} Q (\tau_a - \tau_d) \le \left\|\xi_2^T M_a^{-1} Q (\tau_a - \tau_d)\right\| \le \varepsilon_2\varepsilon_3 \left\|\xi_2\right\| \tag{6.28}$$

Substitute (6.26), (6.27), and (6.28) into (6.25), one has

$$\begin{aligned}
\dot{V}_\xi &= -c_1\xi_1^T \xi_1 - c_2\xi_2^T \xi_2 + \xi_1^T (x_{2c} - x_{2d}) \\
&\quad + \xi_1^T \xi_2 + \xi_2^T M_a^{-1} Q (\tau_a - \tau_d) \\
&\le -\left(c_1 - \frac{1}{2}\right)\xi_1^T \xi_1 - \left(c_2 - \frac{1}{2}\right)\xi_2^T \xi_2 \\
&\quad + \varepsilon_1 \left\|\xi_1\right\| + \varepsilon_2\varepsilon_3 \left\|\xi_2\right\| \\
&\le -c_4 V_\xi + \varepsilon_4\sqrt{V_\xi}
\end{aligned} \tag{6.29}$$

where $c_4 = \min\left\{2\left(c_1 - \frac{1}{2}\right), 2\left(c_2 - \frac{1}{2}\right)\right\}$ and $\varepsilon_4 = \max\left\{\sqrt{2}\varepsilon_1, \sqrt{2}\varepsilon_2\varepsilon_3\right\}$.

Therefore, according to (6.29), the compensation signals ξ_i satisfies the following bound.

$$\left\|\xi_i\right\| \le \frac{2\varepsilon_4}{c_4}\left(1 - e^{-\frac{1}{2}c_4 t}\right) \tag{6.30}$$

From $z_1 = q_{ev} - \xi_1$, one has

$$
\begin{aligned}
\|q_{ev}\| &\leq \|z_1\| + \|\xi_1\| \\
&\leq \sqrt{V_{cf}(0)e^{-c_3 t} + \frac{d_m}{c_3}\left(1 - e^{-c_3 t}\right) + \frac{2\varepsilon_4}{c_4}\left(1 - e^{-\frac{1}{2}c_4 t}\right)}
\end{aligned} \tag{6.31}
$$

Therefore, the closed-loop system is SGUUB, that is, the spacecraft attitude tracking error q_{ev} and the compensator parameters ξ_1 and ξ_2 are SGUUB. □

It is worth noting that the theorem and proof in this subsection are the cases of the closed-loop system under the controller (6.12) considering the disturbances. The above proof covers both the stability analysis of the closed-loop system under controller (6.12) with and without disturbances. When the disturbances are not considered, it is only necessary to set the disturbance item to 0, and the proof process is still valid. Considering the external disturbances, it has more useful practical value. Although the controller is designed under the nominal case, it has a certain disturbance suppression ability through the proof of this subsection.

6.3 Spacecraft Attitude Performance Guaranteed Asymptotic Control Based on Command Filter

6.3.1 Control Design

In the previous section, we designed the traditional command filtering backstepping controller for the spacecraft attitude tracking error system. In this section, the performance guaranteed technique is introduced, and the asymptotically stable spacecraft attitude performance guaranteed control based on command filter is proposed. Same as in the previous section, we still design the control algorithm that does not consider the external disturbances first. In the following theorem and proof, the stability of the algorithm in disturbed environments is further considered.

Same as Sect. 2.4.2, define $\sigma_i(t, t_{ps})$ as performance-constrained auxiliary variable, see (2.60) for its definition. t_{ps} is the predefined maximum settling time, and the relationship between t_{ps} and $\sigma_i(t, t_{ps})$ is analyzed in Sect. 2.4.2. In the following text, the abbreviation of $\sigma_i(t, t_{ps})$ is σ_i.

Same as Sect. 6.2, Considering the spacecraft attitude tracking error model (2.28), for the convenience of unifying symbols, define the system tracking error $\tilde{x}_1 = x_1 = q_{ev}$, and the tracking error of control to virtual control in backstepping method $\tilde{x}_2 = \dot{q}_{ev} - x_{2c}$. where x_{2c} is the signal output by the command filter after x_{2d} has passed through the command filter (2.71), x_{2d} is the expectation $x_2 = \dot{q}_{ev}$ designed by backstepping. The controller designed by theory is defined as τ_d, which passes through the command filter, and the obtained control input τ_a is the actual input of the spacecraft attitude error model (2.28). Define the compensator ξ_1 with respect to $x_{2c} - x_{2d}$ and the compensator ξ_2 with respect to $\tau_a - \tau_d$.

First, the preparatory derivation is carried out. The derivative of σ_i with respect to time is as follows,

$$\dot{\sigma}_i = \frac{\partial \varphi^{-1}(\vartheta_i)}{\partial \vartheta_i} \dot{\vartheta}_i \tag{6.32}$$

$$\dot{\vartheta}_i = \frac{2}{(1+\delta_i)} \frac{\dot{\tilde{x}}_{1i}\rho_i - \tilde{x}_{1i}\dot{\rho}_i}{\rho_i^2} = \frac{2}{(1+\delta_i)\rho_i^2}\left(\rho_i \dot{\tilde{x}}_{1i} - \dot{\rho}_i \tilde{x}_{1i}\right) \tag{6.33}$$

$$\dot{\sigma}_i = \frac{2r_i\rho_i}{(1+\delta_i)\rho_i^2}\dot{\tilde{x}}_{1i} - \frac{2r_i\dot{\rho}_i}{(1+\delta_i)\rho_i^2}\tilde{x}_{1i} \tag{6.34}$$

The situation of this study is that the system state is within the constraint range at the initial, and the controller needs to be designed to keep the corresponding state within the constraint range all the time. Considering that within the entire constraint interval, $\partial\varphi^{-1}(\vartheta_i)/\partial\vartheta_i > 0$ holds, let $r_i = |\partial\varphi^{-1}(\vartheta_i)/\partial\vartheta_i|$, then within the scope of the study there is $\dot{\sigma}_i = r_i\dot{\vartheta}_i$.

And (6.34) is rewritten in a compact form suitable for MIMO systems with $\sigma = [\sigma_1, \sigma_2, \sigma_3]^T$ as follows

$$\dot{\sigma} = R_1\dot{\tilde{x}}_1 - R_2\tilde{x}_1 \tag{6.35}$$

where

$$\begin{aligned} R_1 &= \operatorname{diag}\left(\frac{2r_1}{(1+\delta_1)\rho_1}, \ldots, \frac{2r_n}{(1+\delta_n)\rho_n}\right) \\ R_2 &= \operatorname{diag}\left(\frac{2r_1\dot{\rho}_1}{(1+\delta_1)\rho_1^2}, \ldots, \frac{2r_n\dot{\rho}_n}{(1+\delta_n)\rho_n^2}\right) \end{aligned} \tag{6.36}$$

Remark 6.2 In order to limit overshoot, $\tilde{x}_{1i} = (1-\delta_i)\rho_i(t)/2$ will be the new equilibrium point or the tracking path of the error. Therefore, when $\delta_i \neq 1$, it will bring steady-state error. When $\tilde{x}_{1i}(0) = 0$, δ_i can not be 0.

Remark 6.3 In this research, the symmetric tanh function is used to solve the problem of the asymmetric performance constraints with the transformation as (2.58), which makes it possible to restrict the overshoot and simplify the design process simultaneously.

Different from Sect. 6.2, to achieve the purpose of PMST and performance guaranteed, it needs first to stabilize σ instead of $\tilde{x}_1$. So the desired auxiliary variable x_{2d} is designed. The candidate Lyapunov function for σ is chosen as follows

$$V_\sigma = \frac{1}{2}\sigma^T\sigma \tag{6.37}$$

Deriving the above formula with respect to time, one has

$$\dot{V}_\sigma = \sigma^T\dot{\sigma} \tag{6.38}$$

If $\dot{\sigma} = -c_1\sigma$, then

$$\dot{\sigma} = R_1\dot{\tilde{x}}_1 - R_2\tilde{x}_1 = R_1 x_2 - R_2\tilde{x}_1 = -c_1\sigma \tag{6.39}$$

Thus, x_{2d} can be preliminarily selected as

$$x_{2d} = -c_1 R_1^{-1}\sigma + R_1^{-1}R_2\tilde{x}_1 \tag{6.40}$$

So, it can be guaranteed that $\dot{V}_\sigma = -c_1\sigma^T\sigma \leq 0$. If the system state x_2 can track x_{2d}, then σ can be stable, and the upper bound of the settling time will be further guaranteed. It should be noted that since the current selection of the desired control auxiliary variable x_{2d} only considers the performance constraints and does not consider the command filter, the desired controller here is preliminarily selected. Subsequent design can still be carried out as needed, and the stability of the closed-loop system needs to be based on the candidate Lyapunov function in the subsequent design. Since the errors generated by the filters are involved in the closed-loop system, it also needs to consider the compensators, so the intermediate variables of the backstepping method are re-designed as follows

$$z_1 = \sigma - \xi_1 \tag{6.41}$$

$$z_2 = \tilde{x}_2 - \xi_2 \tag{6.42}$$

Next, the controller is designed through the backstepping method. Similarly, this part is also divided into two-step designs.

Step 1. Design the virtual auxiliary variable x_{2d} and the compensator ξ_1 to stabilize z_1.

The candidate Lyapunov function for z_1 is selected as follows

$$V_1 = \frac{1}{2}z_1^T z_1 \tag{6.43}$$

In order to facilitate the calculation of the derivative of (6.43), some preliminary calculations are made.

$$\begin{aligned}\dot{\tilde{x}}_1 &= \dot{x}_1 \\ &= (x_2 + x_{2d} - x_{2d} + x_{2c} - x_{2c}) \\ &= x_{2d} + (x_{2c} - x_{2d}) + \tilde{x}_2\end{aligned} \tag{6.44}$$

where x_{2c} is the signal obtained after x_{2d} passes through the command filter.

$$\begin{aligned}\dot{\sigma} &= R_1\dot{\tilde{x}}_1 - R_2\tilde{x}_1 \\ &= R_1\left(x_{2d} + (x_{2c} - x_{2d}) + \tilde{x}_2\right) - R_2 x_1\end{aligned} \tag{6.45}$$

$$\dot{z}_1 = \dot{\sigma} - \dot{\xi}_1 = R_1 \left(x_{2d} + (x_{2c} - x_{2d}) + \tilde{x}_2 \right) - R_2 x_1 - \dot{\xi}_1 \tag{6.46}$$

Considering the existence of the filter compensator, (6.40) is designed as $x_{2d} = -c_1 R_1^{-1}\sigma + R_1^{-1} R_2 \tilde{x}_1 - \kappa_R \xi_1$, κ_R is a positive number, Substituting it into (6.46), one has

$$\dot{z}_1 = R_1 \left(-c_1 R_1^{-1}\sigma + R_1^{-1} R_2 \tilde{x}_1 + (x_{2c} - x_{2d}) - \kappa_R \xi_1 + \tilde{x}_2 \right) - R_2 x_1 - \dot{\xi}_1 \tag{6.47}$$

Simplify (6.47), one has

$$\begin{aligned} \dot{z}_1 &= R_1 \left(-c_1 R_1^{-1}\sigma + R_1^{-1} R_2 \tilde{x}_1 + (x_{2c} - x_{2d}) - \kappa_R \xi_1 + \tilde{x}_2 \right) - R_2 x_1 - \dot{\xi}_1 \\ &= -c_1 \sigma + R_1 (x_{2c} - x_{2d}) + R_1 \tilde{x}_2 - \dot{\xi}_1 - \kappa_R R_1 \xi_1 \\ &= -c_1 z_1 - c_1 \xi_1 + R_1 (x_{2c} - x_{2d}) + R_1 z_2 \\ &\quad + R_1 \xi_2 - \dot{\xi}_1 - \kappa_R R_1 \xi_1 \end{aligned} \tag{6.48}$$

If the compensator ξ_1 is taken as

$$\dot{\xi}_1 = -c_1 \xi_1 + R_1 (x_{2c} - x_{2d}) + R_1 \xi_2 - \kappa_R R_1 \xi_1 \tag{6.49}$$

then, (6.48) can be rewritten as

$$\dot{z}_1 = -c_1 z_1 + R_1 z_2 \tag{6.50}$$

So far, the derivative of (6.43) with respect to time is as follows

$$\dot{V}_1 = z_1^T \dot{z}_1 = -c_1 z_1^T z_1 + z_1^T R_1 z_2 \tag{6.51}$$

Step 2. Design the controller τ_d and the compensator ξ_2 to stabilize z_2.

The main function of the command filter is to filter the command signal and obtain a modified command signal. In this section, τ_d is the designed command signal, which generates the signal τ after passing through the filter (2.71). τ is input to the system and becomes the actual control signal of the system. The desired command signal τ_d is now designed.

Without considering V_1, the ideal controller τ_d is obtained through the Lyapunov method and backstepping technique. Select the candidate Lyapunov function only about $\tilde{x}_2$ as follows and derive it, and one has

$$V_\tau = \frac{1}{2} \tilde{x}_2^T \tilde{x}_2 \tag{6.52}$$

$$\dot{V}_\tau = \tilde{x}_2^T \dot{\tilde{x}}_2 \tag{6.53}$$

If it can be made that $\dot{\tilde{x}}_2 = -c_2\tilde{x}_2$, the constant $c_2 > 1/2$, then $\dot{V}_\tau = \tilde{x}_2^T\dot{\tilde{x}}_2 = -c_2\tilde{x}_2^T\tilde{x}_2 \leq 0$. According to

$$M_a^{-1}(-C_a\dot{q}_{ev} - G_a) + M_a^{-1}Q\tau_d - M_a^{-1}d - \dot{x}_{2c} = -c_2\tilde{x}_2 \tag{6.54}$$

where Q is defined as in the previous section, find the desired τ_d as follows

$$\tau_d = Q^{-1}M_a\left(-c_2\tilde{x}_2 - M_a^{-1}(-C_a\dot{q}_{ev} - G_a) + \dot{x}_{2c}\right) \tag{6.55}$$

Considering that there is an item $z_1^T R_1 z_2$ in (6.51) and $z_1^T R_1 z_2 = z_2^T R_1^T z_1$, modify the desired τ_d as follows

$$\tau_d = Q^{-1}M_a\left(-c_2\tilde{x}_2 - M_a^{-1}(-C_a\dot{q}_{ev} - G_a) + \dot{x}_{2c} - R_1^T z_1\right) \tag{6.56}$$

The candidate Lyapunov function for z_2 is selected as follows

$$V_2 = \frac{1}{2}z_2^T z_2 \tag{6.57}$$

Find the derivative of z_2 with respect to time, one has

$$\begin{aligned}
\dot{z}_2 &= \dot{\tilde{x}}_2 - \dot{\xi}_2 \\
&= M_a^{-1}(-C_a\dot{q}_{ev} - G_a) + M_a^{-1}Q\tau_a - M_a^{-1}d - \dot{x}_{2c} - \dot{\xi}_2 \\
&= M_a^{-1}(-C_a\dot{q}_{ev} - G_a) + M_a^{-1}Q\tau_d + M_a^{-1}Q(\tau_a - \tau_d) \\
&\quad - M_a^{-1}d - \dot{x}_{2c} - \dot{\xi}_2
\end{aligned} \tag{6.58}$$

Substitute (6.56) into (6.58),

$$\begin{aligned}
\dot{z}_2 &= M_a^{-1}(-C_a\dot{q}_{ev} - G_a) + M_a^{-1}Q\tau_d + M_a^{-1}Q(\tau_a - \tau_d) \\
&\quad - M_a^{-1}d - \dot{x}_{2c} - \dot{\xi}_2 \\
&= -c_2\tilde{x}_2 - R_1^T z_1 + M_a^{-1}Q(\tau_a - \tau_d) - M_a^{-1}d - \dot{\xi}_2 \\
&= -c_2 z_2 - c_2\xi_2 - R_1^T z_1 + M_a^{-1}Q(\tau_a - \tau_d) \\
&\quad - M_a^{-1}d - \dot{\xi}_2
\end{aligned} \tag{6.59}$$

If the compensator ξ_2 is designed as

$$\dot{\xi}_2 = -c_2\xi_2 + M_a^{-1}Q(\tau_a - \tau_d) \tag{6.60}$$

and substitute it into (6.59), and then one has

$$\dot{z}_2 = -c_2 z_2 - R_1^T z_1 - M_a^{-1}d \tag{6.61}$$

Then it can be found that the derivative of (6.57) with respect to time as follows

$$\dot{V}_2 = z_2^T \dot{z}_2 = -c_2 z_2^T z_2 - z_2^T R_1^T z_1 - z_2^T M_a^{-1} d \tag{6.62}$$

So far, the design of the controller and compensator has been completed. In the next section, the stability and disturbance rejection of the closed-loop system will be analyzed in combination with the proposed controller and compensator.

6.3.2 Stability Analysis

Theorem 6.2 *For the spacecraft attitude tracking error model (2.28) with considering external disturbances, if Assumption 2.1 is satisfied and the controller is chosen as (6.56), the command filter compensators are chosen as (6.49) and (6.60), the command filters are chosen as (2.71), and the differences between the input and output signals of the command filters are bounded, then the spacecraft attitude tracking error is semi-globally uniformly ultimately bounded (SGUUB). Simultaneously, the practical settling time of the spacecraft attitude tracking error is less than the pre-defined maximum settling time* t_{ps}, *and* t_{ps} *can be set in advance according to actual needs. The transient and steady-state behavior of the closed-loop system satisfies performance constraints (2.55). The command filters can sufficiently constrain the rate, amplitude, and bandwidth of the control command or signal.*

Proof The candidate Lyapunov function for V_1 and V_2 is selected as follows

$$V = V_1 + V_2 \tag{6.63}$$

Taking the derivative of (6.63) with respect to time, one has

$$\begin{aligned} \dot{V} &= \dot{V}_1 + \dot{V}_2 \\ &= -c_1 z_1^T z_1 + z_1^T R_1 z_2 - c_2 z_2^T z_2 \\ &\quad - z_2^T R_1^T z_1 - z_2^T M_a^{-1} d \\ &= -c_1 z_1^T z_1 - c_2 z_2^T z_2 - z_2^T M_a^{-1} d \end{aligned} \tag{6.64}$$

According to Young's inequality (Lemma 2.11) and the matrix norm properties, $-z_2^T M_a^{-1} d$ in (6.64) is analyzed.

$$\begin{aligned} -z_2^T M_a^{-1} d &\le \left\| z_2^T M_a^{-1} d \right\| \le \left\| z_2 \right\| \left\| M_a^{-1} d \right\| \\ &\le \frac{1}{2} \left\| z_2 \right\|^2 + \frac{1}{2} \left\| M_a^{-1} d \right\|^2 \\ &= \frac{1}{2} z_2^T z_2 + \frac{1}{2} \left\| M_a^{-1} d \right\|^2 \end{aligned} \tag{6.65}$$

Substitute (6.65) into (6.64), and one has

$$\begin{aligned}\dot{V} &= -c_1 z_1^T z_1 - c_2 z_2^T z_2 - z_2^T M_a^{-1} d \\ &\le -c_1 z_1^T z_1 - c_2 z_2^T z_2 + \frac{1}{2} z_2^T z_2 + \frac{1}{2}\left\|M_a^{-1} d\right\|^2 \\ &= -c_1 z_1^T z_1 - \left(c_2 - \frac{1}{2}\right) z_2^T z_2 + \frac{1}{2}\left\|M_a^{-1} d\right\|^2\end{aligned} \tag{6.66}$$

Because M_a^{-1} and d are both bounded, a new constant d_m is set, which satisfies

$$d_m \ge \frac{1}{2}\left\|M_a^{-1} d\right\|^2 \tag{6.67}$$

Equation (6.66) can be rewritten as

$$\begin{aligned}\dot{V} &\le -c_1 z_1^T z_1 - \left(c_2 - \frac{1}{2}\right) z_2^T z_2 + d_m \\ &\le -c_3 V + d_m\end{aligned} \tag{6.68}$$

where $0 < c_3 \le \min\left\{2c_1, 2\left(c_2 - \frac{1}{2}\right)\right\}$.

From the above formula, one has

$$V(t) \le V(0)e^{-c_3 t} + \frac{d_m}{c_3}\left(1 - e^{-c_3 t}\right) \tag{6.69}$$

$$\|z_1\| \le \sqrt{2V} \le \sqrt{2V(0)e^{-c_3 t} + \frac{2d_m}{c_3}\left(1 - e^{-c_3 t}\right)} \tag{6.70}$$

Considering that the filter is stable [2], that is, if the signal input to the filter is bounded, then the signal output by the filter is also bounded. The difference between the signals before and after filtering is bounded. Then it can be assumed that $\|x_{2c} - x_{2d}\| \le \varepsilon_1$, $\|\tau - \tau_d\| \le \varepsilon_2$, and $\left\|M_a^{-1} Q\right\| \le \varepsilon_3$.

Next, the convergence and stability of the compensators are considered, and the candidate Lyapunov function for ξ_1 and ξ_2 is selected as follows

$$V_\xi = V_{\xi 1} + V_{\xi 2} = \frac{1}{2}\xi_1^T \xi_1 + \frac{1}{2}\xi_2^T \xi_2 \tag{6.71}$$

Considering the stability of $V_{\xi 2}$, take its derivative with respect to time and substitute $\dot{\xi}_2$, one has

$$\dot{V}_{\xi 2} = +\xi_2^T \dot{\xi}_2 = -c_2 \xi_2^T \xi_2 + \xi_2^T M_a^{-1} Q\left(\tau_a - \tau_d\right) \tag{6.72}$$

Considering the boundedness of $M_a^{-1} Q(\tau_a - \tau_d)$, it can be known that

$$\xi_2^T M_a^{-1} Q(\tau_a - \tau_d) \le \left\| \xi_2^T M_a^{-1} Q(\tau_a - \tau_d) \right\| \le \varepsilon_2 \varepsilon_3 \left\| \xi_2 \right\| \tag{6.73}$$

Substitute (6.73) into (6.72), one has

$$\begin{aligned} \dot{V}_{\xi 2} &\le - c_2 \xi_2^T \xi_2 + \varepsilon_2 \varepsilon_3 \left\| \xi_2 \right\| \\ &\le - c_4 V_{\xi 2} + \varepsilon_4 \sqrt{V_{\xi 2}} \end{aligned} \tag{6.74}$$

where $c_4 = 2c_2$ and $\varepsilon_4 = \sqrt{2}\varepsilon_2\varepsilon_3$. Therefore, according to (6.74), the compensation signal ξ_2 satisfies the following bound,

$$\left\| \xi_2 \right\| \le \varepsilon_{\xi 2} \tag{6.75}$$

where $\varepsilon_{\xi 2} = \frac{2\varepsilon_4}{c_4}\left(1 - e^{-\frac{1}{2}c_4 t}\right)$.

Then consider the stability of $V_{\xi 1}$, take its derivative with respect to time and substitute $\dot{\xi}_1$, one has

$$\begin{aligned} \dot{V}_{\xi 1} = \xi_1^T \dot{\xi}_1 &= -c_1 \xi_1^T \xi_1 + \xi_1^T R_1 (x_{2c} - x_{2d}) + \xi_1^T R_1 \xi_2 - \kappa_R \xi_1^T R_1 \xi_1 \\ &= - c_1 \xi_1^T \xi_1 - \xi_1^T R_1 (x_{2d} - x_{2c} - \xi_2) - \kappa_R \xi_1^T R_1 \xi_1 \end{aligned} \tag{6.76}$$

Considering that $(x_{2d} - x_{2c} - \xi_2)$ is bounded, assuming $||x_{2d} - x_{2c} - \xi_2|| \le \varepsilon_1 + \varepsilon_{\xi 2} = \varepsilon_5$, taking into account the inequality $-aX^T M X - X^T M Y \le \frac{1}{4a} Y^T M Y$, where X, M, and Y are the vectors or matrices with matching dimensions, M is positive definite, a is a positive number, then, by (6.76), one has

$$\dot{V}_{\xi 1} \le -2c_1 V_{\xi 1} + \frac{1}{4\kappa_R}\varepsilon_5^2 \tag{6.77}$$

Then ξ_1 is bounded, there is

$$V_{\xi 1} \le V_{\xi 1}(0) e^{-2c_1 t} + \frac{\varepsilon_5^2}{8\kappa_R c_1}\left(1 - e^{-2c_1 t}\right) \tag{6.78}$$

$$\|\xi_1\| \le \sqrt{2V_{\xi 1}} \le \sqrt{2V_{\xi 1}(0) e^{-2c_1 t} + \frac{\varepsilon_5^2}{4\kappa_R c_1}\left(1 - e^{-2c_1 t}\right)} \triangleq \varepsilon_{\xi 1} \tag{6.79}$$

From $z_1 = \sigma - \xi_1$, one has

$$\begin{aligned} \|\sigma\| &\le \|z_1\| + \|\xi_1\| \\ &\le \sqrt{2V(0) e^{-c_3 t} + \frac{2d_m}{c_3}(1 - e^{-c_3 t})} + \varepsilon_{\xi 1} \end{aligned} \tag{6.80}$$

Therefore, σ is SGUUB. According to (2.58), (2.60), and Theorems 2.1 and 2.2, the spacecraft closed-loop system attitude tracking error q_{ev} is SGUUB. Moreover, q_{ev} satisfies the performance constraints (2.55), and the practical settling time that the attitude tracking errors converge to the predefined steady-state error boundary is less than the predefined maximum settling time t_{ps}. By the command filters, the rate, amplitude, and bandwidth of the control command or signal can be sufficiently constrained. □

6.4 Simulation Analysis

6.4.1 Simulation Parameters

This subsection first presents the initial conditions and parameters for the simulation. Considering the background of the on-orbit service spacecraft mission, although this study takes attitude motion as the research object, this section still gives relevant information about the spacecraft's orbit.

Considering that the spacecraft is very close to the target spacecraft in the short-range on-orbit service mission. Therefore, the orbital parameters of the spacecraft to be controlled are also shown in Table 6.1. Next, consider the initial parameters of the spacecraft as shown in Table 6.2. Without loss of generality, the parameters of the Tables 6.1 and 6.2 can be changed according to specific tasks.

Next, consider the desired attitude of the on-orbit service spacecraft. In the numerical simulation of this study, the spacecraft is considered to track the attitude of the

Table 6.1 The orbital parameters of the target spacecraft

Parameters	Values
Semimajor axis	6.85×10^6 m
Eccentricity	0.0073
Inclination	60 deg
Argument of periapsis	0 deg
Longitude of the ascending node	0 deg
Mean anomaly	0 deg

Table 6.2 The initial attitude parameters of the spacecraft

Parameters	Values
q_{v0}	$[0.6, 0.4, -0.3]^T$ rad
ω_{a0}	$[0, 0, 0]^T$ rad/s
J_s	$\begin{bmatrix} 100 & -10 & 5 \\ -10 & 120 & -3 \\ 5 & -3 & 100 \end{bmatrix}$ $\text{kg} \cdot \text{m}^2$

target spacecraft, and the attitude of the target spacecraft is the same as the settings in Sect. 4.4.1.

$$q_{dv} = \begin{bmatrix} (r_{li-23} - r_{li-32})/(4q_{d0}) \\ (r_{li-31} - r_{li-13})/(4q_{d0}) \\ (r_{li-12} - r_{li-21})/(4q_{d0}) \end{bmatrix} \tag{6.81}$$

In the simulation of this chapter, the control input maximum and minimum values of the spacecraft attitude actuator are considered to be limited to ± 1.1 N · m. Simultaneously, through the command filter, this chapter actively limits the actual control input amplitude to $[-1, 1]$ N · m. The active limit amplitude is smaller than the actual input amplitude, mainly to leave a certain margin.

According to the spacecraft attitude tracking model in Sect. 2.3, the tracking error quaternion and the angular velocity tracking error can be obtained. To demonstrate the performance of the proposed controller and to facilitate comparison with similar controllers, the error-index is defined as $I_a = ||q_{ev}|| + ||\omega_e||$. To illustrate the energy consumption of the proposed controller, the energy consumption index $E_a = \sum_{i=1}^{n} \int_0^t |\tau_{ai}(t)|^2 dt$ is defined as in [9]. This index describes the electrical or chemical energy consumed by actuators such as thrusters or reaction flywheels during the stabilization of an attitude closed-loop system. The parameter selection basis for the controller proposed in this study is the same as in the previous chapter.

6.4.2 *Performance and Comparison*

6.4.2.1 Simulation Results of Controller (6.12)

This section presents the parameters selected in the simulation of the controller (6.12) in Sect. 6.2, as shown in Table 6.3, where c_1 and c_2 are controller gain parameters. The magnitude limits in the command filter are ± 1 N · m. Considering the external disturbances to the spacecraft, we take the disturbance variable as follows [10],

$$d_a = \begin{bmatrix} 0.24\sin(0.5t) + 1.44\,\text{rand}_a(t) \\ 0.16\sin(0.5t + \pi/2\,) + 0.96\,\text{rand}_a(t) \\ 0.16\sin(0.5t + \pi/3) + 0.96\,\text{rand}_a(t) \end{bmatrix} \text{N} \cdot \text{m} \tag{6.82}$$

Table 6.3 The initial parameters of the proposed controller (6.12)

Param	Value	Param	Value
c_1	0.18	c_2	2

where $\mathrm{rand}_a(t)$ represents a zero-mean Gaussian random signal, and the variance is set to 0.01. It should be pointed out that, to fully demonstrate the disturbance rejection ability of the controller proposed, the disturbance selected in the simulation is larger than the practical situation, and the selected value is still within the reasonable range in orbit.

In this chapter, according to the traditional command filtering backstepping algorithm, the unconstrained command-filter-based spacecraft attitude asymptotic control algorithm is designed. This algorithm is used as the basic algorithm of the subsequent controllers. Therefore, this section only shows the characteristics of this controller and does not compare the algorithms. The subsequent controllers will be compared with other algorithms.

Figures 6.1, 6.2, 6.3, 6.4, 6.5, 6.6, 6.7, 6.8, 6.9 and 6.10 is the simulation of the controller (6.12). This controller is designed for the application of traditional command filtering backstepping methods to spacecraft attitude tracking. It serves as the technical premise for the subsequent design of the controller.

Figures 6.1 and 6.2 are the simulation results of the auxiliary variables z_1 and z_2 for the spacecraft attitude tracking error and the compensator in the command filter backstepping method without and with considering external disturbances. It can be seen from the curves that the auxiliary variables can converge, and when there are bounded disturbances, the auxiliary variables converge to the neighborhood of their equilibrium point. Figures 6.3 and 6.4 are the simulation diagrams of the compensator variables without and with considering external disturbances, respectively. It can be seen that the compensator variables can converge to the equilibrium point when the disturbance is not considered and converge to the neighborhood of their equilibrium point when the disturbance is considered. Based on the auxiliary and compensator variables, it can be seen that the spacecraft attitude tracking error also converges

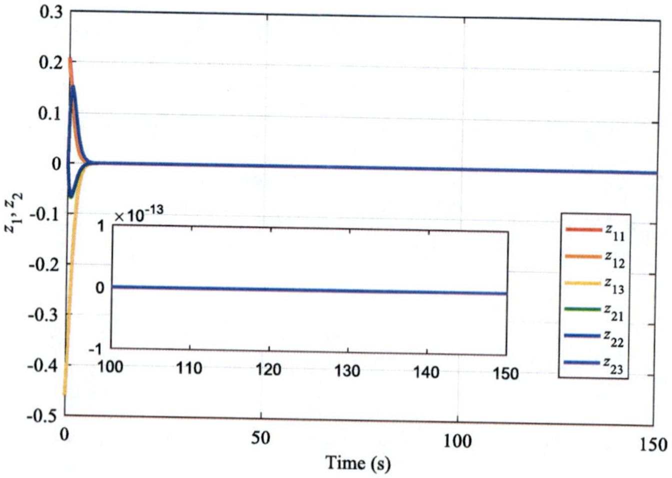

Fig. 6.1 The controller auxiliary variables z_1 and z_2 without disturbances

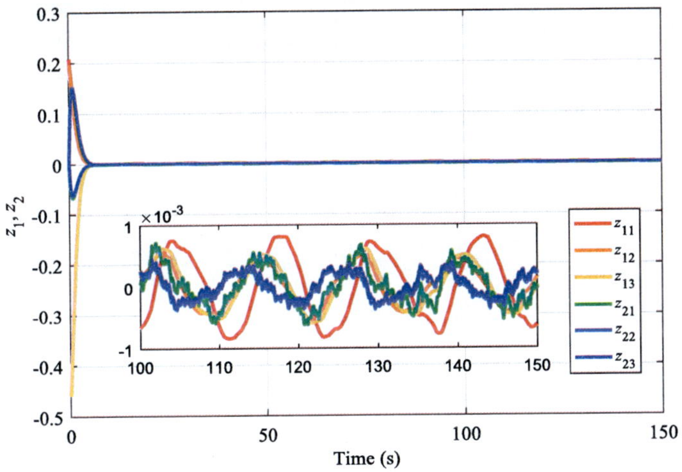

Fig. 6.2 The controller auxiliary variables z_1 and z_2 with disturbances

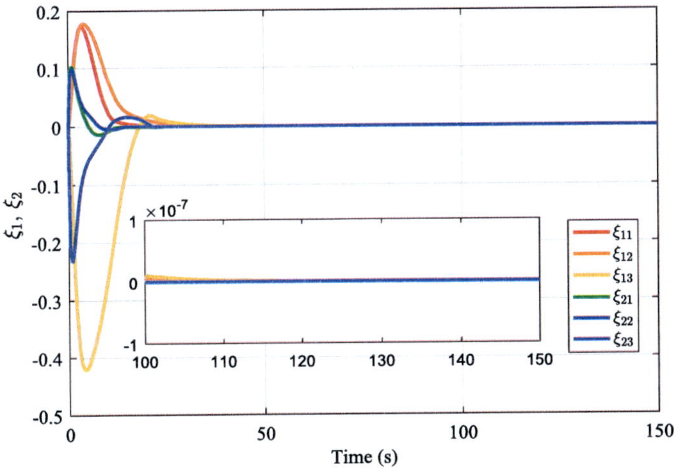

Fig. 6.3 The command filter compensators ξ_1 and ξ_2 without disturbances

to the equilibrium point or the neighborhood of the equilibrium point, as shown in Fig. 6.5. Figure 6.6 shows the spacecraft attitude angular velocity tracking error ω_e under the controller (6.12), which can also converge to the equilibrium point or the neighborhood of the equilibrium point.

Figures 6.7 and 6.8 are the control input τ_a of the spacecraft attitude tracking model and the expected control command τ_d designed by theory, respectively. After τ_d passes through the command filter, the system control input τ_a is obtained. It can be seen from the simulation diagram that the command filter significantly adjusts the

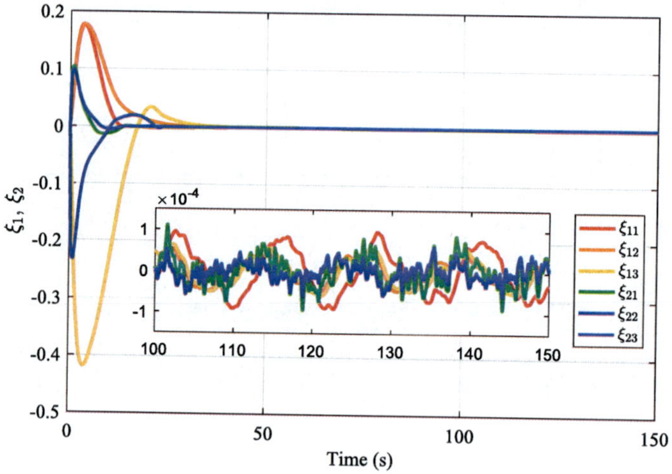

Fig. 6.4 The command filter compensators ξ_1 and ξ_2 with disturbances

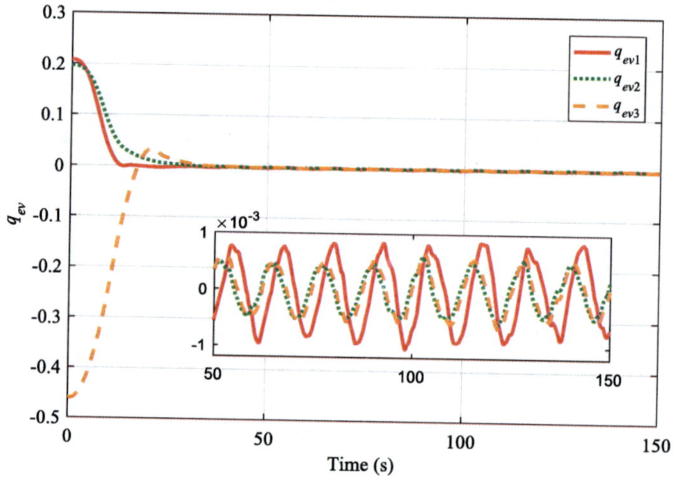

Fig. 6.5 The spacecraft attitude tracking error q_{ev}

amplitude of the control command signal, so that the control input of the spacecraft conforms to the actual situation of the spacecraft. As a comparison, the amplitude of the control input designed by the theory is too large, which is unsuitable for the actual spacecraft actuator. Figure 6.9 is the energy consumption of the controller, and Fig. 6.10 is the external disturbance considered in the numerical simulation.

Through numerical simulation, it is fully demonstrated that the closed-loop system steady-state error converges under the action of the controller in Sect. 6.2. It also shows that the command filter has a strong ability to adjust the theoretically designed

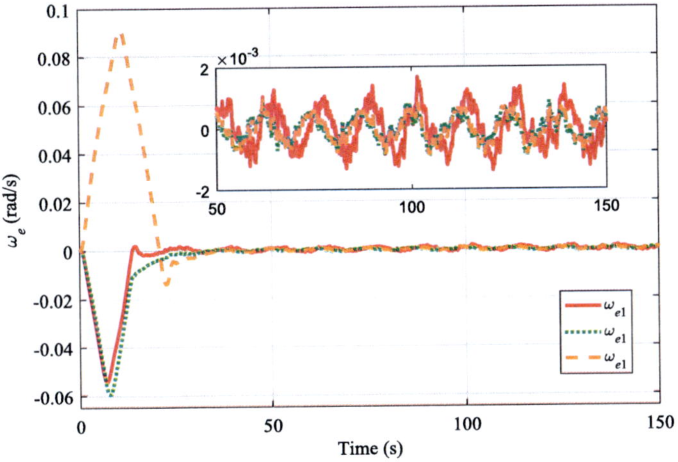

Fig. 6.6 The angular velocity tracking error ω_e

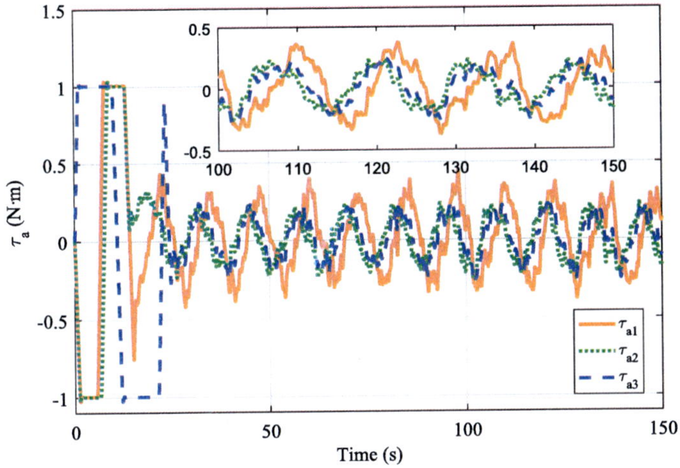

Fig. 6.7 The spacecraft attitude control input τ_a

control command τ_d, which makes the obtained actual control input τ_a more suitable for the practical spacecraft attitude control application.

6.4.2.2 Simulation Results of Controller (6.56)

This section presents the parameters selected in the simulation of the controller (6.56) in Sect. 6.3, as shown in the following table.

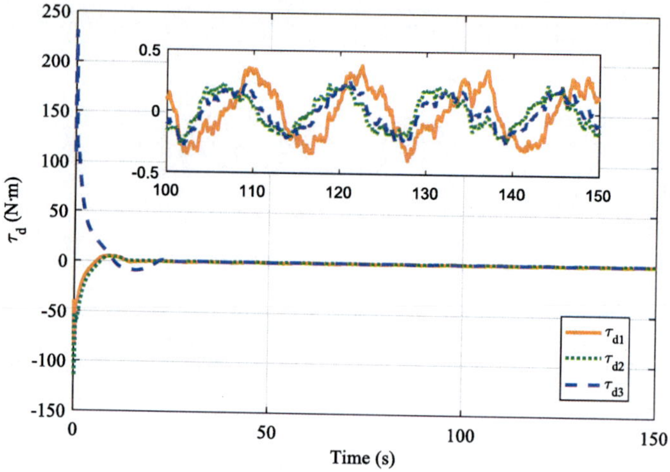

Fig. 6.8 The control command τ_d before filtering

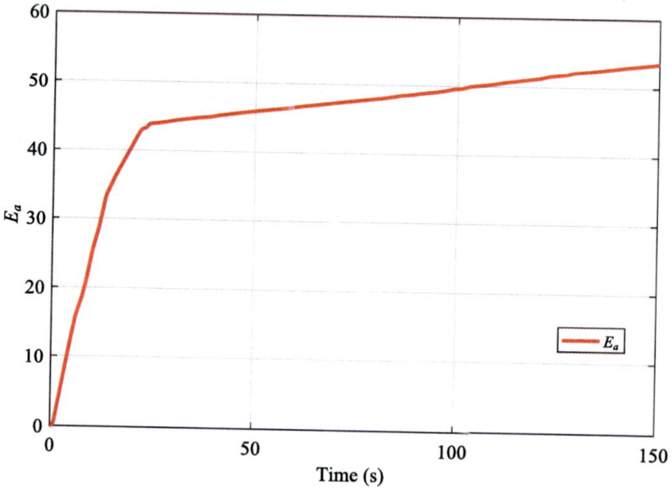

Fig. 6.9 The energy consumption

In Table 6.4, the parameter t_{ps} is the predefined maximum settling time. ρ_0 is the maximum initial value of the predefined performance boundary, which needs to be larger than the initial error quaternion. Considering that all quaternion elements are less than or equal to 1, thus, ρ_0 is taken as 1. ρ_t and ε_ρ provide steady-state error bounds. δ_i is about whether to limit the overshoot. c_1 and c_2 are the controller gain parameters. The magnitude of the saturation functions in the command filter is limited to ± 1 N · m.

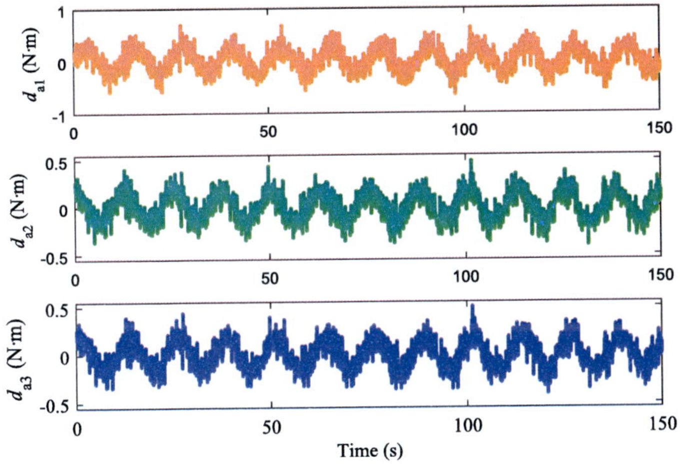

Fig. 6.10 The external disturbances

Table 6.4 The initial parameters of the proposed controller (6.56)

Param	Value	Param	Value	Param	Value
ρ_0	1	t_{ps}	50s	c_1	0.15
ρ_t	0.02	δ_i	1	c_2	1
ε_ρ	0.01	κ_R	0.001		

Considering the external disturbances to the spacecraft, we take the disturbance variable as follows [10],

$$d_a = \begin{bmatrix} 0.08\sin(0.5t) + 0.5\,\text{rand}_a(t) \\ 0.08\sin(0.5t + \pi/2\,) + 0.5\,\text{rand}_a(t) \\ 0.08\sin(0.5t + \pi/3) + 0.5\,\text{rand}_a(t) \end{bmatrix} \text{N}\cdot\text{m} \tag{6.83}$$

where $\text{rand}_a(t)$ represents a zero-mean Gaussian random signal, and the variance is set to 0.01. It should be pointed out that, in order to demonstrate the disturbance rejection ability of the controller proposed fully, the disturbance selected in the simulation is larger than the practical situation, and the selected value is still within the reasonable range in orbit.

The controller proposed in this chapter is compared with existing state-of-the-art controllers. For the convenience of description, this section labels the proposed controller (6.56) in Sect. 6.3 as controller 'A'. Label the asymptotic stabilization control of spacecraft attitude based on the command filter without performance constraints (6.12) proposed in Sect. 6.2 as controller 'B'. Mark the command-filter-based algorithm in [4] as controller 'C'. Label the spacecraft attitude performance constrained asymptotic stabilization controller (3.16) in Sect. 3.2 as controller 'D'. Label the tra-

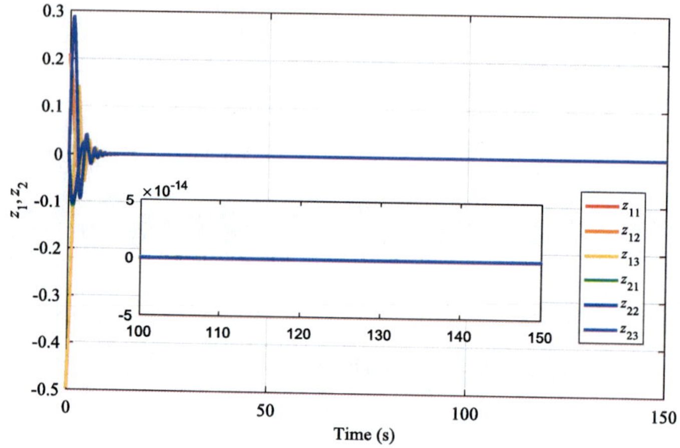

Fig. 6.11 The controller auxiliary variables z_1 and z_2 without disturbances

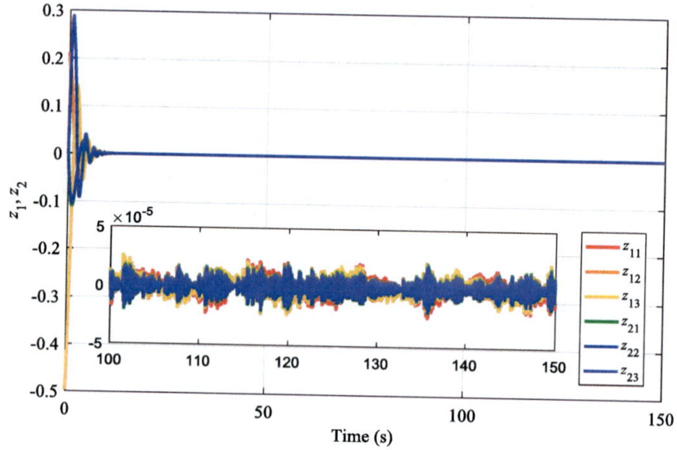

Fig. 6.12 The controller auxiliary variables z_1 and z_2 with disturbances

ditional prescribed performance controller in [11] as controller 'E'. In this section, the comparative simulation of the above controllers will be carried out to illustrate the effectiveness and advantages of this study. In the simulation diagram and analysis, the symbols 'A', 'B', 'C', 'D', and 'E' represent the above five controllers. Similar to the previous chapter, controller 'E' uses an accurate model.

Figures 6.11, 6.12, 6.13, 6.14, 6.15, 6.16, 6.17, 6.18, 6.19, 6.20, 6.21, 6.22, 6.23, 6.24, 6.25, 6.26, 6.27, 6.28, 6.29 and 6.30 are the comparative simulation results of the controller (6.56) in Sect. 6.3. Consider the controller itself first. Figures 6.11 and 6.12 show the simulation results of the auxiliary variables z_1 and z_2 for the spacecraft attitude tracking error and the compensator in the command filter backstepping

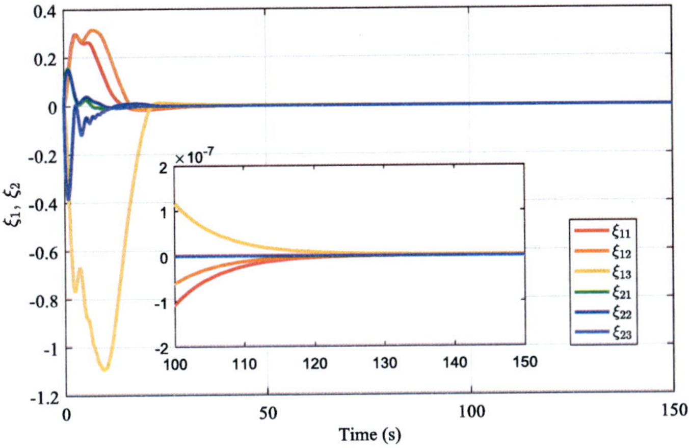

Fig. 6.13 The command filter compensators ξ_1 and ξ_2 without disturbances

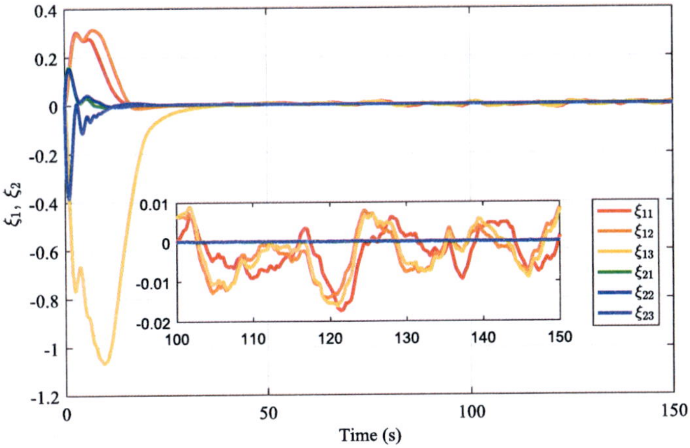

Fig. 6.14 The command filter compensators ξ_1 and ξ_2 with disturbances

method under performance constraints without and with considering external disturbances. It can be seen from the curve that the auxiliary variables can converge. When there are bounded disturbances, the auxiliary variables converge to the neighborhood of their equilibrium point. Figures 6.13 and 6.14 are the simulation diagrams of the compensator variables without and with considering external disturbances, respectively. It can be seen that the compensator variables can converge to their equilibrium point when the disturbance is not considered and converge to the neighborhood of the equilibrium point when the disturbance is considered.

Figures 6.15, 6.16 and 6.17 show the closed-loop system tracking error curve under the action of the five controllers 'A', 'B', 'C', 'D', and 'E'. In contrast, the per-

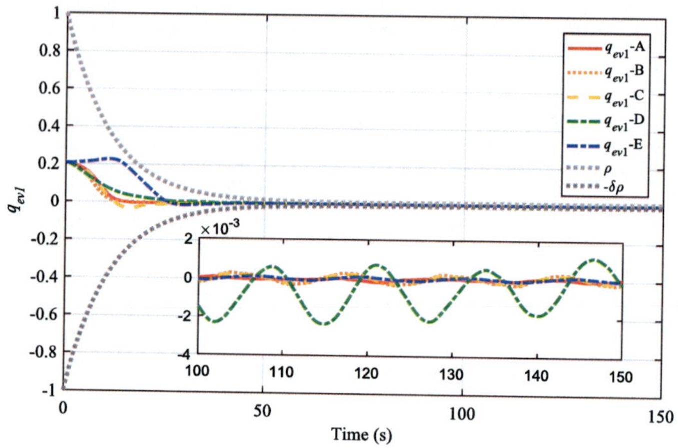

Fig. 6.15 The spacecraft attitude tracking error q_{ev1}

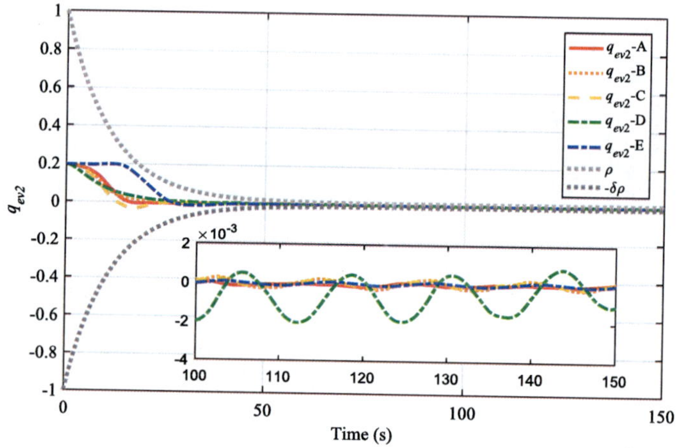

Fig. 6.16 The spacecraft attitude tracking error q_{ev2}

formance guaranteed command-filter-based controller proposed in Sect. 6.3 has better transient and steady-state performance and satisfies the performance constraints. Specifically, the proposed controller has performance constraints compared to the controllers 'B' and 'C' obtained by the traditional command filtering backstepping method. As shown in Fig. 6.17, the tracking error does not meet the performance constraints under the action of controller 'C'. Both controllers can meet the performance constraints compared to controller 'D'. Controller 'A' has a smaller steady-state error, is more suitable for the control input of spacecraft, and has stronger disturbance rejection capability. Controller 'A' has control inputs more suitable for the spacecraft than controller 'E'. Figures 6.18, 6.19 and 6.20 show the tracking error curves of the

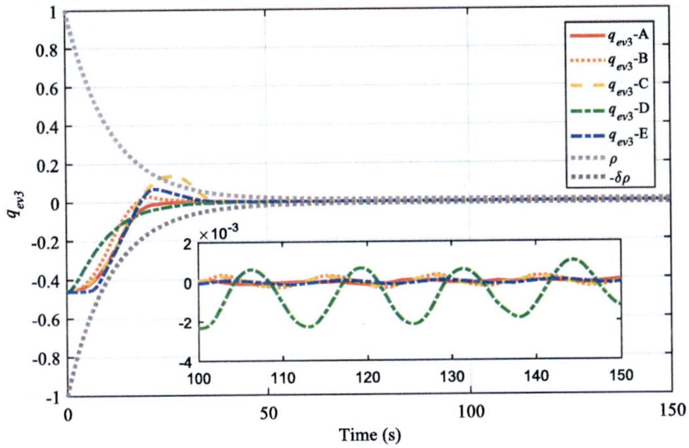

Fig. 6.17 The spacecraft attitude tracking error q_{ev3}

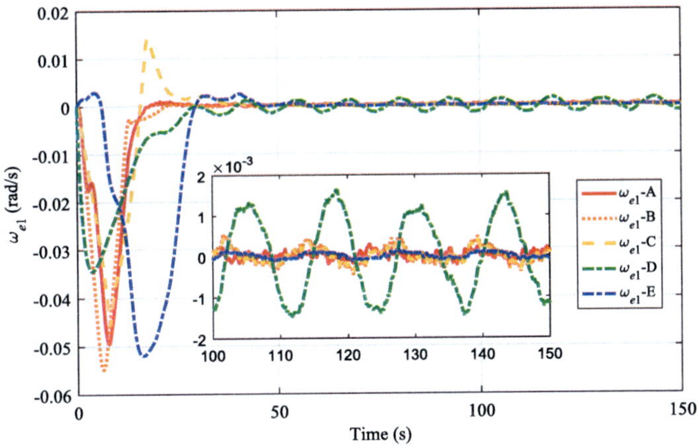

Fig. 6.18 The angular velocity tracking error ω_{e1}

spacecraft attitude angular velocity under the action of the five controls, 'A', 'B', 'C', 'D' and 'E'. It can be seen that the controller proposed has similar or better transient and steady-state performance than similar controllers, that is, the angular velocity amplitude changes slightly, and the process is smooth.

Figures 6.21 and 6.22 show the spacecraft attitude error index $I_a = ||q_{ev}|| + ||\omega_e||$ and the energy consumption E_a. The comparison shows that the controller in Sect. 6.3 has a faster convergence rate and smaller steady-state error. Moreover, it can be seen from the simulation curves that the controller proposed has smaller energy consumption.

Figures 6.23 and 6.24 are the control input τ_a of the spacecraft attitude tracking model and the expected control command τ_d designed by theory, respectively. After

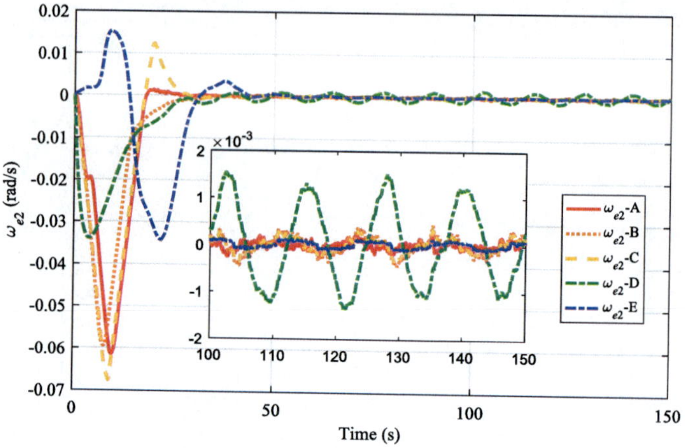

Fig. 6.19 The angular velocity tracking error ω_{e2}

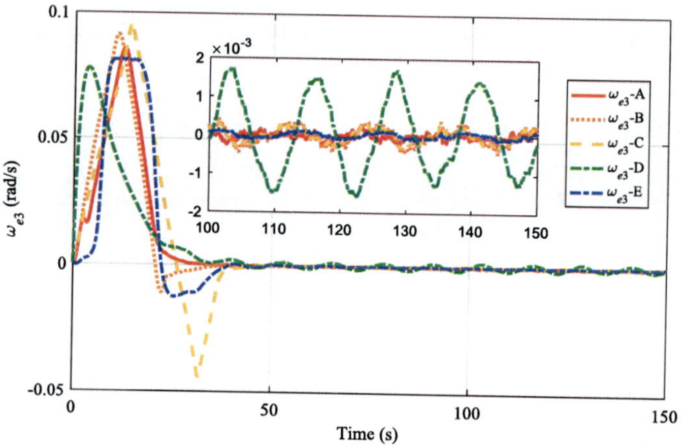

Fig. 6.20 The angular velocity tracking error ω_{e3}

τ_d passes through the command filter, the system control input τ_a is obtained. It can be seen from the simulation diagram that the command filter significantly adjusts the amplitude and frequency of the control command signal, so that the control input τ_a of the spacecraft conforms to the practical situation of the spacecraft. In contrast, the amplitude of the theoretically designed control input τ_d is too large, which is unsuitable for the practical spacecraft actuator. By comparison, the control input amplitude of the proposed controller is more reasonable, and the signal is smoother, which shows the important role of the command filter, that is, to adjust the amplitude, rate, and bandwidth of the command signal τ_d. Since the spacecraft control signal is generally not a high-frequency signal, the main function of the command filter

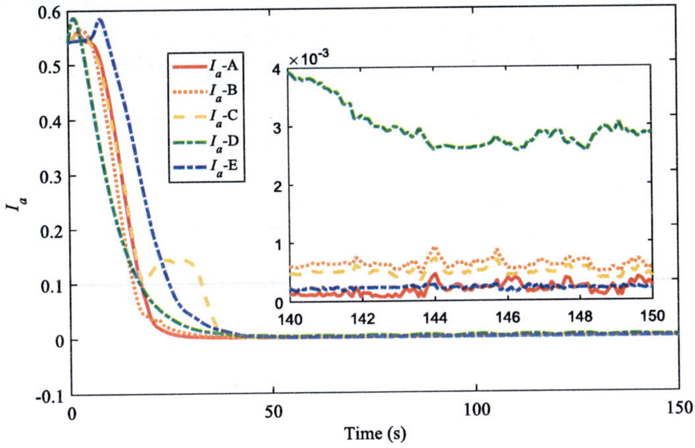

Fig. 6.21 The error index I_a

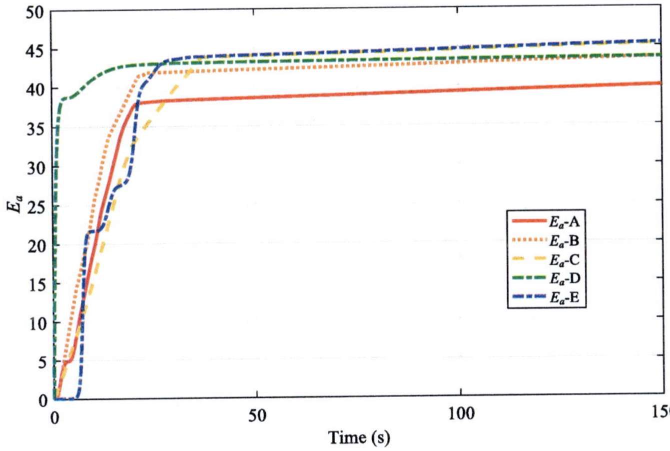

Fig. 6.22 The energy consumption

in this controller is to limit the actual control input amplitude. This also solves the problem of control input saturation.

Regarding the parameter damping ratio η of the filter itself and the natural frequency ω_n, the numerical simulation is shown in Figs. 6.25 and 6.26. When the damping ratio $\eta = 1$ and the natural frequency $\omega_n \geq 10$, the parameter $I_\xi = ||\xi_1|| + ||\xi_2||$ changes slightly. When the natural frequency $\omega_n = 100$ and the damping ratio $\eta \in (0, 1]$, the parameter $I_\xi = ||\xi_1|| + ||\xi_2||$ changes slightly. Therefore, the values can be taken according to these, such as $\eta = 1$ and $\omega_n \geq 10$.

In order to further illustrate the effect of the performance constraint functions with the predefined maximum settling time parameter on the error response of the closed-

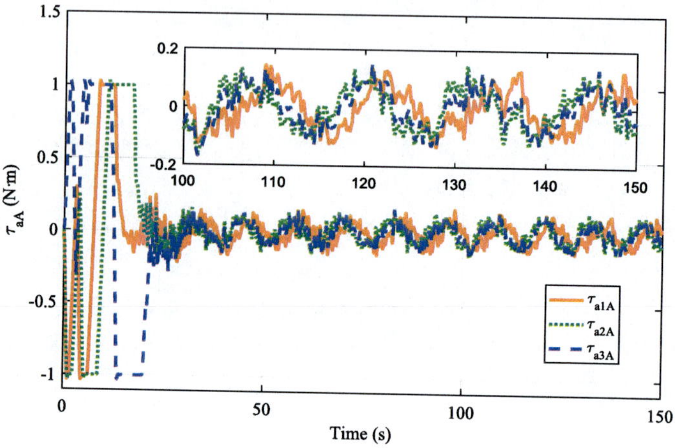

Fig. 6.23 The control input τ_a of 'A'

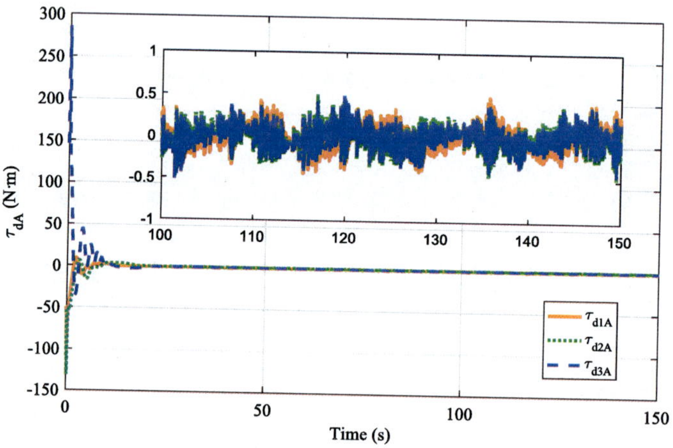

Fig. 6.24 The control command τ_d of 'A' before filtering

loop system in this controller, this section adds the comparison simulations about performance constraint parameters, as shown in Figs. 6.27, 6.28 and 6.29. Figure 6.27 considers the constraint effect of different maximum settling time parameters t_{ps} on the tracking error. It can be seen from the simulation that the smaller the maximum settling time parameter is, the greater the error convergence rate is. Moreover, the error will converge to the predefined steady-state interval in a shorter time. Figure 6.28 considers the constraint effect of different steady-state error parameters ρ_t on the tracking error. It can be seen from the simulation that even if there are external disturbances when the predefined steady-state error parameter is smaller, the steady-state error of the closed-loop system is smaller. Figure 6.29 considers the constraint effect of different overshoot parameters δ on the tracking error. It can be seen from

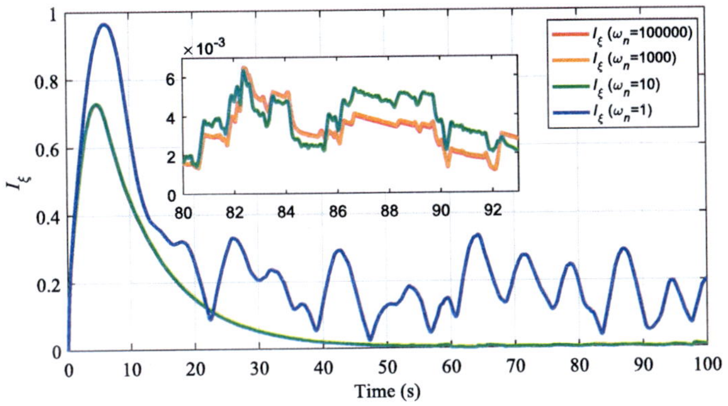

Fig. 6.25 When the damping ratio $\eta = 1$, $I_\xi = ||\xi_1|| + ||\xi_2||$ under different natural frequency ω_n

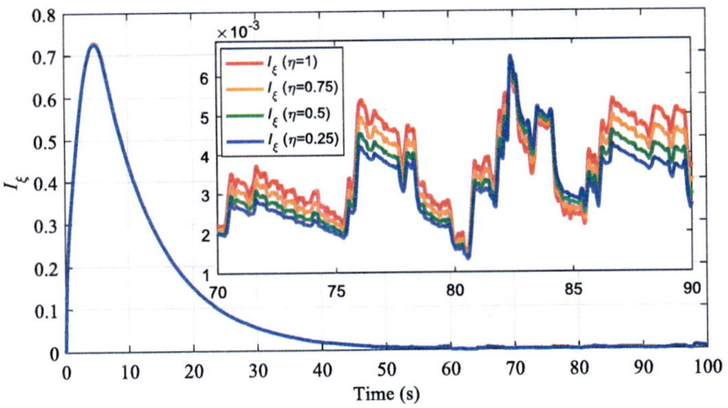

Fig. 6.26 When the natural frequency $\omega_n = 100$, $I_\xi = ||\xi_1|| + ||\xi_2||$ under the different damping ratio η

the simulation that the overshoot parameter has obvious constraints on the overshoot. When $\delta = 0$, there is no overshoot in the closed-loop system, but the steady-state error will increase. Finally, Fig. 6.30 presents the external disturbances considered in the simulation.

Through numerical simulation, it is fully demonstrated that the closed-loop system tracking errors converge under the action of the controller proposed in Sect. 6.3. It shows that the closed-loop system can obtain better transient and steady-state performance by consuming less energy. Furthermore, the performance is guaranteed. That is, the maximum settling time of the tracking error is guaranteed to be less than the predefined maximum settling time, the steady-state error is less than the predefined steady-state error, and the overshoot satisfies the constraints. In addition, the command filter has a strong ability to adjust the theoretically designed control

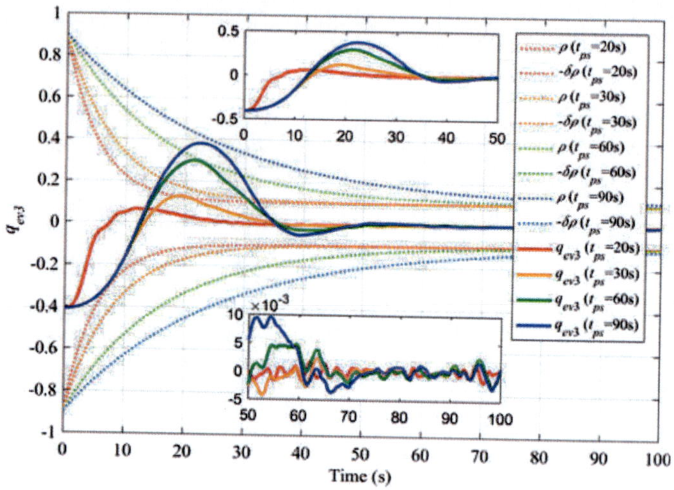

Fig. 6.27 The effect of different maximum settling time parameters t_{ps} on tracking error

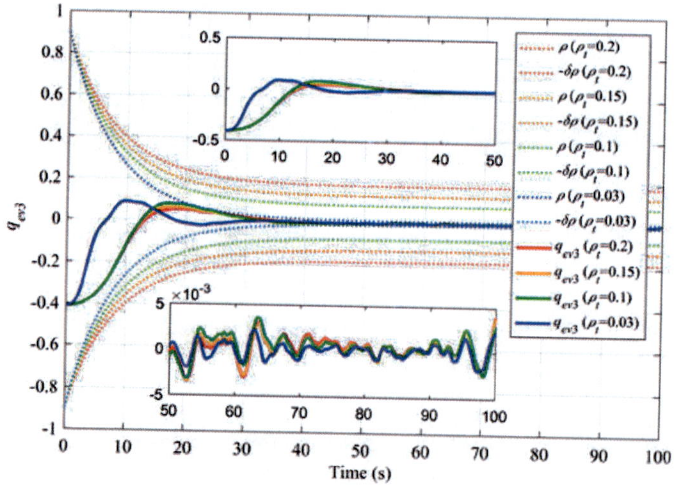

Fig. 6.28 The effect of different steady-state error parameters ρ_t on tracking error

command τ_d, which makes the obtained actual control input τ_a more suitable for the practical spacecraft attitude control application.

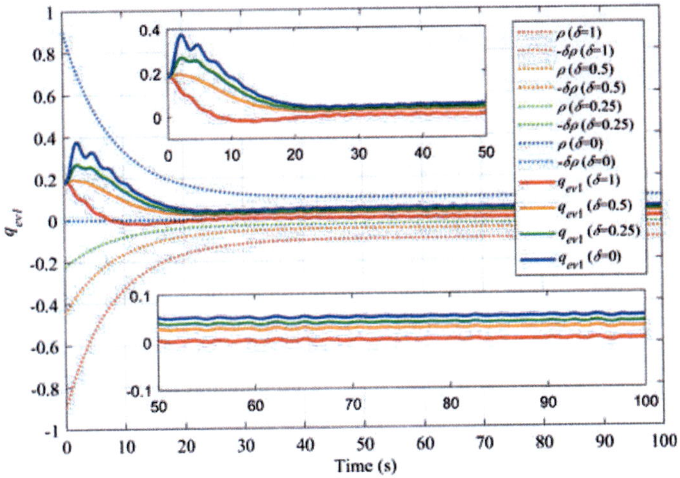

Fig. 6.29 The effect of different overshoot parameters δ on tracking error

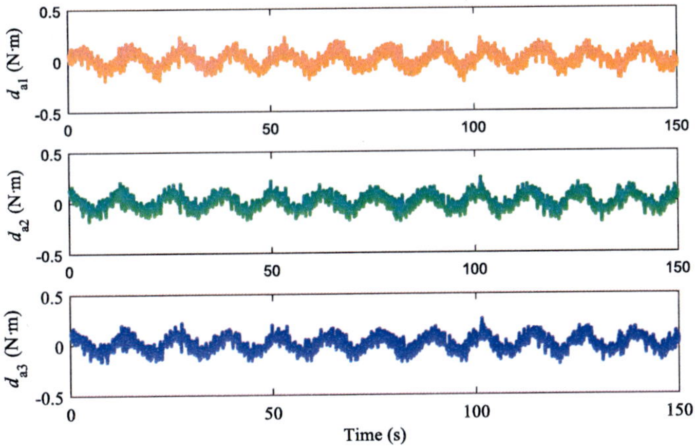

Fig. 6.30 The external disturbances

6.5 Conclusion

Based on the spacecraft performance guaranteed control with considering time parameters in Chaps. 3–5, this chapter further considers the limitation and adjustment of the control input. Specifically, the previous chapters propose the corresponding controllers for the closed-loop system to meet the performance constraints. Although the tracking errors of the spacecraft closed-loop system can meet the performance constraints, the magnitude/amplitude and constraints of the control input still need to be fully considered. However, the capabilities of the attitude control actuators of practical spacecraft are often limited, and even the rate and bandwidth of control commands are limited. Therefore, for the practice of aerospace engineering, the limited control input of the spacecraft attitude control system should also be considered. Therefore, this chapter applies command filters to tune and constrain the theoretically designed desired control commands. When the theoretically designed desired control command passes through the filter, the control input, whose magnitude/amplitude, rate, and bandwidth are adjusted and suitable for the practical actuators, is obtained. By refining the design of the command filter, this chapter first proposes asymptotic stable control of spacecraft based on the command filter. In this design, the controller suitable for spacecraft attitude tracking is designed according to the traditional command filtering backstepping method, which provides a design basis for performance guaranteed control. Furthermore, through the performance-constrained auxiliary variables and combined with the command filtering backstepping method, this chapter designs the spacecraft performance guaranteed control with the command filter. Under the effect of this controller, not only the tracking error of the spacecraft attitude closed-loop system can converge to the predefined steady-state error range within the predefined maximum settling time, but also the practical control input of the spacecraft has the capability of magnitude/amplitude constraints. Finally, the rationality and effectiveness of the control algorithms proposed in this chapter are proved through rigorous theoretical proof and numerical simulation analysis and comparison.

References

1. Farrell, J., Polycarpou, M., Sharma, M.: Adaptive backstepping with magnitude, rate, and bandwidth constraints: Aircraft longitude control. In: Proceedings of the 2003 American Control Conference, 2003., vol. 5, pp. 3898–3904. IEEE (2003)
2. Farrell, J.A., Polycarpou, M., Sharma, M., Dong, W.: Command filtered backstepping. IEEE Transactions on Automatic Control **54**(6), 1391–1395 (2009)
3. Sonneveldt, L., Chu, Q., Mulder, J.: Nonlinear flight control design using constrained adaptive backstepping. Journal of Guidance, Control, and Dynamics **30**(2), 322–336 (2007)
4. Wang, Y., Chen, W., Zhang, S., Zhu, J., Cao, L.: Command-filtered incremental backstepping controller for small unmanned aerial vehicles. Journal of Guidance, Control, and Dynamics **41**(4), 954–967 (2018)
5. Homayoun, B., Arefi, M.M., Vafamand, N., Yin, S.: Neuro-adaptive command filter control of stochastic time-delayed nonstrict-feedback systems with unknown input saturation. Journal of the Franklin Institute **357**(12), 7456–7482 (2020)
6. Xia, J., Zhang, J., Feng, J., Wang, Z., Zhuang, G.: Command filter-based adaptive fuzzy control for nonlinear systems with unknown control directions. IEEE Transactions on Systems, Man, and Cybernetics: Systems **51**(3), 1945–1953 (2019)
7. Qiu, J., Sun, K., Rudas, I.J., Gao, H.: Command filter-based adaptive nn control for mimo nonlinear systems with full-state constraints and actuator hysteresis. IEEE transactions on cybernetics **50**(7), 2905–2915 (2019)
8. Patel, R., Hafiz, F., Swain, A., Ukil, A.: Nonlinear excitation control of diesel generator: A command filter backstepping approach. IEEE Transactions on Industrial Informatics **17**(7), 4809–4817 (2020)

9. Jiang, B., Li, C., Ma, G.: Finite-time output feedback attitude control for spacecraft using "adding a power integrator" technique. Aerospace Science and Technology **66**, 342–354 (2017)
10. Zou, A.M.: Finite-time output feedback attitude tracking control for rigid spacecraft. IEEE Transactions on Control Systems Technology **22**(1), 338–345 (2013)
11. Bechlioulis, C.P., Rovithakis, G.A.: Robust adaptive control of feedback linearizable mimo nonlinear systems with prescribed performance. IEEE Transactions on Automatic Control **53**(9), 2090–2099 (2008)

Chapter 7
Spacecraft Finite-Time Performance Guaranteed Control with Command Filter

Abstract The attitude control system of the on-orbit service spacecraft has control design requirements such as precise motion and safety assurance. Moreover, the actuators in the existing system have practical constraints on the control input signals, such as amplitude, rate, and bandwidth. The performance guaranteed control algorithm based on the command filter can give the closed-loop system the ability to adjust and limit the control command signal while having the performance guaranteed capability. The command filter provides the closed-loop system with the ability to adjust the control input so that the actual control signal meets the control input design constraints required by the actuators. Considering the ability of the finite-time stability theory to accurately describe the settling time and the steady-state error boundary of the closed-loop system, the finite-time performance guaranteed algorithm based on the command filter is further proposed in this chapter. The finite-time stability design reduces the conservative of the performance guaranteed control algorithm. That is, it has a more accurate design value and the estimated settling time upper bound value. Simultaneously, the closed-loop system can converge to a more accurate and specific steady-state error range within the finite time. The spacecraft attitude tracking closed-loop system with desired performance and desired control input is finite-time stable and has strong disturbance suppression ability. Finally, the rigorous theoretical proof verifies the stability of the closed-loop system. The sufficient numerical simulation analysis and comparative analysis verify the rationality, effectiveness, and advantages of the proposed control algorithms.

7.1 Introduction

In Chaps. 3, 4 and 5, the performance guaranteed control algorithms are fully considered for asymptotically stable and finite-time stable closed-loop systems. In order to further adjust the control input/command signal in performance guaranteed control to make the control input more suitable for the practical spacecraft system, the performance guaranteed control algorithm based on the command filter (CF) is designed in Chap. 6, so that the closed-loop system can be asymptotically stable with the control input improvement.

Y. Gao and D. Li, *Spacecraft Maneuver with Performance Guaranteed*,
https://doi.org/10.1007/978-981-99-4653-2_7

Similar to the idea in Chap. 4, the finite-time method based on adding a power integrator (API) technique [1] can provide better transient and steady-state performance of the system. This has been analyzed in Sect. 4.1. For the algorithm design in Chap. 6, we can still use the ideas in Chap. 4 to improve the algorithms. This makes the related algorithms less conservative and more accurate in predicting the settling time. Considering the design of the command filter and the finite-time stability of the closed-loop system, and compared with the traditional command filtering backstepping method, the compensators on the command filter need to be redesigned to accommodate the finite-time stability of the closed-loop system.

In recent years, many scholars have paid attention to the design of finite-time command filtering algorithms and obtained some research results. Intuitively, the sign functions commonly used in finite-time theory are used in the design of command filters and compensators. In Ref. [2], the sign function is applied, and based on finite-time theory, finite-time command filter control is proposed, and at the same time, adaptive fault-tolerant control for uncertain nonlinear systems is further considered. Based on the neural networks, Ref. [3] designs the finite-time command filtering control algorithm for switched nonlinear systems. Among them, the sign function is applied to transform the compensators and the closed-loop system to have the property of finite-time stability. Reference [4] also adopts the first-order Levant differentiator as a filter for the fuzzy control. The corresponding finite-time command filtering output feedback controller is designed based on this. Based on the first-order Levant differentiator, Ref. [5] further considers the input saturation problem in the finite-time command filtering algorithm and designs the fuzzy controller. Based on finite-time command filtering control, Ref. [6] further considers the quantized input problem, Ref. [7] considers the uncertain fractional-order nonlinear systems problem, Ref. [8] considers event-triggered problems for the stochastic nonlinear systems, Ref. [9] considers this problem for the power system. Reference [10] considers the uncertain disturbance problem. In addition, the state-dependent switching method [11] and the immersion and invariance [12] are also applied to the design of the finite-time command filtering control algorithm.

The primary purpose of this chapter is to modify the performance guaranteed control based on the command filter with the finite-time approach. The algorithm is essentially the performance guaranteed control algorithm, with the advantages of the settling time design and transient and steady-state behavior constraints that the performance guaranteed method has. Synchronously, the command filter provides the control input adjustment for the algorithm, the same as in the previous chapter. In addition, finite-time stability makes the closed-loop system more descriptive, especially the actual settling time of the closed-loop system.

Based on the above motivation and analysis, in this chapter, we aim to design the performance guaranteed control based on the command filter that can stabilize the closed-loop system in a finite time. Specifically, the command filter has the ability to actively adjust the amplitude, rate, and bandwidth of the control command signals. By selecting the desired amplitude, rate, bandwidth, and other parameters of the control signal in the filter, the signal adjusted by the filter can meet the capability of the spacecraft actuators. It only needs to input the theoretically designed control command into the filter. That is, the filtered control command signal can be obtained. Using the filtered control signal as the control input of the closed-loop system, the closed-loop system that satisfies the performance guaranteed and the actuator/control input constraints can be obtained. Considering the finite-time stability of the closed-loop system, we need to make the compensators about the filters also stable in finite time, so compared to the previous chapter, we need to make some improvements. Combined with the specific situation of spacecraft attitude tracking, the mission objectives of this chapter are highlighted here:

Problem 7.1 Consider the spacecraft attitude tracking error model (2.28), and select predefined related parameters according to practical mission requirements, including predefined maximum settling time t_{ps}, the maximum steady-state error ρ_t, the maximum overshoot χ_t, design the performance guaranteed control such that:
(1) the conditions in Problem 3.1 are satisfied;
(2) the amplitude and rate constraints on the control input/command signals of the closed-loop system are satisfied;
(3) the closed-loop system is finite-time stable;
(4) the compensators on the command filter are finite-time stable;
(5) the closed-loop system has the property of disturbance suppression. That is, when strong external disturbances exist, the closed-loop system still satisfies the above performance constraints.

The control designs as described above have advantages, including: (1) the performance guaranteed control can provide the active design of the settling time upper bounds, the transient and steady-state performance constraints; (2) the command filter can adjust the amplitude, rate, and bandwidth of the control input signal to adapt to the constraints of the actual system actuators. (3) the finite-time stability may further reduce the conservative of the settling time upper bound that can be described, and further adjust the transient performance of the closed-loop system, etc.

This chapter is arranged as follows. Without considering disturbances, the finite-time control algorithm and compensators are designed and analyzed in Sect. 7.2. Further, the design of the control algorithm considering disturbances is presented in Sect. 7.3. Section 7.4 analyzes and proves the finite-time stability of closed-loop systems. Section 7.5 shows the numerical simulation and analysis of the control algorithms proposed in this chapter. Finally, Sect. 7.6 gives the conclusions of this chapter.

7.2 Control Design and Stability Analysis Without Disturbances

The constraint method of the performance guaranteed technique can make the closed-loop system state or tracking error converge to the desired range within the desired predefined time. Based on the traditional performance constraint framework with the predefined settling time, the range and performance of the tracking error can be limited. The system can only achieve asymptotic stability or bounded stability, which is more conservative and less descriptive. Therefore, the performance guaranteed algorithm based on the command filter designed in the previous chapter also has limitations. As an improvement and stability enhancement (bounded stability to finite time stability) of the algorithms in the previous chapter, to further accurately describe the system performance and also consider the performance guaranteed of the high precision control of spacecraft attitude, the performance-constrained control design based on a finite-time stable command filter is presented in this chapter. The re-consideration of the stability of the closed-loop system based on CF means that, compared with the control algorithm in the previous chapter, in this chapter, it is necessary to design both the finite-time stable controller and the finite-time stable command filter compensators so that the entire closed-loop system is finite-time stable. This is also one of the main innovations of this chapter. In this chapter, the finite-time controller is designed by the method of *adding a power integrator technique* (API). The controller without considering the external disturbances is designed first, and the controller with considering disturbance rejection is designed in the next section. In addition, it should be noted that this section simplifies the control structure based on the command filter. That is, only the command filter is used on the control signal side.

Before we start designing the controller, we define some calculation symbols as follows

$$\mathrm{sig}(x)^{\gamma} = [|x_1|^{\gamma_1}\,\mathrm{sign}(x_1), \ldots, |x_n|^{\gamma_n}\,\mathrm{sign}(x_n)]^T \tag{7.1}$$

$$x^{\gamma} = [{x_1}^{\gamma_1}, \ldots, {x_n}^{\gamma_n}]^T, |x|^{\gamma} = [|x_1|^{\gamma_1}, \ldots, |x_n|^{\gamma_n}]^T \tag{7.2}$$

where $\mathrm{sign}(\cdot)$ stands for a symbolic function, $\gamma = [\gamma_1, \ldots, \gamma_n]^T$, $1 < \gamma_1, \ldots, \gamma_n < 2$.

In this section, the controller design is first carried out for the nominal system (2.28) with $d = 0$, which does not consider the external disturbances. First, considering the relevant links of the performance guaranteed technique as in the previous chapters, we list the function terms and auxiliary systems required for performance constraints as follows. The definition of σ_i is the same as (2.60), if σ_i converges or is bounded, then the closed-loop system satisfies the performance constraints. To facilitate the derivation, we first find $\dot{\sigma}_i$, the same as the previous chapter, $r_i = |\partial\varphi^{-1}(\vartheta_i)/\partial\vartheta_i|$, $\dot{\vartheta}_i = \frac{2\dot{q}_{evi}(t)}{(1+\delta)\rho(t)} - \frac{2\dot{\rho}(t)q_{evi}(t)}{(1+\delta)\rho^2(t)}$, then $\dot{\sigma}_i = r_i\dot{\vartheta}_i$.

$$\dot{\sigma}_i = \frac{\partial\varphi^{-1}(\vartheta_i)}{\partial\vartheta_i}\dot{\vartheta}_i \tag{7.3}$$

$$\ddot{\sigma}_i = \dot{r}_i\dot{\vartheta}_i + r_i\ddot{\vartheta}_i = \alpha_i + \beta_i\ddot{q}_{evi}(t) \tag{7.4}$$

where

$$\begin{aligned}\alpha_i = \dot{r}_i\dot{\vartheta}_i &- \frac{2r_i\dot{q}_{evi}(t)\dot{\rho}_i(t)}{(1+\delta_i)\rho_i{}^2(t)} + \frac{4r_iq_{evi}(t)\dot{\rho}_i^2(t)}{(1+\delta_i)\rho_i{}^3(t)} \\ &- \frac{2r_i\ddot{\rho}_i(t)q_{evi}(t) + 2r_i\dot{\rho}_i(t)\dot{q}_{evi}(t)}{(1+\delta_i)\rho_i{}^2(t)}\end{aligned} \tag{7.5}$$

$$\beta_i = \frac{2r_i}{(1+\delta_i)\rho_i(t)} \tag{7.6}$$

By expressing (7.4) into a compact form, the available auxiliary systems for the spacecraft attitude tracking error to meet the performance constraints are as follows

$$\ddot{\sigma} = A + B\ddot{q}_{ev} \tag{7.7}$$

where $\sigma = [\sigma_1, \ldots, \sigma_n]^T$ $A = [\alpha_1, \ldots, \alpha_n]^T$ $B = \text{diag}(\beta_1, \ldots, \beta_n)$.

Substitute the system equation into the auxiliary system as follows

$$\ddot{\sigma} = A + B\ddot{q}_{ev} = A - BM_a^{-1}(C_a\dot{q}_{ev} + G_a) + BM_a^{-1}Q\tau_a - BM_a^{-1}d \tag{7.8}$$

where Q is defined as the previous section.

Same as the previous chapter, because the CF is designed in the controller, it is necessary to design parameters to compensate for the signal attenuation caused by filtering [13, 14]. Unlike the previous chapter, the compensator in this chapter needs to settle in the finite time. Likewise, two compensators are designed in this chapter, φ_1 and φ_2. Different from the previous chapter, the compensator φ_2 in this chapter is used to compensate for the error of the control signal passing through the filter. The compensator φ_1 does not do signal compensation but converts the signal of the compensator φ_2 to the controller Therefore, we design the following auxiliary variables,

$$z_1 = \sigma - \varphi_1 \tag{7.9}$$

$$\alpha = \dot{\sigma} - \varphi_2. \tag{7.10}$$

The derivative of (7.9) with respect to time is as follows

$$\begin{aligned}\dot{z}_1 &= \dot{\sigma} - \dot{\varphi}_1 \\ &= \alpha + \varphi_2 - \dot{\varphi}_1 - \kappa_1\varphi_1 + \kappa_1\sigma \\ &\quad - \kappa_1 z_1 - \kappa_2\,\text{sig}(\varphi_1)^{p_1} + \kappa_2\,\text{sig}(\sigma - z_1)^{p_1}\end{aligned} \tag{7.11}$$

where $\kappa_1 \geq \varrho^2/2$, ϱ and $\kappa_2 \geq 0$ are positive constants.

We design the finite-time compensator φ_1 as follows to use it as a first-order compensation signal.

$$\dot{\varphi}_1 = -\kappa_1 \varphi_1 - \kappa_2 \operatorname{sig}(\varphi_1)^{p_1} + \varphi_2 \tag{7.12}$$

where $p_1 \in (0, 1)$ is a power number. Also, $p_2 = (1 + p_1)/2$ is defined in advance to be used in convergence proof of φ_1.

In order to simplify the subsequent formula, we take some variables instead of complicated function terms as

$$\Psi_1 = \kappa_1 \sigma - \kappa_1 z_1 + \kappa_2 \operatorname{sig}(\sigma - z_1)^{p_1} \tag{7.13}$$

We can simplify (7.11) to

$$\dot{z}_1 = \Psi_1 + \alpha \tag{7.14}$$

Taking the derivative of (7.10) with respect to time, we have

$$\begin{aligned}
\dot{\alpha} &= \ddot{\sigma} - \dot{\varphi}_2 \\
&= A - BM_a^{-1}(C_a \dot{q}_{ev} + G_a) - BM_a^{-1} d + BM_a^{-1} Q \tau_a - \dot{\varphi}_2 \\
&= A - BM_a^{-1}(C_a \dot{q}_{ev} + G_a) - BM_a^{-1} d + BM_a^{-1} Q \tau_d \\
&\quad + BM_a^{-1} Q (\tau_a - \tau_d) - \dot{\varphi}_2 - \kappa_3 \varphi_2 - \kappa_4 \operatorname{sig}(\varphi_2)^{p_1} \\
&\quad + \kappa_3 (\dot{\sigma} - \alpha) + \kappa_4 \operatorname{sig}(\dot{\sigma} - \alpha)^{p_1}
\end{aligned} \tag{7.15}$$

where $\kappa_3 \geq 0$ and $\kappa_4 \geq 0$ are constants.

We take the second compensator φ_2 as follows and make some simplifications as before, and we can get

$$\dot{\varphi}_2 = -\kappa_3 \varphi_2 - \kappa_4 \operatorname{sig}(\varphi_2)^{p_1} + BM_a^{-1} Q (\tau_a - \tau_d) - \kappa_B B \varphi_2 \tag{7.16}$$

$$\Psi_2 = A - BM_a^{-1}(C_a \dot{q}_{ev} + G_a) + \kappa_B B \varphi_2 + \kappa_3 (\dot{\sigma} - \alpha) + \kappa_4 \operatorname{sig}(\dot{\sigma} - \alpha)^{p_1} \tag{7.17}$$

$$\dot{\alpha} = \Psi_2 + BM_a^{-1} Q \tau_d - BM_a^{-1} d \tag{7.18}$$

where κ_B is a constant. In (7.16), we find that φ_2 is able to compensate for $\tau_a - \tau_d$ accordingly, which is the signal loss caused by the command filter.

To obtain a system form suitable for applying the API technique to design the controller, we further perform a coordinate transformation as

$$z_2 = \dot{z}_1 = \Psi_1 + \alpha \tag{7.19}$$

Taking the derivative of (7.19), we have

$$\dot{z}_2 = \Psi_3 + BM_a^{-1}Q\tau_d - BM_a^{-1}d \tag{7.20}$$

where $\Psi_3 = \dot{\Psi}_1 + \Psi_2$.

Remark 7.1 There is a singularity problem in $\dot{\Psi}_1$ because the derivative of $\mathrm{sig}(\beta)^{p_1}$ is infinite when $\beta_i = 0$ and $\dot{\beta}_i \neq 0$ in $\mathrm{d}(\kappa_2 \mathrm{sig}(\beta)^{p_1})/\mathrm{d}t$, $\beta = \sigma - z_1$. To avoid this problem, the threshold value ϖ is set to judge singularity so that the following definition is given [15]

$$\frac{\mathrm{d}}{\mathrm{d}t}(\kappa_2\, \mathrm{sig}(\beta_i)^{p_1}) = \begin{cases} \kappa_2 p_1 |\beta_i|^{p_1-1}\dot{\beta}_i, & \text{if } |\beta_i| \geq \varpi,\ \dot{\beta}_i \neq 0 \\ \kappa_2 p_1 |\iota_i|^{p_1-1}\dot{\beta}_i, & \text{if } |\beta_i| < \varpi,\ \dot{\beta}_i \neq 0 \\ 0, & \text{if } \dot{\beta}_i = 0 \end{cases} \tag{7.21}$$

where ϖ and ι_i are both small positive constants.

Without considering external disturbances, we first design the finite-time controller as follows, and we will analyze the situation with disturbances later.

$$\tau_d = -\kappa_5 Q^{-1} M_a B^{-1}\left(z_2^{p_3} + \kappa_6^{p_3} z_1\right)^{2/p_3 - 1} - Q^{-1} M_a B^{-1}\Psi_3 \tag{7.22}$$

where $1 < p_3 = p_4/p_5 < 2$, p_4, p_5 are positive odd integers, κ_5 and κ_6 will be defined later. The following theorem is then proposed.

Theorem 7.1 *For the spacecraft attitude tracking error model (2.28) with a known smooth reference trajectory and without considering external disturbances, the control input τ_a can be obtained after τ_d passes the CF (2.71), and $||\tau_a - \tau_d||$ is bounded. If the controller τ_d is chosen as (7.22), and the command filter compensators are selected as (7.12) and (7.16), then z_1, z_2 can converge to the equilibrium point in finite time, and the finite settling time T_1 of z_1, z_2 satisfies*

$$T_1 \leq \frac{V(0)^{1-p_6/2}}{\kappa_9(1 - p_6/2\,)} < +\infty \tag{7.23}$$

where κ_9 and p_6 are defined later.

Proof We choose the candidate Lyapunov function about z_1 as

$$V_0 = \frac{1}{2} z_1^T z_1 \tag{7.24}$$

To stabilize V_0, taking the derivative of (7.24), we get

$$\dot{V}_0 = z_1^T \dot{z}_1 \tag{7.25}$$

To stabilize V_0, if $z_2 = \dot{z}_1 = -\kappa_6 z_1^{1/p_3}$, $\kappa_6 > 0$ can be achieved, which can stabilize z_1, then the ideal z_2^* can be designed as

$$z_2^* = -\kappa_6 z_1^{1/p_3} \tag{7.26}$$

It should be noted, in order to make z_1 converge within finite time, which also means that V_0 is stable, we can design $z_2 = \dot{z}_1 = -\kappa_6 z_1^{1/p_3}$. If this condition is satisfied, then z_1 is finite-time stable. However, this z_2 is the z_2 we expect, not the z_2 in the actual system, so we denote it $z_2^* = -\kappa_6 z_1^{1/p_3}$ as a desired virtual control variable. In order to show that under the proposed controller, such conditions can be satisfied, we need to analyze the following further so that z_2 in the actual system tracks z_2^*, that is, $z_2 \to z_2^*$ within finite time.

Further, we select the complete candidate Lyapunov function as follows

$$V = V_0 + \sum\nolimits_{i=1}^{n} V_i \tag{7.27}$$

$$V_i = \frac{1}{(2 - 1/p_3)\, 2^{1-1/p_3} \kappa_6^{1+p_3}} \int_{z_{2i}^*}^{z_{2i}} \left(s^{p_3} - z_{2i}^{*p_3}\right)^{2-1/p_3} \mathrm{d}s \tag{7.28}$$

By Propositions B1 and B2 in [16], $\int_{z_{2i}^*}^{z_{2i}} \left(s^{p_3} - z_{2i}^{*p_3}\right)^{2-1/p_3} \mathrm{d}s$ are differentiable and positive definite. It is worth noting that the spacecraft attitude system used is a vector form system. For the traditional API technique, we will make some adjustments. That is, we design V_i for each dimension and sum them as shown in (7.27) and (7.28). In this way, for the vector form system, the API technique can be used to analyze its finite-time stability.

To facilitate the derivation, a new auxiliary variable is defined as

$$\xi_i = z_{2i}^{p_3} - z_{2i}^{*p_3} \tag{7.29}$$

Step 1: Analyze $\dot{V}_0$.

After taking the derivative of (7.24), we have

$$\dot{V}_0 = z_1^T \dot{z}_1 = z_1^T z_2 = \sum\nolimits_{i=1}^{n} z_{1i} z_{2i}^* + \sum\nolimits_{i=1}^{n} z_{1i}(z_{2i} - z_{2i}^*) \tag{7.30}$$

By Lemma 2.8, we obtain

$$z_{1i}(z_{2i} - z_{2i}^*) \le 2^{1-1/p_3} |z_{1i}| |z_{2i}^{p_3} - z_{2i}^{*p_3}|^{1/p_3} \tag{7.31}$$

By Lemma 2.9, we have

$$z_{1i}(z_{2i} - z_{2i}^*) \le 2^{1-1/p_3} \left(\frac{p_3}{1+p_3} |z_{1i}|^{p_6} + \frac{1}{1+p_3} |\xi_i|^{p_6} \right) \tag{7.32}$$

where $p_6 = 1 + 1/p_3$.

Substituting (7.32) into (7.30) results in

$$\begin{aligned}\dot{V}_0 &= -\kappa_6 \sum\nolimits_{i=1}^{n} z_{1i} z_{1i}^{1/p_3} + \sum\nolimits_{i=1}^{n} z_{1i}\left(z_{2i} - z_{2i}^*\right) \\ &\le -\kappa_6 \sum\nolimits_{i=1}^{n} z_{1i}^{1+1/p_3} + \sum\nolimits_{i=1}^{n} 2^{1-1/p_3}\left(\frac{p_3}{1+p_3}|z_{1i}|^{p_6} + \frac{1}{1+p_3}|\xi_i|^{p_6}\right)\end{aligned} \tag{7.33}$$

$$\begin{aligned}\dot{V}_0 &\le -\kappa_6 \sum\nolimits_{i=1}^{n} z_{1i}^{p_6} + 2^{1-1/p_3}\frac{p_3}{1+p_3}\sum\nolimits_{i=1}^{n} |z_{1i}|^{p_6} + \frac{2^{1-1/p_3}}{1+p_3}\sum\nolimits_{i=1}^{n} |\xi_i|^{p_6} \\ &= -\left(\kappa_6 - 2^{1-1/p_3}\frac{p_3}{1+p_3}\right)\sum\nolimits_{i=1}^{n} |z_{1i}|^{p_6} + \frac{2^{1-1/p_3}}{1+p_3}\sum\nolimits_{i=1}^{n} |\xi_i|^{p_6}\end{aligned} \tag{7.34}$$

Step 2: Calculate $\sum_{i=1}^{n} \dot{V}_i$.

After taking the derivative of V_i in (7.28), we have

$$\begin{aligned}\dot{V}_i = &-\frac{1}{2^{1-1/p_3}\kappa_6^{1+p_3}}\frac{\partial z_{2i}^{*p_3}}{\partial t}\int_{z_{2i}^*}^{z_{2i}}\left(s^{p_3} - z_{2i}^{*p_3}\right)^{1-1/p_3}\mathrm{d}s \\ &+ \frac{1}{(2-1/p_3)\,2^{1-1/p_3}\kappa_6^{1+p_3}}\xi_i^{2-1/p_3}\bar{\tau}_i, i = 1, 2, \ldots, n\end{aligned} \tag{7.35}$$

where $\bar{\tau}_i = \partial z_{2i}/\partial t$. According to $z_2^* = -\kappa_6 z_1^{1/p_3}$ and (7.19), we have

$$\frac{\partial z_2^{*p_3}}{\partial t} = \left[\frac{\partial z_{21}^{*p_3}}{\partial t}, \ldots, \frac{\partial z_{2i}^{*p_3}}{\partial t}, \ldots, \frac{\partial z_{2n}^{*p_3}}{\partial t}\right]^T = -\kappa_6^{p_3} z_2 \tag{7.36}$$

Noticing that

$$\left|\frac{\partial z_{2i}^{*p_3}}{\partial t}\right| \le \kappa_6^{p_3}\left(|z_{21}| + \cdots + |z_{2i}| + \cdots + |z_{2n}|\right) = \kappa_6^{p_3}\sum\nolimits_{i=1}^{n} |z_{2i}| \tag{7.37}$$

By imposing the result in (7.37), (7.35) can be written as follows

$$\dot{V}_i \le \frac{\left(\sum_{i=1}^{n} |z_{2i}|\right)\left|z_{2i} - z_{2i}^*\right|}{2^{1-1/p_3}\kappa_6}|\xi_i|^{1-1/p_3} + \frac{\xi_i^{2-1/p_3}\bar{\tau}_i}{(2-1/p_3)\,2^{1-1/p_3}\kappa_6^{1+p_3}} \tag{7.38}$$

By Lemmas 2.8 and 2.9, for any $i, m = 1, 2, \ldots, n$, we have

$$\begin{aligned}&|z_{2m}|\left|z_{2i} - z_{2i}^*\right||\xi_i|^{1-1/p_3} \le 2^{1-1/p_3}\,|z_{2m}|\,|\xi_i|^{1/p_3}\,|\xi_i|^{1-1/p_3} \\ &\le 2^{1-1/p_3}\left(\frac{\left(2^{1-1/p_3} + \kappa_6\right)p_3}{1+p_3}|\xi_i|^{p_6} + \frac{2^{1-1/p_3}}{1+p_3}|\xi_m|^{p_6} + \frac{\kappa_6}{1+p_3}|z_{1m}|^{p_6}\right)\end{aligned} \tag{7.39}$$

Substituting (7.39) into (7.38) results in

$$\dot{V}_i \le \left(\frac{2^{1-1/p_3} + \kappa_6}{\kappa_6}\frac{np_3}{1+p_3}|\xi_i|^{p_6} + \frac{1}{\kappa_6}\frac{2^{1-1/p_3}}{1+p_3}\sum\nolimits_{m=1}^{n}|\xi_m|^{p_6} + \frac{1}{1+p_3}\sum\nolimits_{m=1}^{n}|z_{1m}|^{p_6}\right) + \frac{1}{(2-1/p_3)\,2^{1-1/p_3}\kappa_6^{1+p_3}}\xi_i^{2-1/p_3}\bar{\tau}_i \tag{7.40}$$

Step 3: Calculate $\dot{V}$.

Taking the derivative of (7.27), we have

$$\dot{V} = \dot{V}_0 + \sum\nolimits_{i=1}^{n}\dot{V}_i \tag{7.41}$$

Substituting (7.33) and (7.40) into (7.41) and considering $|z_{1i}|^{p_6} = |z_{1i}|^{(1+p_3)/p_3} = z_{1i}^{p_6}$, $|\xi_i|^{p_6} = \xi_i^{p_6}$ we have

$$\begin{aligned}\dot{V} \le& -\left(\kappa_6 - \frac{2^{1-1/p_3}p_3}{1+p_3}\right)\sum\nolimits_{i=1}^{n}|z_{1i}|^{p_6} + \frac{2^{1-1/p_3}}{1+p_3}\sum\nolimits_{i=1}^{n}|\xi_i|^{p_6}\\ &+\sum\nolimits_{i=1}^{n}\frac{\left(2^{1-1/p_3} + \kappa_6\right)np_3}{\kappa_6(1+p_3)}|\xi_i|^{p_6}\\ &+\sum\nolimits_{i=1}^{n}\frac{2^{1-1/p_3}}{\kappa_6(1+p_3)}\sum\nolimits_{m=1}^{n}|\xi_m|^{p_6} + \sum\nolimits_{i=1}^{n}\frac{1}{1+p_3}\sum\nolimits_{m=1}^{n}|z_{1m}|^{p_6}\\ &+\sum\nolimits_{i=1}^{n}\frac{\xi_i^{2-1/p_3}\bar{\tau}_i}{(2-1/p_3)\,2^{1-1/p_3}\kappa_6^{1+p_3}}\end{aligned} \tag{7.42}$$

$$\begin{aligned}\dot{V} \le& -\left(\kappa_6 - \frac{2^{1-1/p_3}p_3 + n}{1+p_3}\right)\sum\nolimits_{i=1}^{n}|z_{1i}|^{p_6} + \frac{\sum\nolimits_{i=1}^{n}\xi_i^{2-1/p_3}\bar{\tau}_i}{(2-1/p_3)\,2^{1-1/p_3}\kappa_6^{1+p_3}}\\ &+\left(\frac{2^{1-1/p_3} + np_3}{1+p_3} + \frac{2^{1-1/p_3}n}{\kappa_6}\right)\sum\nolimits_{i=1}^{n}|\xi_i|^{p_6}\end{aligned} \tag{7.43}$$

According to Theorem 7.1 proposed, if we take the controller as (7.22), then

$$\dot{z}_2 = -\kappa_5\left(z_2^{p_3} + \kappa_6^{p_3}z_1\right)^{2/p_3-1} \tag{7.44}$$

Then we have

$$\bar{\tau}_i = \partial z_{2i}/\partial t = -\kappa_5\left(z_{2i}^{p_3} + \kappa_6^{p_3}z_{1i}\right)^{2/p_3-1} = -\kappa_5\xi_i^{2/p_3-1} \tag{7.45}$$

Substituting (7.45) into (7.43), we can obtain

$$\begin{aligned}\dot{V} \leq & -\left(\kappa_6 - \frac{2^{1-1/p_3} p_3 + n}{1+p_3}\right)\sum_{i=1}^{n} |z_{1i}|^{p_6} \\ & + \left(\frac{2^{1-1/p_3} + n p_3}{1+p_3} + \frac{2^{1-1/p_3} n}{\kappa_6}\right)\sum_{i=1}^{n} |\xi_i|^{p_6} \\ & - \frac{\kappa_5 \sum_{i=1}^{n} \xi_i^{2-1/p_3} \xi_i^{2/p_3 - 1}}{(2-1/p_3)\, 2^{1-1/p_3} \kappa_6^{1+p_3}}\end{aligned} \tag{7.46}$$

So that (7.46) can be rewritten as

$$\begin{aligned}\dot{V} \leq & -\left(\kappa_6 - \frac{2^{1-1/p_3} p_3 + n}{1+p_3}\right)\sum_{i=1}^{n} |z_{1i}|^{p_6} \\ & + \left(\frac{2^{1-1/p_3} + n p_3}{1+p_3} + \frac{2^{1-1/p_3} n}{\kappa_6}\right)\sum_{i=1}^{n} |\xi_i|^{p_6} \\ & - \frac{\kappa_5 \sum_{i=1}^{n} \xi_i^{p_6}}{(2-1/p_3)\, 2^{1-1/p_3} \kappa_6^{1+p_3}}\end{aligned} \tag{7.47}$$

To simplify the above inequality, we have

$$\begin{aligned}\dot{V} \leq & -\left(\kappa_6 - \frac{2^{1-1/p_3} p_3 + n}{1+p_3}\right)\sum_{i=1}^{n} z_{1i}^{p_6} \\ & -\left(\frac{\kappa_5}{(2-1/p_3)\, 2^{1-1/p_3} \kappa_6^{1+p_3}} - \frac{2^{1-1/p_3} + n p_3}{1+p_3} - \frac{2^{1-1/p_3} n}{\kappa_6}\right)\sum_{i=1}^{n} \xi_i^{p_6}\end{aligned} \tag{7.48}$$

Take $\kappa_7 > 0$, the above inequality becomes

$$\dot{V} \leq -\kappa_7 \sum_{i=1}^{n} z_{1i}^{p_6} - \kappa_7 \sum_{i=1}^{n} \xi_i^{p_6} \tag{7.49}$$

where κ_5, κ_6 and κ_7 need to satisfy the following conditions

$$\begin{aligned}& \kappa_6 - \frac{2^{1-1/p_3} p_3 + n}{1+p_3} > \kappa_7 \\ & \kappa_5 \frac{1}{(2-1/p_3)\, 2^{1-1/p_3} \kappa_6^{1+p_3}} - \frac{2^{1-1/p_3} + n p_3}{1+p_3} - \frac{2^{1-1/p_3} n}{\kappa_6} > \kappa_7\end{aligned} \tag{7.50}$$

The value ranges of κ_5, κ_6 are obtained

$$\begin{aligned}& \kappa_6 > \frac{2^{1-1/p_3} p_3 + n}{1+p_3} + \kappa_7 \\ & \kappa_5 > (2-1/p_3)\, 2^{1-1/p_3} \kappa_6^{1+p_3} \left(\frac{2^{1-1/p_3} + n p_3}{1+p_3} + \frac{2^{1-1/p_3} n}{\kappa_6} + \kappa_7\right)\end{aligned} \tag{7.51}$$

From the above derivation, we know that the closed-loop system is stable. Next, we prove that $V(t)$ can converge to the origin within a finite time.

Through the calculation of (7.35) and Lemma 2.8, we can get

$$V_i \leq \frac{|z_{2i} - z_{2i}^*||z_{2i}^{p_3} - z_{2i}^{*p_3}|^{2-1/p_3}}{(2-1/p_3)2^{1-1/p_3}\kappa_6^{1+p_3}} \leq \frac{\xi_i^2}{(2-1/p_3)\kappa_6^{1+p_3}} \tag{7.52}$$

Hence,

$$V = V_0 + \sum_{i=1}^{n} V_i \leq \kappa_8\left(\sum_{i=1}^{n} z_{1i}^2 + \sum_{i=1}^{n} \xi_i^2\right) \tag{7.53}$$

where $\kappa_8 = \max\{1/2, 1/(2-1/p_3)\kappa_6^{1+p_3}\}$.

By Lemma 2.10, we have

$$V^{p_6/2} \leq \kappa_8^{p_6/2}\left(\sum_{i=1}^{n} z_{1i}^{p_6} + \sum_{i=1}^{n} \xi_i^{p_6}\right) \tag{7.54}$$

Noticing

$$\begin{aligned}\dot{V} + \kappa_9 V^{p_6/2} \leq & -\kappa_7 \sum_{i=1}^{n} z_{1i}^{p_6} - \kappa_7 \sum_{i=1}^{n} \xi_i^{p_6} \\ & + \kappa_9\kappa_8^{p_6/2} \sum_{i=1}^{n} z_{1i}^{p_6} + \kappa_9\kappa_8^{p_6/2} \sum_{i=1}^{n} \xi_i^{p_6}\end{aligned} \tag{7.55}$$

Simplify the above inequality, we get

$$\begin{aligned}\dot{V} + \kappa_9 V^{p_6/2} \leq & -\left(\kappa_7 - \kappa_9\kappa_8^{p_6/2}\right)\sum_{i=1}^{n} z_{1i}^{p_6} \\ & -\left(\kappa_7 - \kappa_9\kappa_8^{p_6/2}\right)\sum_{i=1}^{n} \xi_i^{p_6}\end{aligned} \tag{7.56}$$

Because of $\kappa_7 - \kappa_9\kappa_8^{p_6/2} \geq 0, \kappa_9 > 0, \ \kappa_9 \leq \kappa_7/\kappa_8^{p_6/2}$, we can take $\kappa_9 = \kappa_7/(2\kappa_8^{p_6/2})$, then

$$\dot{V} + \kappa_9 V^{p_6/2} \leq 0 \tag{7.57}$$

Note that $0 < p_6/2 < 1$, according to Lemma 2.3, the closed-loop system is finite-time stable with the finite settling time

$$T_1 \leq \frac{V(0)^{1-p_6/2}}{\kappa_9(1-p_6/2)} < +\infty \tag{7.58}$$

This implies $z_1 \to 0, z_2 \to 0$ in finite time and $z_1 = z_2 = 0 \ \forall t \geq T_1$. □

7.3 Control Design and Stability Analysis with Disturbances

Consider the finite-time control based on the performance guaranteed and command filter designed in the previous section. If the external disturbances are considered, we need to consider the nonlinear terms $-BM_a{}^{-1}d$ about disturbances in the model (7.8). To enhance the robustness of the system, in this section, the robust finite-time controller is further designed.

According to Lemma 2.6 and the analysis in the previous section, the general form of the observer for the system $\dot{x} = f + gv + d$ is as follows,

$$\dot{y}_0 = -k_1 \frac{y_0 - x}{||y_0 - x||^{1/2}} + y_1 - k_2 (y_0 - x) + f + gv \tag{7.59}$$

$$\dot{y}_1 = -k_3 \frac{y_0 - x}{||y_0 - x||} - k_4 (y_0 - x) \tag{7.60}$$

where k_1, k_2, k_3, k_4 represent the positive constant gain, y_0 and y_1 are observations for the state x and the unknown disturbance d.

So far, we design the following observer to observe the uncertainty term $BM_a^{-1}d$ in $\ddot{\sigma} = A - BM_a^{-1}(C_a\dot{q}_{ev} + G_a) + BM_a^{-1}Q\tau_a - BM_a^{-1}d$.

$$\begin{aligned}\dot{y}_0 = & -\kappa_{y1} \frac{y_0 - \dot{\sigma}}{||y_0 - \dot{\sigma}||^{1/2}} + y_1 - \kappa_{y2} (y_0 - \dot{\sigma}) \\ & + A - BM_a^{-1}(C_a\dot{q}_{ev} + G_a) + BM_a^{-1}Q\tau_a\end{aligned} \tag{7.61}$$

$$\dot{y}_1 = -\kappa_{y3} \frac{y_0 - \dot{\sigma}}{||y_0 - \dot{\sigma}||} - \kappa_{y4} (y_0 - \dot{\sigma}) \tag{7.62}$$

where $\kappa_{y1}, \kappa_{y2}, \kappa_{y3}, \kappa_{y4}$ represent the positive constant gain, y_0 and y_1 are observations for the state $\dot{\sigma}$ and the unknown disturbance $BM_a^{-1}d$, i.e. $y_0 \to \dot{\sigma}$, $y_1 \to BM_a^{-1}d$. Considering that in this controller, the model is precisely known, so only y_1 is needed. According to the observer property Lemma 2.6, there exists the finite time T_y such that the tracking error of y_1 for $BM_a^{-1}d$ converges in T_y.

According to the function term y_1 designed by the above observer, the modified controller (7.22) is as follows,

$$\tau_d = -\kappa_5 Q^{-1} M_a B^{-1} \left(z_2^{p_3} + \kappa_6^{p_3} z_1\right)^{2/p_3 - 1} - Q^{-1} M_a B^{-1} \Psi_3 + Q^{-1} M_a B^{-1} y_1 \tag{7.63}$$

Considering the observation error of the observer, define $F_y = y_1 - BM_a^{-1}d$, according to the nature of the observer, the observation error is bounded by $||F_y|| \le \varepsilon_d$, where ε_d is a positive constant.

Theorem 7.2 *For the spacecraft attitude tracking error model (2.28) with a known smooth reference trajectory and with considering external disturbances, the control input τ_a can be obtained after τ_d passes the CF (2.71), and $||\tau_a - \tau_d||$ is bounded.*

If the Assumption 2.1 is satisfied and the controller τ_d is chosen as (7.63), and the command filter compensators are selected as (7.12) and (7.16), then z_1, z_2 can converge into the neighborhood Ω_z containing the equilibrium point within finite time T_2,

$$\begin{aligned}\Omega_z = \{(z_1^T, z_2^T) : |z_{1i}| \le \kappa_{12}(2\kappa_{10}/\kappa_7)^{p_3/(2-p_3)}, \\ |z_{2i}| \le \kappa_{13}(2\kappa_{10}/\kappa_7)^{1/(2-p_3)}\}\end{aligned} \tag{7.64}$$

where T_2, κ_{10}, κ_{12}, and κ_{13} are defined later.

Proof We substitute the controller (7.63) into the definition of $\bar{\tau}_i$, then

$$\bar{\tau}_i = -\kappa_5 \xi_i^{2/p_3 - 1} + F_{yi} \tag{7.65}$$

We rewrite (7.43) as

$$\begin{aligned}\dot{V} \le &-\left(\kappa_6 - \frac{2^{1-1/p_3} p_3 + n}{1 + p_3}\right)\sum\nolimits_{i=1}^{n} |z_{1i}|^{p_6} \\ &+ \left(\frac{2^{1-1/p_3} + n p_3}{1 + p_3} + \frac{2^{1-1/p_3} n}{\kappa_6}\right)\sum\nolimits_{i=1}^{n} |\xi_i|^{p_6} \\ &+ \frac{1}{(2 - 1/p_3)\, 2^{1-1/p_3} \kappa_6^{1+p_3}} \sum\nolimits_{i=1}^{n} \xi_i^{2-1/p_3} \left(-\kappa_5 \xi_i^{2/p_3 - 1} + F_{yi}\right)\end{aligned} \tag{7.66}$$

Then, we can obtain

$$\begin{aligned}\dot{V} \le &-\left(\kappa_6 - \frac{2^{1-1/p_3} p_3 + n}{1 + p_3}\right)\sum\nolimits_{i=1}^{n} z_{1i}^{p_6} + \kappa_{10}\sum\nolimits_{i=1}^{n} |\xi_i|^{2-1/p_3} \\ &-\left(\frac{\kappa_5}{(2 - 1/p_3)\, 2^{1-1/p_3} \kappa_6^{1+p_3}} - \frac{2^{1-1/p_3} + n p_3}{1 + p_3} - \frac{2^{1-1/p_3} n}{\kappa_6}\right)\sum\nolimits_{i=1}^{n} \xi_i^{p_6}\end{aligned} \tag{7.67}$$

Simplify (7.67), we can get

$$\dot{V} \le -\kappa_7 \sum\nolimits_{i=1}^{n} z_{1i}^{p_6} - \kappa_7 \sum\nolimits_{i=1}^{n} \xi_i^{p_6} + \kappa_{10}\sum\nolimits_{i=1}^{n} |\xi_i|^{2-1/p_3} \tag{7.68}$$

where $\kappa_{10} = \frac{\varepsilon_d}{(2-1/p_3)2^{1-1/p_3}\kappa_6^{1+p_3}}$.

In order to prove the stability of a disturbed system, we will prove it through the following three steps. We first define two sets with $\xi = [\xi_1, ..., \xi_n]$ as follows.

$$\begin{aligned}\Omega_1 = \Big\{(z_1^T, \xi^T) : \sum\nolimits_{i=1}^{n} z_{1i}^{p_6} + \sum\nolimits_{i=1}^{n} \xi_i^{p_6} \\ < \left(n^{(1+p_3)/(2p_3-1)} + n^{n/(1+p_3)}\right)(2\kappa_{10}/\kappa_7)^{(1+p_3)/(2-p_3)}\Big\}\end{aligned} \tag{7.69}$$

$$\begin{aligned}\Omega_2 = \Big\{\big(z_1^T, \xi^T\big) : V(t) < \varepsilon_m, \\ \varepsilon_m = \kappa_8\big(n^{(1+p_3)/(2p_3-1)} + n^{n/(1+p_3)}\big)^{2/p_6}(2\kappa_{10}/\kappa_7)^{2p_3/(2-p_3)}\Big\}\end{aligned} \tag{7.70}$$

We first prove in Step 1 that when $(z_1^T, \xi^T) \notin \Omega_1$, $\dot{V} < 0$ holds. However, this step of proof does not guarantee that when $(z_1^T, \xi^T) \in \Omega_1$, $\dot{V} < 0$ holds. In other words, Ω_1 is not guaranteed to be an attractive region, which means that (z_1^T, ξ^T) may escape again after entering Ω_1. Fortunately, we can design an area Ω_2, which is larger than Ω_1 ($\Omega_1 \subset \Omega_2$). In Step 2, we prove that Ω_2 is the attractive region for any initial states. Simultaneously, we prove that all states will converge to Ω_2 within finite time T_2, that is, $(z_1^T, \xi^T) \in \Omega_2$ for $\forall t > T_2$. In Step 3, we will find the corresponding boundaries.

Step 1: Prove $\dot{V} < 0$ when $(z_1^T, \xi^T) \notin \Omega_1$.

When $(z_1^T, \xi^T) \notin \Omega_1$, we have

$$\begin{aligned}\sum\nolimits_{i=1}^{n} z_{1i}^{p_6} + \sum\nolimits_{i=1}^{n} \xi_i^{p_6} \\ \geq \big(n^{(1+p_3)/(2p_3-1)} + n^{n/(1+p_3)}\big)(2\kappa_{10}/\kappa_7)^{(1+p_3)/(2-p_3)}\end{aligned} \tag{7.71}$$

To facilitate analysis, we discuss two cases.

Case 1: $\sum\nolimits_{i=1}^{n} \xi_i^{p_6} \geq n^{(1+p_3)/(2p_3-1)}(2\kappa_{10}/\kappa_7)^{(1+p_3)/(2-p_3)}$.

We notice that $1 < p_3 < 2$ and $p_6 = 1 + 1/p_3$, so we have $0 < (2 - 1/p_3)/p_6 < 1$, and further, according to Lemma 2.10, we get

$$\begin{aligned}\sum\nolimits_{i=1}^{n} |\xi_i|^{2-1/p_3} &= \sum\nolimits_{i=1}^{n} \big(|\xi_i|^{p_6}\big)^{(2-1/p_3)/p_6} \\ &\leq n^{(2-p_3)/(1+p_3)}\Big(\sum\nolimits_{i=1}^{n} \xi_i^{p_6}\Big)^{(2p_3-1)/(1+p_3)}\end{aligned} \tag{7.72}$$

After that,

$$\sum\nolimits_{i=1}^{n} \xi_i^{p_6} \geq n^{(p_3-2)/(2p_3-1)}\Big(\sum\nolimits_{i=1}^{n} |\xi_i|^{2-1/p_3}\Big)^{(1+p_3)/(2p_3-1)} \tag{7.73}$$

The following inequality holds

$$\begin{aligned}\frac{\kappa_7}{2}\sum\nolimits_{i=1}^{n} \xi_i^{p_6} - \kappa_{10}\sum\nolimits_{i=1}^{n} |\xi_i|^{2-1/p_3} \geq \Big(\sum\nolimits_{i=1}^{n} |\xi_i|^{2-1/p_3}\Big)\cdot \\ \left[\frac{\kappa_7}{2} n^{(p_3-2)/(2p_3-1)}\Big(\sum\nolimits_{i=1}^{n} |\xi_i|^{2-1/p_3}\Big)^{(1+p_3)/(2p_3-1)-1} - \kappa_{10}\right]\end{aligned} \tag{7.74}$$

Simultaneously, we found that

$$\sum\nolimits_{i=1}^{n} \xi_i^{p_6} \geq n^{(1+p_3)/(2p_3-1)}(2\kappa_{10}/\kappa_7)^{(1+p_3)/(2-p_3)} \tag{7.75}$$

By Lemma 2.10, we have

$$\begin{aligned}\sum_{i=1}^{n}|\xi_i|^{2-1/p_3} &= \sum_{i=1}^{n}\left(\xi_i^{p_6}\right)^{(2-1/p_3)/p_6} \\ &\geq \left(\sum_{i=1}^{n}\xi_i^{p_6}\right)^{(2-1/p_3)/p_6} \geq n\left(\frac{2\kappa_{10}}{\kappa_7}\right)^{(2p_3-1)/(2-p_3)}\end{aligned} \tag{7.76}$$

Substituting inequality (7.76) into (7.74), we have

$$\frac{\kappa_7}{2}\sum_{i=1}^{n}\xi_i^{p_6}-\kappa_{10}\sum_{i=1}^{n}|\xi_i|^{2-1/p_3}\geq 0 \tag{7.77}$$

Combining (7.68), we have

$$\begin{aligned}\dot{V} &\leq -\kappa_7\sum_{i=1}^{n}z_{1i}^{p_6}-\kappa_7\sum_{i=1}^{n}\xi_i^{p_6}+\kappa_{10}\sum_{i=1}^{n}|\xi_i|^{2-1/p_3} \\ &\leq -\kappa_7\sum_{i=1}^{n}z_{1i}^{p_6}-\frac{\kappa_7}{2}\sum_{i=1}^{n}\xi_i^{p_6}<0\end{aligned} \tag{7.78}$$

Case 2: $\sum_{i=1}^{n}\xi_i^{p_6}<n^{(1+p_3)/(2p_3-1)}\left(2\kappa_{10}/\kappa_7\right)^{(1+p_3)/(2-p_3)}$.
Under such conditions, when $(z_1^T,\xi^T)\notin\Omega_1$, it is easy to know that

$$\sum_{i=1}^{n}z_{1i}^{p_6}>n^{n/(1+p_3)}\left(2\kappa_{10}/\kappa_7\right)^{(1+p_3)/(2-p_3)} \tag{7.79}$$

By (7.72), we have

$$\sum_{i=1}^{n}|\xi_i|^{2-1/p_3}\leq n^{n/(1+p_3)}\left(\frac{2\kappa_{10}}{\kappa_7}\right)^{(2p_3-1)/(2-p_3)} \tag{7.80}$$

Substituting the inequalities (7.79) and (7.80) into (7.68), we obtain

$$\dot{V}\leq-\frac{\kappa_7}{2}\sum_{i=1}^{n}z_{1i}^{p_6}-\kappa_7\sum_{i=1}^{n}\xi_i^{p_6}<0 \tag{7.81}$$

Therefore, we can know from Cases 1 and 2 that $\dot{V}<0$ when $(z_1^T,\xi^T)\notin\Omega_1$.
Step 2: Prove $\dot{V}<0$ when $(z_1^T,\xi^T)\notin\Omega_2$.
By (7.54), for any $(z_1^T,\xi^T)\in\Omega_1$, we have

$$\begin{aligned}V^{p_6/2} &\leq \kappa_8^{p_6/2}\left(\sum_{i=1}^{n}z_{1i}^{p_6}+\sum_{i=1}^{n}\xi_i^{p_6}\right) \\ &< \kappa_8^{p_6/2}\left(n^{(1+p_3)/(2p_3-1)}+n^{n/(1+p_3)}\right)\left(\frac{2\kappa_{10}}{\kappa_7}\right)^{(1+p_3)/(2-p_3)} \\ &= \varepsilon_m^{p_6/2}\end{aligned} \tag{7.82}$$

From the above formula, it is easy to know that $(z_1^T,\xi^T)\in\Omega_2$, i.e., $\Omega_1\subset\Omega_2$. Thus, when $(z_1^T,\xi^T)\notin\Omega_2$, we have $\dot{V}<0$. For Ω_2 is a level set of the Lyapunov function, so there exists finite time $T_2 \geq T_y$, such that $(z_1^T, \xi^T) \in \Omega_2$, i.e., $V(t) < \varepsilon_m$, $\forall t > T_2$.

Next, we analyze the value range of T_2.

We noticed that, in fact, in (7.78) and (7.81), not only $\dot{V} < 0$, but also within finite time, we now analyze it. According to (7.78) and (7.81), we have $\dot{V} \leq -\frac{\kappa_7}{2}\sum_{i=1}^n z_{1i}^{p_6} - \frac{\kappa_7}{2}\sum_{i=1}^n \xi_i^{p_6} < 0$, and then combined with (7.54), we have $\dot{V} + \kappa_{11} V^{p_6/2} \leq -(\kappa_7/2 - \kappa_{11}\kappa_8^{p_6/2})\left(\sum_{i=1}^n z_{1i}^{p_6} + \sum_{i=1}^n \xi_i^{p_6}\right) < 0$, where $0 < \kappa_{11} < \kappa_7/(2\kappa_8^{p_6/2})$ so that $\kappa_7/2 - \kappa_{11}\kappa_8^{p_6/2} > 0$. According to Lemma 2.3, we can find the corresponding finite time as $T^* \leq V(0)^{1-(p_6/2)}/(\kappa_{11}(1 - p_6/2)) < +\infty$. Considering $\Omega_1 \subset \Omega_2$, the time for the state to enter Ω_2 from the initial value must be less than T^*, i.e. $T_2 \leq T^*$.

Step 3: Find the corresponding boundaries.

When $t \geq T_2$, we have

$$|z_{1i}| \leq \sqrt{z_1^T z_1} = \sqrt{2V_0(t)} \leq \sqrt{2V(t)} \leq \sqrt{2\varepsilon_m} \leq \kappa_{12}(2\kappa_{10}/\kappa_7)^{p_3/(2-p_3)} \tag{7.83}$$

where $\kappa_{12} = \sqrt{2\kappa_8}\left(n^{(1+p_3)/(2p_3-1)} + n^{n/(1+p_3)}\right)^{p_3/(1+p_3)}$.

By Lemma 2.8 and the definition of the candidate Lyapunov function, we have noticed that

$$\left|s - z_{2i}^*\right| \leq 2^{1-1/p_3}\left|s^{p_3} - z_{2i}^{*p_3}\right|^{1/p_3} \tag{7.84}$$

then

$$\left|s^{p_3} - z_{2i}^{*p_3}\right| \geq \left(2^{1/p_3 - 1}\left|s - z_{2i}^*\right|\right)^{p_3} \tag{7.85}$$

If $z_{2i} \geq z_{2i}^*$, we have

$$V_i \geq \frac{2^{1-2p_3}}{(2p_3 - 1)\kappa_6^{1+p_3}}\left(z_{2i} - z_{2i}^*\right)^{2p_3} \tag{7.86}$$

When $z_{2i} < z_{2i}^*$, the analysis process is similar. Because $V_i \leq V \leq \varepsilon_m$, $\forall t \geq T_2$, when $t \geq T_2$, the following formula holds

$$\left|z_{2i} - z_{2i}^*\right| \leq 2\varepsilon_m^{\frac{1}{2p_3}}\left[\left(p_3 - \frac{1}{2}\right)\kappa_6^{1+p_3}\right]^{\frac{1}{2p_3}} \tag{7.87}$$

Through the above derivation, when $t \geq T_2$, we can get the boundary of z_{2i} as follows

$$|z_{2i}| \leq \left|z_{2i}^*\right| + \left|z_{2i} - z_{2i}^*\right| = \kappa_6\left|z_{1i}^{1/p_3}\right| + \left|z_{2i} - z_{2i}^*\right| \leq \kappa_{13}(2\kappa_{10}/\kappa_7)^{1/(2-p_3)} \tag{7.88}$$

where

$$\kappa_{13} = \left[2^{\frac{1}{2p_3}}\kappa_6 + 2\left(\left(p_3 - \frac{1}{2}\right)\kappa_6^{1+p_3}\right)^{\frac{1}{2p_3}}\right]\kappa_8^{\frac{1}{2p_3}} \cdot \left(n^{(1+p_3)/(2p_3-1)} + n^{n/(1+p_3)}\right)^{1/(1+p_3)} \tag{7.89}$$

which completes the proof. □

7.4 Finite-Time Convergence Analysis of Command Filter Compensators

The above analysis can prove that z_1, z_2 converges in finite time with or without disturbances, which means that σ converges to φ_1 in finite time. This part proves that φ_1, φ_2 also converges in finite time. It can be further explained that closed-loop systems converge in finite time. The upper bound of the closed-loop system settling time is also given.

Theorem 7.3 *For the spacecraft attitude tracking error model (2.28) with a known smooth reference trajectory and with considering external disturbances, the control input τ_a can be obtained after τ_d passes the CF (2.71), and $||\tau_a - \tau_d||$ is bounded. If the Assumption 2.1 is satisfied and the controller τ_d is chosen as (7.63), and the command filter compensators are selected as (7.12) and (7.16), then the compensators φ_1 and φ_2 converge to the compact set Ω_φ (the neighborhood of the equilibrium point) within the finite time T_3,*

$$\Omega_\varphi = \{\varphi_1, \varphi_2 : ||\varphi_1|| \le \varepsilon_{\varphi_1}, ||\varphi_2|| \le \varepsilon_{\varphi_2}\} \tag{7.90}$$

where T_3 is the finite settling time, and its value is analyzed later, ε_{φ_2} and ε_{φ_2} are positive constants, the value range is analyzed later. The closed-loop system performance-constrained auxiliary variable σ is finite-time stable. Without considering the external bounded disturbances, σ converges to $\Omega_a = \Omega_\varphi$ within the finite time $T_4 = \max\{T_1, T_3\}$. When considering external bounded disturbances, σ within the finite time $T_5 = \max\{T_y, T_2, T_3\}$ converges to

$$\Omega_b = \left\{\sigma : \sigma = z_1 + \varphi_1, z_1 \in \Omega_z, \varphi_1 \in \Omega_\varphi\right\} \cup \Omega_z \cup \Omega_\varphi.$$

The spacecraft attitude tracking error is FTUUB, and its settling time is $T_s \le \min\{T_4, T_{ps}\}$ (disturbances are not considered), $T_s \le \min\{T_5, T_{ps}\}$ (disturbances are considered), where T_{ps} is the predefined maximum settling time, and the attitude tracking error bound is given by Theorem 2.2.

Proof Select the candidate Lyapunov function for φ_1, φ_2.

$$V_\varphi = \frac{1}{2}\varphi_1^T\varphi_1 + \frac{1}{2}\varphi_2^T\varphi_2 \tag{7.91}$$

where $V_{\varphi 1} = \frac{1}{2}\varphi_1^T\varphi_1$, $V_{\varphi 2} = \frac{1}{2}\varphi_2^T\varphi_2$.

Taking the derivative of the above two equations with respect to time and substituting the compensators (7.12) and (7.16) into them, we get

$$\dot{V}_{\varphi 1} = \varphi_1^T \dot{\varphi}_1 = -\kappa_1 \varphi_1^T \varphi_1 - \kappa_2 \varphi_1^T \operatorname{sig}(\varphi_1)^{p_1} + \varphi_1^T \varphi_2 \tag{7.92}$$

$$\begin{aligned} \dot{V}_{\varphi 2} = \varphi_2^T \dot{\varphi}_2 &= -\kappa_3 \varphi_2^T \varphi_2 - \kappa_4 \varphi_2^T \operatorname{sig}(\varphi_2)^{p_1} \\ &- \varphi_2^T B M_a^{-1} Q\,(\tau_d - \tau_a) - \kappa_B \varphi_2^T B \varphi_2 \end{aligned} \tag{7.93}$$

As in the previous section, according to the property of the command filter and the boundedness of Q, it is assumed that $||\tau_a - \tau_d|| \le \varepsilon_\tau$, $||M_a^{-1} Q|| \le \varepsilon_{MQ}$, where ε_{MQ} is a positive constant. Considering the inequality $-aX^T MX - X^T MY \le \frac{1}{4a} Y^T MY$ [17], where X, M, Y are vectors or matrices with matching dimensions, M is positive definite, a is a positive number, then,

$$\begin{aligned} \dot{V}_{\varphi 2} \le &- \kappa_3 \varphi_2^T \varphi_2 - \kappa_4 \varphi_2^T \operatorname{sig}(\varphi_2)^{p_1} + \frac{1}{4\kappa_B} \varepsilon_\tau^2 \varepsilon_{MQ}^2 \\ = &- 2\kappa_3 V_{\varphi 2} - 2^{p_2} \kappa_4 V_{\varphi 2}^{p_2} + \frac{1}{4\kappa_B} \varepsilon_\tau^2 \varepsilon_{MQ}^2 V_{\varphi 2}^0 \end{aligned} \tag{7.94}$$

According to Lemma 2.5, φ_2 is finite-time uniformly ultimately bounded (FTUUB), that is, there is the finite time T_{φ_2} makes $||\varphi_2|| \le \varepsilon_{\varphi_2}$.

According to Young's inequality and the matrix norm properties, the following equations hold:

$$\varphi_1^T \varphi_2 \le \left\| \varphi_1^T \varphi_2 \right\| \le \|\varphi_1\| \, \|\varphi_2\| \le \frac{\varrho^2}{2} \|\varphi_1\|^2 + \frac{1}{2\varrho^2} \|\varphi_2\|^2 \tag{7.95}$$

When $t \ge T_{\varphi_2}$, according to (7.95), there are

$$\begin{aligned} \dot{V}_{\varphi 1} &\le - \kappa_1 \varphi_1^T \varphi_1 - \kappa_2 \varphi_1^T \operatorname{sig}(\varphi_1)^{p_1} + \frac{\varrho^2}{2} \|\varphi_1\|^2 + \frac{1}{2\varrho^2} \|\varphi_2\|^2 \\ &\le - (\kappa_1 - \frac{\varrho^2}{2}) \varphi_1^T \varphi_1 - \kappa_2 \varphi_1^T \operatorname{sig}(\varphi_1)^{p_1} + \frac{1}{2\varrho^2} \varepsilon_{\varphi_2}^2 \\ &\le - \kappa_{14} V_{\varphi 1} - \kappa_{15} V_{\varphi 1}^{p_2} + \kappa_{16} V_{\varphi 1}^0 \end{aligned} \tag{7.96}$$

where $\kappa_{14} = 2(\kappa_1 - \frac{\varrho^2}{2})$, $\kappa_{15} = 2^{p_2} \kappa_2$, $\kappa_{16} = \frac{\varepsilon_{\varphi_2}^2}{2\varrho^2}$, $p_2 = (1 + p_1)/2$. According to Lemma 2.5, φ_1 is finite-time uniformly ultimately bounded (FTUUB), that is, there is the finite time T_{φ_1} makes $||\varphi_1|| \le \varepsilon_{\varphi_1}$.

By Lemma 2.5, the states φ_1, φ_2 can converge to the compact set Ω_φ (a region of equilibrium point) within finite time $T_3 \le T_{\varphi_1} + T_{\varphi_2}$. Together with Theorems 7.1 and 7.2, we know that the tracking error of the closed-loop system is stable in finite time, as described in Theorem 7.3. Combining Theorems 2.1 and 2.2, it can be seen that the spacecraft attitude tracking error is FTUUB. The settling time of the closed-loop system tracking error converges to the predefined steady-state error range is less than the predefined maximum settling time T_{ps}, transient and steady-state performance satisfy the constraints (2.55).

The proof is completed. □

Remark 7.2 (*A note on finite-time convergence*) The control algorithm design proposed in this chapter considers the finite-time stability of the closed-loop system. This finite-time stability is manifested in that the auxiliary variables z_1 and z_2 and the command filter compensators φ_1 and φ_2 are finite-time convergent or bounded. This finite-time stability relies on an assumption or precondition that when $||\tau_a - \tau_d|| \leq \varepsilon_\tau$, the closed-loop system is finite-time stable or bounded. That is to say, when the errors between the input and output of the command filter are bounded, the closed-loop system under these algorithms is finite-time stable. That means the algorithm is partially finite-time stable. These algorithms mainly consider the algorithm framework of the finite-time closed-loop system applying the command filter, so the finite-time design of the command filter itself has not been considered yet. Because it is only necessary to select different command filters or by changing the finite-time stability of the filter itself, the finite-time stable conclusion about the closed-loop system state can be obtained. In fact, under the framework of the proposed algorithms, as long as the command filter itself is stable, then $||\tau_a - \tau_d||$ is always bounded, then the closed-loop system is finite-time stable. The control command signal errors caused by the command filter only affect the steady-state error boundary described by the finite-time stability.

7.5 Simulation Analysis

7.5.1 Simulation Parameters

This subsection first presents the initial conditions and parameters for the simulation. Considering the background of the on-orbit service spacecraft mission, although this study takes attitude motion as the research object, this section still gives the relevant information about the spacecraft orbit. The orbital parameters of the target spacecraft are as follows.

Considering that in the short-range on-orbit service mission, the spacecraft is very close to the target spacecraft. Therefore, the orbital parameters of the spacecraft to be controlled are also shown in Table 7.1. Next, consider the initial parameters of the spacecraft as shown in Table 7.2. Without loss of generality, the parameters of the Tables 7.1 and 7.2 can be changed according to specific tasks.

Next, consider the desired attitude of the on-orbit service spacecraft. In the numerical simulation of this study, the spacecraft is considered to track the attitude of the

Table 7.1 The orbital parameters of the target spacecraft

Parameters	Values
Semimajor axis	6.85×10^6 m
Eccentricity	0.0073
Inclination	60 deg
Argument of periapsis	0 deg
Longitude of the ascending node	0 deg
Mean anomaly	0 deg

Table 7.2 The initial attitude parameters of the spacecraft

Parameters	Values
q_{v0}	$[0.6, 0.4, -0.3]^T$ rad
ω_{a0}	$[0, 0, 0]^T$ rad/s
J_s	$\begin{bmatrix} 100 & -10 & 5 \\ -10 & 120 & -3 \\ 5 & -3 & 100 \end{bmatrix}$ $\text{kg} \cdot \text{m}^2$

target spacecraft, and the attitude of the target spacecraft is the same as the settings in Sect. 4.4.1.

$$q_{dv} = \begin{bmatrix} (r_{li-23} - r_{li-32})/(4q_{d0}) \\ (r_{li-31} - r_{li-13})/(4q_{d0}) \\ (r_{li-12} - r_{li-21})/(4q_{d0}) \end{bmatrix} \tag{7.97}$$

In the simulation of this chapter, the control input maximum and minimum values of the spacecraft attitude actuator are considered to be limited to ± 1.1 N · m. Simultaneously, through the command filter, this chapter actively limits the actual control input amplitude to $[-1, 1]$ N · m. The active limit amplitude is smaller than the actual input amplitude, mainly to leave a certain margin.

According to the spacecraft attitude tracking model in Sect. 2.3, the tracking error quaternion and the angular velocity tracking error can be obtained. To demonstrate the performance of the proposed controller and to facilitate comparison with similar controllers, the error-index is defined as $I_a = ||q_{ev}|| + ||\omega_e||$. To illustrate the energy consumption of the proposed controller, the energy consumption index $E_a = \sum_{i=1}^{n} \int_0^t |\tau_{ai}(t)|^2 dt$ is defined as in [18]. This index describes the electrical or chemical energy consumed by actuators such as thrusters or reaction flywheels during the stabilization of an attitude closed-loop system. For the controller proposed in this study, the parameter selection basis is the same as in the previous chapter.

This section presents the parameters selected in the simulation of the controller (7.63) in this chapter, as shown in the following table.

In Table 7.3, the parameter t_{ps} is the predefined maximum settling time. ρ_0 is the maximum initial value of the predefined performance boundary, which needs to be larger than the initial error quaternion. Considering that all quaternion elements are

Table 7.3 The initial parameters of the proposed controller (7.63)

Param	Value	Param	Value	Param	Value
t_{ps}	50s	κ_1	0.2	κ_6	1
ρ_0	1	κ_2	0.1	κ_B	0.001
ρ_t	0.02	κ_3	0.15	p_1	0.99
ε_ρ	0.01	κ_4	0.1	p_2	0.995
δ_i	1	κ_5	1	p_3	11/9

less than or equal to 1, thus, ρ_0 is taken as 1. ρ_t and ε_ρ provide steady-state error bounds. δ_i is about whether to limit the overshoot. κ_1, κ_2, κ_3, κ_4, κ_5, and κ_6 are the controller gain parameters. p_1, p_2, and p_3 are the relevant power exponents in the controller, respectively. The magnitude of the saturation function in the command filter is limited to $\pm 1\,\mathrm{N}\cdot\mathrm{m}$.

Considering the external disturbances to the spacecraft, we take the disturbance variable as follows [19],

$$d_a = \begin{bmatrix} 0.24\sin(0.5t) + 1.44\,\mathrm{rand}_a(t) \\ 0.16\sin(0.5t + \pi/2) + 0.96\,\mathrm{rand}_a(t) \\ 0.16\sin(0.5t + \pi/3) + 0.96\,\mathrm{rand}_a(t) \end{bmatrix} \mathrm{N}\cdot\mathrm{m} \tag{7.98}$$

where $\mathrm{rand}_a(t)$ represents a zero-mean Gaussian random signal, and the variance is set to 0.01. It should be pointed out that, to fully demonstrate the disturbance rejection ability of the controller proposed, the disturbance selected in the simulation is larger than the practical situation, and the selected value is still within the reasonable range in orbit.

7.5.2 Performance and Comparison

The controller proposed in this chapter is compared with existing state-of-the-art controllers. For the convenience of representation, this section labels the proposed controller (7.63) as controller 'A'. Label the spacecraft attitude performance constrained finite-time controller (4.41) proposed in Sect. 4.3 without the filter as controller 'B'. Mark the command filtering algorithm in [13] as controller 'C'. Mark the improved command filtering algorithm based on low-pass filter in [13] as controller 'D'. Label the traditional prescribed performance controller in [20] as controller 'E'. Label the finite-time controller in [21] as controller 'F'. In this section, a comparative simulation of the above controllers will be carried out to illustrate the effectiveness and advantages of this study. In the simulation diagram and analysis, the symbols or subscripts 'A', 'B', 'C', 'D', 'E', and 'F' represent the above six controllers. Like the previous section, controller 'E' uses the accurate model.

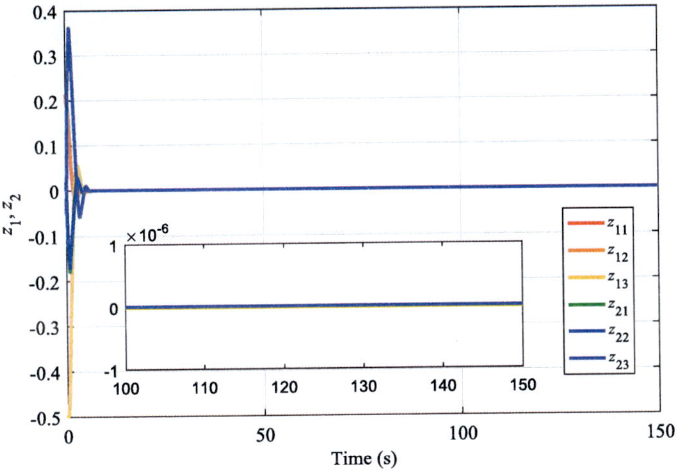

Fig. 7.1 The controller auxiliary variables z_1 and z_2 without disturbances

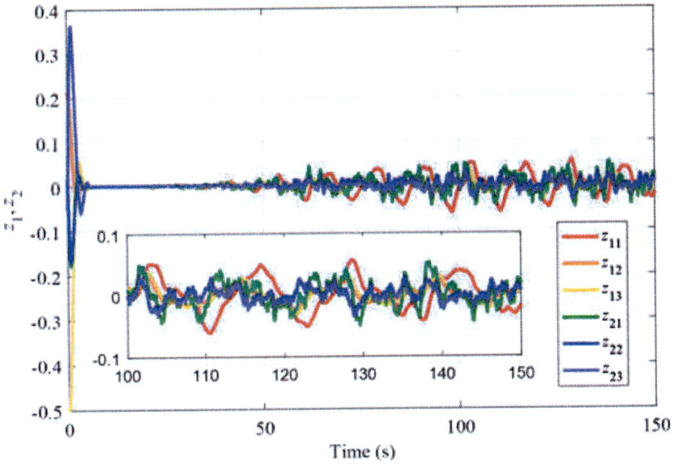

Fig. 7.2 The controller auxiliary variables z_1 and z_2 with disturbances

Figures 7.1, 7.2, 7.3, 7.4, 7.5, 7.6, 7.7, 7.8, 7.9, 7.10, 7.11, 7.12, 7.13, 7.14 and 7.15 are the comparative simulation results of the controller (7.63) proposed in this chapter. Compared with the controller in Chap. 6, the controller proposed in this chapter is the finite-time stable performance guaranteed controller.

Figures 7.1 and 7.2 show the simulation results of the auxiliary variables z_1 and z_2 for the spacecraft attitude tracking error and the compensator in the command filter backstepping method under performance constraints without and with considering external disturbances. It can be seen from the curves that the auxiliary variables can

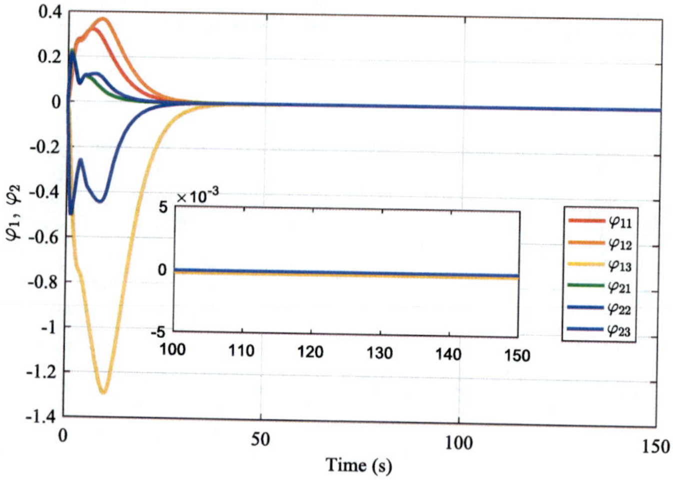

Fig. 7.3 The command filter compensators φ_1 and φ_2 without disturbances

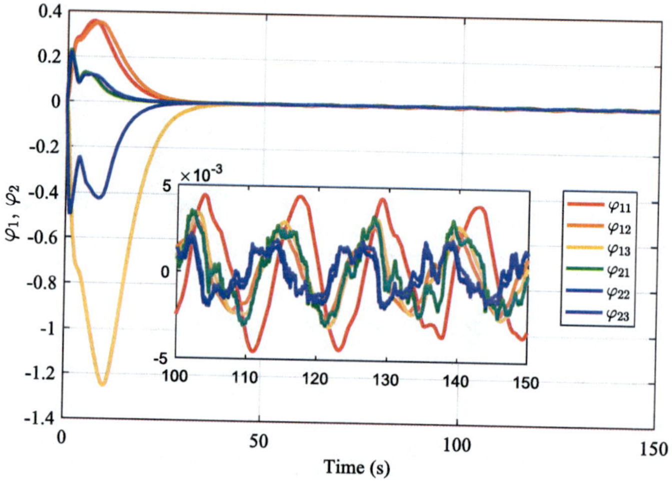

Fig. 7.4 The command filter compensators φ_1 and φ_2 with disturbances

converge. When there is a bounded disturbance, the auxiliary variables converge to the neighborhood of the equilibrium point. Figures 7.3 and 7.4 are the simulation diagrams of the compensator variables without and with considering external disturbances, respectively. It can be seen that the compensator variables can converge to the equilibrium point when the disturbance is not considered and converge to the neighborhood of the equilibrium point when the disturbance is considered.

Figures 7.5, 7.6 and 7.7 show the closed-loop system tracking error curves under the action of six controllers, 'A', 'B', 'C', 'D', 'E', and 'F'. The proposed controller

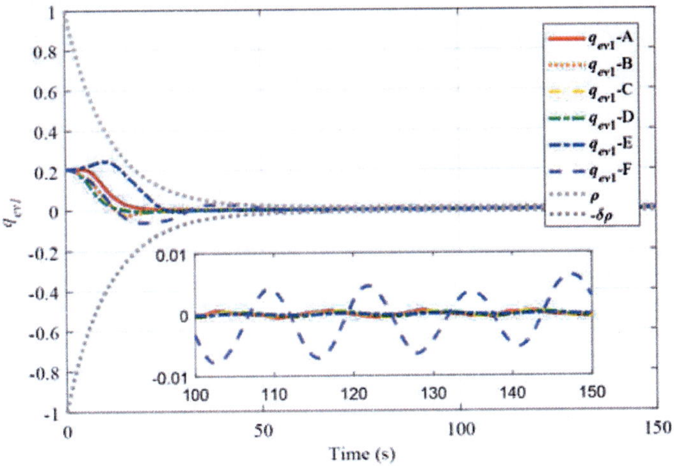

Fig. 7.5 The spacecraft attitude tracking error q_{ev1}

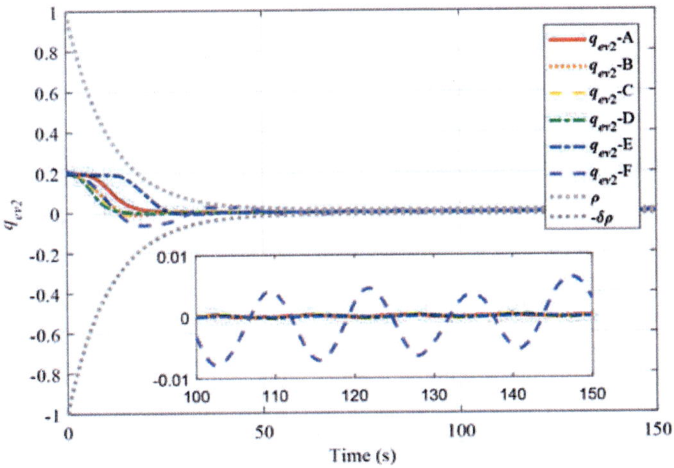

Fig. 7.6 The spacecraft attitude tracking error q_{ev2}

adds a filter structure compared to controller 'B'. Therefore, although the tracking error convergence rate of the closed-loop system is slower than that under the action of controller 'B', the control input is smaller than that of 'B'. Moreover, considering the actuator's saturation, it is more suitable for the practical spacecraft system. Compared with the command filter controller 'C' and the command filter controller 'D' based on a low-pass filter, the controller proposed in this chapter can achieve similar or better transient performance and guarantee similar or less steady-state error. Compared with the traditional prescribed performance controller 'E', the controller proposed in this

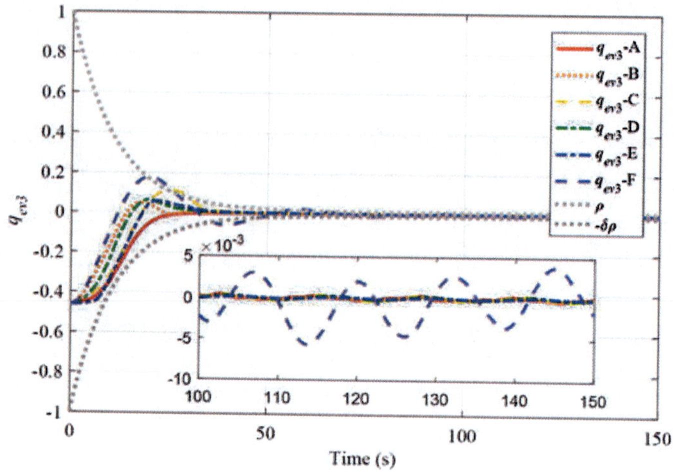

Fig. 7.7 The spacecraft attitude tracking error q_{ev3}

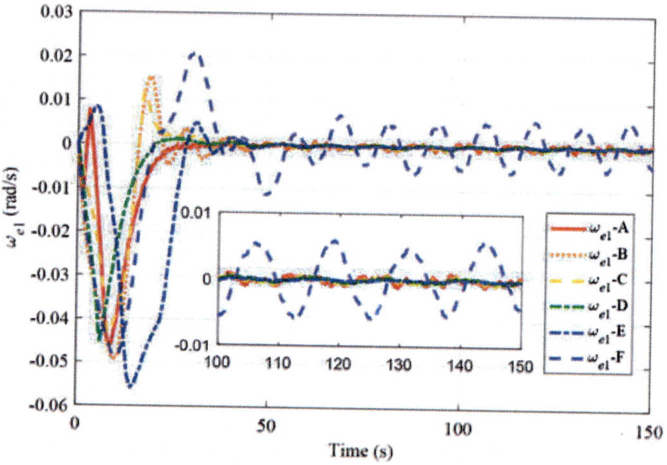

Fig. 7.8 The angular velocity tracking error ω_{e1}

chapter has the control input, whose amplitude and frequency are more suitable for the spacecraft system. Compared with the finite-time controller 'F', the controller proposed in this chapter has explicit performance constraints and guarantees, as well as smaller steady-state errors. Figures 7.8, 7.9 and 7.10 show the curves of the spacecraft attitude angular velocity tracking errors under the action of six controllers, 'A', 'B', 'C', 'D', 'E' and 'F'. It can be seen that the controller proposed in this chapter has a smoother angular velocity error curve, a smaller steady-state error, and better disturbance suppression performance.

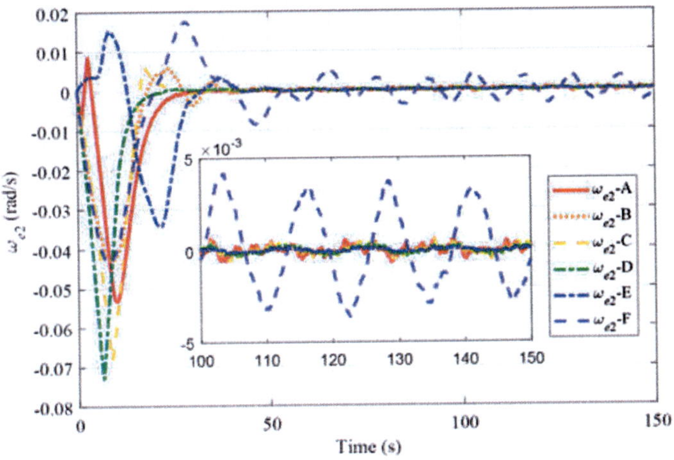

Fig. 7.9 The angular velocity tracking error ω_{e2}

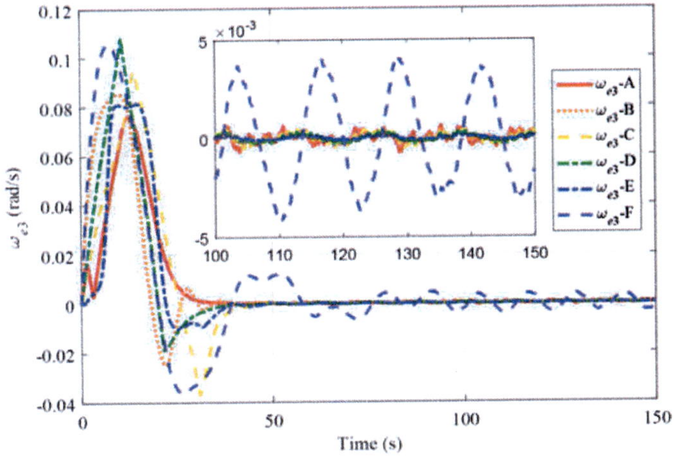

Fig. 7.10 The angular velocity tracking error ω_{e3}

Figures 7.11 and 7.12 respectively show the spacecraft attitude error index $I_a = ||q_{ev}|| + ||\omega_e||$ and the energy consumption. The comparison shows that the proposed controller has a faster convergence rate and smaller steady-state error. Moreover, through the simulation curves, it can be seen that the proposed controller has smaller energy consumption when the closed-loop system achieves similar transient and steady-state performance as the comparison controllers. Figures 7.13 and 7.14 are the control input τ_a of the spacecraft attitude tracking model and the expected control command τ_d designed by theory, respectively. After τ_d passes through the command filter, the system control input τ_a is obtained. It can be seen from the simulation

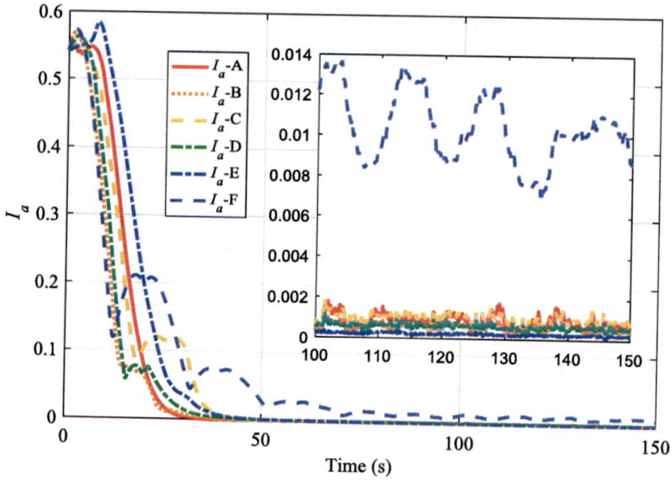

Fig. 7.11 The error index I_a

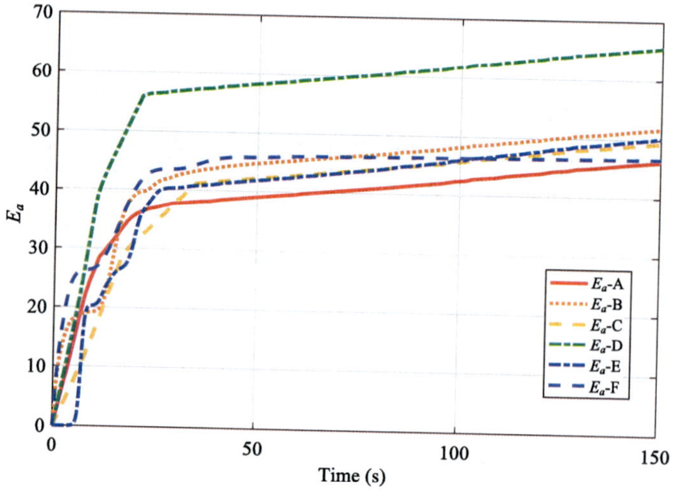

Fig. 7.12 The energy consumption

diagram that the command filter significantly adjusts the amplitude of the control command signal, so that the control input of the spacecraft conforms to the practical situation of the spacecraft. In contrast, the amplitude of the control input designed by the theory is too large, which is unsuitable for the practical spacecraft actuator. Figure 7.15 presents the external disturbances considered in the numerical simulation.

Through numerical simulation, it is fully demonstrated that under the action of the controller proposed in this chapter, the closed-loop system can obtain better transient and steady-state performance by consuming less energy. The performance

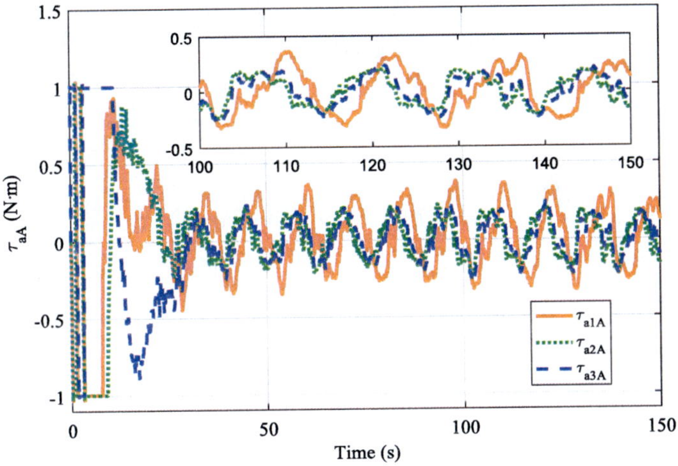

Fig. 7.13 The control input τ_a of 'A'

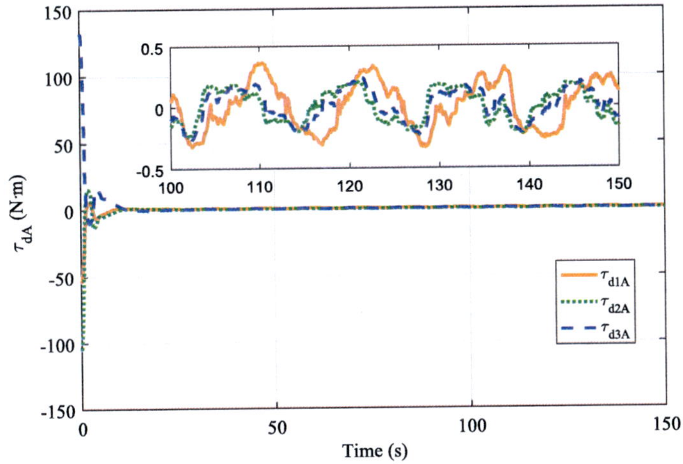

Fig. 7.14 The control command τ_d of 'A' before filtering

is guaranteed. That is, the maximum settling time of the tracking error is guaranteed to be less than the predefined maximum settling time, the steady-state error is less than the predefined steady-state error interval, and the overshoot satisfies the constraints. Considering that the controller proposed in this chapter has the ability to converge in the finite time, the transient performance can be further improved by further tuning the controller parameters. Simultaneously, the command filter has the strong ability to adjust the theoretically designed control command τ_d, so that the obtained actual control input τ_a is more suitable for the practical spacecraft attitude control application.

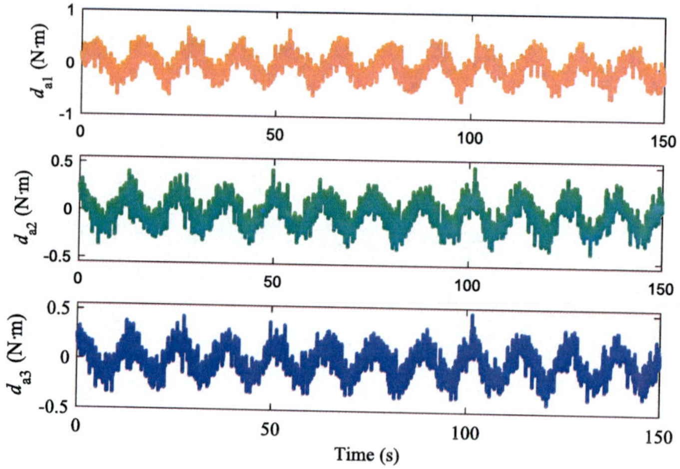

Fig. 7.15 The external disturbances

7.6 Conclusion

Similar to the idea in Chap. 4, considering the conservativeness of constrained asymptotic stability, based on Chap. 6, this chapter designs the spacecraft finite-time performance guaranteed control with command filter. On the one hand, the performance improvement brought by finite-time stability to the closed-loop system further enhances the transient and steady-state performance of spacecraft attitude control. On the other hand, finite-time controllers provide a more detailed description of the constraint interval and settling time. Simultaneously, this chapter designs finite-time stable compensators for the command filters. Considering external disturbances, this chapter further refines the proposed controller. The rationality of the proposed controllers is demonstrated through rigorous theoretical analysis and proof. Under the proposed controllers' effect, the tracking error of the spacecraft attitude closed-loop system can converge to the predefined steady-state error range within the predefined maximum settling time. Furthermore, the practical control input of the spacecraft attitude tracking system satisfies the amplitude constraints, etc. Finally, through the numerical simulation analysis and comparison with the advanced controller, the effectiveness of the controller proposed in this chapter is verified.

References

1. Huang, X., Lin, W., Yang, B.: Global finite-time stabilization of a class of uncertain nonlinear systems. Automatica **41**(5), 881–888 (2005)
2. Li, Y.X.: Finite time command filtered adaptive fault tolerant control for a class of uncertain nonlinear systems. Automatica **106**, 117–123 (2019)

3. Fu, C., Wang, Q.G., Yu, J., Lin, C.: Neural network-based finite-time command filtering control for switched nonlinear systems with backlash-like hysteresis. IEEE Transactions on Neural Networks and Learning Systems **32**(7), 3268–3273 (2020)
4. Wang, L., Wang, H., Liu, P.X., Ling, S., Liu, S.: Fuzzy finite-time command filtering output feedback control of nonlinear systems. IEEE transactions on fuzzy systems **30**(1), 97–107 (2020)
5. Yu, J., Zhao, L., Yu, H., Lin, C., Dong, W.: Fuzzy finite-time command filtered control of nonlinear systems with input saturation. IEEE transactions on cybernetics **48**(8), 2378–2387 (2017)
6. Ma, J., Park, J.H., Xu, S.: Command-filter-based finite-time adaptive control for nonlinear systems with quantized input. IEEE Transactions on Automatic Control **66**(5), 2339–2344 (2020)
7. You, X., Dian, S., Liu, K., Guo, B., Xiang, G., Zhu, Y.: Command filter-based adaptive fuzzy finite-time tracking control for uncertain fractional-order nonlinear systems. IEEE Transactions on Fuzzy Systems (2022)
8. Xia, J., Li, B., Su, S.F., Sun, W., Shen, H.: Finite-time command filtered event-triggered adaptive fuzzy tracking control for stochastic nonlinear systems. IEEE Transactions on Fuzzy Systems **29**(7), 1815–1825 (2020)
9. Xu, W., Li, Z., Cui, G., Wang, C., Hu, F.: Fuzzy adaptive finite time command filter backstepping control of power system. International Journal of Control, Automation and Systems **19**(11), 3812–3822 (2021)
10. Zhang, J., Xia, J., Sun, W., Wang, Z., Shen, H.: Command filter-based finite-time adaptive fuzzy control for nonlinear systems with uncertain disturbance. Journal of the Franklin Institute **356**(18), 11270–11284 (2019)
11. Li, S., Ahn, C.K., Xiang, Z.: Command-filter-based adaptive fuzzy finite-time control for switched nonlinear systems using state-dependent switching method. IEEE Transactions on Fuzzy Systems **29**(4), 833–845 (2020)
12. Yu, J., Shi, P., Chen, X., Cui, G.: Finite-time command filtered adaptive control for nonlinear systems via immersion and invariance. Science China information sciences **64**(9), 1–14 (2021)
13. Wang, Y., Chen, W., Zhang, S., Zhu, J., Cao, L.: Command-filtered incremental backstepping controller for small unmanned aerial vehicles. Journal of Guidance, Control, and Dynamics **41**(4), 954–967 (2018)
14. Farrell, J.A., Polycarpou, M., Sharma, M., Dong, W.: Command filtered backstepping. IEEE Transactions on Automatic Control **54**(6), 1391–1395 (2009)
15. Liu, H., Zhang, T.: Adaptive neural network finite-time control for uncertain robotic manipulators. Journal of Intelligent & Robotic Systems **75**(3), 363–377 (2014)
16. Qian, C., Lin, W.: A continuous feedback approach to global strong stabilization of nonlinear systems. IEEE Transactions on Automatic Control **46**(7), 1061–1079 (2001)
17. Liu, T., Gao, Y., Buss, M.: Adaptive output tracking control of piecewise affine systems with prescribed performance. IEEE Transactions on Systems, Man, and Cybernetics: Systems pp. 1–13 (2021). 10.1109/TSMC.2021.3122982
18. Jiang, B., Li, C., Ma, G.: Finite-time output feedback attitude control for spacecraft using "adding a power integrator" technique. Aerospace Science and Technology **66**, 342–354 (2017)
19. Zou, A.M.: Finite-time output feedback attitude tracking control for rigid spacecraft. IEEE Transactions on Control Systems Technology **22**(1), 338–345 (2013)
20. Bechlioulis, C.P., Rovithakis, G.A.: Robust adaptive control of feedback linearizable mimo nonlinear systems with prescribed performance. IEEE Transactions on Automatic Control **53**(9), 2090–2099 (2008)
21. Du, H., Li, S., Qian, C.: Finite-time attitude tracking control of spacecraft with application to attitude synchronization. IEEE Transactions on Automatic Control **56**(11), 2711–2717 (2011)

Chapter 8
Spacecraft Performance Guaranteed Optimal Control with Adaptive Dynamic Programming

Abstract Optimal control is crucial for spacecraft missions, with limited spacecraft fuel or energy storage and few opportunities for refueling after launch. The performance guaranteed control can provide transient or steady-state performance constraints for spacecraft closed-loop systems and even the settling time upper bounds. However, it cannot guarantee the optimization of the control inputs. To optimize the control input under the condition of satisfying the performance constraints, considering the optimal control method based on adaptive dynamic programming (ADP), this chapter proposes the spacecraft performance guaranteed optimal control and its improved algorithm based on experience replay. Detailed theoretical derivation, analysis, and rigorous theoretical proofs of the algorithms are given. By applying these algorithms, the transient and steady-state behavior of the spacecraft closed-loop system can meet the desired constraints, and the control input can be optimized when the constraints are considered. Finally, comprehensive numerical simulation analysis and comparative simulation verify the effectiveness, feasibility, and advantages of the proposed algorithms.

8.1 Introduction

Chapters 3–7 design the performance guaranteed control algorithms for the spacecraft with practical constraints. Although the closed-loop system satisfies the performance constraints, including the desired settling time, some controllers are constrained by the command filter to adjust the control signal, but the aforementioned control algorithms are not optimal or suboptimal.

For spacecraft control tasks, the optimization problems of tracking error and energy consumption are critical. Although we expect higher accuracy and tighter boundary constraints for spacecraft motion through performance guaranteed control, the optimization problems of tracking error and energy consumption still need to be seriously considered. Consider the high cost of fuel storage and the utility of fuel and energy in practical aerospace engineering. This chapter seeks to design the performance guaranteed optimal or suboptimal control algorithms for spacecraft tracking control.

Y. Gao and D. Li, *Spacecraft Maneuver with Performance Guaranteed*,
https://doi.org/10.1007/978-981-99-4653-2_8

It can be seen from the foregoing that the attitude tracking error model of the spacecraft considering the guaranteed performance constraints is such as (3.6). In this chapter, the optimal control algorithms based on adaptive dynamic programming are designed based on this model so that the spacecraft attitude tracking process not only satisfies the predefined maximum settling time and transient and steady-state performance constraints but also achieves the control input optimization.

Specifically, according to the related spacecraft attitude tracking model and the performance constraints with the predefined maximum settling time in Chap. 2, combined with the performance guaranteed control in Chap. 3, we can see that the key technique of the performance guaranteed control lies in the range mapping of constrained intervals. Through performance-constrained auxiliary variables, new auxiliary systems can be obtained, such as system (3.5), (6.35), (7.8) and so on. In the previous chapter's control design, the auxiliary system's stability is used to obtain the stability and performance guaranteed of the spacecraft closed-loop system. On this basis, the optimal or sub-optimal control algorithms can be further designed for the spacecraft through the same idea. Based on the description in Sect. 2.4.4, for nonlinear systems, design the cost function (2.75) and a Hamiltonian function (2.76). Substituting the designed optimal control law (2.80) back into the conditional Eq. (2.79), the Hamilton–Jacobi–Bellman (HJB) equation (2.81) about the gradient of the value function can be obtained. Solving this equation to obtain the gradient expression of the value function, then the optimal control law can be obtained. When the Hamiltonian function is bounded, suboptimal control can be obtained.

Combined with the specific situation of spacecraft attitude tracking, the problems or control objectives studied in this chapter are described as follows:

Problem 8.1 Consider the spacecraft attitude tracking error model (2.28), which does not consider the small environmental disturbance moment suffered by the spacecraft. Design predefined performance guaranteed parameters according to the task requirements, including the predefined maximum settling time t_{ps}, the maximum steady-state error ρ_t, the maximum overshoot χ_t. Let the system $\dot{x} = f + g\tau_a$ be the performance-constrained auxiliary system, where $x \in \mathbb{R}^n$ is the performance-constrained auxiliary variable, that is, if x is bounded, then the tracking error of the spacecraft attitude closed-loop system satisfies all the predefined performance constraints. Design the optimal control τ_a to minimize the value function of the auxiliary system as (2.75). To achieve the objective that the Hamiltonian function is zero (the HJB equation has a solution) or the Hamiltonian function is bounded (the HJB equation has an approximately optimal suboptimal solution).

For linear systems, the solution of the HJB equation can be obtained by solving the algebraic Riccati equation, and the optimal control can be obtained by substituting the solution into the control. However, the HJB equation is often challenging to solve for nonlinear systems. Therefore, this chapter adopts the adaptive dynamic programming (ADP) method to approximate the solution of the HJB equation through the neural networks to obtain the optimal control. Due to the approximation error, the obtained approximate solution and the actual solution often have bounded errors. Therefore, the solution of the HJB equation obtained by the neural network is often a suboptimal solution that approximates the optimal solution, and the obtained control is suboptimal. It should be noted that through the ADP method, the control is not optimal in the process of neural network weight fitting, which also means that in the online training process of control input, it is not optimal control. However, the optimal or sub-optimal control can be obtained after the weight estimation error has converged and bounded. Therefore, in the transient process, the control is sub-optimal, and when the tracking error converges, the control is closer to the optimal.

In practical applications, the selection of predefined performance constraints often needs to consider the practical situation. According to the practical requirements of the task, it is necessary to give the corresponding expected transient and steady-state constraints that are more accurate than traditional expected values. The expected conditions cannot be arbitrarily given beyond the practical capabilities of the space-craft itself. For the controllable spacecraft attitude tracking system, this chapter aims to achieve the optimization objective under the transient and steady-state guaranteed performance constraints of the closed-loop system described in the Problem 8.1 through the design of the performance guaranteed control, and further improve the spacecraft control performance, etc. Instead, it is impossible to directly improve the capabilities of the spacecraft actuator itself only through algorithm design.

In addition, the problem proposed above is an optimal control problem under performance constraints. Considering the mapping or transformation of the system with respect to performance constraints, the definition of the corresponding value function will change according to the mapping system. Since the system is performance guaranteed, the optimal control here is no longer the optimality of the original spacecraft system, but the optimality of the performance-constrained system and the appropriate optimization of the original system.

This chapter is arranged as follows. In Sect. 8.2, the spacecraft performance guaranteed optimal control algorithm is proposed based on the online training adaptive dynamic programming method. Section 8.3 relaxes the persistent excitation that the control algorithm in Sect. 8.2 must satisfy and proposes spacecraft performance guaranteed optimal control based on experience replay adaptive dynamic programming. Section 8.4 shows the simulation analysis results, followed by concluding remarks in Sect. 8.5.

8.2 Spacecraft Performance Guaranteed Optimal Control Based on Adaptive Dynamic Programming

To realize the optimal control algorithm design of spacecraft attitude under the performance guaranteed or constraints, according to the model and the performance guaranteed technique in Chap. 2, the original system is transformed to obtain the auxiliary system about the performance-constrained auxiliary variable $\sigma_i(t, t_{ps})$ in (2.60). The optimal control algorithm is designed for the auxiliary system and mapped to the spacecraft attitude control input, and the optimal algorithm for the guaranteed attitude performance is obtained.

8.2.1 Control Design

Considering the spacecraft attitude tracking error model (2.28), define auxiliary variables $\sigma_i(t, t_{ps})$, which is defined in (2.60), with respect to the spacecraft attitude tracking error q_{ev}, where t_{ps} is the PMST. The relationship between t_{ps} and $\sigma_i(t, t_{ps})$ is shown in Sect. 2.4.2. In the following text, the abbreviation σ_i is $\sigma_i(t, t_{ps})$. If σ_i is bounded, then according to Theorem 2.1, the closed-loop system tracking error q_{ev} will converge to the pre-defined steady-state error range within the pre-defined time t_{ps}, and satisfy the transient and steady-state performance constraints (2.55).

Considering that the spacecraft attitude system is a second-order system, we design a sliding mode surface ζ_i as follows with respect to auxiliary variable σ_i in (2.60), where λ_i is a positive constant. It is easy to know that when ζ_i converges or is bounded, σ_i also converges or is bounded.

$$\zeta_i = \dot{\sigma}_i + \lambda_i \sigma_i \tag{8.1}$$

Taking the derivative of (8.1) with respect to time and substituting (2.61), we have

$$\dot{\zeta}_i = \alpha_i + \beta_i \ddot{q}_{evi}(t) \tag{8.2}$$

where

$$\begin{aligned}\alpha_i = &- \frac{2r_i \dot{q}_{evi}(t)\dot{\rho}_i(t)}{(1+\delta_i)\rho_i{}^2(t)} + \frac{4r_i q_{evi}(t)\dot{\rho}_i^2(t)}{(1+\delta_i)\rho_i{}^3(t)} + \dot{r}_i\dot{\vartheta}_i \\ &- \frac{2r_i\ddot{\rho}_i(t)q_{evi}(t) + 2r_i\dot{\rho}_i(t)\dot{q}_{evi}(t)}{(1+\delta_i)\rho_i{}^2(t)} + \lambda_i r_i \dot{\vartheta}_i\end{aligned} \tag{8.3}$$

$$\beta_i = \frac{2r_i}{(1+\delta_i)\rho_i(t)} \tag{8.4}$$

$$r_i = \partial\varphi^{-1}(\vartheta_i)/\partial\vartheta_i \tag{8.5}$$

Considering the requirements of the pre-defined settling time and pre-defined performance, and at the same time, in order to facilitate the design of the optimized control algorithm, we design the auxiliary system as follows. By expressing (8.2) into a compact form, we have

$$\dot{\zeta} = \ddot{\sigma} + \Gamma\dot{\sigma} = A + B\ddot{q}_{ev}(t) \tag{8.6}$$

where $\zeta = [\zeta_1, \ldots, \zeta_n]^T$, $\sigma = [\sigma_1, \ldots, \sigma_n]^T$, $A = [\alpha_1, \ldots, \alpha_n]^T$, $B = \text{diag}(\beta_1, \ldots, \beta_n)$, $\Gamma = \text{diag}(\lambda_1, \ldots, \lambda_n)$.

If we can design the control algorithm τ_d so that the auxiliary system (8.6) is stable or bounded, then σ is stable or bounded, and the closed-loop system satisfies the constraint condition (2.55). Therefore, the performance of the closed-loop system based on the desired parameters is guaranteed. Substituting (2.28) into (8.6), where external disturbances are not considered for this control algorithm, the spacecraft attitude tracking error model related to the performance constraints can be obtained as follows:

$$\dot{\zeta} = f + g\tau_d \tag{8.7}$$

where $f = A + BM_a^{-1}[-C_a\dot{q}_{ev} - G_a]$, $g = \frac{1}{2}M_a^{-1}P^T\kappa_B^{-1}$, $\tau_d = \kappa_B(P^T)^{-1}M_a BM_a^{-1}P^T\tau_a$, $\kappa_B = \text{diag}(\frac{2\delta_1\rho_1(t)}{1+\delta_1}, \ldots, \frac{2\delta_n\rho_n(t)}{1+\delta_n})$, δ_i is the overshoot adjustment parameter in the performance constraint. Considering that when the error converges, $\kappa_B(P^T)^{-1}M_a BM_a^{-1}P^T \to 1$. Therefore, if τ_d is the optimal control, then τ_a can converge to the optimal control. According to the spacecraft quaternion model, it can be known that g is bounded, that is, $||g|| \le \varepsilon_{b_g}$, where ε_{b_g} is a positive number. Considering that f in the auxiliary system is related to the state of the system, there exists a $\mathcal{K}$-like function on $||f||$ such that $||f|| \le \varepsilon_{bf}||\zeta||$, where ε_{bf} is a positive constant. The auxiliary system (8.7) was obtained.

Next, the optimal control algorithm with the guaranteed performance constraints can be achieved simply by designing the optimal control algorithm for the above auxiliary system (8.7) such that ζ converges or is bounded. Specifically, that is, the tracking error q_{ev} of the spacecraft attitude closed-loop system converges within a pre-defined range within a pre-defined time, satisfying the transient and steady-state performance constraints (2.55), and the controller is optimal.

Next, we complete the design of the optimal control algorithm.

The value function used in this algorithm is defined as

$$V(\zeta) \triangleq \int_0^t \left(\zeta^T Q_\zeta \zeta + \tau_d^T R_\zeta \tau_d\right) \mathrm{d}\tau \tag{8.8}$$

where Q_ζ and R_ζ are positive definite symmetric matrices. The Hamiltonian function related to the value function is defined as

$$H(\zeta, \tau_d, \nabla V) = \zeta^T Q_\zeta \zeta + \tau_d^T R_\zeta \tau_d + (\nabla V)^T \cdot (f + g\tau_d) \tag{8.9}$$

where $\nabla V = \frac{\partial V}{\partial \zeta}$ represents the partial derivative of V with respect to ζ. According to the optimal control theory, the optimal control strategy can be obtained as

$$\tau_d{}^* = -\frac{1}{2}R_\zeta^{-1}g^T\nabla V \tag{8.10}$$

Then the Hamilton–Jacobi–Bellman (HJB) equation associated with the optimal control problem is

$$\zeta^T Q_\zeta \zeta + (\nabla V)^T f - \frac{1}{4}(\nabla V)^T g R_\zeta^{-1} g^T \nabla V = 0 \tag{8.11}$$

According to the Weierstrass higher-order approximation theorem [1–3], there exist complete independent basis sets such that the solution to V in (8.11) and their gradients ∇V are uniformly approximated [4, 5]. So we write the value functions based on neural networks (NN) as

$$V(\zeta) = W^T \phi(\zeta) + \varepsilon(\zeta) \tag{8.12}$$

where $W \in \mathbb{R}^K$ represents the unknown bounded ideal weights of the critic neural network of the value function, and $||W|| \leq W_{\max}$. $\phi(\zeta) \in \mathbb{R}^K$ represents the NN activation function vector, K indicates the number of hidden layer neurons, $\varepsilon(\zeta) \in \mathbb{R}$ is the weight estimation error. In addition, the partial derivative of the value function with respect to ζ, that is, the gradient of V with respect to ζ is

$$\nabla V(\zeta) = \nabla \phi^T(\zeta) W + \nabla \varepsilon \tag{8.13}$$

where $\nabla \phi(\zeta)$ and $\nabla \varepsilon$ are the bounded gradients of the activation function and estimation error, respectively. According to the current research results of NN approximation, when $K \to \infty$, the estimated error $\varepsilon \to 0$, the gradient of the estimated error $\nabla \varepsilon \to 0$. When K is a finite fixed value, $\nabla \phi(\zeta)$ and $\nabla \varepsilon$ are bounded [6, 7].

The ideal fixed weights W are unknown, and it is difficult to obtain the actual W for nonlinear systems. Therefore, in this section, the ideal weights are estimated by the adaptive neural network. Define the estimated weight as $\hat{W}$, and the estimated error as $\tilde{W} = W - \hat{W}$. Then the estimated value of the value function $\hat{V}(\zeta)$ is as follows

$$\hat{V} = \hat{W}^T \phi \tag{8.14}$$

According to the theory of adaptive dynamic programming, such as in Sect. 2.4.4, the Hamiltonian function becomes,

$$H = \zeta^T Q_\zeta \zeta + \tau_d^T R_\zeta \tau_d + W^T \nabla \phi \left(f + g\tau_d\right) \triangleq \varepsilon_H \tag{8.15}$$

where the residual error is

$$\varepsilon_H = -(\nabla \varepsilon)^T \left(f + g\tau_d\right) \tag{8.16}$$

Substituting the estimated value of the value function into the Hamiltonian equation, we have

$$H = \zeta^T Q_\zeta \zeta + \tau_d^T R_\zeta \tau_d + \hat{W}^T \nabla \phi \left(f + g\tau_d\right) \triangleq e \tag{8.17}$$

It can be seen from the above formula that if there is $e \to 0$, then the control algorithm (8.10) obtained by $\hat{W}$ is the optimal control. If e is bounded and stable, then a suboptimal control can be obtained. Therefore, it is desired to select $\hat{W}$ to minimize the square residual error

$$E = \frac{1}{2} e^T e \tag{8.18}$$

The common adaptive law based on gradient descent algorithm for the critic neural network weight parameter $\hat{W}$ is as follows [4, 8, 9],

$$\dot{\hat{W}} = -a \frac{\partial E}{\partial \hat{W}} = -a\theta e \tag{8.19}$$

where $a > 0$ is a constant gain, $\theta = \nabla\phi\,(f + g\tau_d)$. And there exists a positive constant θ_M such that $||\theta|| \le \theta_M$ [6]. Based on the above adaptive law (8.19), in this section, the adaptive law of the following form is adopted [7]

$$\dot{\hat{W}} = -a \frac{\theta}{\left(1 + \theta^T\theta\right)^2} e \tag{8.20}$$

It should be noted that $\frac{\theta}{\left(1+\theta^T\theta\right)^2}$ is the modified Levenberg Marquardt algorithm. The normalization term in its denominator is raised to the second power [7, 10, 11].

According to (8.15) and (8.17), it can be seen that $e = \varepsilon_H - \tilde{W}^T\theta$. According to the adaptive law (8.20) and the Hamiltonian equation, the dynamic equation about the weight estimation error $\tilde{W}$ can be obtained as follows

$$\dot{\tilde{W}} = a \frac{\theta}{\left(1 + \theta^T\theta\right)^2} e = a \frac{\theta}{\left(1 + \theta^T\theta\right)^2} (\varepsilon_H - \tilde{W}^T\theta) \tag{8.21}$$

To make the neural network fully trained, it is necessary to make θ satisfy the persistent excitation (PE) conditions, that is, $||\theta|| \ge \theta_m$, where θ_m is a positive number [4]. In order to satisfy PE conditions, the noises can be introduced into the system, and the system state can be reset [6].

So far, the design for $\hat{W}$ is completed. Since the system model and control form considered in this section are known, only the critic neural network needs to be trained. Substituting (8.13) into (8.10), the optimal control can be obtained as

$$\tau_d = -\frac{1}{2} R_\zeta^{-1} g^T \nabla V = -\frac{1}{2} R_\zeta^{-1} g^T (\nabla\phi^T(\zeta) W + \nabla\varepsilon) \tag{8.22}$$

Then, the estimated optimal control policy $\hat{\tau}_d$ is

$$\hat{\tau}_d = -\frac{1}{2} R_\zeta^{-1} g^T \nabla\phi^T \hat{W} \tag{8.23}$$

8.2.2 Stability Analysis

Theorem 8.1 *Considering the spacecraft attitude tracking error model (2.28) (disturbances are not considered) with the predefined maximum settling time T_{ps} and predefined transient and steady-state constraints (2.55), if the spacecraft performance guaranteed optimal control strategy (8.23) and the adaptive update law of the critic neural network weights (8.20) are selected, and the PE conditions are satisfied, then the critic neural network weight estimation error $\tilde{W}$ is uniformly ultimately bounded.*

Proof Select the candidate Lyapunov function with respect to $\tilde{W}$ as

$$L_W = \frac{1}{2}a^{-1}\mathrm{tr}(\tilde{W}^T\tilde{W}) \tag{8.24}$$

Find the derivative of (8.24) with respect to time, and substitute (8.21) to get

$$\begin{aligned} \dot{L}_W &= a^{-1}\mathrm{tr}(\tilde{W}^T\dot{\tilde{W}}) \\ &= a^{-1}\mathrm{tr}\left(\tilde{W}^T a\frac{\theta}{(1+\theta^T\theta)^2}(\varepsilon_H - \tilde{W}^T\theta)\right) \end{aligned} \tag{8.25}$$

Considering $\theta_m \leq ||\theta|| \leq \theta_M$, then $\frac{\theta}{(1+\theta^T\theta)^2} \leq \frac{\theta_M}{(1+\theta_m^T\theta_m)^2} \triangleq \varepsilon_\theta$, where ε_θ is a positive constant. Then, we have

$$\begin{aligned} \dot{L}_W &\leq a^{-1}\mathrm{tr}\left(\tilde{W}^T a\varepsilon_\theta(\varepsilon_H - \tilde{W}^T\theta)\right) \\ &\leq a^{-1}\mathrm{tr}\left(\tilde{W}^T a\varepsilon_\theta\varepsilon_H - \tilde{W}^T a\varepsilon_\theta\tilde{W}^T\theta\right) \end{aligned} \tag{8.26}$$

According to Young's inequality, taking into account $a^{-1}\tilde{W}^T a\varepsilon_\theta\varepsilon_H \leq \frac{a^2}{2}\varepsilon_\theta^2||\tilde{W}||^2 + \frac{\varepsilon_H^2}{2a^2}$, then

$$\begin{aligned} \dot{L}_W &\leq \frac{a^2}{2}\varepsilon_\theta^2||\tilde{W}||^2 + \frac{\varepsilon_H^2}{2a^2} - \varepsilon_\theta\theta_m||\tilde{W}||^2 \\ &\leq -(\varepsilon_\theta\theta_m - \frac{a^2}{2}\varepsilon_\theta^2)||\tilde{W}||^2 + \frac{\varepsilon_H^2}{2a^2} \end{aligned} \tag{8.27}$$

Choose an appropriate $a > 0$ such that $\varepsilon_\theta\theta_m - \frac{a^2}{2}\varepsilon_\theta^2 > 0$, i.e. satisfy the condition $a < \sqrt{\frac{2\varepsilon_\theta\theta_m}{\varepsilon_\theta^2}}$. Then, when the following inequality is satisfied,

$$||\tilde{W}|| > \sqrt{\frac{\varepsilon_H^2}{a^2(2\varepsilon_\theta\theta_m - a^2\varepsilon_\theta^2)}} \tag{8.28}$$

$\dot{L}_W < 0$ holds. Then the neural network weight estimation error $||\tilde{W}||$ is uniformly ultimately bounded, and the boundary is $||\tilde{W}|| \leq \sqrt{\frac{\varepsilon_H^2}{a^2(2\varepsilon_\theta\theta_m - a^2\varepsilon_\theta^2)}}$. □

Theorem 8.2 *Considering the spacecraft attitude tracking error model (2.28) (disturbances are not considered) with the predefined maximum settling time T_{ps} and predefined transient and steady-state constraints (2.55), if the spacecraft performance guaranteed optimal control strategy (8.23) and the adaptive update law of the critic neural network weights (8.20) are selected, the PE conditions are satisfied, and Theorem 8.1 holds, For any initial conditions t_0 and ζ_0, there exists the time $T(\zeta_0, \kappa_B)$ such that $\zeta(t)$ is uniformly ultimately bounded with the bound κ_B described as*

$$\|\zeta(t)\| \leq \kappa_\beta, t \geq t_0 + T \tag{8.29}$$

where $\kappa_\beta \triangleq \sqrt{\kappa_\varpi \lambda_{\min}^{-1}(Q_1)}$, $\kappa_\varpi \in \mathbb{R}^+$. The performance-constrained auxiliary variable σ_i and the spacecraft attitude tracking error q_{ev} are uniformly ultimately bounded. And the maximum settling time $T_{q_{ev}}$ satisfies $T_{q_{ev}} \leq \min\{T_{ps}, T(\zeta_0, \kappa_B)\}$, where T_{ps} is the predefined maximum settling time. The final bound of the steady-state error is given by Theorem 2.2.

Proof We take the derivative of the value function and substitute the optimal control policy to get

$$\dot{V}(t) = (\nabla V(\zeta))^T \left(f + g\hat{\tau}_d\right) \tag{8.30}$$

Considering the HJB equation, we have

$$(\nabla V)^T f = +\frac{1}{4}(\nabla V)^T g R_\zeta^{-1} g^T \nabla V - \zeta^T Q_\zeta \zeta \tag{8.31}$$

Substituting (8.31) into (8.30) yields

$$\dot{V}(t) = +(\nabla V)^T g\hat{\tau}_d + \frac{1}{4}(\nabla V)^T g R_\zeta^{-1} g^T \nabla V - \zeta^T Q_\zeta \zeta \tag{8.32}$$

and

$$\dot{V}(t) = -\zeta^T Q_\zeta \zeta - \frac{1}{4}(\nabla V)^T g R_\zeta^{-1} g^T \nabla V - (\nabla V_i)^T g\left(\tau_d - \hat{\tau}_d\right) \tag{8.33}$$

Substituting (8.22) and (8.13) into (8.33), we have

$$\dot{V}(t) = -\zeta^T Q_\zeta \zeta - \frac{1}{4}(\nabla V)^T g R_\zeta^{-1} g^T \nabla V + \varpi \tag{8.34}$$

where $\varpi = \frac{1}{2}(\nabla\phi^T W + \nabla\varepsilon)^T g R_\zeta^{-1} g^T (\nabla\phi^T(\zeta)\tilde{W} + \nabla\varepsilon(\zeta))$.

Considering that the items in ϖ are all UUB, then ϖ has a boundary as follows

$$\varpi \leq \left\| \frac{1}{2}(\nabla\phi^T W + \nabla\varepsilon)^T g R_\zeta^{-1} g^T \left(\nabla\phi^T(\zeta)\tilde{W} + \nabla\varepsilon(\zeta)\right) \right\| \leq \kappa_\varpi \tag{8.35}$$

where κ_ϖ is constant.

Considering that R_ζ is the positive definite symmetric matrix, then the following term is also a positive definite

$$\frac{1}{4}(\nabla V)^T g R_\zeta^{-1} g^T \nabla V > 0 \tag{8.36}$$

Then we have

$$\dot{V}(t) \leq -\zeta^T Q_\zeta \zeta + \kappa_\varpi \leq -\lambda_{\min}(Q_\zeta)||\zeta||^2 + \kappa_\varpi \tag{8.37}$$

where $\lambda_{\min}(Q_\zeta)$ represents the least eigenvalue of the matrix Q_ζ. Define the compact set

$$\Omega_\zeta \triangleq \{\zeta : ||\zeta|| \leq \sqrt{\kappa_\varpi \lambda_{\min}^{-1}(Q_\zeta)}\} \tag{8.38}$$

If ζ is inside Ω_ζ, ζ is bounded. If ζ is outside Ω_ζ, then all the derivatives $\dot{V}(t)$ are negative. So $||\zeta||$ is UUB.

Considering that in (8.1), $\zeta_i = \dot{\sigma}_i + \lambda_i \sigma_i$, similar to the derivation in the previous section, according to $||\zeta_i||$ is UUB, we know that σ_i is also UUB. According to Theorems 2.1 and 2.2, it can be seen that the spacecraft attitude tracking error q_{ev} is UUB, and its boundary is given by the Theorem 2.2. Its practical settling time is less than the predefined maximum settling time T_{ps} and satisfies the performance constraint (2.55).

□

8.3 Spacecraft Performance Guaranteed Optimal Control Based on Experience Replay Adaptive Dynamic Programming

8.3.1 *Control Design*

In the control design of the previous section, it is necessary to make the system meet the PE condition, which is relatively strict. To relax this condition, this section adopts the experience replay technique. That is, the sampling data of the past time is applied to evaluate the update law of the critic NN. In this way, the PE condition can be relaxed. Specifically, the optimal control policy based on experience replay for the auxiliary system (8.7) is designed in this section. Based on the design in the previous section, the value function (8.8), Hamiltonian function (8.9), and ideal optimal control policy (8.10) is still defined in this section. The optimal control policy can be obtained by solving the HJB equation and obtaining the gradient expression of the value function. In this section, the optimal control policy is obtained by estimating the gradient of the value function through an adaptive neural network.

Similar to the previous section, we again design the estimated value function through the neural network as follows,

$$V(\zeta) = W^T\phi(\zeta) + \varepsilon(\zeta) \tag{8.39}$$

where $W \in \mathbb{R}^K$ represents the critic NN weight of the value function, $\phi(\zeta) \in \mathbb{R}^K$ represents the activation function vector, K is the number of neurons in the hidden layer, $\varepsilon(\zeta) \in \mathbb{R}$ is defined as the weight estimation error. Therefore, the partial derivative with respect to ζ is

$$\nabla V(\zeta) = \nabla\phi^T(\zeta)W + \nabla\varepsilon \tag{8.40}$$

Through the optimal control theory, the Hamiltonians become

$$H = \zeta^T Q_\zeta \zeta + \tau_d^T R_\zeta \tau_d + W^T \nabla\phi\,(f + g\tau_d) = \varepsilon_H \tag{8.41}$$

where the residual errors are

$$\varepsilon_H = -(\nabla\varepsilon)^T\,(f + g\tau_d) \tag{8.42}$$

The optimal control policy for spacecraft attitude tracking, considering performance constraints, is

$$\tau_d = -\frac{1}{2}R_\zeta^{-1}g^T\nabla V \tag{8.43}$$

Substituting τ_d into the HJB equation (8.11), we have

$$\zeta^T Q_\zeta \zeta - \frac{1}{4}W^T R_g W + W^T\nabla\phi f = \varepsilon_{HJB} \tag{8.44}$$

where $R_g = \nabla\phi g R_\zeta^{-1} g^T \nabla\phi^T$, $\varepsilon_{HJB} = -\nabla\varepsilon^T f + \frac{1}{2}W^T\nabla\phi g R_\zeta^{-1} g^T \nabla\varepsilon + \frac{1}{4}\nabla\varepsilon^T g R_\zeta^{-1} g^T \nabla\varepsilon$ is the estimation error of the HJB equation. According to [1, 12], if the number of neurons in the hidden layer $K \to \infty$, then ε_{HJB} will converge to 0, if K is finite, then ε_{HJB} is bounded, and has $\|\varepsilon_{HJB}\| \le \varepsilon_{b_{HJB}}$, where $\varepsilon_{b_{HJB}}$ is a constant.

For nonlinear systems, it is tough, even impossible, to obtain the actual W, so neural networks are needed to estimate these values. Define the estimated value of W as $\hat{W}$, and the estimated errors are $\tilde{W} = W - \hat{W}$. Therefore, the estimated value function is

$$\hat{V} = \hat{W}^T\phi \tag{8.45}$$

Then, the approximate optimal control policy for the system is given by

$$\hat{\tau}_d = -\frac{1}{2}R_\zeta^{-1}g^T\nabla\phi^T\hat{W} \tag{8.46}$$

Substituting the approximated control policy into (8.9), we get the approximated Hamiltonian equation with the definition of the residual errors e as follows

$$\begin{aligned} H &= \zeta^T Q_\zeta\zeta + \hat{\tau}_d^T R_\zeta\hat{\tau}_d + \hat{W}^T\nabla\phi\left(f + g\hat{\tau}_d\right) \\ &= \zeta^T Q_\zeta\zeta + \frac{1}{4}\hat{W}^T R_g\hat{W} + \hat{W}^T\nabla\phi f - \frac{1}{2}\hat{W}^T R_g\hat{W} = e \end{aligned} \tag{8.47}$$

If $e = 0$, then the control policy is optimal. If $e \to 0$, then the control policy tends to be optimal, and if e is bounded, then the control policy is suboptimal. It is desired to select $\hat{W}$ to minimize the square residual error as

$$E = \frac{1}{2}e^T e \tag{8.48}$$

The adaptive update law of gradient descent for $\hat{W}$ designed in the previous section is as follows,

$$\dot{\hat{W}} = -a\frac{\partial E}{\partial\hat{W}} = -a\frac{\theta}{\left(1 + \theta^T\theta\right)^2}e \tag{8.49}$$

where $\theta = \nabla\phi\left(f + g\hat{\tau}_d\right)$, $a > 0$ is the adaptive rate of critic networks. It should be noted that here $\frac{\theta}{\left(1+\theta^T\theta\right)^2}$ is the modified Levenberg Marquardt algorithm, and the normalization term in the denominator is raised to the second power [7]. According to the current research results, there is a constant θ_M such that $\|\theta\| \le \theta_M$ [5].

In the traditional adaptive dynamic programming method, θ in the adaptive update law (8.49) needs to satisfy the persistence of excitation (PE) conditions, so that $\|\theta\| \ge \theta_m$ holds, where θ_m is a positive constant, especially in online update algorithms. Inspired by [8], the critic NN update law based on experience replay is designed in this section to relax the PE condition.

First, according to the experience replay technique, the weights of the critic NN can be updated by the stored past data, and the current weight data [13]. Define the residuals about the current estimated weights $\hat{W}(t)$ and stored data at t_k as

$$e(t_k) = r_k(t_k) + \hat{W}^T(t)\theta_k \tag{8.50}$$

where $r_k(t_k) = \zeta^T(t_k)Q_\zeta\zeta(t_k) + \tau_d^T(t_k)R_\zeta\tau_d(t_k)$, $\theta_k = \nabla\phi(t_k)\left(f(t_k) + g(t_k)\tau_d(t_k)\right)$ is the sample data stored at time t_k, $k = 1, ..., p$, p is the number of samples.

According to the following conditions, the PE conditions in the update process can be relaxed through the experience replay technique. That is, the system can meet the PE conditions through the experience replay technique [13]: Let $\Xi = [\bar{\theta}_1, ..., \bar{\theta}_p]$ be the weight estimate record data for the critic NN $\hat{W}$, where $\bar{\theta}_k = \frac{\theta_k}{1+\theta_k^T\theta_k}$. Then Ξ

contains as many linear independent elements as the number of hidden neurons in the corresponding critic NN, i.e., rank(Ξ) = K.

Therefore, for the critic NN, the number of data stores is a fixed constant, and $p > K$, where K is the number of neurons. The experience replay used in this section collects past bounded values. Although in a reasonable control process, the past data must be bounded, for the sake of rigor, this limitation is still added when defining the experience playback data collection.

Based on the above analysis, the update law of the critic neural network based on experience replay is designed as follows:

$$\begin{aligned}\dot{\hat{W}} = & -a\frac{\theta}{\left(1+\theta^T\theta\right)^2}\left(e - \frac{1}{4}\hat{W}^T R_g \hat{W}\right) \\ & - a(\kappa_W + \kappa_{W\theta}\bar{\theta}^T)\hat{W} - a\sum\nolimits_{k=1}^{p}\frac{\theta_k}{\left(1+\theta_k^T\theta_k\right)^2}e_k\end{aligned} \tag{8.51}$$

where $\bar{\theta} = \frac{\theta}{1+\theta^T\theta}$, κ_W, and $\kappa_{W\theta}$ are adjustable parameters, p is the number of past parameters required. Considering that the model used in this study and the optimal controller in the form of a closed-loop system is known, it is only necessary to train the critic network.

According to the facts stated in [7], for a given compact set $\Omega \subset \mathbb{R}^n$: 1) The critic NN estimation errors and their gradients are bounded so that $||\varepsilon|| < \varepsilon_{b_\varepsilon}$, $||\nabla\varepsilon|| < \varepsilon_{b_{\nabla\varepsilon}}$. 2) The NN activation function and its gradient are bounded so that $||\phi|| < \varepsilon_{b_\phi}$, $||\nabla\phi|| < \varepsilon_{b_{\nabla\phi}}$. 3) The critic NN weights are bounded, and there is a positive constant $W_{\max}$ such that $||W|| < W_{\max}$.

8.3.2 Stability Analysis

Theorem 8.3 *Considering the spacecraft attitude tracking error model (2.28) (disturbances are not considered) with the predefined maximum settling time T_{ps} and predefined transient and steady-state constraints (2.55), if the spacecraft performance guaranteed optimal control strategy (8.46) and the experience-replay-based adaptive update law of the critic neural network weights (8.51) are selected, and the PE conditions are satisfied, then the critic neural network weight estimation error $\tilde{W}$ is uniformly ultimately bounded. The performance-constrained auxiliary variable σ_i is uniformly ultimately bounded. The spacecraft attitude tracking error q_{ev} is uniformly ultimately bounded. The tracking error practical settling time is less than the predefined maximum settling time T_{ps} and satisfies the performance constraint (2.55).*

Proof The candidate Lyapunov function for the value function V and the critic NN estimation error $\tilde{W}$ are as follows,

$$L = L_V + L_W \tag{8.52}$$

$$L_V = V \tag{8.53}$$

$$L_W = \frac{1}{2}a^{-1}\tilde{W}^T\tilde{W} \tag{8.54}$$

Step 1. Calculate $\dot{L}_V$.

Differentiating V in (8.39) along the trajectories generated by the approximate control policies $\hat{\tau}_d$, we have

$$\begin{aligned}\dot{V} &= W^T\nabla\phi\left(f + g\hat{\tau}_d\right) + \varepsilon_V \\ &= W^T\left(\nabla\phi f - \frac{1}{2}R_g W\right) + \frac{1}{2}W^T R_g\tilde{W} + \varepsilon_V\end{aligned} \tag{8.55}$$

where $\varepsilon_V = \nabla\varepsilon^T(f - \frac{1}{2}gR_\zeta^{-1}g^T\nabla\phi^T\hat{W})$.

According to (8.44), we have

$$W^T\nabla\phi f - \frac{1}{2}W^T R_g W = \varepsilon_{HJB} - \zeta^T Q_\zeta\zeta - \frac{1}{4}W^T R_g W \tag{8.56}$$

Equation (8.55) can be rewritten as

$$\dot{V} = -\zeta^T Q_\zeta\zeta - \frac{1}{4}W^T R_g W + \frac{1}{2}W^T R_g\tilde{W} + \varepsilon_V + \varepsilon_{HJB} \tag{8.57}$$

Substituting (8.57) into the derivative of (8.53) with respect to time, we have

$$\dot{L}_V = \dot{V} = -\zeta^T Q_\zeta\zeta - \frac{1}{4}W^T R_g W + \frac{1}{2}W^T R_g\tilde{W} + \varepsilon_V + \varepsilon_{HJB} \tag{8.58}$$

For the item ε_V, we scale it as follows. From the definition of ε_V, we have

$$||\varepsilon_V|| \leq \varepsilon_{b\nabla_\varepsilon}\varepsilon_{bf}||\zeta|| + \frac{1}{2}\varepsilon_{b\nabla\varepsilon 1}\varepsilon_{b_g}^2\varepsilon_{b\nabla\phi 1}\lambda_{\max}(R_\zeta^{-1})(||W|| + ||\tilde{W}||) \tag{8.59}$$

Step 2. Calculate $\dot{L}_W$.

According to (8.44) and (8.47), we can get

$$\begin{aligned} e &= +\hat{W}^T\nabla\phi f - W^T\nabla\phi f + \frac{1}{4}W^T R_g W - \frac{1}{4}\hat{W}^T R_g \hat{W} + \varepsilon_{HJB} \\ &= -\tilde{W}^T\theta + \frac{1}{4}\tilde{W}^T R_g \tilde{W} + \frac{1}{4}W^T R_g \tilde{W} - \frac{1}{4}\tilde{W}^T R_g W + \varepsilon_{HJB} \\ &= -\tilde{W}^T\theta + \Upsilon \end{aligned} \tag{8.60}$$

where $\Upsilon = \frac{1}{4}\tilde{W}^T R_g \tilde{W} + \frac{1}{4}W^T R_g \tilde{W} - \frac{1}{4}\tilde{W}^T R_g W + \varepsilon_{HJB}$. Similarly, the experience replay term can be written as

$$e_k(t_k) = -\tilde{W}^T\theta_k(t_k) + \Upsilon_k(t_k) \tag{8.61}$$

where $\Upsilon_k(t_k)$ is the value of $\Upsilon(t)$ at time t_k.

Substituting (8.60) and (8.61) into (8.51), we have

$$\begin{aligned} \dot{\hat{W}} = & -a\frac{\theta}{\left(1+\theta^T\theta\right)^2}\left(e - \frac{1}{4}\hat{W}^T R_g \hat{W}\right) - a(\kappa_W + \kappa_{W\theta}\bar{\theta}^T)\hat{W} \\ & - a\sum\nolimits_{k=1}^{p}\frac{\theta_k}{\left(1+\theta_k^T\theta_k\right)^2}e_k \\ = & -a\frac{\theta}{\left(1+\theta^T\theta\right)^2}\left(-\tilde{W}^T\theta + \frac{1}{4}\tilde{W}^T R_g \tilde{W} + \varepsilon_{HJB} - \frac{1}{4}\hat{W}^T R_g \hat{W}\right) \\ & - a(\kappa_W + \kappa_{W\theta}\bar{\theta}^T)\hat{W} + a\sum\nolimits_{k=1}^{p}\frac{\theta_k}{\left(1+\theta_k^T\theta_k\right)^2}\theta_k^T\tilde{W} - a\sum\nolimits_{k=1}^{p}\frac{\theta_k\Upsilon_k}{\left(1+\theta_k^T\theta_k\right)^2} \end{aligned} \tag{8.62}$$

Substitute the above formula into the derivative of $\tilde{W}$ with respect to time, and simplify, we can get

$$\begin{aligned} \dot{\tilde{W}} = & -\dot{\hat{W}} = -a\frac{\theta}{\left(1+\theta^T\theta\right)^2}\tilde{W}^T\theta + a\frac{\theta}{\left(1+\theta^T\theta\right)^2}\varepsilon_{HJB} \\ & - a\frac{1}{4}\frac{\theta}{\left(1+\theta^T\theta\right)^2}W^T R_g W + a\frac{1}{2}\frac{\theta}{\left(1+\theta^T\theta\right)^2}W^T R_g \tilde{W} \\ & + a\kappa_W W - a\kappa_W \tilde{W} + a\kappa_{W\theta}\bar{\theta}^T W - a\kappa_{W\theta}\bar{\theta}^T\tilde{W} \\ & - a\left(\sum\nolimits_{k=1}^{p}\frac{\theta_k}{\left(1+\theta_k^T\theta_k\right)^2}\theta_k^T\right)\tilde{W} \\ & + a\sum\nolimits_{k=1}^{p}\frac{\theta_k\Upsilon_k}{\left(1+\theta_k^T\theta_k\right)^2} \end{aligned} \tag{8.63}$$

Considering that $\bar{\theta} = \frac{\theta}{(1+\theta^T\theta)}, m_s = 1 + \theta^T\theta, \frac{\bar{\theta}}{m_s} = \frac{\theta}{(1+\theta^T\theta)^2}$, in addition, $\bar{\theta}_k$ and m_{sk} is the data of $\bar{\theta}$ and m_s at time t_k respectively, and the above formula is simplified to,

$$\begin{aligned}
\dot{\tilde{W}} &= \dot{W} - \dot{\hat{W}} \\
&= -a\bar{\theta}\bar{\theta}^T\tilde{W} \\
&\quad -a\kappa_{W\theta}\bar{\theta}^T\tilde{W} + a\frac{1}{2}\frac{\bar{\theta}}{m_s}W^T R_g\tilde{W} \\
&\quad -a\sum\nolimits_{k=1}^{p}\bar{\theta}_k\bar{\theta}_k^T\tilde{W} - a\kappa_W\tilde{W} \\
&\quad -a\frac{1}{4}\frac{\bar{\theta}}{m_s}W^T R_g W + a\kappa_W W + a\kappa_{W\theta}\bar{\theta}^T W \\
&\quad +a\sum\nolimits_{k=1}^{p}\frac{\bar{\theta}_k}{m_{sk}}\Upsilon_k + a\frac{\bar{\theta}}{m_s}\varepsilon_{HJB}
\end{aligned} \tag{8.64}$$

Substitute the above formula into the derivative of (8.54) with respect to time, we can get

$$\begin{aligned}
\dot{L}_W &= a^{-1}\tilde{W}^T\dot{\tilde{W}} \\
&= -\tilde{W}^T\bar{\theta}\bar{\theta}^T\tilde{W} \\
&\quad -\tilde{W}^T\bar{\theta}\kappa_{W\theta}\tilde{W} + \tilde{W}^T\bar{\theta}\frac{W^T R_g}{2m_s}\tilde{W} \\
&\quad +\tilde{W}^T\bar{\theta}\frac{\varepsilon_{HJB}}{m_s} - \tilde{W}^T\bar{\theta}\frac{W^T R_g W}{4m_s} \\
&\quad -\tilde{W}^T\sum\nolimits_{k=1}^{p}\bar{\theta}_k\bar{\theta}_k^T\tilde{W} - \tilde{W}^T\kappa_W\tilde{W} \\
&\quad +\tilde{W}^T\bar{\theta}\kappa_{W\theta}^T W \\
&\quad +\tilde{W}^T\kappa_W W + \tilde{W}^T\sum\nolimits_{k=1}^{p}\frac{\bar{\theta}_k}{m_{sk}}\Upsilon_k
\end{aligned} \tag{8.65}$$

Step 3. Calculate $\dot{L}$.

Combining Steps 1 and 2, we write the derivative of L as

$$
\begin{aligned}
\dot{L} &= \dot{L}_V + \dot{L}_W \\
&= -\zeta^T Q_\zeta \zeta - \tilde{W}^T \bar{\theta}\bar{\theta}^T \tilde{W} \\
&\quad - \tilde{W}^T \bar{\theta}\kappa_{W\theta}\tilde{W} + \tilde{W}^T \bar{\theta}\frac{W^T R_g}{2m_s}\tilde{W} \\
&\quad + \tilde{W}^T \bar{\theta}\frac{\varepsilon_{HJB}}{m_s} - \tilde{W}^T \bar{\theta}\frac{W^T R_g W}{4m_s} \\
&\quad - \tilde{W}^T \sum_{k=1}^{p} \bar{\theta}_k \bar{\theta}_k^T \tilde{W} - \tilde{W}^T \kappa_W \tilde{W} \\
&\quad + \tilde{W}^T \bar{\theta}\kappa_{W\theta}^T W \\
&\quad + \frac{1}{2}\tilde{W}^T R_g W + \tilde{W}^T \kappa_W W + \tilde{W}^T \sum_{k=1}^{p} \frac{\bar{\theta}_k}{m_{sk}} \Upsilon_k \\
&\quad - \frac{1}{4} W^T R_g W + \varepsilon_V + \varepsilon_{HJB} \\
&= -\zeta^T Q_\zeta \zeta - \tilde{W}^T \bar{\theta}\bar{\theta}^T \tilde{W} \\
&\quad - \tilde{W}^T \left(\sum_{k=1}^{p} \bar{\theta}_k \bar{\theta}_k^T + \kappa_W \right) \tilde{W} \\
&\quad - \tilde{W}^T \bar{\theta} \left(\kappa_{W\theta} - \frac{W^T R_g}{2m_s} \right) \tilde{W} \\
&\quad + \tilde{W}^T \bar{\theta} \left(\frac{\varepsilon_{HJB}}{m_s} - \frac{W^T R_g W}{4m_s} + \kappa_{W\theta}^T W \right) \\
&\quad + \tilde{W}^T \left(\frac{R_g W}{2} + \kappa_W W + \sum_{k=1}^{p} \frac{\bar{\theta}_k}{m_{sk}} \Upsilon_k \right) \\
&\quad - \frac{1}{4} W^T R_g W + \varepsilon_V + \varepsilon_{HJB}
\end{aligned} \tag{8.66}
$$

Substituting (8.59) into the above formula, we have

$$
\begin{aligned}
\dot{L} \le &- \zeta^T Q_\zeta \zeta - \tilde{W}^T \bar{\theta}\bar{\theta}^T \tilde{W} \\
&- \tilde{W}^T \left(\sum_{k=1}^{p} \bar{\theta}_k \bar{\theta}_k^T + \kappa_W \right) \tilde{W} \\
&- \tilde{W}^T \bar{\theta} \left(\kappa_{W\theta} - \frac{W^T R_g}{2m_s} \right) \tilde{W} \\
&+ \varepsilon_{b_{\nabla\varepsilon}} \varepsilon_{bf} ||\zeta|| \\
&+ \tilde{W}^T \bar{\theta} \left(\frac{\varepsilon_{HJB}}{m_s} - \frac{W^T R_g W}{4m_s} + \kappa_{W\theta}^T W \right) \\
&+ \tilde{W}^T \left(\frac{R_g W}{2} + \kappa_W W + \sum_{k=1}^{p} \frac{\bar{\theta}_k}{m_{sk}} \Upsilon_k + \frac{1}{2} \varepsilon_{b_{\nabla\varepsilon 1}} \varepsilon_{b_g}^2 \varepsilon_{b_{\nabla\phi 1}} \lambda_{\max}(R_\zeta^{-1}) \right) \\
&- \frac{1}{4} W^T R_g W + \frac{1}{2} \varepsilon_{b_{\nabla\varepsilon 1}} \varepsilon_{b_g}^2 \varepsilon_{b_{\nabla\phi 1}} \lambda_{\max}(R_\zeta^{-1}) W_{\max} + \varepsilon_{HJB}
\end{aligned} \tag{8.67}
$$

where $\lambda_{\min}(\cdot)$ and $\lambda_{\max}(\cdot)$ represent the minimum and maximum eigenvalues of the matrix, respectively.

Select the variable $Z = [\zeta, \bar{\theta}^T \tilde{W}, \tilde{W}]$, then the above formula can be written in a compact form as follows,

$$\dot{L} < -Z^T M Z + Z^T D + \varepsilon_c \tag{8.68}$$

where

$$M = \begin{bmatrix} m_{11} & 0 & 0 \\ 0 & m_{22} & m_{23} \\ 0 & m_{32} & m_{33} \end{bmatrix}, D = \begin{bmatrix} d_1 \\ d_2 \\ d_3 \end{bmatrix},$$

with

$$\begin{aligned}
m_{11} &= Q_\zeta, \\
m_{22} &= I, \\
m_{33} &= \sum\nolimits_{k=1}^{p} \bar{\theta}_k \bar{\theta}_k^T + \kappa_W, \\
m_{23} = m_{32}^T &= \frac{1}{2}\kappa_{W\theta} - \frac{W^T R_g}{4m_s}, \\
d_1 &= \varepsilon_{b_{\nabla_\varepsilon}} \varepsilon_{bf}, \\
d_2 &= \frac{\varepsilon_{HJB}}{m_s} - \frac{W^T R_g W}{4m_s} + \kappa_{W\theta}^T W, \\
d_3 &= \frac{R_g W}{2} + \kappa_W W + \sum\nolimits_{k=1}^{p} \frac{\bar{\theta}_k}{m_{sk}} \Upsilon_k + \frac{1}{2}\varepsilon_{b_{\nabla\varepsilon 1}} \varepsilon_{b_g}^2 \varepsilon_{b_{\nabla\phi 1}} \lambda_{\max}(R_\zeta^{-1}), \\
\varepsilon_c &= -\frac{1}{4} W^T R_g W + \frac{1}{2}\varepsilon_{b_{\nabla\varepsilon 1}} \varepsilon_{b_g}^2 \varepsilon_{b_{\nabla\phi 1}} \lambda_{\max}(R_\zeta^{-1}) W_{\max} + \varepsilon_{HJB}.
\end{aligned} \tag{8.69}$$

According to the aforementioned boundedness analysis, D and ε_c are both bounded, and the maximum boundary is defined as $D_{\max}$ and $\varepsilon_{c\max}$, respectively. Choose the appropriate κ_W and $\kappa_{W\theta}$ parameters so that M is positive definite, and then L is UUB. To make M positive definite, analyze the block matrix in M, where $m_{11} > 0$. Next, the Schur complements of the diagonal elements m_{22} and m_{33} can be obtained as $S_{22} = m_{22} - m_{23} m_{33}^{-1} m_{32}$ and $S_{33} = m_{33} - m_{32} m_{22}^{-1} m_{23}$, respectively. Choose appropriate parameters κ_W and $\kappa_{W\theta}$ such that $S_{22} > 0$ and $S_{33} > 0$, then a positive definite M can be obtained [7]. Thus, (8.68) can be rewritten as

$$\dot{L} < -\lambda_{\min}(M)||Z||^2 + D_{\max}||Z|| + \varepsilon_{c\max} \tag{8.70}$$

According to (8.70), we have $\dot{L} < 0$, if

$$||Z|| > \frac{D_{\max}}{2\lambda_{\min}(M)} + \sqrt{\frac{D_{\max}^2}{4\lambda_{\min}^2(M)} + \frac{\varepsilon_{c\max}}{\lambda_{\min}(M)}} = \varepsilon_Z. \tag{8.71}$$

It can be seen from (8.71) that when $||Z||$ is less than ε_Z, the system state and the estimated weight errors are bounded, and when $||Z|| > \varepsilon_Z$, $\dot{L}$ is negative definite. In summary, the system state and the weight errors are UUB.

Considering that in (8.1), $\zeta_i = \dot{\sigma}_i + \lambda_i \sigma_i$, similar to the derivation in the previous section, according to ζ_i is UUB, we know that σ_i is UUB. According to Theorems 2.1 and 2.2, it can be seen that the spacecraft attitude tracking error q_{ev} is UUB, and its boundary is given by Theorem 2.2. The spacecraft attitude tracking error practical settling time is less than the predefined maximum settling time T_{ps}, which satisfies the performance guaranteed constraint (2.55), which completes the proof. The control policy (8.23) is the optimal control policy for the system to satisfy the predefined performance guaranteed constraints after the neural network weights converge. □

8.4 Simulation Analysis

8.4.1 Simulation Parameters

This subsection first presents the initial conditions and parameters for the simulation. Considering the background of the on-orbit service spacecraft mission, although this study takes attitude motion as the research object, this section still gives relevant information about the spacecraft's orbit.

Considering that in the short-range on-orbit service mission, the spacecraft is very close to the target spacecraft. Therefore, the orbital parameters of the spacecraft to be controlled are also shown in Table 8.1. Next, consider the initial parameters of the spacecraft as shown in Table 8.2. Without loss of generality, the parameters of the Tables 8.1 and 8.2 can be changed according to specific tasks.

Next, consider the desired attitude of the on-orbit service spacecraft. In the numerical simulation of this study, the spacecraft is considered to track the attitude of the target spacecraft, and the attitude of the target spacecraft is the same as the settings in Sect. 4.4.1.

Table 8.1 The orbital parameters of the target spacecraft

Parameters	Values
Semimajor axis	7.35×10^6 m
Eccentricity	0.0068
Inclination	60 deg
Argument of periapsis	0 deg
Longitude of the ascending node	0 deg
Mean anomaly	0 deg

Table 8.2 The initial attitude parameters of the spacecraft

Parameters	Values
q_{v0}	$[0.5, 0.2, -0.3]^T$ rad
ω_{a0}	$[0, 0, 0]^T$ rad/s
J_s	$\begin{bmatrix} 150 & -20 & 10 \\ -20 & 150 & -10 \\ 10 & -10 & 120 \end{bmatrix}$ kg · m^2

$$q_{dv} = \begin{bmatrix} (r_{li-23} - r_{li-32})/(4q_{d0}) \\ (r_{li-31} - r_{li-13})/(4q_{d0}) \\ (r_{li-12} - r_{li-21})/(4q_{d0}) \end{bmatrix} \tag{8.72}$$

According to the spacecraft attitude tracking model in Sect. 2.3, the tracking error quaternion and the angular velocity tracking error can be obtained. To demonstrate the performance of the proposed controller and to facilitate comparison with similar controllers, the error-index is defined as $I_a = ||q_{ev}|| + ||\omega_e||$. To illustrate the energy consumption of the proposed controller, the energy consumption index $E_a = \sum_{i=1}^{n} \int_0^t |\tau_{ai}(t)|^2 dt$ is defined as in [14]. This index describes the electrical or chemical energy consumed by actuators such as thrusters or reaction flywheels during the stabilization of an attitude closed-loop system. The parameter selection basis for the controller proposed in this study is the same as in the previous chapter.

8.4.2 *Performance and Comparison*

8.4.2.1 Simulation Results of Controller (8.23)

This section presents the parameters selected in the simulation of the controller (8.23) in Sect. 8.2, as shown in the following table.

In Table 8.3, the parameter t_{ps} is the predefined maximum settling time. ρ_0 is the maximum initial value of the predefined performance boundary, which needs to be larger than the initial error quaternion. Considering that all quaternion elements are less than or equal to 1, thus, ρ_0 is taken as 1. ρ_t and ε_ρ provide steady-state error

Table 8.3 The initial parameters of the proposed controller (8.23)

Param	Value	Param	Value	Param	Value
t_{ps}	100s	ε_ρ	0.01	Q_ς	$0.01 I_{3\times3}$
ρ_0	1	δ_i	1	R_ς	$0.02 I_{3\times3}$
ρ_t	0.02	λ_i	1	a	0.05

bounds. δ_i is about whether to limit the overshoot, and λ_i is the sliding mode surface scale coefficient in the controller. Q_ζ and R_ζ are the constant matrices in the value function respectively, $I_{3\times3}$ is the 3×3 unit matrix, a is the ratio coefficient of the neural network weight adaptive law. In addition, the neural network basis set used in this section is $\phi = [\zeta_1^2, \zeta_2^2, \zeta_3^2, \zeta_1\zeta_2, \zeta_2\zeta_3, \zeta_3\zeta_1]^T$. The controller in this chapter does not consider external disturbances.

The controller proposed in this study is compared with existing state-of-the-art controllers. For the convenience of representation, this section labels the optimal performance guaranteed control based on adaptive dynamic programming (8.23) proposed in Sect. 8.2 as controller 'A'. If the controller (8.23) removes the corresponding performance constraints, then a controller that does not consider the predefined performance constraints can be obtained, marked as controller 'B'. Label the optimal control algorithm based on adaptive dynamic programming in [4] as controller 'C'. In order to illustrate the optimization effect of the optimal controller designed in this chapter, the controller proposed is compared with the traditional prescribed performance controller in [15], marked as controller 'D', and the finite-time controller in [16], labeled as 'E'. In this section, the comparative simulations of the above controllers will be carried out to illustrate the effectiveness and advantages of this study. In the simulation diagram and analysis, the symbols or subscripts 'A', 'B', 'C', 'D', and 'E' represent the above controllers.

Figures 8.1, 8.2, 8.3, 8.4, 8.5, 8.6, 8.7, 8.8, 8.9, 8.10, 8.11, 8.12 and 8.13 are the simulation of the controller (8.23) in Sect. 8.2. Figures 8.1, 8.2 and 8.3 show the tracking error curves of the spacecraft attitude closed-loop system. The error elements of each dimension are plotted separately to compare the performance of the controllers. From

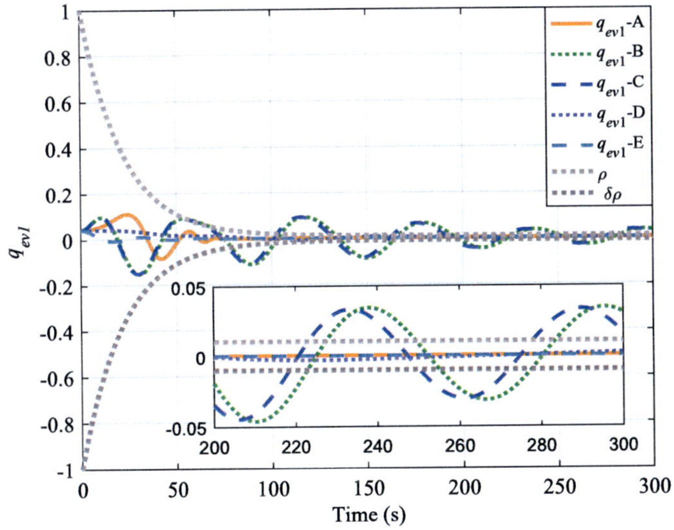

Fig. 8.1 The spacecraft attitude tracking error q_{ev1}

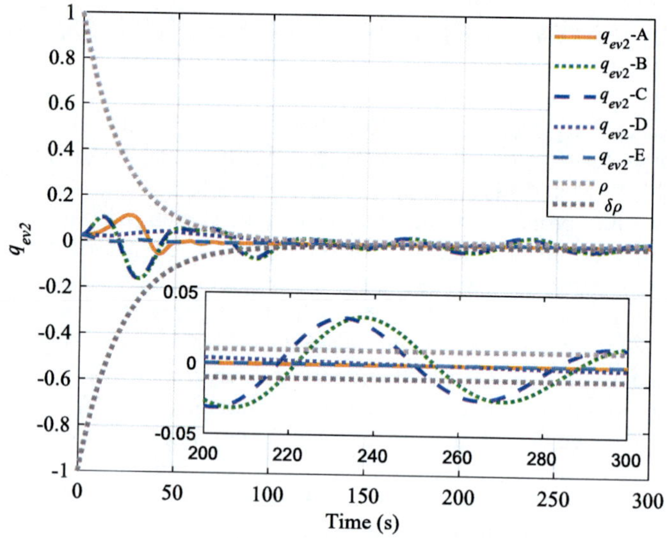

Fig. 8.2 The spacecraft attitude tracking error q_{ev2}

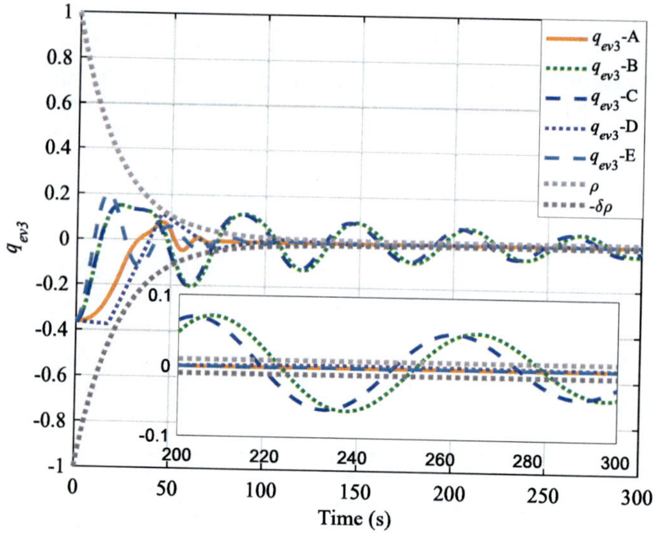

Fig. 8.3 The spacecraft attitude tracking error q_{ev3}

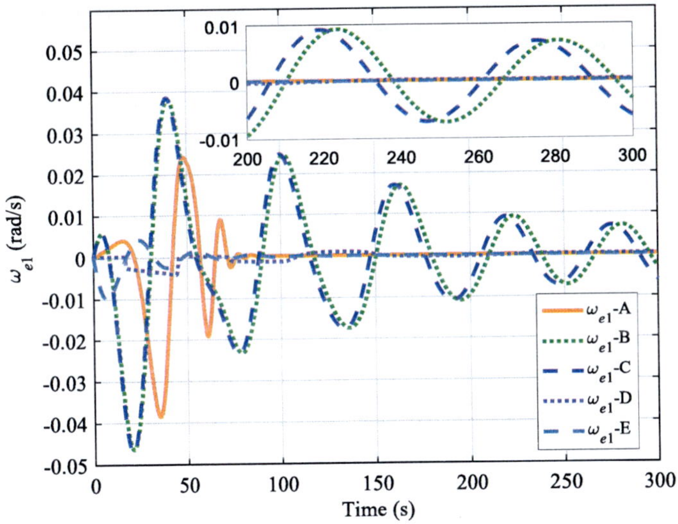

Fig. 8.4 The angular velocity tracking error ω_{e1}

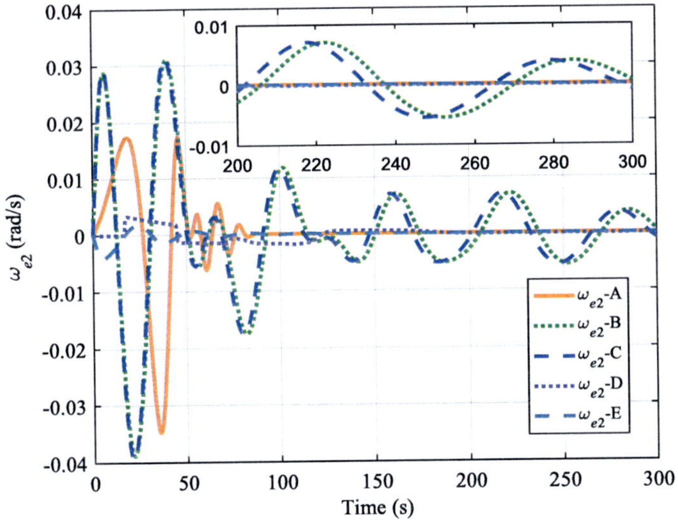

Fig. 8.5 The angular velocity tracking error ω_{e2}

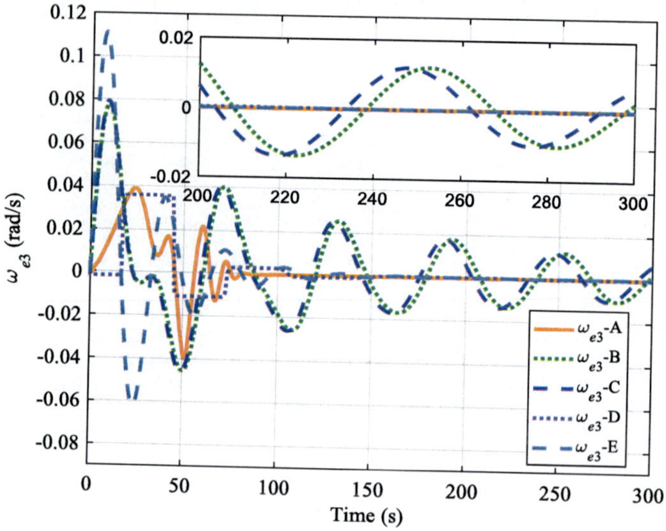

Fig. 8.6 The angular velocity tracking error ω_{e3}

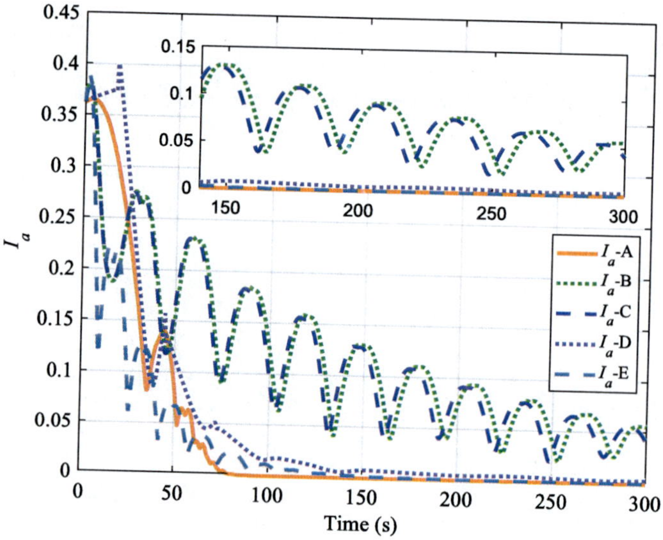

Fig. 8.7 The error index I_a

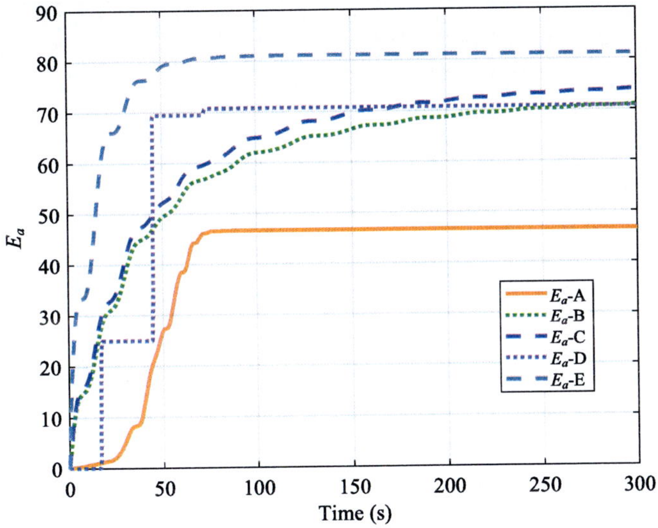

Fig. 8.8 The energy consumption

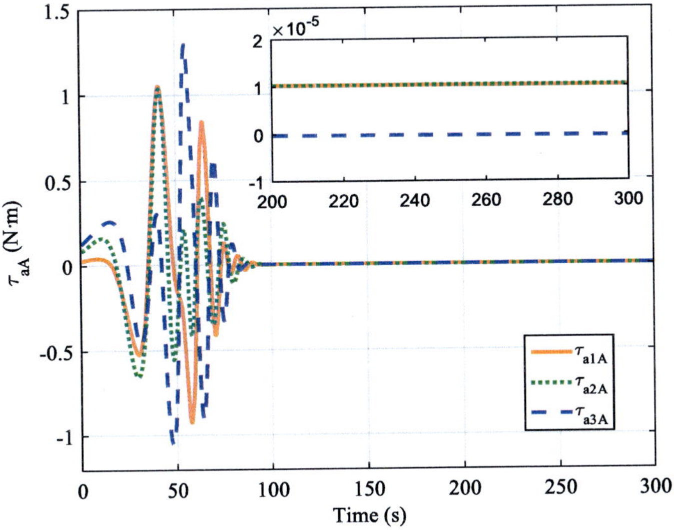

Fig. 8.9 The control input τ_a of 'A'

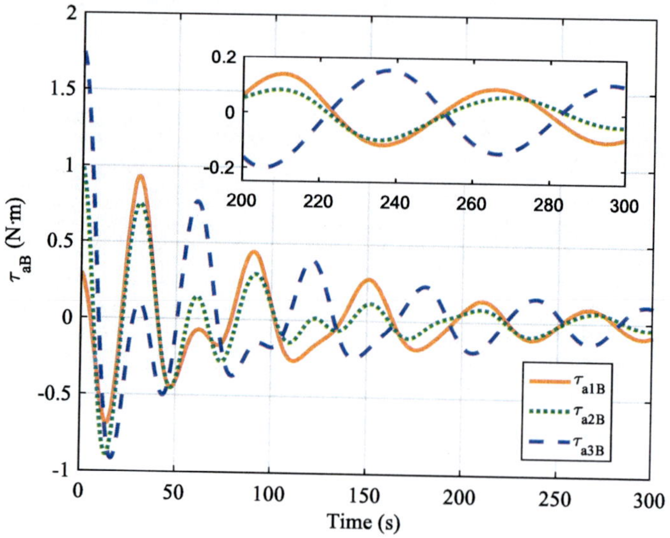

Fig. 8.10 The control input τ_a of 'B'

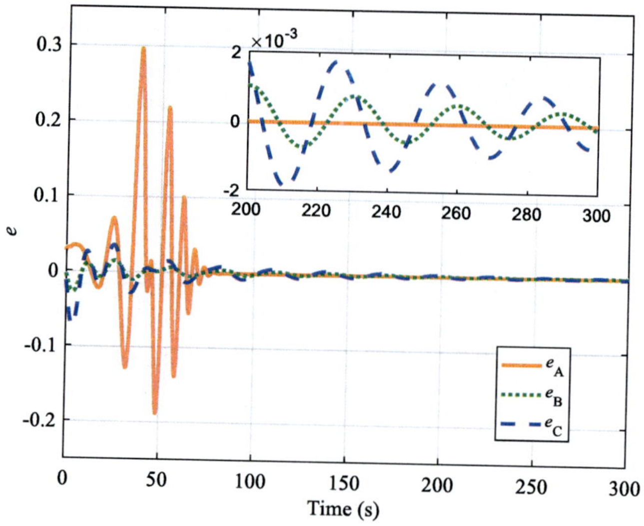

Fig. 8.11 The residual errors of the HJB equations

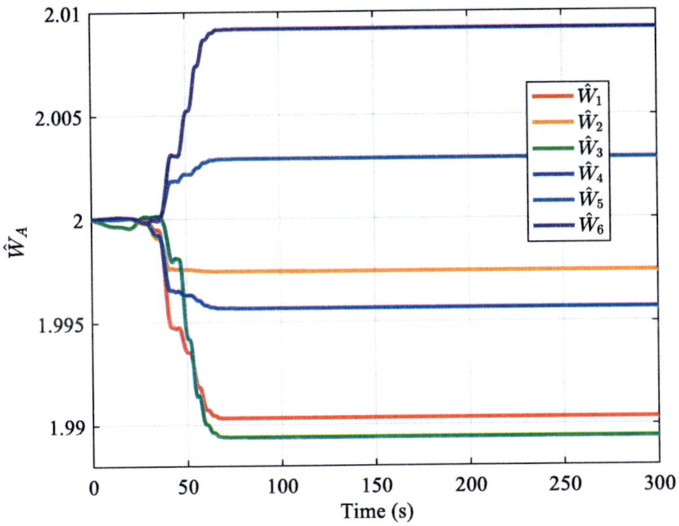

Fig. 8.12 The adaptive weights of the controller 'A'

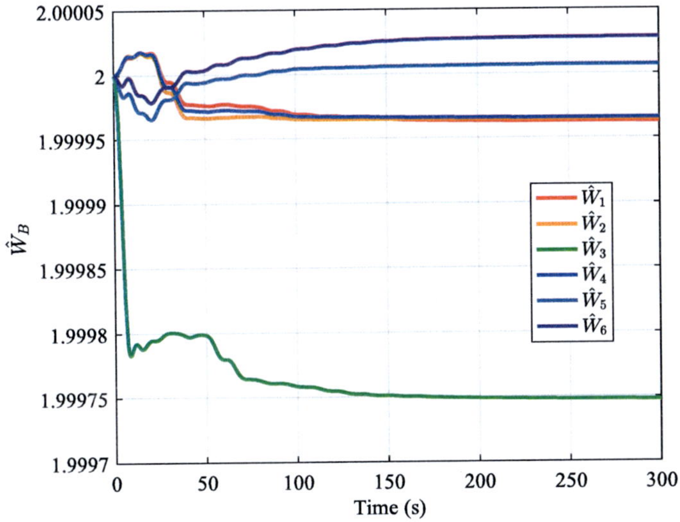

Fig. 8.13 The adaptive weights of the controller 'B'

these three graphs, it can be seen that the tracking error converges under the action of the controller proposed with performance constraints. Moreover, the tracking error of the closed-loop system satisfies the performance constraints, which can converge to the predefined steady-state error interval more quickly within the predefined maximum settling time, with better transient performance and predefined performance guarantees. Under the action of controller 'B' without considering performance constraints and optimal controller 'C' based on adaptive dynamic programming, the error can be converged, but the performance constraints are not guaranteed. This means that the transient and steady-state performance of their closed-loop systems needs to be accomplished through parameter adjustment, and there is no theoretically strong constraint on errors. Compared with the traditional prescribed performance controller 'D' and the unconstrained finite-time controller 'E', the tracking error transient performance under the action of the 'A' controller is better, and the convergence rate is faster. Figures 8.4, 8.5 and 8.6 show the spacecraft attitude angular velocity tracking error curves under the action of the controllers. It can be seen from the curves that due to the existence of performance constraints, the angular velocity errors have a faster convergence rate under the action of the controller proposed. This means better transient performance. After converging to the equilibrium point, the angular velocity error remains at the equilibrium point.

Figures 8.7 and 8.8 respectively show that under the action of the controllers, the spacecraft attitude error index $I_a = ||q_{ev}|| + ||\omega_e||$ and the energy consumption. Through the comparison, under the action of the proposed controller, the transient performance of the closed-loop system is better. Compared with other controllers, the error index I_a converges faster, and the steady-state error is smaller. To achieve such transient and steady-state performance, the controller proposed consumes less energy. Compared with controllers 'B' and 'C', the optimal control with performance constraints shows an increasing trend of 'slow-fast-slow' energy consumption. This reflects the impact of performance constraints on tracking errors. That is, when the error is close to the constraint boundary, the control input increases, and the energy consumption increases. It is precise because of the performance constraints that the tracking error is constrained to be within the range close to the equilibrium point, which slows down the subsequent energy consumption growth and saves more energy than the controllers 'B' and 'C'. It should be noted that 'B' and 'C' are also online updated adaptive optimal controllers. Therefore, the energy consumption is mainly consumed in the process of the controller approximating the optimal controller. However, the controller is not optimal in the approximating process, that is, in the process of HJB equation residual convergence. After the HJB equation residuals converge, the optimal controller is obtained. Compared to the controllers 'D' and 'E', in similar performance due to optimal design considerations, the controller proposed has less energy consumption even in the process of controller approximating to the optimal controller.

Figures 8.9 and 8.10 show the control input curves of the controllers respectively. Figure 8.11 shows the residual values of the HJB equations associated with the three controllers.

Figures 8.12 and 8.13 show the adaptive weight parameters of the corresponding controllers, respectively. It can be seen from the figures that the adaptive weights can be stabilized around a constant after a period of time.

Through the comparative simulation, it is fully demonstrated that under the action of the controller (8.23) in Sect. 8.2, the steady-state error of the closed-loop system can obtain better transient and steady-state performance by consuming less energy. Moreover, the performance is guaranteed, that is, the maximum settling time of the tracking error is guaranteed to be less than the predefined maximum settling time, the steady-state error is less than the predefined steady-state error, and the overshoot satisfies the constraints. When the residuals of the HJB equation converge, the optimal controller is obtained.

8.4.2.2 Simulation Results of Controller (8.46)

This section presents the parameters selected in the simulation of the controller (8.46) in Sect. 8.3, as shown in the following table.

In Table 8.4, the parameter t_{ps} is the predefined maximum settling time. ρ_0 is the maximum initial value of the predefined performance boundary, which needs to be larger than the initial error quaternion. Considering that all quaternion elements are less than or equal to 1, thus, ρ_0 is taken as 1. ρ_t and ε_ρ provide steady-state error bounds. δ_i is about whether to limit the overshoot, and λ_i is the sliding mode surface scale coefficient in the controller. Q_ζ and R_ζ are the constant matrices in the value function respectively, $I_{3\times3}$ is the 3×3 unit matrix, a is the ratio coefficient of the neural network weight adaptive law. κ_W and $\kappa_{W\theta}$ are the corresponding scale coefficients, $I_{6\times6}$ is the 6×6 identity matrix, $I_{6\times1} = [1, 1, 1, 1, 1, 1]^T$ is a vector whose all elements are 1. In addition, the neural network basis set used by this controller is $\phi = [\zeta_1^2, \zeta_2^2, \zeta_3^2, \zeta_1\zeta_2, \zeta_2\zeta_3, \zeta_3\zeta_1]^T$. This controller does not consider external disturbances.

The controller proposed in this study is compared with existing state-of-the-art controllers. For the convenience of representation, this section labels the proposed optimal control with experience replay (8.46) as controller 'A'. If the controller (8.46) removes the corresponding performance constraints, then a controller that does not consider the predefined performance constraints can be obtained, marked as controller 'B'. The controller, after adding predefined performance constraints to the

Table 8.4 The initial parameters of the proposed controller (8.46)

Param	Value	Param	Value	Param	Value
t_{ps}	100s	δ_i	1	a	0.1
ρ_0	1	λ_i	1	κ_W	$0.0001 I_{6\times6}$
ρ_t	0.02	Q_ζ	$0.02 I_{3\times3}$	$\kappa_{W\theta}$	$0.0001 I_{6\times1}$
ε_ρ	0.01	R_ζ	$0.01 I_{3\times3}$		

optimal control algorithm based on adaptive dynamic programming in [4], is marked as controller ‘C’. Label the optimal control algorithm based on adaptive dynamic programming in [4] as controller ‘D’. In this section, the comparative simulation of the above controllers will be carried out to illustrate the effectiveness and advantages of this study. In the simulation diagram and analysis, the symbols or subscripts ‘A’, ‘B’, ‘C’, and ‘D’ represent the above four controllers. The controller (8.46) in Sect. 8.3 is an extension of the controller (8.23) in Sect. 8.2, and the two controllers have similar effects. The comparison of the optimized controller with the traditional prescribed performance controller in [15] and the finite-time controller in [16] has been given in the previous section, so it will not be repeated in this section.

Figures 8.14, 8.15, 8.16, 8.17, 8.18, 8.19, 8.20, 8.21, 8.22, 8.23, 8.24, 8.25 and 8.26 show the simulation of the controller (8.46) in Sect. 8.3. Compared with the previous Sect. 8.2, the controller proposed in Sect. 8.3 relaxes the PE condition through experience replay in theory so that it can be regarded as an improvement on the controller in Sect. 8.2. Figures 8.14, 8.15 and 8.16 are the tracking error curves of the spacecraft attitude closed-loop system. The error elements of each dimension are plotted separately to compare the performance of the four controllers. From these three graphs, it can be seen that under the action of the proposed controller (8.46), considering the performance constraints, the tracking error converges. The performance constraints are satisfied, which can more quickly converge to the predefined steady-state error interval within the predefined maximum settling time with better transient performance and predefined performance guarantees. Controller ‘B’ is the result of removing performance constraints from controller ‘A’. It can be seen that the tracking error of the closed-loop system under its action is similar to that

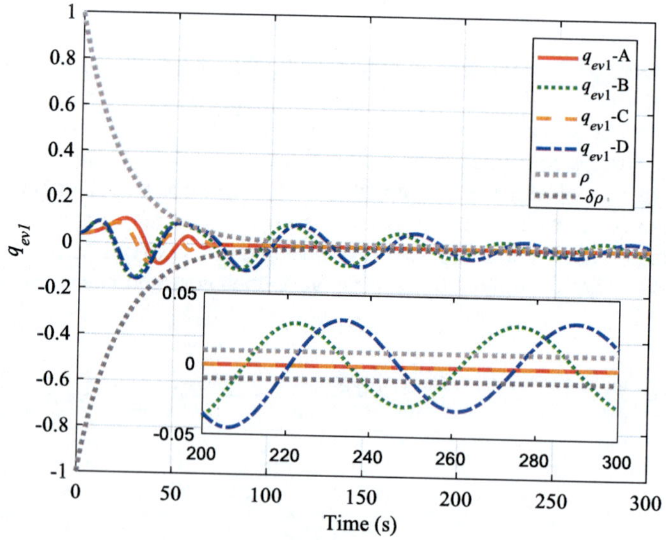

Fig. 8.14 The spacecraft attitude tracking error q_{ev1}

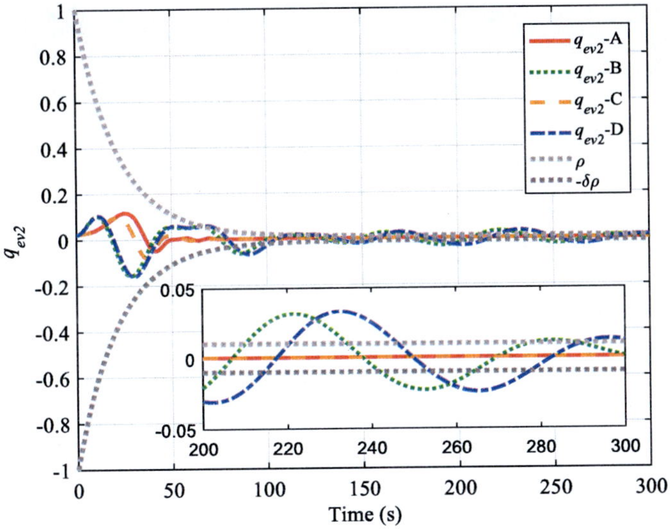

Fig. 8.15 The spacecraft attitude tracking error q_{ev2}

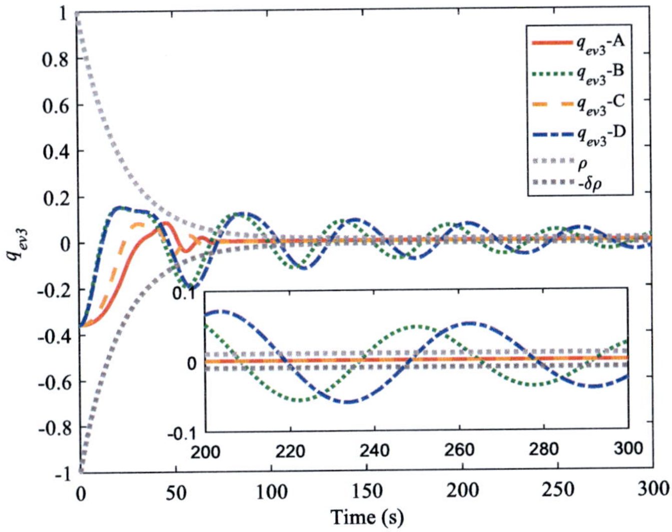

Fig. 8.16 The spacecraft attitude tracking error q_{ev3}

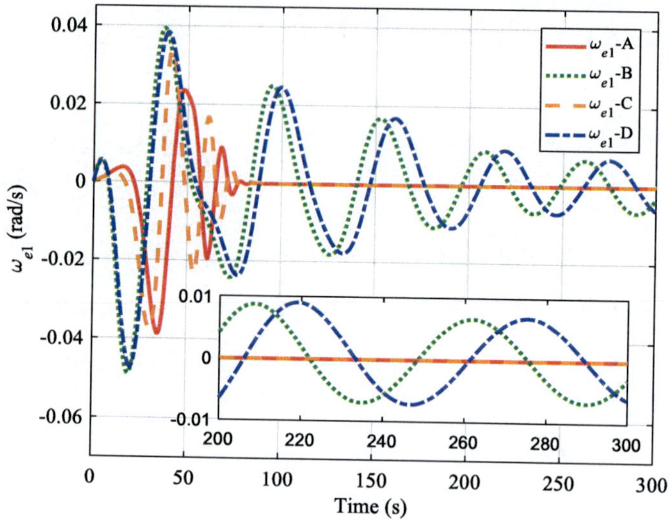

Fig. 8.17 The angular velocity tracking error ω_{e1}

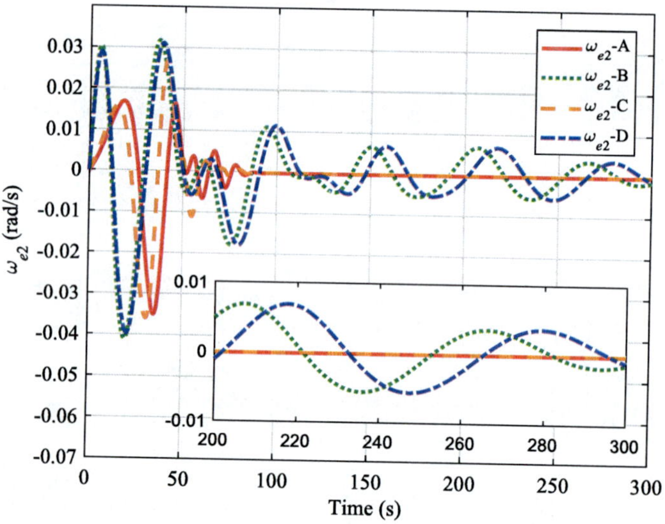

Fig. 8.18 The angular velocity tracking error ω_{e2}

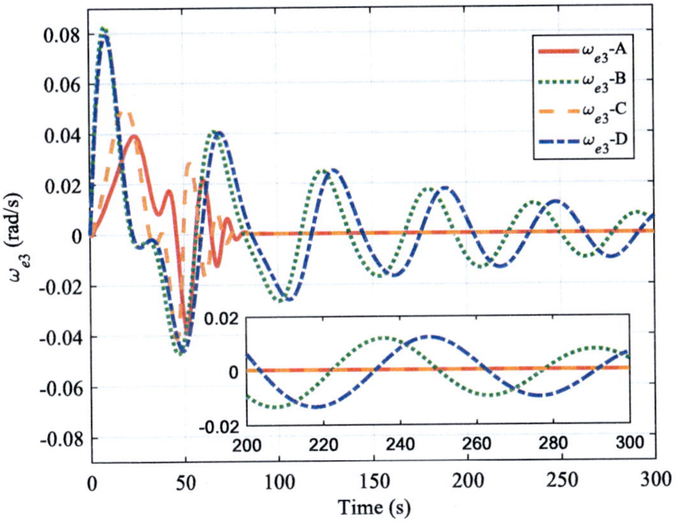

Fig. 8.19 The angular velocity tracking error ω_{e3}

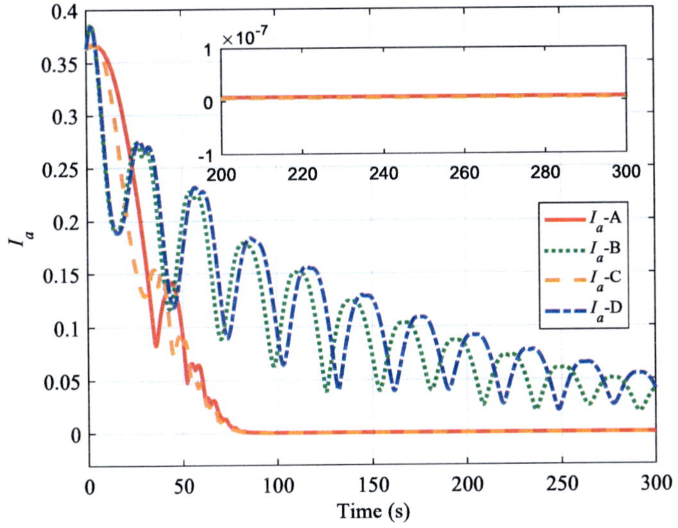

Fig. 8.20 The error index I_a

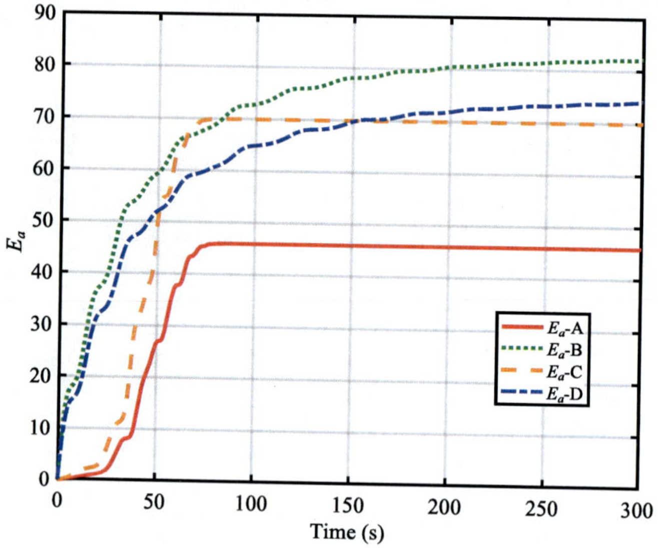

Fig. 8.21 The energy consumption

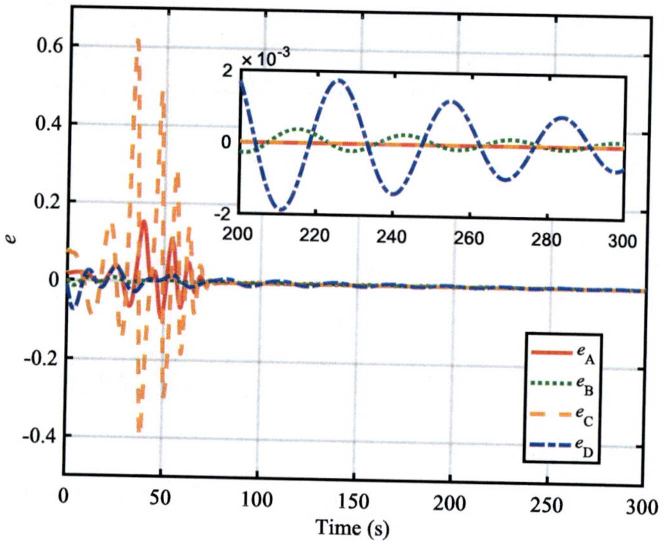

Fig. 8.22 The residual errors of the HJB equations

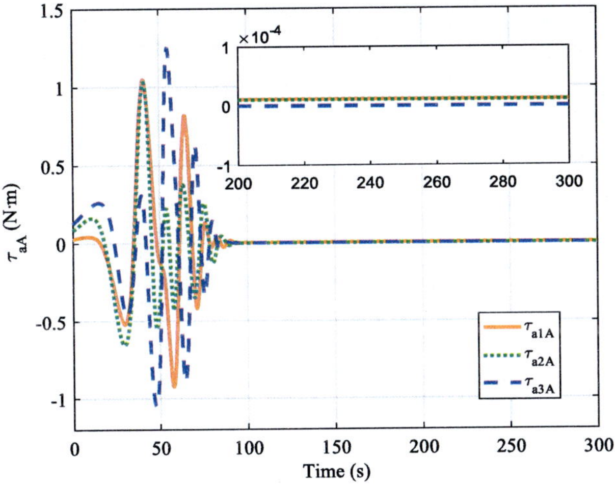

Fig. 8.23 The control input τ_a of 'A'

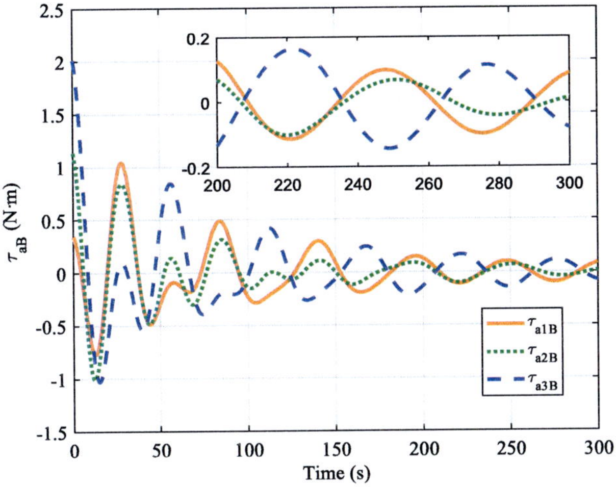

Fig. 8.24 The control input τ_a of 'B'

of the optimal controller 'D' based on adaptive dynamic programming, and neither can satisfy the performance constraints well. For more precise space missions, the optimal controllers 'B' and 'D' require a large number of parameter adjustments to achieve precise control. The controller (8.46) can satisfy the performance constraints in principle and, at the same time, can achieve optimal control under the performance constraints. The advantages of the proposed controller (8.46) in the performance-constrained precise control can be seen. Controller 'B' is the result of

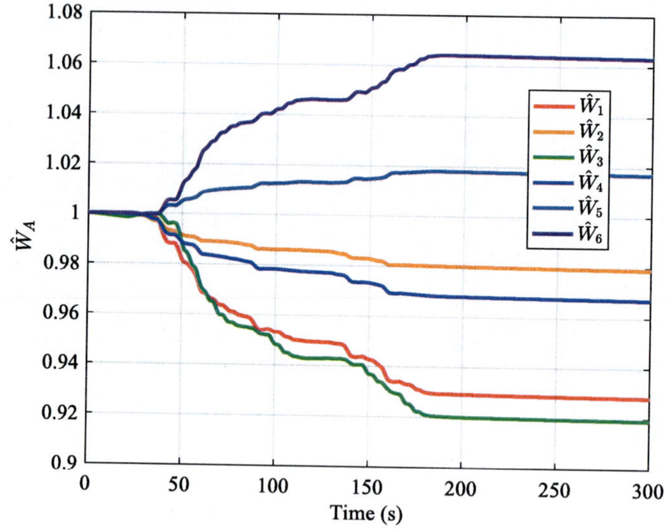

Fig. 8.25 The adaptive weights of the controller 'A'

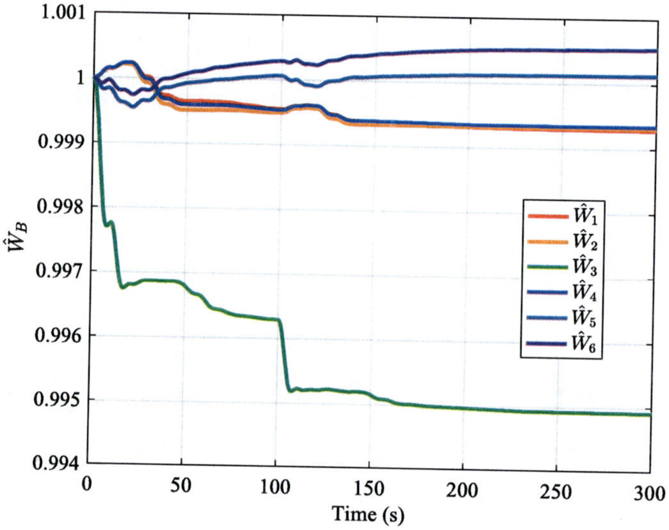

Fig. 8.26 The adaptive weights of the controller 'B'

adding performance constraints to controller 'D'. Applying performance constraints can impose strong performance guarantees on tracking errors.

Figures 8.17, 8.18 and 8.19 are the spacecraft attitude angular velocity tracking error curves. Similar to the tracking error analysis, under the condition of performance constraints, the spacecraft attitude angular velocity tracking error can converge quickly, and the steady-state error is small so that the spacecraft can obtain better transient and steady-state performance.

Figures 8.20 and 8.21 respectively show that under the action of the controllers, the spacecraft attitude error index $I_a = ||q_{ev}|| + ||\omega_e||$ and the energy consumption. Through the controller comparison, it can be seen that the proposed controller (8.46) has smaller energy consumption, faster convergence rate, and smaller steady-state error when the other controllers to be compared have similar energy consumption. Figure 8.22 shows the residual values of the HJB equations associated with the four controllers. Figures 8.23 and 8.24 show the control input curves of the controllers respectively. Figures 8.25 and 8.26 show the adaptive weight parameters of the corresponding controllers, respectively. It can be seen from the figures that the adaptive weights can be stabilized around a constant after a period of time.

Through comparative simulation, it is fully demonstrated that under the action of the controller (8.46) proposed in Sect. 8.3, the steady-state error of the closed-loop system can obtain better transient and steady-state performance by consuming less energy. Moreover, the performance is guaranteed, that is, the maximum settling time of the tracking error is guaranteed to be less than the predefined maximum settling time, the steady-state error is less than the predefined steady-state error, and the overshoot satisfies the constraints. When the residuals of the HJB equation converge, the optimal controller is obtained.

8.5 Conclusion

In this chapter, the optimal control problem is further considered based on the spacecraft's performance guaranteed control considering settling time parameters. The optimal or sub-optimal design of the control input is of critical significance to aerospace engineering due to the limited fuel or energy that the spacecraft can carry. Specifically, although the performance guaranteed control can bring the performance to meet the expected value range, the optimal control input may not be obtained. Based on spacecraft mission requirements and practical conditions, this chapter proposes the spacecraft performance guaranteed optimal control based on adaptive dynamic programming and the improved optimal control algorithm based on experience replay. First, based on the auxiliary system considering performance constraints, this chapter designs the value function, and the corresponding HJB equation is obtained. The approximate solution of the HJB equation is obtained by the adaptive dynamic programming method. That is, neural networks fit the solution of the HJB equation. Further, the proposed optimal control is obtained through the approximate solution of the HJB equation. Considering that the proposed optimal control needs to satisfy the persistent excitation conditions, the improved algorithm based on experience replay is proposed to relax this condition. Simultaneously, considering the role of the performance guaranteed constraint function, the spacecraft closed-loop system also satisfies the predefined constraints, including the maximum settling time, overshoot, and steady-state error. Applying the algorithms proposed in this chapter, the spacecraft closed-loop system has the desired transient and steady-state performance, and the control input is optimized with considering the constraints. This chapter systematically proposes and analyzes the above two algorithms and expounds on the rationality of the proposed algorithms through theoretical analysis and proof. Finally, the rationality and effectiveness of the control algorithms proposed in this chapter are proved by the comparisons and numerical simulation experiments with advanced control algorithms in similar achievements.

References

1. Abu-Khalaf, M., Lewis, F.L.: Nearly optimal control laws for nonlinear systems with saturating actuators using a neural network hjb approach. Automatica **41**(5), 779–791 (2005)
2. Finlayson, B.A.: The method of weighted residuals and variational principles. SIAM (2013)
3. Hornik, K., Stinchcombe, M., White, H.: Universal approximation of an unknown mapping and its derivatives using multilayer feedforward networks. Neural networks **3**(5), 551–560 (1990)
4. Vamvoudakis, K.G., Lewis, F.L.: Online actor–critic algorithm to solve the continuous-time infinite horizon optimal control problem. Automatica **46**(5), 878–888 (2010)
5. Liu, D., Li, H., Wang, D.: Online synchronous approximate optimal learning algorithm for multi-player non-zero-sum games with unknown dynamics. IEEE Transactions on Systems, Man, and Cybernetics: Systems **44**(8), 1015–1027 (2014)
6. Liu, D., Li, H., Wang, D.: Online synchronous approximate optimal learning algorithm for multi-player non-zero-sum games with unknown dynamics. IEEE Transactions on Systems, Man, and Cybernetics: Systems **44**(8), 1015–1027 (2014). 10.1109/TSMC.2013.2295351
7. Vamvoudakis, K.G., Lewis, F.L.: Multi-player non-zero-sum games: Online adaptive learning solution of coupled hamilton–jacobi equations. Automatica **47**(8), 1556–1569 (2011)
8. Modares, H., Lewis, F.L., Naghibi-Sistani, M.B.: Integral reinforcement learning and experience replay for adaptive optimal control of partially-unknown constrained-input continuous-time systems. Automatica **50**(1), 193–202 (2014)
9. Yasini, S., Naghibi Sitani, M.B., Kirampor, A.: Reinforcement learning and neural networks for multi-agent nonzero-sum games of nonlinear constrained-input systems. International Journal of Machine Learning and Cybernetics **7**(6), 967–980 (2016)
10. Ioannou, P., Fidan, B.: Adaptive control tutorial. SIAM (2006)
11. Tao, G.: Adaptive control design and analysis, vol. 37. John Wiley & Sons (2003)
12. Zhang, H., Cui, L., Luo, Y.: Near-optimal control for nonzero-sum differential games of continuous-time nonlinear systems using single-network adp. IEEE transactions on cybernetics **43**(1), 206–216 (2012)
13. Zhao, D., Zhang, Q., Wang, D., Zhu, Y.: Experience replay for optimal control of nonzero-sum game systems with unknown dynamics. IEEE transactions on cybernetics **46**(3), 854–865 (2015)
14. Jiang, B., Li, C., Ma, G.: Finite-time output feedback attitude control for spacecraft using "adding a power integrator" technique. Aerospace Science and Technology **66**, 342–354 (2017)
15. Bechlioulis, C.P., Rovithakis, G.A.: Robust adaptive control of feedback linearizable mimo nonlinear systems with prescribed performance. IEEE Transactions on Automatic Control **53**(9), 2090–2099 (2008)
16. Du, H., Li, S., Qian, C.: Finite-time attitude tracking control of spacecraft with application to attitude synchronization. IEEE Transactions on Automatic Control **56**(11), 2711–2717 (2011)

Part II
Performance Guaranteed Games for Multiple Spacecraft

This part focuses on the cooperative control game problem among several spacecraft. Of course, consistent with the topic of this book, such game problems need to be under the framework of performance guaranteed. For multiple spacecraft, they can be separated, such as in formation, and each spacecraft is unique in non-contact. The performance guaranteed cooperative game strategy for this situation is presented in Chap. 9. Multiple spacecraft can also be combined, appearing as a combined spacecraft, with each spacecraft becoming the *sub-spacecraft*. The cooperative game control strategy with performance guaranteed is designed for each sub-spacecraft so that the sub-spacecraft jointly drives the attitude motion of the whole combined spacecraft. This research will be introduced in Chap. 10.

Chapter 9
Performance Guaranteed Non-Zero-Sum Games for Multiple Spacecraft

Abstract In the on-orbit service spacecraft rendezvous and docking mission or the multi-spacecraft attitude coordination mission, the service spacecraft and the target spacecraft often need to reach an attitude consistency or the desired attitude orientation. To address this issue, feasible ideas are considered in this chapter. The cooperative target spacecraft can be maneuvered together with the service spacecraft or multiple spacecraft to complete the maneuver control objectives of eliminating relative attitude errors. In response to this idea, the performance guaranteed non-zero-sum game algorithms ars designed in this chapter. The algorithms are introduced into the multi-spacecraft system. Due to the utilization of the performance guaranteed techniques, the transient and steady-state behavior of the system, such as settling time, overshoot, steady-state error, etc., will meet predefined expectations. Based on the performance guaranteed or constraint transformation system, the non-zero-sum game strategies provide each member spacecraft with attitude maneuver policies that achieve Nash equilibrium. This allows all spacecraft to share the entire control task in an optimized way, reducing the control cost of individual spacecraft. The detailed neural network-based performance guaranteed non-zero-sum game control strategy design and the rigorous theoretical proofs are given. Finally, comprehensive numerical simulation analysis and comparative simulation verify the effectiveness, feasibility, and advantages of the proposed algorithms.

9.1 Introduction

In the mission or system consisting of multiple spacecraft, such as rendezvous and docking [1], on-orbit service [2, 3], formation flight [4, 5], satellite constellation [6, 7], etc., we usually need to maneuver multiple spacecraft to achieve attitude consensus [8–10], common pointing [11, 12], or other desired objectives. For such mission objectives, we can use the suitable control methods in the multi-agent system theory to make all spacecraft in the system have consistent attitude attributes. As described in [13–15], this problem will be further presented in Part III of this book.

For multi-spacecraft systems with fewer individuals, such as dual-spacecraft systems, one spacecraft can track the attitude of the other spacecraft so that they have

Y. Gao and D. Li, *Spacecraft Maneuver with Performance Guaranteed*,
https://doi.org/10.1007/978-981-99-4653-2_9

the common desired attitude. However, when only two or more spacecraft attitude consistency is considered, such an approach implies that the maneuvering spacecraft bears the total cost of control. For this problem, we can consider another implementation scheme: the two or more spacecraft move together to have the same or desired attitude. In other words, by designing their own control strategies for multiple spacecraft, the multiple cooperative spacecraft are respectively driven to perform attitude maneuvers, so the attitude errors between them are eliminated. This is the main problem studied in this chapter. This way, the cost of eliminating systematic errors will be shared by dual or multiple spacecraft. This not only shares energy consumption but also makes it faster. Therefore, the control allocation algorithm can be used to allocate the attitude motion of the spacecraft in advance to realize the attitude maneuver tasks of all spacecraft [16–18]. However, in the process above, the motion of each spacecraft in the system is allocated by allocation algorithms, which is not necessarily optimal. If the optimal control allocation algorithm is used for design, the algorithm of a single spacecraft may be optimal. However, the overall strategy of multiple individuals is not necessarily optimal and Nash-equilibrium. This means unnecessary energy loss may be caused between individuals due to the non-Nash equilibrium state.

For the above problems, in fact, the multiple spacecraft system has already constituted a multiple-player game system. According to the Non-Zero-Sum (NZS) Game theory [19] and Nash equilibrium theory [20], the individual control problems of the aforementioned multi-spacecraft system can be solved. References [21–26] address the design and solution methods of the non-zero-sum game problem through Adaptive Dynamic Programming (ADP). This provides us with a state-of-the-art method for solving multi-spacecraft system-level optimization. By applying advanced non-zero-sum game strategies, all spacecraft can achieve the optimal control input with considering the states of other spacecraft, and the strategies of all spacecraft together constitute a nash equilibrium solution. This algorithm will appear in the algorithm comparison simulation in this chapter. However, this simple application of the game strategy brings new problems: (1) in order to make the control input optimal or sub-optimal, the settling time of the system will be increased; (2) in order to reduce the settling time of the system, the control parameter adjustment may increase the energy consumption; (3) during the stabilization process, each spacecraft may generate overshoots, and these overshoots will cause accumulated errors in other spacecraft, resulting in system jitter and more energy consumption; (4) after the system enters the steady state, due to information sharing among all spacecraft, there may also be excessive system jitters that are not beneficial to the mission and increase energy consumption; (5) more transient additional motion and more complex steady-state jitter will lose the control precision and accuracy of the system, which is not conducive to high-precision attitude control.

In response to the above problems, we face new challenges: (1) it is hoped to achieve the optimal control of each spacecraft and the Nash equilibrium of all spacecraft control inputs through the non-zero-sum game strategy, that is, system optimization; (2) it is hoped that the system has customized convergence or the descriptive transient process; (3) it is hoped that the system has slight steady-state errors. In response to these challenges, this chapter adopts the performance guaranteed algorithm proposed in the previous chapters to realize the performance constraints and control enhancement of the system's transient and steady-state behavior. Specifically, this chapter focuses on the practical problems of multi-spacecraft control and proposes the performance guaranteed non-zero-sum games strategy and specific implementation algorithm based on the non-zero-sum game theory. It is applied to the multi-spacecraft attitude control system, and rigorous theoretical proof and detailed simulation analysis are given to prove the feasibility, effectiveness, and advantages of the proposed algorithm.

The main contributions of the algorithm proposed in this chapter are listed as follows: (1) The non-zero-sum game control strategy based on the performance guaranteed method is proposed. This strategy extends the ADP-based optimal control method for the performance-constrained control field so that the input optimization problem can be solved while the system has performance constraints. This strategy extends the performance constraints for game theory. The algorithm proposed in this chapter realizes the optimization of the original controller through performance-constrained system mapping and auxiliary controller design, thus enriching the field of differential game theory. (2) For the joint attitude control problem of multi-spacecraft, the controllers are designed based on the proposed control strategy so that the control inputs related to each member spacecraft's system are optimal, and the control strategies of all spacecraft constitute the Nash equilibrium of the system. The performance guaranteed algorithm is adopted so that the system has pre-definable settling time, steady-state error, overshoot, etc. The filters are used to make the system satisfy the Finite Excitation (FE) conditions. (3) Excessive jitter and more energy consumption are avoided. Due to the performance constraints, the system settles quickly, thus reducing the excessive jitter of the overshoot of the member spacecraft and thus avoiding more energy consumption.

It should be noted that to make the system satisfy both performance constraints and optimal conditions, this chapter proposes a compromise solution. This optimality can also be considered a moderate optimality. The optimality of the system is reflected in the adjustment of the control input to make the value function optimal. In order to meet the system performance constraints, the algorithm adopts the optimal design of the auxiliary control variables, which will be reflected in Sect. 9.3.1, which means that the auxiliary control variables are optimal. When the system is stabilized, the auxiliary control variables will converge to the actual control inputs of the system, so the proposed control design is optimal when the system has converged. In addition, although the controller proposed in this chapter takes spacecraft attitude control as an example, it is not limited to this application. After abstracting the spacecraft systems as the Euler-Lagrange (EL) System, it can also be used for other EL systems such as relative orbital motion, robot control, aircraft control, aircraft formation, etc.

The remainder of this study is organized as follows. Section 9.2 summarizes the two-spacecraft attitude game model and proposes the control objectives.

Section 9.3 provides algorithm design, stability analysis, Nash equilibrium analysis, and multiple spacecraft situations. Section 9.4 shows the simulation results, followed by concluding remarks in Sect. 9.5.

9.2 Spacecraft Game Model and Preliminaries

9.2.1 *Two-Spacecraft Attitude Game Model*

We describe the attitude kinematics of the spacecraft under the Earth-Centered Inertial (ECI) frame $\mathfrak{F}_i$ [27, 28] and the body-fixed Cartesian frame. For the two spacecraft, we use $\mathfrak{F}_s$ and $\mathfrak{F}_t$ to represent their body coordinate systems, which are located at centroid on the two spacecraft, respectively. The attitude movement between the two spacecraft can be described by the rotation of their body-fixed reference frames relative to the $\mathfrak{F}_i$. The game of two spacecraft means that both spacecraft have control inputs. Our control objective is the convergence of the attitude tracking errors of the two spacecraft. Therefore, we now begin to derive the game model.

Considering the relative attitude movement of the rigid spacecraft, we can use the unit quaternion $q_a = [q_0, q_v^T]^T \in \mathbb{R}^4$, $q_v \in \mathbb{R}^3$ to describe the rotation of the body-fixed frame $\mathfrak{F}_s$ with respect to the inertial frame $\mathfrak{F}_i$, which is subject to the constraint [29]

$$q_v^T q_v + q_0^2 = 1 \tag{9.1}$$

Based on the dynamics and kinematics models of the spacecraft [28, 30], the attitude kinematics and dynamics of the first spacecraft are as follows

$$J_a \dot{\omega}_a = -\omega_a^{\times} J_a \omega_a + \tau_a \tag{9.2}$$

$$\dot{q}_v = \frac{1}{2}\left(q_v^{\times} \omega_a + q_0 \omega_a\right) \tag{9.3}$$

$$\dot{q}_0 = -\frac{1}{2} q_v^T \omega_a \tag{9.4}$$

$$R_a = (q_0^2 - q_v^T q_v) I_3 + 2 q_v q_v^T - 2 q_0 q_v^{\times} \tag{9.5}$$

$$\omega_a = 2(q_0 \dot{q}_v - q_v \dot{q}_0) - 2 q_v^{\times} \dot{q}_v \tag{9.6}$$

where $J_a \in \mathbb{R}^{3\times 3}$ is the constant, positive-definite, symmetric inertia matrix of the first spacecraft, $\omega_a \in \mathbb{R}^3$ is the angular velocity of the body-fixed frame $\mathfrak{F}_s$ with respect to the inertial reference frame $\mathfrak{F}_i$ expressed in $\mathfrak{F}_s$, $\tau_a \in \mathbb{R}^3$ is a vector of control torques in $\mathfrak{F}_s$, $R_a(q_v, q_0) \in SO(3)$ is the rotation matrix that brings $\mathfrak{F}_i$ onto $\mathfrak{F}_s$, and the notation $a^{\times}$, $\forall a = [a_1, a_2, a_3]^T$, denotes the following skew-symmetric matrix: $a^{\times} \triangleq [0, -a_3, a_2; a_3, 0, -a_1; -a_2, a_1, 0]$.

Similarly, we list the attitude kinematics and dynamics of the second spacecraft as follows, where the definition of variables is the same as that of the first spacecraft, and we use the index d to distinguish.

$$J_d\dot{\omega}_d = -\omega_d^{\times} J_d\omega_d + \tau_d \tag{9.7}$$

$$\dot{q}_{dv} = \frac{1}{2}\left(q_{dv}^{\times}\omega_d + q_{d0}\omega_d\right) \tag{9.8}$$

$$\dot{q}_{d0} = -\frac{1}{2}q_{dv}^{T}\omega_d \tag{9.9}$$

$$R_d = (q_{d0}^2 - q_{dv}^T q_{dv})I_3 + 2q_{dv}q_{dv}^T - 2q_{d0}q_{dv}^{\times} \tag{9.10}$$

$$\omega_d = 2(q_{d0}\dot{q}_{dv} - q_{dv}\dot{q}_{d0}) - 2q_{dv}^{\times}\dot{q}_{dv} \tag{9.11}$$

Through the rotation matrix of R_a and R_d, we can find their error matrix and the tracking error of the angular velocity as follows

$$R_e = R_a R_d^T = (q_{e0}^2 - q_{ev}^T q_{ev})I_3 + 2q_{ev}q_{ev}^T - 2q_{e0}q_{ev}^{\times} \tag{9.12}$$

$$q_{ev} = q_{d0}q_v - q_0 q_{dv} + q_v^{\times} q_{dv} \tag{9.13}$$

$$q_{e0} = q_0 q_{d0} + q_v^T q_{dv} \tag{9.14}$$

$$\omega_e = \omega_a - R_e\omega_d \tag{9.15}$$

Find the derivative of (9.13), (9.14), and (9.15) with respect to time, we have

$$\dot{\omega}_e = \dot{\omega}_a - \dot{R}_e\omega_d - R_e\dot{\omega}_d \tag{9.16}$$

$$\dot{q}_{ev} = \frac{1}{2}(q_{ev}^{\times} + q_{e0}I_3)\omega_e = \frac{1}{2}T\omega_e \tag{9.17}$$

$$\dot{q}_{e0} = -\frac{1}{2}q_{ev}^{T}\omega_e \tag{9.18}$$

where $\dot{R}_e = -{\omega_e}^{\times} R_e$ and $T = q_{ev}^{\times} + q_{e0}I_3$. Taking the derivative of (9.17) with respect to time, we have

$$\ddot{q}_{ev} = \frac{1}{2}\dot{T}\omega_e + \frac{1}{2}T\dot{\omega}_e \tag{9.19}$$

Multiplying both sides of the above formula by $T^{-T}J_aT^{-1}$ at the same time, we have

$$T^{-T}J_aT^{-1}\ddot{q}_{ev} = \frac{1}{2}T^{-T}J_aT^{-1}\dot{T}\omega_e + \frac{1}{2}T^{-T}J_a\dot{\omega}_e \tag{9.20}$$

Substituting the error dynamics and the dynamics of the two spacecraft into (9.20), after sorting out, the game model of the two spacecraft attitude tracking is obtained.

$$M\ddot{q}_{ev} + C\dot{q}_{ev} + G = +G_a\tau_a + G_d\tau_d \tag{9.21}$$

where $M = T^{-T}J_aT^{-1}$, $C = -T^{-T}J_aT^{-1}\dot{T}T^{-1}$, $G = \frac{1}{2}T^{-T}(+\omega_a^{\times}J_a\omega_a - J_a{\omega_e}^{\times}$ $R_e\omega_d - J_aR_eJ_d^{-1}\omega_d^{\times}J_d\omega_d)$, $G_a = \frac{1}{2}T^{-T}$, and $G_d = -\frac{1}{2}T^{-T}J_aR_eJ_d^{-1}$.

In this chapter, we consider the performance guaranteed attitude tracking control game problem of two cooperative spacecraft, that is, design the optimal control strategy of the two spacecraft respectively, so that the attitude tracking error of the two spacecraft converges and satisfies the pre-defined performance requirements. The strategy group reaches the Nash equilibrium. We give the control objectives of this chapter as follows.

Problem 9.1 Given the initial game system (9.21) of two spacecraft and appropriate tracking control indicators include the maximum settling time t_{ps}, the maximum steady-state error ρ_t, and the maximum overshoot χ_t, design the optimal control algorithm τ_a and τ_d, such that the following conditions are satisfied: (1) the closed-loop system is stable, and $t_s \leq t_{ps}$, where t_s is the actual settling time of the closed-loop tracking error; (2) $-\delta\rho_t < q_{ev\,i}|_{t\geq t_{ps}} < \rho_t$ when $q_{ev\,i}(0) > 0$ or $-\rho_t < q_{ev\,i}|_{t\geq t_{ps}} < \delta\rho_t$ when $q_{ev\,i}(0) < 0$, where $q_{ev\,i}|_{t\geq t_{ps}}$ represents the actual steady-state error when $t \geq t_{ps}$, $\delta \in [0, 1]$ is a constant to be set; (3) the actual maximum overshoot less than χ_t; (4) τ_a and τ_d are optimal, and their control strategy group reaches the Nash equilibrium.

9.2.2 *Performance Guaranteed with Pre-defined Maximum Settling Time Function*

In order to ensure the transient performance and steady-state accuracy of the spacecraft closed-loop system, we need to impose strong constraints on the attitude tracking error.

Inspired by [31, 32], we make some improvements to the performance constraint function in the prescribed performance so that the maximum settling time can be explicitly included in the constraint function to achieve the purpose of defining the settling time in advance. In this chapter, we introduce the so-called Pre-defined Maximum Settling Time constraint Function (PMSTF) that can obviously contain the maximum settling time as follows.

Definition 9.1 A smooth function $\rho(t) : \mathbb{R}^+ \rightarrow \mathbb{R}^+\backslash 0$ will be called a PMSTF if: (1) $\rho(t)$ is positive and decreasing; (2) $\rho(t_{ps}) = \rho_t$, where $t_{ps} > 0$ is the expected upper bound of settling time set in advance according to actual needs and $\rho_t > 0$ is the upper bound of the expected steady-state error specified in advance; (3) $\lim\limits_{t\rightarrow\infty} \rho(t) = \rho_t - \varepsilon_\rho$, where $0 < \varepsilon_\rho < \rho_t$ is a small constant to provide an asymptote.

According to Definition 9.1, the exponential PMSTF can be designed as follows.

$$\rho(t) = \left(\rho_0 - \rho_t + \varepsilon_\rho\right) \exp\left(\frac{t}{t_{ps}} \ln \frac{\varepsilon_\rho}{\rho_0 - \rho_t + \varepsilon_\rho}\right) + \rho_t - \varepsilon_\rho \tag{9.22}$$

For the spacecraft attitude tracking system with multiple inputs and multiple outputs, we apply (9.22) as the constraint functions to force the tracking errors of the closed-loop system to be within the allowable range described by (9.22). In this way, the closed-loop system can obtain the desired settling time, and transient and steady-state performance. Therefore, we have the following constraints.

$$\begin{array}{ll} -\delta_i \rho_i(t) < q_{evi}(t) < \rho_i(t) & q_{evi}(0) \geq 0 \\ -\rho_i(t) < q_{evi}(t) < \delta_i \rho_i(t) & q_{evi}(0) < 0 \end{array} \tag{9.23}$$

We define the variable ϑ_i as

$$\vartheta_i = \begin{cases} \dfrac{2q_{evi}(t)}{(1+\delta_i)\,\rho_i(t)} + \dfrac{\delta_i - 1}{1+\delta_i},\ q_{evi}(0) \geq 0 \\ \dfrac{2q_{evi}(t)}{(1+\delta_i)\,\rho_i(t)} + \dfrac{1-\delta_i}{1+\delta_i},\ q_{evi}(0) < 0 \end{cases}. \tag{9.24}$$

Then, the constraint inequality (9.23) can be described as $-1 < \vartheta_i < 1$. In other words, if ϑ_i can always remain in $(-1, 1)$, then the tracking error of the closed-loop system will always be constrained in (9.23). If we map $(-1, 1)$ to $(-\infty, +\infty)$, then if the mapped variable associated with ϑ_i is stable or bounded in $(-\infty, +\infty)$, then (9.23) holds. In this way, we solve the constraint problem by solving the stability problem.

To map ϑ_i, we define a smooth function $\varphi : \mathbb{R} \to \mathbb{R}$ that satisfies $\lim\limits_{\sigma \to +\infty} \varphi(\sigma_i) = 1$, $\lim\limits_{\sigma \to -\infty} \varphi(\sigma_i) = -1$, where σ_i is considered to be the transform parameter. We take a function that satisfies the above conditions, for example, $\varphi(\sigma_i) = \tanh(\sigma_i)$. We then take $\vartheta_i = \varphi(\sigma_i)$ and $\sigma_i = \varphi^{-1}(\vartheta_i)$. If σ_i is stable or bounded, then $-1 < \vartheta_i < 1$ is true.

If we can design the controller so that σ_i is stable or bounded, then the closed-loop system satisfies the constraint condition (9.23).

9.2.3 *Optimal Non-Zero-Sum Games and Nash Equilibrium*

We consider the Non-zero-sum (NZS) games problem of the following form [22, 33, 34]

$$\dot{x} = f(x) + \sum\nolimits_{j=1}^{N} g_j(x) u_j(t) \tag{9.25}$$

with the state vector $x \in \mathbb{R}^n$, N control input $u_j(t) \in \mathbb{R}^{m_j}$, and the smooth nonlinear dynamics $f(x) \in \mathbb{R}^n$ and $g_j(x) \in \mathbb{R}^{n \times m_j}$, which are also assumed to satisfy the Lipschitz condition on a compact set $\Omega \subseteq \mathbb{R}^n$.

For each spacecraft, we define the following value functions as

$$J_i(x_0, u_1, \ldots, u_N) = \int_0^{\infty} \left(Q_i(x) + \sum\nolimits_{j=1}^{N} u_j^T R_{ij} u_j \right) \mathrm{d}t \tag{9.26}$$

where Q_i is positive definite, R_{ii} and R_{ij} are positive definite symmetric matrices, x_0 is the initial state, $i, j = 1, ..., N$.

If we can find a feedback control policy pair $\mu = \{\mu_1, ..., \mu_N\}$, μ can stabilize (9.25) on the set Ω, (9.26) is finite, and μ_i is continuous on the set Ω, then we defined it as admissible on the set Ω, and denote it by $\mu_i \in \Psi(\Omega)$ [22]. Then we define the value function for the admissible control policies as

$$V_i(x, \mu_1, ..., \mu_N) = \int_t^{\infty} \left(Q_i(x(\tau)) + \sum_{j=1}^{N} \mu_j^T(\tau) R_{ij} \mu_j(\tau)\right) \mathrm{d}\tau \tag{9.27}$$

The optimal admissible control policy pair $\mu^* = \{\mu_1^*, ..., \mu_N^*\}$ means that such a control strategy can minimize the value function (9.27) for each spacecraft. If this optimal strategy μ^* can satisfy the following definition, then such a strategy is a Nash equilibrium solution.

Definition 9.2 [35] An N-tuple of policies $\{\mu_1^*, \ldots, \mu_N^*\}$ with $\mu_i^* \in \Psi(\Omega)$ is said to constitute a Nash equilibrium for an N-player games, if the following N inequalities are satisfied for all $\mu_i \in \Psi(\Omega)$:

$$J_i^*(\mu_1^*, ..., \mu_i^*, ..., \mu_N^*) \leq J_i(\mu_1^*, ..., \mu_i, ..., \mu_N^*), i \in \mathbf{N} \tag{9.28}$$

Taking the derivative of (9.27), we can get the derivative equation of the value function

$$0 = r_i(x, \mu_1, ..., \mu_N) + (\nabla V_i)^T \left(f + \sum_{j=1}^{N} g_j \mu_j\right) \tag{9.29}$$

where $r_i(x, u_1, ..., u_N) = x^T Q_i x + \sum_{j=1}^{N} u_j^T R_{ij} u_j$, $\nabla V_i = \partial V_i / \partial x$, and $V_i(0) = 0$. So the Hamiltonian functions and the optimal value functions can be expressed as follows

$$H_i(x, \nabla V_i, \mu_1, ..., \mu_N) = r_i(x, \mu_1, ..., \mu_N) + (\nabla V_i)^T \left(f + \sum_{j=1}^{N} g_j \mu_j\right) \tag{9.30}$$

$$V_i^*(x) = \min_{\mu_i} \int_t^{\infty} \left(x^T(\tau) Q_i x(\tau) + \sum_{j=1}^{N} \mu_j^T(\tau) R_{ij} \mu_j(\tau)\right) \mathrm{d}\tau \tag{9.31}$$

By utilizing the stationarity conditions $\partial H_i / \partial \mu_i = 0$, we can get the optimal control strategy for each spacecraft as

$$\mu_i^* = -\frac{1}{2} R_{ii}^{-1} g_i^T \nabla V_i^* \tag{9.32}$$

And the coupled HJ equation can be described as

$$\begin{aligned} 0 =& r_i(x, \mu_1^*, ..., \mu_N^*) + (\nabla V_i^*)^T \left(f + \sum\nolimits_{j=1}^{N} g_j \mu_j^*\right) \\ =& x^T Q_i x + \frac{1}{4} \sum\nolimits_{j=1}^{N} (\nabla V_j^*)^T g_j R_{jj}^{-T} R_{ij} R_{jj}^{-1} g_j^T \nabla V_j^* \\ &+ (\nabla V_i^*)^T f - \frac{1}{2} (\nabla V_i^*)^T \sum\nolimits_{j=1}^{N} g_j R_{jj}^{-1} g_j^T \nabla V_j^* \end{aligned} \tag{9.33}$$

where $V_i^*(0) = 0$.

By solving the above HJ equation, we can obtain V_i^*, so we can obtain the optimal control policies (9.32). At the same time, this pair of optimal controls also reach a Nash equilibrium solution. However, for nonlinear systems, it is often difficult to obtain the solution of the HJ equation. Therefore, we can use the neural network approximation method to obtain the corresponding solution.

9.3 Performance Guaranteed NZS Game with Experience Replay for Multiple Spacecraft

9.3.1 Algorithm Design

We first design the algorithms between two spacecraft, and then extend it to multi-spacecraft. In order to achieve the pre-defined performance of the spacecraft attitude cooperative control, by introducing the parameter σ_i associated with performance constraints, considering that the spacecraft attitude system is a second-order system, we design an auxiliary variable as follows, where λ_i is a positive constant.

$$\zeta_i = \dot{\sigma}_i + \lambda_i \sigma_i \tag{9.34}$$

Here the new state variable ζ_i is an auxiliary variable. The primary purpose is to use ζ_i to connect the converted σ_i with the tracking error of the spacecraft attitude game system, and to re-establish a second-order system that can satisfy the constraints through simple controller design. Therefore, the constraint problem is transformed into a controller design problem.

Substituting (9.24) into the derivative of (9.34) results in

$$\dot{\zeta}_i = \alpha_i + \beta_i \ddot{q}_{evi}(t) \tag{9.35}$$

where

$$\begin{aligned} \alpha_i =& -\frac{2r_i \dot{q}_{evi}(t) \dot{\rho}_i(t)}{(1+\delta_i)\rho_i{}^2(t)} + \frac{4r_i q_{evi}(t) \dot{\rho}_i^2(t)}{(1+\delta_i)\rho_i{}^3(t)} + \dot{r}_i \dot{\vartheta}_i \\ &- \frac{2r_i \ddot{\rho}_i(t) q_{evi}(t) + 2r_i \dot{\rho}_i(t) \dot{q}_{evi}(t)}{(1+\delta_i)\rho_i{}^2(t)} + \lambda_i r_i \dot{\vartheta}_i \end{aligned} \tag{9.36}$$

$$\beta_i = \frac{2r_i}{(1+\delta_i)\,\rho_i\,(t)} \tag{9.37}$$

$$r_i = \partial\varphi^{-1}(\vartheta_i)/\partial\vartheta_i \tag{9.38}$$

Express the above formula into a compact form, we have

$$\dot{\zeta} = \ddot{\sigma} + \Gamma\dot{\sigma} = A + B\ddot{q}_{ev}\,(t) \tag{9.39}$$

where $\zeta = [\zeta_1, \ldots, \zeta_n]^T$, $\sigma = [\sigma_1, \ldots, \sigma_n]^T$, $A = [\alpha_1, \ldots, \alpha_n]^T$, $B = \mathrm{diag}(\beta_1, \ldots, \beta_n)$, $\Gamma = \mathrm{diag}(\lambda_1, \ldots, \lambda_n)$.

Remark 9.1 Since $-1 < \vartheta_i < 1$ is always satisfied during the control process, $\beta_i > 0$ is always authentic, that is, B^{-1} exists. Also, both matrices A and B can be utilized in controller design.

Substituting the game model (9.21) of the two spacecraft into (9.39), we have

$$\begin{aligned}\dot{\zeta} = {} & A - BM^{-1}C\dot{q}_{ev} - BM^{-1}G \\ & + BM^{-1}G_a\tau_a + BM^{-1}G_d\tau_d\end{aligned} \tag{9.40}$$

In order to simplify the symbolic expression in the subsequent derivation process, we rewrite the system (9.40) into the general form as follows

$$\dot{x} = f + g_1u_1 + g_2u_2 \tag{9.41}$$

where $x = \zeta$, $f = A + BM^{-1}(-C\dot{q}_{ev} - G)$, $g_1 = M^{-1}G_a\kappa_g^{-1}$, $u_1 = \kappa_g G_a^{-1}M \cdot BM^{-1}G_a\tau_a$, $g_2 = M^{-1}G_d\kappa_g^{-1}$, $u_2 = \kappa_g G_d^{-1}M \cdot BM^{-1}G_d\tau_d$, $\kappa_g = \mathrm{diag}(\frac{2\delta_1\rho_1(t)}{1+\delta_1}, \ldots, \frac{2\delta_n\rho_n(t)}{1+\delta_n})$, $\delta_1, \ldots, \delta_n$ are the overshoot adjustment parameter in the performance constraint. When the system error converges, that is, when r_i converges, there is $\kappa_g G_i^{-1}MBM^{-1}G_i \to 1$, so when u_1 and u_2 are the optimal control, τ_a and τ_d can converge to the optimal control. Considering that κ_g, M^{-1}, G_a, and G_d are all bounded, then g_1 and g_2 are bounded by constants $||g_1|| < \varepsilon_{b_{g1}}$ and $||g_2|| < \varepsilon_{b_{g2}}$. In addition, we have $||f|| \leq \varepsilon_{bf}||x||$, where ε_{bf} is a constant.

Design the optimal control algorithm to make the closed-loop system state $x = \zeta$ converge or UUB and the two objectives that pre-defined performance and optimal control with game solutions can be achieved.

According to the Weierstrass higher-order approximation theorem, there exist complete independent basis sets such that the solution to (9.33) and their gradients are uniformly approximated [36–38]. So we write the value functions based on neural networks as

$$V_1(x) = W_1^T\phi_1(x) + \varepsilon_1(x) \tag{9.42}$$

$$V_2(x) = W_2^T\phi_2(x) + \varepsilon_2(x) \tag{9.43}$$

where $W_1, W_2 \in \mathbb{R}^K$ represent the critic NNs weights, $\phi_1(x), \phi_2(x) \in \mathbb{R}^K$ are the activation function vectors, K shows the number of hidden neurons in neural networks, and $\varepsilon_1(x), \varepsilon_2(x) \in \mathbb{R}$ are the approximation errors. Therefore, their derivative with respect to x can be written as

$$\nabla V_1(x) = \nabla\phi_1^T(x)W_1 + \nabla\varepsilon_1 \tag{9.44}$$

$$\nabla V_2(x) = \nabla\phi_2^T(x)W_2 + \nabla\varepsilon_2 \tag{9.45}$$

Through the NZS game theory, the Hamiltonians become

$$\begin{aligned} H_1 = {} & x^T Q_1 x + u_1^T R_{11} u_1 + u_2^T R_{12} u_2 \\ & + W_1^T \nabla\phi_1 (f + g_1 u_1 + g_2 u_2) = \varepsilon_{H_1} \end{aligned} \tag{9.46}$$

$$\begin{aligned} H_2 = {} & x^T Q_2 x + u_1^T R_{21} u_1 + u_2^T R_{22} u_2 \\ & + W_2^T \nabla\phi_2 (f + g_1 u_1 + g_2 u_2) = \varepsilon_{H_2} \end{aligned} \tag{9.47}$$

where the residual errors are

$$\varepsilon_{H_1} = -(\nabla\varepsilon_1)^T (f + g_1 u_1 + g_2 u_2) \tag{9.48}$$

$$\varepsilon_{H_2} = -(\nabla\varepsilon_2)^T (f + g_1 u_1 + g_2 u_2) \tag{9.49}$$

Consider the controller as

$$u_1 = -\frac{1}{2} R_{11}^{-1} g_1^T \nabla V_1 \tag{9.50}$$

$$u_2 = -\frac{1}{2} R_{22}^{-1} g_2^T \nabla V_2 \tag{9.51}$$

Substituting u_1 and u_2 into the HJ equations (9.33), we have

$$\begin{aligned} & x^T Q_1 x + \frac{1}{4} W_1^T E_{11} W_1 + \frac{1}{4} W_2^T E_{12} W_2 \\ & \quad + W_1^T \nabla\phi_1 f - \frac{1}{2} W_1^T D_{11} W_1 - \frac{1}{2} W_1^T D_{12} W_2 = \varepsilon_{HJ_1} \end{aligned} \tag{9.52}$$

$$\begin{aligned} & x^T Q_2 x + \frac{1}{4} W_1^T E_{21} W_1 + \frac{1}{4} W_2^T E_{22} W_2 \\ & \quad + W_2^T \nabla\phi_2 f - \frac{1}{2} W_2^T D_{21} W_1 - \frac{1}{2} W_2^T D_{22} W_2 = \varepsilon_{HJ_2} \end{aligned} \tag{9.53}$$

where $E_{11} = \nabla\phi_1 g_1 R_{11}^{-T} R_{11} R_{11}^{-1} g_1^T \nabla\phi_1^T$, $E_{12} = \nabla\phi_2 g_2 R_{22}^{-T} R_{12} R_{22}^{-1} g_2^T \nabla\phi_2^T$, $E_{21} = \nabla\phi_1 g_1 R_{11}^{-T} R_{21} R_{11}^{-1} g_1^T \nabla\phi_1^T$, $E_{22} = \nabla\phi_2 g_2 R_{22}^{-T} R_{22} R_{22}^{-1} g_2^T \nabla\phi_2^T$, $D_{11} = \nabla\phi_1 g_1 R_{11}^{-1} g_1^T \nabla\phi_1^T$, $D_{12} = \nabla\phi_1 g_2 R_{22}^{-1} g_2^T \nabla\phi_2^T$, $D_{21} = \nabla\phi_2 g_1 R_{11}^{-1} g_1^T \nabla\phi_1^T$, $D_{22} = \nabla\phi_2 g_2 R_{22}^{-1} g_2^T \nabla\phi_2^T$, ε_{HJ_1}, ε_{HJ_2} are the coupled HJ approximation errors.

If the hidden layer neuron $K \to \infty$, $\varepsilon_{HJ_1}, \varepsilon_{HJ_2}$ converge to 0, if is finite, then $\varepsilon_{HJ_1}, \varepsilon_{HJ_2}$ are bounded that $\|\varepsilon_{HJ_1}\| \le \varepsilon_{b_{HJ_1}}$, $\|\varepsilon_{HJ_2}\| \le \varepsilon_{b_{HJ_2}}$, where $\varepsilon_{b_{HJ_1}}, \varepsilon_{b_{HJ_2}}$ are constants.

For the nonlinear systems, it is extremely difficult or even impossible to get the actual W_1, W_2, so we need to use the neural networks to estimate these values. Here we define the estimated value $\hat{W}_1, \hat{W}_2$ for W_1, W_2 and the estimated error as $\tilde{W}_1 = W_1 - \hat{W}_1$ and $\tilde{W}_2 = W_2 - \hat{W}_2$. So the estimated value functions are

$$\hat{V}_1 = \hat{W}_1^T \phi_1 \tag{9.54}$$

$$\hat{V}_2 = \hat{W}_2^T \phi_2 \tag{9.55}$$

Then, the approximate optimal control policies for the system are given by

$$\hat{u}_1 = -\frac{1}{2} R_{11}^{-1} g_1^T \nabla\phi_1^T \hat{W}_1 \tag{9.56}$$

$$\hat{u}_2 = -\frac{1}{2} R_{22}^{-1} g_2^T \nabla\phi_2^T \hat{W}_2 \tag{9.57}$$

Substituting the approximated controller into (9.30), we get the approximated Hamiltonian equations with the definition of the residual errors e_1 and e_2 as follows

$$\begin{aligned} H_1 =& x^T Q_1 x + \hat{u}_1^T R_{11} \hat{u}_1 + \hat{u}_2^T R_{12} \hat{u}_2 \\ &+ \hat{W}_1^T \nabla\phi_1 \left(f + g_1 \hat{u}_1 + g_2 \hat{u}_2\right) \\ =& x^T Q_1 x + \frac{1}{4} \hat{W}_1^T E_{11} \hat{W}_1 + \frac{1}{4} \hat{W}_2^T E_{12} \hat{W}_2 \\ &+ \hat{W}_1^T \nabla\phi_1 f - \frac{1}{2} \hat{W}_1^T D_{11} \hat{W}_1 - \frac{1}{2} \hat{W}_1^T D_{12} \hat{W}_2 = e_1 \end{aligned} \tag{9.58}$$

$$\begin{aligned} H_2 =& x^T Q_2 x + \hat{u}_1^T R_{21} \hat{u}_1 + \hat{u}_2^T R_{22} \hat{u}_2 \\ &+ \hat{W}_2^T \nabla\phi_2 \left(f + g_1 \hat{u}_1 + g_2 \hat{u}_2\right) \\ =& x^T Q_2 x + \frac{1}{4} \hat{W}_1^T E_{21} \hat{W}_1 + \frac{1}{4} \hat{W}_2^T E_{22} \hat{W}_2 \\ &+ \hat{W}_2^T \nabla\phi_2 f - \frac{1}{2} \hat{W}_2^T D_{21} \hat{W}_1 - \frac{1}{2} \hat{W}_2^T D_{22} \hat{W}_2 = e_2 \end{aligned} \tag{9.59}$$

It is desired to select $\hat{W}_1, \hat{W}_2$ to minimize the square residual error

$$E = \frac{1}{2} e_1^T e_1 + \frac{1}{2} e_2^T e_2 \tag{9.60}$$

We utilize the update rate for $\hat{W}_1$ and $\hat{W}_2$ as follows [22]

$$\dot{\hat{W}}_1 = -a_1 \frac{\partial E}{\partial \hat{W}_1} = -a_1 \frac{\theta_1}{\left(1+\theta_1^T \theta_1\right)^2} e_1 \tag{9.61}$$

$$\dot{\hat{W}}_2 = -a_2 \frac{\partial E}{\partial \hat{W}_2} = -a_2 \frac{\theta_2}{\left(1+\theta_2^T \theta_2\right)^2} e_2 \tag{9.62}$$

where $\theta_i = \nabla\phi_i \left(f + g_1\hat{u}_1 + g_2\hat{u}_2\right), i = 1, 2.$

According to [39], the error $\tilde{W}$ is bounded under the above updating laws. However, in this updating method, $\bar{\theta}_i = \theta_i/(1+\theta_i^T\theta_i)$ needs to satisfy the PE conditions. In order to relax the PE conditions, we use the experience replay technique to make $\bar{\theta}_i$ satisfy the Finite Excitation (FE) conditions.

Definition 9.3 [40, 41] (*FE condition*) A bounded signal $s(\cdot) : \mathbb{R} \to \mathbb{R}^{n\times m}$ is said to be finite exciting over an interval $[t, t+T]$, where $t \geq 0$ is a fixed time index, if there exist finite constants $T > 0$ and $c > 0$ such that

$$\int_t^{t+T} s^T(\tau)s(\tau)\mathrm{d}\tau \geq cI_{m\times m} \tag{9.63}$$

where $I_{m\times m}$ is the m-dimensional identity matrix.

In order to achieve the above FE conditions, we utilize the following auxiliary systems as the related item of experience replay [41],

$$\Upsilon_i\left(t, t_{r1}, t_{r2}\right) = \xi_{i1}\left(t_{r1}, t_{r2}\right)\hat{W}_i\left(t\right) + \xi_{i2}\left(t_{r1}, t_{r2}\right) \tag{9.64}$$

$$\dot{\xi}_{i1}\left(t_{r1}, t_{r2}\right) = -\kappa_{\xi i}\xi_{i1}\left(t_{r1}, t_{r2}\right) + \varphi_{i1}\left(t\right) \tag{9.65}$$

$$\dot{\xi}_{i2}\left(t_{r1}, t_{r2}\right) = -\kappa_{\xi i}\xi_{i2}\left(t_{r1}, t_{r2}\right) + \varphi_{i2}\left(t\right) \tag{9.66}$$

where $0 < t_{r1} < t_{r2}$ show the time interval of the data used for experience replay, $\xi_{i1}\left(t_{r1}\right) = 0_{p\times p}$, $\xi_{i2}\left(t_{r1}\right) = 0_{p\times 1}$, $\varphi_{i1} = \bar{\theta}_i\bar{\theta}_i^T$, $\varphi_{i2} = \frac{\bar{\theta}_i}{m_{si}}\left(x^T Q_i x + \hat{u}_1^T R_{i1}\hat{u}_1 + \hat{u}_2^T R_{i2}\hat{u}_2\right)$, $\bar{\theta}_i = \theta_i/(1+\theta_i^T\theta_i)$, $m_{si} = 1+\theta_i^T\theta_i$, $\tilde{u}_i = u_i - \hat{u}_i$, $i = 1, 2$.

Based on the above experience replay technique, we propose a modified update law of estimated weights for critic neural networks with experience replay as follows

$$\begin{aligned}\dot{\hat{W}}_i = &- a_i \frac{\theta_i}{\left(1+\theta_i^T\theta_i\right)^2}\left(e_i - \frac{1}{4}\sum\nolimits_{j=1}^{2} \hat{W}_j^T E_{ij}\hat{W}_j\right) \\ &- a_i(\kappa_{Wi} + \kappa_{W\theta i}\bar{\theta}_i^T)\hat{W}_i - a_i\Upsilon_i\left(t, t_{r1}, t_{r2}\right), i = 1, 2\end{aligned} \tag{9.67}$$

where κ_{Wi} and $\kappa_{W\theta i}$ are adjustable parameters. Because our model is known and the optimal controller in the form of a closed-loop system is known, we only need to train the critic network instead of the critic-action network.

For a given compact set $\Omega \subset \mathbb{R}^n$ [22]: (1) The NN approximation errors and their gradients are bounded so that $||\varepsilon_1|| < \varepsilon_{b_{\varepsilon 1}}$, $||\varepsilon_2|| < \varepsilon_{b_{\varepsilon 2}}$, $||\nabla\varepsilon_1|| < \varepsilon_{b_{\nabla\varepsilon 1}}$, $||\nabla\varepsilon_2|| < \varepsilon_{b_{\nabla\varepsilon 2}}$. (2) The NN activation functions and their gradients are bounded so that $||\phi_1|| <$

$\varepsilon_{b_{\phi1}}$, $||\phi_2|| < \varepsilon_{b_{\phi2}}$, $||\nabla\phi_1|| < \varepsilon_{b_{\nabla\phi1}}$, $||\nabla\phi_2|| < \varepsilon_{b_{\nabla\phi2}}$. (3) The critic NN weights are bounded by known constants $||W_1|| < W_{1\max}$ and $||W_2|| < W_{2\max}$.

9.3.2 Stability Analysis

Theorem 9.1 *Considering the two spacecraft relative attitude game system (9.21), we choose the admissible performance guaranteed bounded control policies (9.56) and (9.57) with the weight updating laws (9.67) of the critic neural networks, where the experience replay technique is used to make $\bar{\theta}_i$ satisfy the FE condition. Then, the weight estimate errors of the critic neural networks are UUB.*

Proof Select the candidate Lyapunov function as

$$L = L_V + L_W \tag{9.68}$$

$$L_V = V_1 + V_2 \tag{9.69}$$

$$L_W = \sum_{i=1}^{2} L_{Wi} = \frac{1}{2}\sum_{i=1}^{2} a_i^{-1}\tilde{W}_i^T\tilde{W}_i \tag{9.70}$$

Step 1. Calculate $\dot{L}_V$.

Differentiating V_1 in (9.42) along the trajectories generated by the approximate control policies $\hat{u}_1$, $\hat{u}_2$, we have

$$\begin{aligned}
\dot{V}_1 &= W_1^T\nabla\phi_1\left(f + g_1\hat{u}_1 + g_2\hat{u}_2\right) + \varepsilon_{V_1} \\
&= W_1^T\left(\nabla\phi_1 f - \frac{1}{2}D_{11}\hat{W}_1 - \frac{1}{2}D_{12}\hat{W}_2\right) + \varepsilon_{V_1} \\
&= W_1^T\left(\nabla\phi_1 f - \frac{1}{2}D_{11}W_1 - \frac{1}{2}D_{12}W_2\right) \\
&\quad + \frac{1}{2}W_1^T D_{11}\tilde{W}_1 + \frac{1}{2}W_1^T D_{12}\tilde{W}_2 + \varepsilon_{V_1}
\end{aligned} \tag{9.71}$$

where $\varepsilon_{V_1} = \nabla\varepsilon_1^T(f - \frac{1}{2}g_1R_{11}^{-1}g_1^T\nabla\phi_1^T\hat{W}_1 - \frac{1}{2}g_2R_{22}^{-1}g_2^T\nabla\phi_2^T\hat{W}_2)$. According to (9.52), we have

$$\begin{aligned}
&W_1^T\nabla\phi_1 f - \frac{1}{2}W_1^T D_{11}W_1 - \frac{1}{2}W_1^T D_{12}W_2 \\
&= \varepsilon_{HJ_1} - x^T Q_1 x - \frac{1}{4}W_1^T E_{11}W_1 - \frac{1}{4}W_2^T E_{12}W_2
\end{aligned} \tag{9.72}$$

Equation (9.71) can be rewritten as

$$\begin{aligned}
\dot{V}_1 = &- x^T Q_1 x - \frac{1}{4}W_1^T E_{11}W_1 - \frac{1}{4}W_2^T E_{12}W_2 \\
&+ \frac{1}{2}W_1^T D_{11}\tilde{W}_1 + \frac{1}{2}W_1^T D_{12}\tilde{W}_2 + \varepsilon_{V_1} + \varepsilon_{HJ_1}
\end{aligned} \tag{9.73}$$

In the same way, we have

$$\begin{aligned}\dot{V}_2 =& -x^T Q_2 x - \frac{1}{4}W_1^T E_{21} W_1 - \frac{1}{4}W_2^T E_{22} W_2 \\ &+ \frac{1}{2}W_2^T D_{21}\tilde{W}_1 + \frac{1}{2}W_2^T D_{22}\tilde{W}_2 + \varepsilon_{V_2} + \varepsilon_{HJ_2}\end{aligned} \tag{9.74}$$

where $\varepsilon_{V_2} = \nabla\varepsilon_2^T(f - \frac{1}{2}g_1 R_{11}^{-1} g_1^T \nabla\phi_1^T \hat{W}_1 - \frac{1}{2}g_2 R_{22}^{-1} g_2^T \nabla\phi_2^T \hat{W}_2)$.

Substituting (9.73) and (9.74) into the derivative of (9.69) with respect to time, we have

$$\begin{aligned}\dot{L}_V =& \dot{V}_1 + \dot{V}_2 \\ =& -x^T Q_1 x - x^T Q_2 x \\ &- \frac{1}{4}W_1^T E_{11} W_1 - \frac{1}{4}W_2^T E_{12} W_2 \\ &- \frac{1}{4}W_1^T E_{21} W_1 - \frac{1}{4}W_2^T E_{22} W_2 \\ &+ \varepsilon_{V_1} + \varepsilon_{HJ_1} + \varepsilon_{V_2} + \varepsilon_{HJ_2} \\ &+ \frac{1}{2}W_1^T D_{11}\tilde{W}_1 + \frac{1}{2}W_1^T D_{12}\tilde{W}_2 \\ &+ \frac{1}{2}W_2^T D_{21}\tilde{W}_1 + \frac{1}{2}W_2^T D_{22}\tilde{W}_2\end{aligned} \tag{9.75}$$

For item $\varepsilon_{V_1} + \varepsilon_{V_2}$, we scale it as follows. From the definition of ε_{V_1} and ε_{V_2}, we have

$$\begin{aligned}||\varepsilon_{V_1}|| \le& + \varepsilon_{b_{\nabla\varepsilon 1}}\varepsilon_{bf}||x|| \\ &+ \frac{1}{2}\varepsilon_{b_{\nabla\varepsilon 1}}\varepsilon_{b_{g1}}^2\varepsilon_{b_{\nabla\phi 1}}\lambda_{\max}(R_{11}^{-1})(||W_1|| + ||\tilde{W}_1||) \\ &+ \frac{1}{2}\varepsilon_{b_{\nabla\varepsilon 1}}\varepsilon_{b_{g2}}^2\varepsilon_{b_{\nabla\phi 2}}\lambda_{\max}(R_{22}^{-1})(||W_2|| + ||\tilde{W}_2||)\end{aligned} \tag{9.76}$$

$$\begin{aligned}||\varepsilon_{V_2}|| \le& + \varepsilon_{b_{\nabla\varepsilon 2}}\varepsilon_{bf}||x|| \\ &+ \frac{1}{2}\varepsilon_{b_{\nabla\varepsilon 2}}\varepsilon_{b_{g1}}^2\varepsilon_{b_{\nabla\phi 1}}\lambda_{\max}(R_{11}^{-1})(||W_1|| + ||\tilde{W}_1||) \\ &+ \frac{1}{2}\varepsilon_{b_{\nabla\varepsilon 2}}\varepsilon_{b_{g2}}^2\varepsilon_{b_{\nabla\phi 2}}\lambda_{\max}(R_{22}^{-1})(||W_2|| + ||\tilde{W}_2||)\end{aligned} \tag{9.77}$$

Step 2. Calculate $\dot{L}_W$.

For the item Υ_1 associated with experience replay, according to the definition of φ_{11} and φ_{12}, we have

$$\begin{aligned}\varphi_{12} &= \frac{\bar{\theta}_1}{m_{s1}}\left(x^T Q_1 x + \hat{u}_1^T R_{11}\hat{u}_1 + \hat{u}_2^T R_{12}\hat{u}_2\right)\\ &= \frac{\bar{\theta}_1}{m_{s1}}(\varepsilon_{H_1} + \varepsilon_{u1} + \varepsilon_{\tilde{u}} - W_1^T \nabla\phi_1\left(f + g_1\hat{u}_1 + g_2\hat{u}_2\right))\\ &= \frac{\bar{\theta}_1}{m_{s1}}(\varepsilon_{H_1} + \varepsilon_{u1} + \varepsilon_{\tilde{u}} - m_{s1}\bar{\theta}_1^T W_1)\end{aligned} \tag{9.78}$$

where $\varepsilon_{u1} = +\hat{u}_1^T R_{11}\hat{u}_1 + \hat{u}_2^T R_{12}\hat{u}_2 - u_1^T R_{11}u_1 - u_2^T R_{12}u_2$, $\varepsilon_{\tilde{u}1} = -W_1^T\nabla\phi_1$ $(+g_1\tilde{u}_1 + g_2\tilde{u}_2)$. Considering (9.65) and (9.66), we have $\xi_{11} = \mathrm{e}^{-\kappa_{\xi i}t_{r2}}$ $\int_{t_{r1}}^{t_{r2}} \varphi_{11}(\tau)\,\mathrm{e}^{\kappa_{\xi 1}\tau}\mathrm{d}\tau$, $\xi_{11} = \mathrm{e}^{-\kappa_{\xi i}t_{r2}}\int_{t_{r1}}^{t_{r2}} \varphi_{11}(\tau)\,\mathrm{e}^{\kappa_{\xi 1}\tau}\mathrm{d}\tau$. Therefore, (9.64) can be expressed as

$$\begin{aligned}\Upsilon_1(t, t_{r1}, t_{r2}) &= \xi_{11}(t_{r1}, t_{r2})\,\hat{W}_1(t) + \xi_{12}(t_{r1}, t_{r2})\\ &= \mathrm{e}^{-\kappa_{\xi 1}t_{r2}}\int_{t_{r1}}^{t_{r2}} \mathrm{e}^{\kappa_{\xi 1}\tau}(\varphi_{11}(\tau)\hat{W}_1(t) + \varphi_{12}(\tau))\,\mathrm{d}\tau\\ &= \mathrm{e}^{-\kappa_{\xi 1}t_{r2}}\int_{t_{r1}}^{t_{r2}} \mathrm{e}^{\kappa_{\xi 1}\tau}(-\bar{\theta}_1\bar{\theta}_1^T\tilde{W}_1(t)\\ &\quad + \frac{\bar{\theta}_1}{m_{s1}}(\varepsilon_{H_1} + \varepsilon_{u1} + \varepsilon_{\tilde{u}1}))\,\mathrm{d}\tau\\ &= -\int_{t_{r1}}^{t_{r2}} \mathrm{e}^{-\kappa_{\xi 1}(t_{r2}-\tau)}\bar{\theta}_1\bar{\theta}_1^T\mathrm{d}\tau\cdot\tilde{W}_1(t)\\ &\quad + \int_{t_{r1}}^{t_{r2}} \mathrm{e}^{-\kappa_{\xi 1}(t_{r2}-\tau)}\frac{\bar{\theta}_1}{m_{s1}}(\varepsilon_{H_1} + \varepsilon_{u1} + \varepsilon_{\tilde{u}1})\,\mathrm{d}\tau\\ &= -\xi_{11}\tilde{W}_1(t) + \varepsilon_{\xi_1}\end{aligned} \tag{9.79}$$

where $\varepsilon_{\xi_1} = \int_{t_{r1}}^{t_{r2}} \mathrm{e}^{-\kappa_{\xi 1}(t_{r2}-\tau)}\frac{\bar{\theta}_1}{m_{s1}}(\varepsilon_{H_1} + \varepsilon_{u1} + \varepsilon_{\tilde{u}1})\,\mathrm{d}\tau$.

Similarly, we have

$$\Upsilon_2(t, t_{r1}, t_{r2}) = -\xi_{21}\tilde{W}_2(t) + \varepsilon_{\xi_2} \tag{9.80}$$

where $\varepsilon_{\xi_2} = \int_{t_{r1}}^{t_{r2}} \mathrm{e}^{-\kappa_{\xi 2}(t_{r2}-\tau)}\frac{\bar{\theta}_2}{m_{s2}}(\varepsilon_{H_2} + \varepsilon_{u2} + \varepsilon_{\tilde{u}})\,\mathrm{d}\tau$, $\varepsilon_{u2} = +\hat{u}_1^T R_{21}\hat{u}_1 + \hat{u}_2^T R_{22}\hat{u}_2 - u_1^T R_{21}u_1 - u_2^T R_{22}u_2$, $\varepsilon_{\tilde{u}2} = -W_2^T\nabla\phi_2(+g_1\tilde{u}_1 + g_2\tilde{u}_2)$.

Substituting (9.79) and (9.80) into (9.67), we have

$$\begin{aligned}\dot{\hat{W}}_i &= -a_i\frac{\theta_i}{\left(1+\theta_i^T\theta_i\right)^2}\left(e_i - \frac{1}{4}\hat{W}_1^T E_{i1}\hat{W}_1 - \frac{1}{4}\hat{W}_2^T E_{i2}\hat{W}_2\right)\\ &\quad - a_i(\kappa_{Wi} + \kappa_{W\theta i}\bar{\theta}_i^T)\hat{W}_i + a_i\xi_{i1}\tilde{W}_i(t) - a_i\varepsilon_{\xi_i}\end{aligned} \tag{9.81}$$

Find the derivative of L_i with respect to time as

$$
\begin{aligned}
\dot{L}_i =& a_i^{-1}\tilde{W}_i^T\dot{\tilde{W}}_i = -a_i^{-1}\tilde{W}_i^T\dot{\hat{W}}_i \\
=& \tilde{W}_i^T\frac{\bar{\theta}_i}{m_{si}}\left(e_i - \frac{1}{4}\hat{W}_1^T E_{i1}\hat{W}_1 - \frac{1}{4}\hat{W}_2^T E_{i2}\hat{W}_2\right) \\
&- \tilde{W}_i^T(\kappa_{Wi} + \kappa_{W\theta i}\bar{\theta}_i^T)\hat{W}_i - \tilde{W}_i^T\xi_{i1}\tilde{W}_i + \varepsilon_{\xi_i}^T\tilde{W}_i
\end{aligned}
\tag{9.82}
$$

From (9.52) and (9.58), we have

$$
\begin{aligned}
e_1 - \frac{1}{4}\hat{W}_1^T E_{11}\hat{W}_1 - \frac{1}{4}\hat{W}_2^T E_{12}\hat{W}_2 =& \varepsilon_{HJ_1} - \theta_1^T\tilde{W}_1 \\
&+ \frac{1}{2}W_1^T D_{11}\tilde{W}_1 + \frac{1}{2}W_1^T D_{12}\tilde{W}_2 - \frac{1}{4}W_1^T E_{11}W_1 - \frac{1}{4}W_2^T E_{12}W_2
\end{aligned}
\tag{9.83}
$$

$$
\begin{aligned}
e_2 - \frac{1}{4}\hat{W}_1^T E_{21}\hat{W}_1 - \frac{1}{4}\hat{W}_2^T E_{22}\hat{W}_2 =& \varepsilon_{HJ_2} - \theta_2^T\tilde{W}_2 \\
&+ \frac{1}{2}W_2^T D_{21}\tilde{W}_1 + \frac{1}{2}W_2^T D_{22}\tilde{W}_2 - \frac{1}{4}W_1^T E_{21}W_1 - \frac{1}{4}W_2^T E_{22}W_2
\end{aligned}
\tag{9.84}
$$

Substituting (9.83) and (9.84) into (9.82), we have

$$
\begin{aligned}
\dot{L}_W =& \sum\nolimits_{i=1}^{2}\dot{L}_{Wi} \\
=& -\tilde{W}_1^T\bar{\theta}_1\theta_1^T\tilde{W}_1 - \tilde{W}_2^T\bar{\theta}_2\theta_2^T\tilde{W}_2 - \tilde{W}_1^T\xi_{11}\tilde{W}_1 - \tilde{W}_2^T\xi_{21}\tilde{W}_2 \\
&+ \tilde{W}_1^T\bar{\theta}_1\frac{W_1^T D_{11}}{2m_{s1}}\tilde{W}_1 + \tilde{W}_1^T\bar{\theta}_1\frac{W_1^T D_{12}}{2m_{s1}}\tilde{W}_2 \\
&+ \tilde{W}_2^T\bar{\theta}_2\frac{W_2^T D_{21}}{2m_{s2}}\tilde{W}_1 + \tilde{W}_2^T\bar{\theta}_2\frac{W_2^T D_{22}}{2m_{s2}}\tilde{W}_2 \\
&+ \tilde{W}_1^T\bar{\theta}_1(-\kappa_{W\theta 1}W_1 + \frac{\varepsilon_{HJ_1}}{m_{s1}} - \frac{1}{4m_{s1}}W_1^T E_{11}W_1 \\
&- \frac{1}{4m_{s1}}W_2^T E_{12}W_2) + \tilde{W}_2^T\bar{\theta}_2(-\kappa_{W\theta 2}W_2 + \frac{\varepsilon_{HJ_2}}{m_{s2}} \\
&- \frac{1}{4m_{s2}}W_1^T E_{21}W_1 - \frac{1}{4m_{s2}}W_2^T E_{22}W_2) \\
&+ \left(\varepsilon_{\xi_1}^T - \kappa_{W1}W_1^T\right)\tilde{W}_1 + \left(\varepsilon_{\xi_2}^T - \kappa_{W2}W_2^T\right)\tilde{W}_2 \\
&+ \tilde{W}_1^T\kappa_{W1}\tilde{W}_1 + \tilde{W}_1^T\kappa_{W\theta 1}\bar{\theta}_1^T\tilde{W}_1 \\
&+ \tilde{W}_2^T\kappa_{W2}\tilde{W}_2 + \tilde{W}_2^T\kappa_{W\theta 2}\bar{\theta}_2^T\tilde{W}_2
\end{aligned}
\tag{9.85}
$$

Step 3. Calculate $\dot{L}$.
Combining Steps 1 and 2, we write the derivative of L as

$$
\begin{aligned}
\dot{L} &= \dot{L}_V + \dot{L}_W \\
&\le -x^T (Q_1 + Q_2) x - \tilde{W}_1^T \bar{\theta}_1 \bar{\theta}_1^T \tilde{W}_1 - \tilde{W}_2^T \bar{\theta}_2 \bar{\theta}_2^T \tilde{W}_2 \\
&\quad - \tilde{W}_1^T (\xi_{11} - \kappa_{W1}) \tilde{W}_1 - \tilde{W}_2^T (\xi_{21} - \kappa_{W2}) \tilde{W}_2 \\
&\quad + \tilde{W}_1^T \bar{\theta}_1 \left(\frac{W_1^T D_{11}}{2m_{s1}} + \kappa_{W\theta 1}^T \right) \tilde{W}_1 + \tilde{W}_1^T \bar{\theta}_1 \frac{W_1^T D_{12}}{2m_{s1}} \tilde{W}_2 \\
&\quad + \tilde{W}_2^T \bar{\theta}_2 \left(\frac{W_2^T D_{22}}{2m_{s2}} + \kappa_{W\theta 2}^T \right) \tilde{W}_2 + \tilde{W}_2^T \bar{\theta}_2 \frac{W_2^T D_{21}}{2m_{s2}} \tilde{W}_1 \\
&\quad + (\varepsilon_{b_{\nabla\varepsilon 1}} + \varepsilon_{b_{\nabla\varepsilon 2}}) \varepsilon_{bf} ||x|| + \varepsilon_c \\
&\quad + \tilde{W}_1^T \bar{\theta}_1 (-\kappa_{W\theta 1} W_1 + \frac{\varepsilon_{HJ_1}}{m_{s1}} - \frac{1}{4m_{s1}} W_1^T E_{11} W_1 \\
&\quad - \frac{1}{4m_{s1}} W_2^T E_{12} W_2) \\
&\quad + \tilde{W}_2^T \bar{\theta}_2 (-\kappa_{W\theta 2} W_2 + \frac{\varepsilon_{HJ_2}}{m_{s2}} - \frac{1}{4m_{s2}} W_1^T E_{21} W_1 \\
&\quad - \frac{1}{4m_{s2}} W_2^T E_{22} W_2) \\
&\quad + \frac{1}{2} (W_1^T D_{11} + W_2^T D_{21} + 2\varepsilon_{\xi_1}^T - 2\kappa_{W1} W_1^T \\
&\quad + (\varepsilon_{b_{\nabla\varepsilon 1}} + \varepsilon_{b_{\nabla\varepsilon 2}}) \varepsilon_{b_{g1}}^2 \varepsilon_{b_{\nabla\phi 1}} \lambda_{\max}(R_{11}^{-1})) ||\tilde{W}_1|| \\
&\quad + \frac{1}{2} (W_2^T D_{22} + W_1^T D_{12} + 2\varepsilon_{\xi_2}^T - 2\kappa_{W2} W_2^T \\
&\quad + (\varepsilon_{b_{\nabla\varepsilon 1}} + \varepsilon_{b_{\nabla\varepsilon 2}}) \varepsilon_{b_{g2}}^2 \varepsilon_{b_{\nabla\phi 2}} \lambda_{\max}(R_{22}^{-1})) ||\tilde{W}_2||
\end{aligned}
\tag{9.86}
$$

where $\lambda_{\min}(\cdot)$ and $\lambda_{\max}(\cdot)$ represent the minimum and maximum eigenvalues of the matrix, respectively, and

$$
\begin{aligned}
\varepsilon_c > &+ \frac{1}{4} (||E_{11}|| + ||E_{21}||) ||W_1||^2 + ||\varepsilon_{HJ_1}|| + ||\varepsilon_{HJ_2}|| \\
&+ \frac{1}{4} (||E_{12}|| + ||E_{22}||) ||W_2||^2 \\
&+ \frac{1}{2} (\varepsilon_{b_{\nabla\varepsilon 1}} + \varepsilon_{b_{\nabla\varepsilon 2}}) \varepsilon_{b_{g1}}^2 \varepsilon_{b_{\nabla\phi 1}} \lambda_{\max}(R_{11}^{-1}) ||W_1|| \\
&+ \frac{1}{2} (\varepsilon_{b_{\nabla\varepsilon 1}} + \varepsilon_{b_{\nabla\varepsilon 2}}) \varepsilon_{b_{g2}}^2 \varepsilon_{b_{\nabla\phi 2}} \lambda_{\max}(R_{22}^{-1}) ||W_2||
\end{aligned}
\tag{9.87}
$$

We select the variable $Z = [x, \bar{\theta}_1^T \tilde{W}_1, \bar{\theta}_2^T \tilde{W}_2, \tilde{W}_1, \tilde{W}_2]^T$, then the above formula can be written in a compact form as follows.

$$
\dot{L} < -Z^T M Z + Z^T D + \varepsilon_c \tag{9.88}
$$

where

$$M = \begin{bmatrix} m_{11} & 0 & 0 & 0 & 0 \\ 0 & m_{22} & 0 & m_{24} & m_{25} \\ 0 & 0 & m_{33} & m_{34} & m_{35} \\ 0 & m_{42} & m_{43} & m_{44} & 0 \\ 0 & m_{52} & m_{53} & 0 & m_{55} \end{bmatrix}, D = \begin{bmatrix} d_1 \\ d_2 \\ d_3 \\ d_4 \\ d_5 \end{bmatrix},$$

with

$$\begin{aligned}
m_{11} &= Q_1 + Q_2, \\
m_{22} &= m_{33} = I, \\
m_{44} &= \xi_{11} - \kappa_{W1}, \\
m_{55} &= \xi_{21} - \kappa_{W2}, \\
m_{25} &= m_{52}^T = -\frac{W_1^T D_{12}}{4m_{s1}}, \\
m_{34} &= m_{43}^T = -\frac{W_2^T D_{21}}{4m_{s2}}, \\
m_{24} &= m_{42}^T = -\frac{W_1^T D_{11}}{4m_{s1}} - \frac{1}{2}\kappa_{W\theta 1}^T, \\
m_{35} &= m_{53}^T = -\frac{W_2^T D_{22}}{4m_{s2}} - \frac{1}{2}\kappa_{W\theta 2}^T, \\
d_1 &= (\varepsilon_{b_{\nabla\varepsilon 1}} + \varepsilon_{b_{\nabla\varepsilon 2}})\varepsilon_{bf}, \\
d_2 &= (-\kappa_{W\theta 1} W_1 + \frac{\varepsilon_{HJ_1}}{m_{s1}} - \frac{1}{4m_{s1}} W_1^T E_{11} W_1 \\
&\quad - \frac{1}{4m_{s1}} W_2^T E_{12} W_2), \\
d_3 &= (-\kappa_{W\theta 2} W_2 + \frac{\varepsilon_{HJ_2}}{m_{s2}} - \frac{1}{4m_{s2}} W_1^T E_{21} W_1 \\
&\quad - \frac{1}{4m_{s2}} W_2^T E_{22} W_2), \\
d_4 &= \frac{1}{2}(W_1^T D_{11} + W_2^T D_{21} + 2\varepsilon_{\xi_1}^T - 2\kappa_{W1} W_1^T \\
&\quad + (\varepsilon_{b_{\nabla\varepsilon 1}} + \varepsilon_{b_{\nabla\varepsilon 2}})\varepsilon_{b_{g1}}^2 \varepsilon_{b_{\nabla\phi 1}} \lambda_{\max}(R_{11}^{-1})), \\
d_5 &= \frac{1}{2}(W_2^T D_{22} + W_1^T D_{12} + 2\varepsilon_{\xi_2}^T - 2\kappa_{W2} W_2^T \\
&\quad + (\varepsilon_{b_{\nabla\varepsilon 1}} + \varepsilon_{b_{\nabla\varepsilon 2}})\varepsilon_{b_{g2}}^2 \varepsilon_{b_{\nabla\phi 2}} \lambda_{\max}(R_{22}^{-1})).
\end{aligned}$$

According to the boundedness statement at the end of the previous subsection, D and ε_c are both bounded, and the maximum boundary is defined as $D_{\max}$ and $\varepsilon_{c\max}$, respectively. Choose the appropriate κ_{W1}, κ_{W2}, $\kappa_{W\theta 1}$, and $\kappa_{W\theta 2}$ parameters so that M is positive definite, and then L is UUB. In order to make M positive definite, we write M as a block matrix form as

$$M = \begin{bmatrix} M_{11} & 0 & 0 \\ 0 & M_{22} & M_{23} \\ 0 & M_{32} & M_{33} \end{bmatrix},$$

where $M_{11} = m_{11} > 0$, $M_{22} = \text{diag}(m_{22}, m_{33})$, $M_{33} = \text{diag}(m_{44}, m_{55})$, $M_{23} = [m_{24}, m_{25}; m_{34}, m_{35}]$, and $M_{32} = [m_{42}, m_{43}; m_{52}, m_{53}]$. And find the Schur complements for the diagonal elements M_{22} and M_{33} as $S_{22} = M_{22} - M_{23}M_{33}^{-1}M_{32}$ and $S_{33} = M_{33} - M_{32}M_{22}^{-1}M_{23}$, respectively. Choose appropriate parameters $\kappa_{W1}, \kappa_{W2}, \kappa_{W\theta 1}$, and $\kappa_{W\theta 2}$ such that $S_{22} > 0$ and $S_{33} > 0$, then a positive definite M can be obtained [22]. Thus, (9.88) can be rewritten as

$$\dot{L} < -\lambda_{\min}(M)||Z||^2 + D_{\max}||Z|| + \varepsilon_{c\max} \tag{9.89}$$

According to (9.89), we have $\dot{L} < 0$, if

$$||Z|| > \frac{D_{\max}}{2\lambda_{\min}(M)} + \sqrt{\frac{D_{\max}^2}{4\lambda_{\min}^2(M)} + \frac{\varepsilon_{c\max}}{\lambda_{\min}(M)}} = \varepsilon_Z. \tag{9.90}$$

It can be seen from (9.90) that when $||Z||$ is less than ε_Z, the system state and the estimated weight errors are bounded, and when $||Z|| > \varepsilon_Z$, $\dot{L}$ is negative definite. In summary, the system state and the weight errors are UUB, which completes the proof. □

9.3.3 *Nash Equilibrium Analysis*

Theorem 9.2 (Nash Solution of the Game) *Considering the two spacecraft relative attitude game system (9.21), we choose the admissible performance guaranteed bounded control policies (9.56) and (9.57) with the experience replay weight updating laws (9.67) of the critic neural networks. Then, H_1 and H_2 are UUB. That is, $\hat{W}_1$ and $\hat{W}_2$ converge to the approximate coupled HJ solution. Equations (9.56) and (9.57) converge to the approximate Nash solution of the game.*

Proof According to proof of Theorem 9.1, $\tilde{W}_i$ are UUB. Substituting Eqs. (9.52) and (9.53) into (9.58) and (9.59), respectively, the approximate coupled HJ equations are obtained as follows

$$
\begin{aligned}
H_1 = & -\frac{1}{2}W_1^T E_{11}\tilde{W}_1 + \frac{1}{4}\tilde{W}_1^T E_{11}\tilde{W}_1 - \frac{1}{2}W_2^T E_{12}\tilde{W}_2 \\
& + \frac{1}{4}\tilde{W}_2^T E_{12}\tilde{W}_2 - \tilde{W}_1^T \nabla\phi_1 f + W_1^T D_{11}\tilde{W}_1 + \varepsilon_{HJ_1} \\
& - \frac{1}{2}\tilde{W}_1^T D_{11}\tilde{W}_1 - \frac{1}{2}\hat{W}_1^T D_{12}\hat{W}_2 + \frac{1}{2}W_1^T D_{12}W_2
\end{aligned} \tag{9.91}
$$

$$
\begin{aligned}
H_2 = & -\frac{1}{2}W_1^T E_{21}\tilde{W}_1 + \frac{1}{4}\tilde{W}_1^T E_{21}\tilde{W}_1 - \frac{1}{2}W_2^T E_{22}\tilde{W}_2 \\
& + \frac{1}{4}\tilde{W}_2^T E_{22}\tilde{W}_2 - \tilde{W}_2^T \nabla\phi_2 f + W_2^T D_{22}\tilde{W}_2 + \varepsilon_{HJ_2} \\
& - \frac{1}{2}\tilde{W}_2^T D_{22}\tilde{W}_2 + \frac{1}{2}W_2^T D_{21}W_1 - \frac{1}{2}\hat{W}_2^T D_{21}\hat{W}_1
\end{aligned} \tag{9.92}
$$

Considering $\hat{W}_1 = W_1 - \tilde{W}_1$ and $\hat{W}_2 = W_2 - \tilde{W}_2$, and letting $||W_1|| \leq \varepsilon_{W_1\max}$ and $||W_2|| \leq \varepsilon_{W_2\max}$, we have

$$
||\hat{W}_1|| = ||W_1 - \tilde{W}_1|| \leq ||W_1|| + ||\tilde{W}_1|| \leq ||\tilde{W}_1|| + \varepsilon_{W_1\max} \tag{9.93}
$$

$$
||\hat{W}_2|| = ||W_2 - \tilde{W}_2|| \leq ||W_2|| + ||\tilde{W}_2|| \leq ||\tilde{W}_2|| + \varepsilon_{W_2\max} \tag{9.94}
$$

Taking norms in (9.91) and (9.92), we have

$$
\begin{aligned}
||H_1|| \leq & + \frac{1}{2}\varepsilon_{W_1\max}||E_{11}||||\tilde{W}_1|| + \frac{1}{4}||E_{11}||||\tilde{W}_1||^2 \\
& + \frac{1}{2}\varepsilon_{W_2\max}||E_{12}||||\tilde{W}_2|| + \frac{1}{4}||E_{12}||||\tilde{W}_2||^2 \\
& + \varepsilon_{b_{\nabla\phi_1}}\varepsilon_{bf}||x||||\tilde{W}_1|| + \varepsilon_{W_1\max}||D_{11}||||\tilde{W}_1|| + \varepsilon_{b_{HJ_1}} \\
& + \frac{1}{2}||D_{11}||||\tilde{W}_1||^2 + \frac{1}{2}\varepsilon_{W_1\max}\varepsilon_{W_2\max}||D_{12}|| \\
& + \frac{1}{2}||D_{12}||\left(||\tilde{W}_1|| + \varepsilon_{W_1\max}\right)\left(||\tilde{W}_2|| + \varepsilon_{W_2\max}\right)
\end{aligned} \tag{9.95}
$$

$$
\begin{aligned}
||H_2|| \leq & + \frac{1}{2}\varepsilon_{W_1\max}||E_{21}||||\tilde{W}_1|| + \frac{1}{4}||E_{21}||||\tilde{W}_1||^2 \\
& + \frac{1}{2}\varepsilon_{W_2\max}||E_{22}||||\tilde{W}_2|| + \frac{1}{4}||E_{22}||||\tilde{W}_2||^2 \\
& + \varepsilon_{b_{\nabla\phi_2}}\varepsilon_{bf}||x||||\tilde{W}_2|| + \varepsilon_{W_2\max}||D_{22}||||\tilde{W}_2|| + \varepsilon_{b_{HJ_2}} \\
& + \frac{1}{2}||D_{22}||||\tilde{W}_2||^2 + \frac{1}{2}\varepsilon_{W_2\max}\varepsilon_{W_1\max}||D_{21}|| \\
& + \frac{1}{2}||D_{21}||\left(||\tilde{W}_2|| + \varepsilon_{W_2\max}\right)\left(||\tilde{W}_1|| + \varepsilon_{W_1\max}\right)
\end{aligned} \tag{9.96}
$$

It can be found that the terms on the right-hand side of (9.95) and (9.96) are UUB, then H_1 and H_2 are UUB and converge to the approximate HJ equation solution.

For $||\hat{u}_1 - u_1||$ and $||\hat{u}_2 - u_2||$, considering that $\tilde{W}_1$ and $\tilde{W}_2$ are UUB, then there are

$$\begin{aligned} ||\hat{u}_1 - u_1|| &= \frac{1}{2}||R_{11}^{-1} g_1^T \nabla\phi_1^T \tilde{W}_1|| \\ &\le \frac{1}{2}\varepsilon_{b_{g1}}\varepsilon_{b_{\nabla\phi 1}}\lambda_{\max}(R_{11}^{-1})||\tilde{W}_1|| \end{aligned} \tag{9.97}$$

$$\begin{aligned} ||\hat{u}_2 - u_2|| &= \frac{1}{2}||R_{22}^{-1} g_2^T \nabla\phi_2^T \tilde{W}_2|| \\ &\le \frac{1}{2}\varepsilon_{b_{g2}}\varepsilon_{b_{\nabla\phi 2}}\lambda_{\max}(R_{22}^{-1})||\tilde{W}_2|| \end{aligned} \tag{9.98}$$

Therefore, $||\hat{u}_1 - u_1||$ and $||\hat{u}_2 - u_2||$ are UUB. Thus, the approximate Nash equilibrium solution of the game is given by $\hat{u}_1$ and $\hat{u}_2$, which completes the proof. □

9.3.4 *Multiple Spacecraft Situation*

In the previous sections of this chapter, we have detailed the design and analysis of the performance guaranteed game strategy for two spacecraft. The proposed strategy can be easily extended to the multi-spacecraft situation. For the multi-spacecraft system, when we only consider the attitude consistency problem or the expected attitude consistency problem, all the spacecraft can actively maneuver to eliminate the common attitude error of the member spacecraft. Their maneuver strategies satisfy the Nash equilibrium, and all spacecraft deallocate control tasks according to the Nash equilibrium solution and assume part of the maneuver control obligations, respectively.

There are two options for extending the control strategies of dual spacecraft to the multi-spacecraft system: (1) In the multi-spacecraft system, any two spacecraft can be selected to form a dual-spacecraft system. The control strategies of the two spacecraft are designed first, and another spacecraft is added to the system after the system is stabilized. The dual spacecraft control strategies are executed again. At this point, the original two spacecraft will use the same strategies to synchronize with the third spacecraft. This cycle continues until all spacecraft achieve consistency. Although this design can achieve the attitude consistency of multiple spacecraft, it has many defects, such as that unnecessary redundant attitude motions are increased, the process is cumbersome, time-consuming is increased, and the energy consumption is increased. (2) Directly expand the system scale, and design the performance guaranteed game strategies for all control inputs of multiple spacecraft simultaneously. The method of designing game strategies for multiple members will be reflected in the next chapter and will not be repeated here. In this chapter, the case of two spacecraft is used as an example.

9.4 Simulation Analysis

The numerical example is shown in this section to demonstrate the performance of the developed the performance guaranteed non-zero-sum game strategies. The multi-spacecraft case can be extended from the dual-spacecraft case, and the algorithm design in this chapter takes the dual-spacecraft case as an example. Simultaneously, in order to correspond to the algorithm design process, the simulation example is the dual-spacecraft case. To illustrate the rationality, effectiveness, and advantages of the strategies proposed in this chapter, this section introduces some similar algorithms for comparison. Furthermore, the practicability of the proposed algorithm is further illustrated by comparing the performance, error suppression, and energy consumption of the different algorithm results.

For detailed algorithm comparison, we mark different algorithms with different letters. We mark the simulation results of control policies (9.56) and (9.57) with the weight updating laws (9.67) proposed in this chapter as 'A'. The simulation results using an algorithm similar to this chapter, but with the performance constraint strategy removed, are marked as 'B'. Compared with 'A', 'B' just lacks the performance guaranteed ability, and this comparison highlights the utility of performance guarantees to the game algorithm. The simulation results of the combination of the performance guaranteed algorithm and the traditional non-zero-sum game algorithm, marked as 'C', mainly characterize the effect of the additional structural design in this chapter. To show the critical role of the guaranteed performance constraints, the algorithm for multi-player non-zero-sum games in [22] mentioned in the introduction is also cited for comparison and labeled as 'D'.

9.4.1 Simulation Parameters

For two spacecraft at close range, the orbital parameters are given in Table 9.1. The attitude maneuver is taken as an example for the simulation in this section, so the spacecraft orbit maneuver is not considered for the time being. The orbital parameters here are only used to reveal the simulation environment. In the control task designed in this chapter, we use the controllers τ_a and τ_d to drive the two cooperative spacecraft to perform attitude maneuvers so that the attitude errors between them are eliminated. Therefore, we first give the initial values of the respective attitude parameters of the two spacecraft. It is worth noting that the simulation analysis provides a working condition display so that the initial attitude parameters can be changed under different working conditions. Without loss of generality, the initial parameters of the two spacecraft are directly given here, $q_v(0) = [0.6, 0.3, -0.3]^T$, $q_0(0) = \sqrt{1 - q_v^T(0)q_v(0)}$, $\omega_a(0) = [0, 0, 0]^T \text{rad/s}$, $q_{dv}(0) = [0.1, 0.1, -0.1]^T$, $q_{d0}(0) = \sqrt{1 - q_{dv}^T(0)q_{dv}(0)}$, $\omega_d(0) = [0, 0, 0]^T \text{rad/s}$.

Table 9.1 The orbital parameters of two spacecraft

Parameters	Values
Semimajor axis	7.25×10^6m
Eccentricity	0.0069
Inclination	60°
Argument of periapsis	0°
Longitude of the ascending node	0°
Mean anomaly	0°

Table 9.2 The initial parameters of the proposed control policies (9.56) and (9.57)

Param	Value	Param	Value	Param	Value
t_{ps}	100s	ε_ρ	0.01	a_1	0.1
ρ_0	1	δ_i	1	a_2	0.1
ρ_t	0.1	λ_i	1		

$$J_a = J_d = \begin{bmatrix} 100 & -10 & 5 \\ -10 & 120 & -3 \\ 5 & -3 & 100 \end{bmatrix} \text{kg} \cdot \text{m}^2.$$

Further, the relevant parameters of the two spacecraft control policies in the simulation are given, as shown in Table 9.2. We first describe the key parameters regarding the performance guaranteed. In Table 9.2, the definable desired settling time of the system was chosen to be $t_{ps} = 100s$, this parameter suitable for the usual spacecraft attitude maneuver; ρ_0 is the initial value of performance constraints. Considering that it needs to be greater than the initial quaternion errors, it is selected as 1 so that it can be greater than any quaternion element; ρ_t and ε_ρ provide the steady-state error boundary; δ_i is about whether to suppress overshoot; λ_i is the proportional coefficient of the sliding-mode surface in the controller. We simplify these parameters so that the simulation results are close to the algorithmic nature of the control strategy. For the gain parameters of the experience replay auxiliary system in (9.65) and (9.66), we simply take them as $\kappa_{\xi 1} = \kappa_{\xi 2} = 1$. As the gain constants in (9.67), we choose them as $a_1 = a_2 = 0.1$. The adjustable parameters κ_{Wi} and $\kappa_{W\theta i}$ can be taken as $\kappa_{W1} = \kappa_{W2} = 10^{-6} I_{6\times 6}$ and $\kappa_{W\theta 1} = \kappa_{W\theta 2} = 10^{-6} I_{6\times 1}$, respectively. The matrix in the value function is temporarily taken as $R_{11} = R_{22} = 0.004 I_{3\times 3}$, $R_{12} = R_{21} = I_{3\times 3}$, $Q_1 = Q_2 = 0.05 I_{3\times 3}$, where $I_{3\times 3}$ is the unit matrix of 3×3. The neural network basis set used in this chapter is $\phi_1 = \phi_2 = [\zeta_1^2, \zeta_2^2, \zeta_3^2, \zeta_1\zeta_2, \zeta_2\zeta_3, \zeta_3\zeta_1]^T$. The controller in this chapter does not consider external disturbances throughout the process.

In order to show the system error in the whole process of two spacecraft attitude maneuvers, we define the error-index for the controllers as $I_a = ||q_{ev}|| + ||\omega_e||$. This index is also used to compare performance between different control policies. Similarly, the energy consumption between different control policies is also a critical comparison index. For the problem studied in this chapter, we define the energy index as $E_a = \int_0^T \left(||\tau_a(t)||^2 + ||\tau_d(t)||^2\right) dt$. This index represents the total energy consumed by the attitude maneuvers of the two spacecraft during the closed-loop system stabilization.

9.4.2 Performance and Comparison

Figures 9.1, 9.2, 9.3, 9.4, 9.5, 9.6, 9.7, 9.8, 9.9, 9.10, 9.11 and 9.12 comprehensively present the numerical simulation results of the algorithm proposed in this chapter and its comparison results with similar algorithms.

Figures 9.1, 9.2 and 9.3 reveal the tracking errors of the closed-loop systems under control algorithms 'A', 'B', 'C', and 'D'. In these three figures, the behavior of each element of the attitude tracking error $q_{ev} = [q_{ev1}, q_{ev2}, q_{ev3}]^T$ under different control algorithms are presented separately. The system errors can all converge with the participation of these four algorithms. However, under algorithms 'A' and 'C' applying the performance guaranteed technique: the stabilization rate of the system error is faster; the actual settling time is less than or equal to the predefined settling time t_{ps}; the steady-state error is smaller. The results of 'B' and 'C' show that sometimes even the desired performance constraints are not met, which may bring risks in space missions where high control accuracy is required. Figures 9.4, 9.5 and 9.6 show the angular velocity tracking error under the algorithms. The results show that under the proposed algorithm, the system converges faster. There is less error

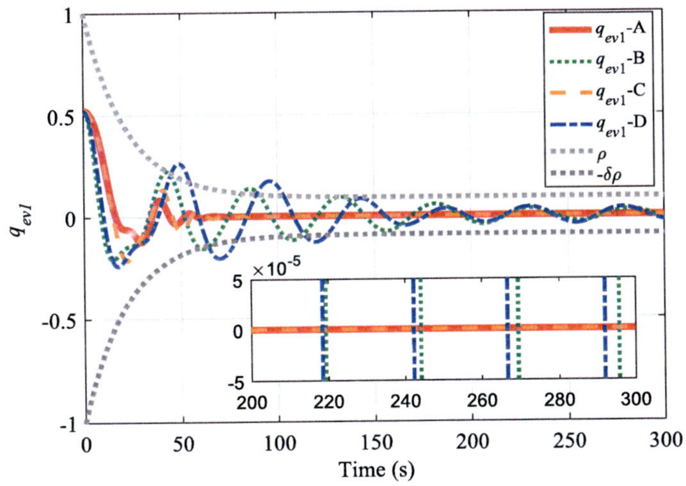

Fig. 9.1 The spacecraft attitude tracking error q_{ev1}

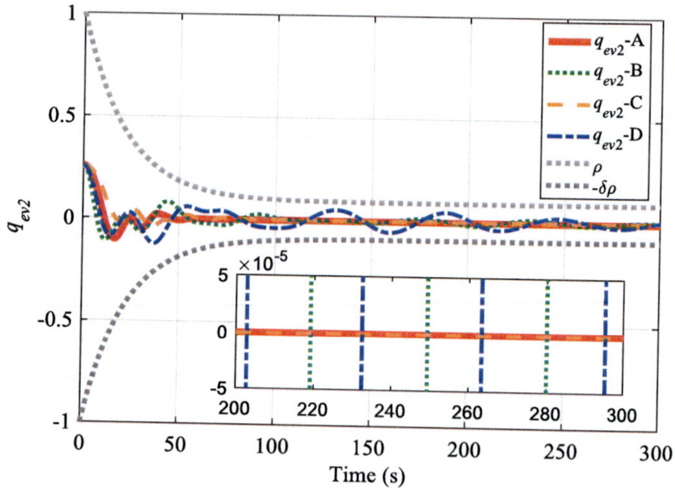

Fig. 9.2 The spacecraft attitude tracking error q_{ev2}

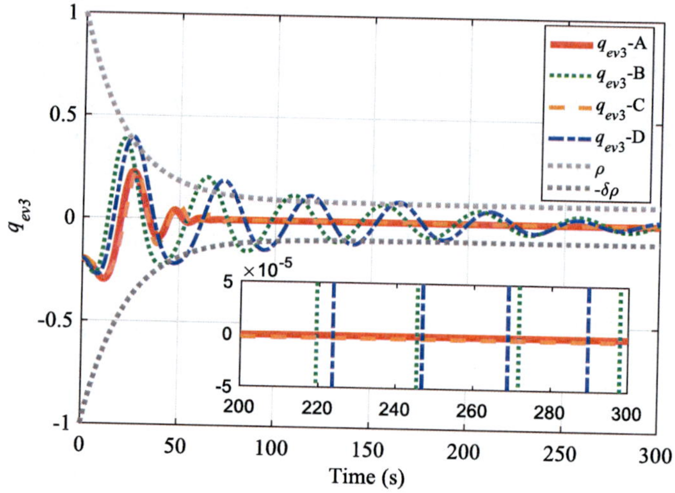

Fig. 9.3 The spacecraft attitude tracking error q_{ev3}

and less redundancy motion in the steady state. Figures 9.7 and 9.8 reflect the control inputs of the two spacecraft under the proposed algorithm, respectively. From the numerical analysis, such control inputs are reasonable and suitable for practical applications. Figures 9.9 and 9.10 are the performance of the neural network weights, and it can be seen that all the weights converge to constants. Figure 9.11 shows the Euclidean norms of the tracking errors $I_a = ||q_{ev}|| + ||\omega_e||$. It is evident from the results that the proposed algorithm provides the system with a faster conver-

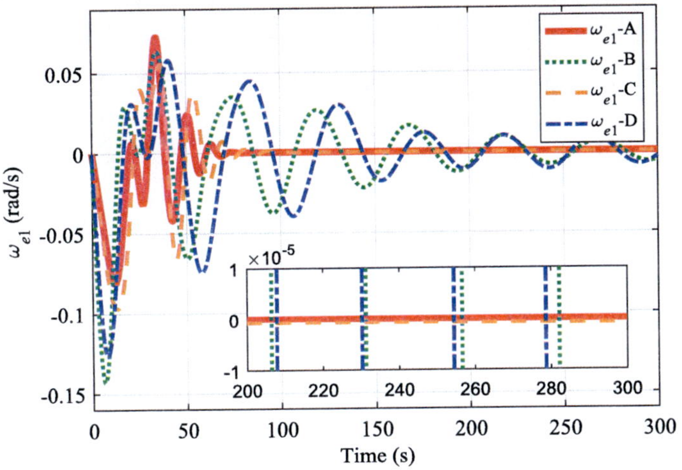

Fig. 9.4 The angular velocity tracking error ω_{e1}

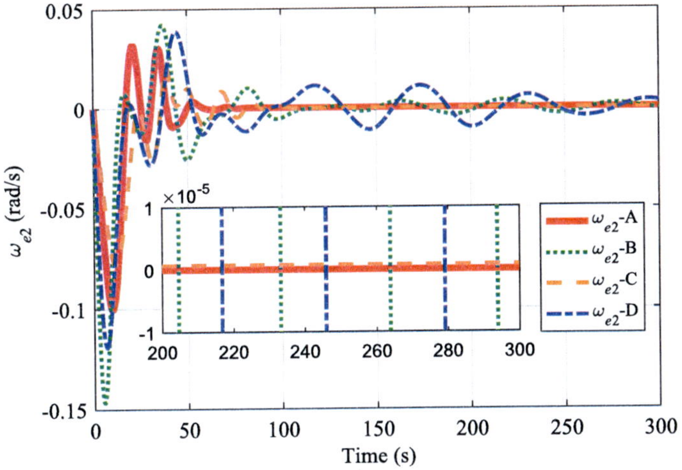

Fig. 9.5 The angular velocity tracking error ω_{e2}

gence rate, smaller steady-state error, and higher control quality, which is reflected in the curve smoothness. Figure 9.12 shows the energy consumption required by the four algorithms. The proposed algorithm provides a smaller energy consumption for the system, which benefits from the rapid stabilization of the system. In the neural network-based game algorithm, the control inputs of members may jitter during the stabilization process due to errors that have not yet converged. Such a jitter may introduce redundant motion, thus producing unintended energy consumption. Due to performance constraints, the system can be stabilized faster, thus reducing

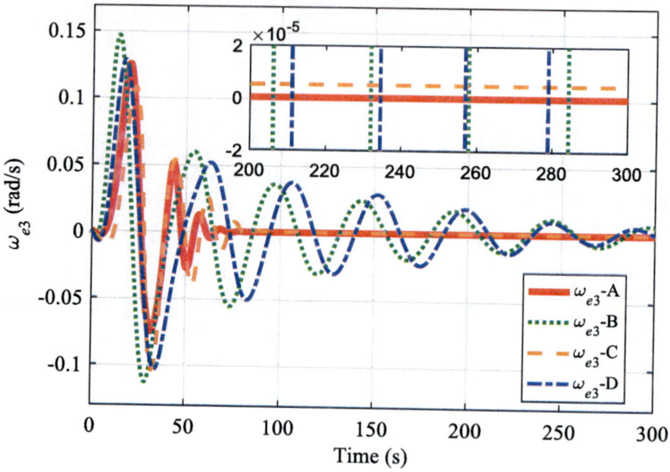

Fig. 9.6 The angular velocity tracking error ω_{e3}

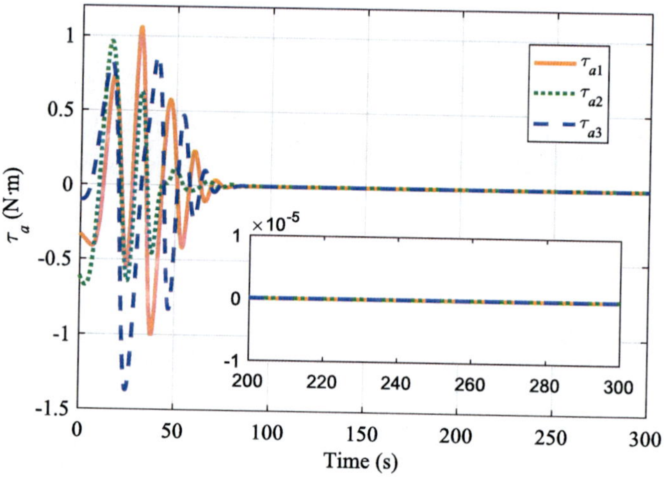

Fig. 9.7 The control input τ_a of the spacecraft 1

redundant movements without changing the game strategy parameters. Thus, energy consumption is also reduced. The energy consumption of 'C' is large because 'C' is derived from simply combining performance constraints with the game strategy and does not adopt the control input transformation as in (9.41). Therefore the energy consumption is greater than 'A'. This reflects the advantages and advanced of the proposed algorithm.

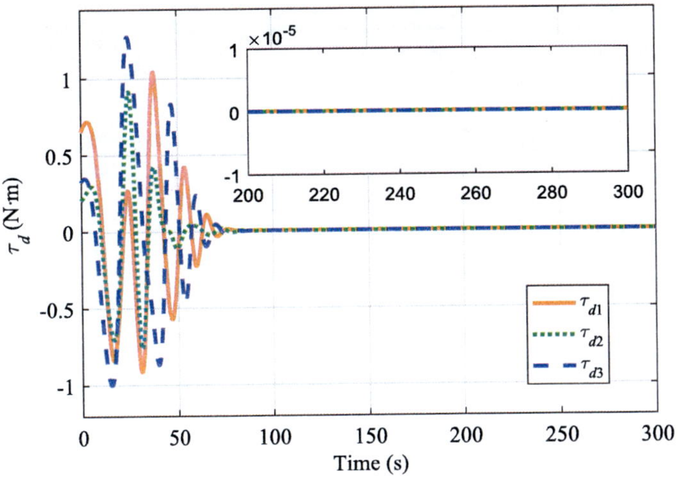

Fig. 9.8 The control input τ_d of the spacecraft 2

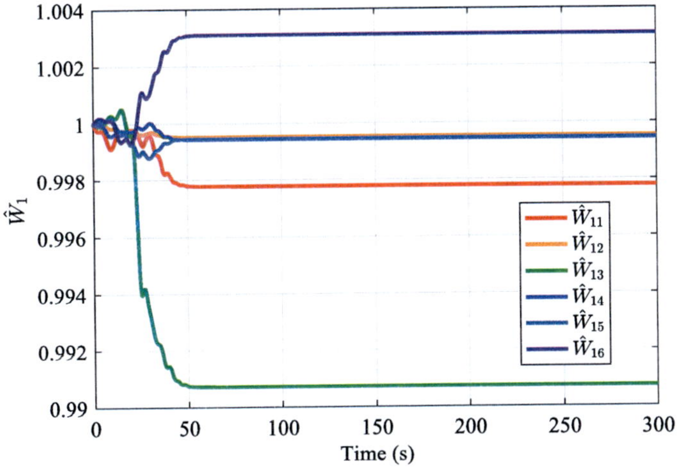

Fig. 9.9 The critic network adaptive weights of the control input τ_a

9.5 Conclusion

This chapter studies the problem of co-maneuvering multiple spacecraft to achieve attitude consistency or expectation. Unlike traditional tracking or multi-agent control design ideas, this chapter studies the algorithm scheme based on non-zero-sum games. By designing the attitude control inputs of all spacecraft in the system, each individual undertakes a part of the control task, and all control inputs reach Nash equilibrium. In this way, the attitude consistency or attitude tracking problem will

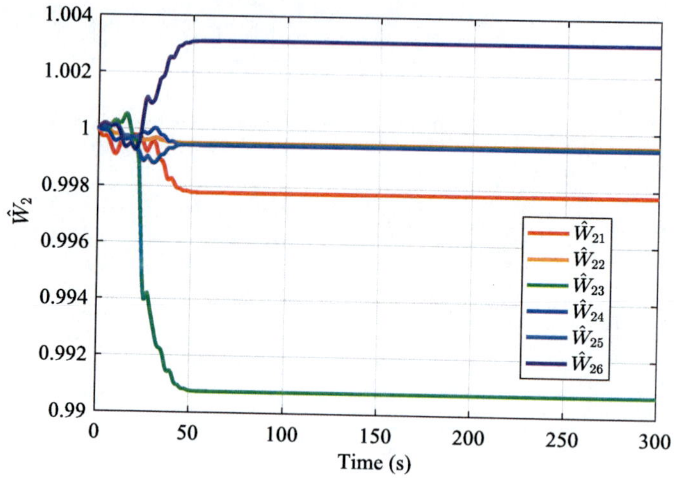

Fig. 9.10 The critic network adaptive weights of control input τ_d

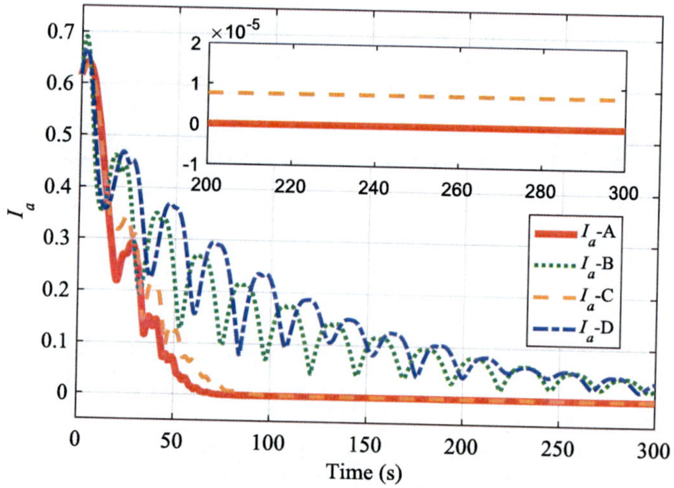

Fig. 9.11 The error index I_a

be transformed into an attitude cooperation problem to reduce the control cost of each spacecraft. Synchronously, based on the performance guaranteed algorithm proposed in previous chapters, the transient and steady-state performance of the system can be described and guaranteed. Therefore, we can reasonably design the expected performance parameters according to the mission requirements so that the comprehensive control effect of the multi-spacecraft system can meet the expected performance constraints. In addition, through the auxiliary system, the adaptive neural network satisfies the Finite Excitation conditions. For the algorithm proposed

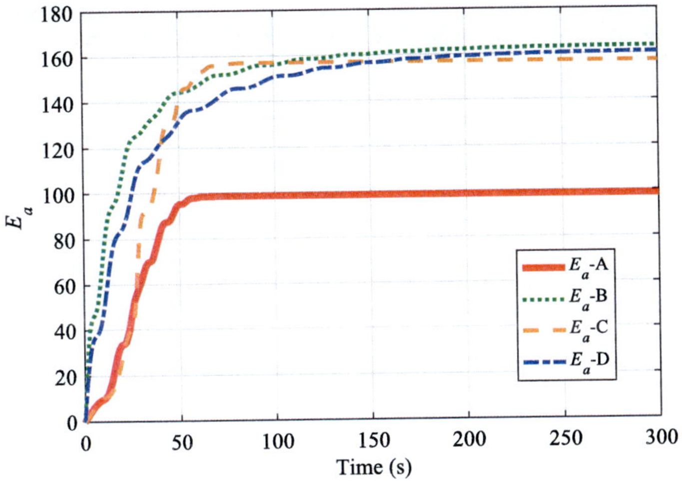

Fig. 9.12 The energy consumption

in this chapter, rigorous theoretical analysis, and numerical simulation analysis are given. The effectiveness, applicability, and advantages of the proposed algorithms are verified.

References

1. Fehse, W.: Automated rendezvous and docking of spacecraft. Cambridge university press (2003)
2. Waltz, D.M.: On-orbit servicing of space systems. Krieger Publishing Company (1993)
3. Li, W., Cheng, D., Liu, X., Wang, Y., Shi, W., Tang, Z., Gao, F., Zeng, F., Chai, H., Luo, W., et al.: On-orbit service (oos) of spacecraft: A review of engineering developments. Progress in Aerospace Sciences **108**, 32–120 (2019)
4. Kapila, V., Sparks, A.G., Buffington, J.M., Yan, Q.: Spacecraft formation flying: Dynamics and control. Journal of Guidance, Control, and Dynamics **23**(3), 561–564 (2000)
5. Kristiansen, R., Nicklasson, P.J.: Spacecraft formation flying: a review and new results on state feedback control. Acta Astronautica **65**(11-12), 1537–1552 (2009)
6. Alvarez, J., Walls, B.: Constellations, clusters, and communication technology: Expanding small satellite access to space. In: 2016 IEEE aerospace conference, pp. 1–11. IEEE (2016)
7. Del Portillo, I., Cameron, B.G., Crawley, E.F.: A technical comparison of three low earth orbit satellite constellation systems to provide global broadband. Acta astronautica **159**, 123–135 (2019)
8. Ren, W.: Formation keeping and attitude alignment for multiple spacecraft through local interactions. Journal of Guidance, Control, and Dynamics **30**(2), 633–638 (2007)
9. Xu, C., Wu, B., Wang, D., Zhang, Y.: Distributed fixed-time output-feedback attitude consensus control for multiple spacecraft. IEEE Transactions on Aerospace and Electronic Systems **56**(6), 4779–4795 (2020)

10. Du, H., Chen, M.Z., Wen, G.: Leader–following attitude consensus for spacecraft formation with rigid and flexible spacecraft. Journal of Guidance, Control, and Dynamics **39**(4), 944–951 (2016)
11. Pittelkau, M.E.: Optimal periodic control for spacecraft pointing and attitude determination. Journal of Guidance, Control, and Dynamics **16**(6), 1078–1084 (1993)
12. Wie, B., Bailey, D., Heiberg, C.: Rapid multitarget acquisition and pointing control of agile spacecraft. Journal of Guidance, Control, and Dynamics **25**(1), 96–104 (2002)
13. Wang, P.K.C., Hadaegh, F., Lau, K.: Synchronized formation rotation and attitude control of multiple free-flying spacecraft. Journal of Guidance, Control, and Dynamics **22**(1), 28–35 (1999)
14. VanDyke, M.C., Hall, C.D.: Decentralized coordinated attitude control within a formation of spacecraft. Journal of Guidance, Control, and Dynamics **29**(5), 1101–1109 (2006)
15. Li, S., Du, H., Shi, P.: Distributed attitude control for multiple spacecraft with communication delays. IEEE Transactions on Aerospace and Electronic Systems **50**(3), 1765–1773 (2014)
16. Servidia, P.A., Pena, R.S.: Spacecraft thruster control allocation problems. IEEE Transactions on Automatic Control **50**(2), 245–249 (2005)
17. Durham, W.C.: Constrained control allocation. Journal of Guidance, control, and Dynamics **16**(4), 717–725 (1993)
18. Hu, Q., Li, B., Xiao, B., Zhang, Y.: Spacecraft attitude fault-tolerant control based on iterative learning observer and control allocation. In: Control Allocation for Spacecraft Under Actuator Faults, pp. 133–155. Springer (2021)
19. Gillies, D.B.: Solutions to general non-zero-sum games. Contributions to the Theory of Games **4**, 47–85 (1959)
20. Kreps, D.M.: Nash equilibrium. In: Game Theory, pp. 167–177. Springer (1989)
21. Vamvoudakis, K.G., Lewis, F.L.: Online actor–critic algorithm to solve the continuous-time infinite horizon optimal control problem. Automatica **46**(5), 878–888 (2010)
22. Vamvoudakis, K.G., Lewis, F.L.: Multi-player non-zero-sum games: Online adaptive learning solution of coupled hamilton–jacobi equations. Automatica **47**(8), 1556–1569 (2011)
23. Vamvoudakis, K.G., Lewis, F.L., Hudas, G.R.: Multi-agent differential graphical games: Online adaptive learning solution for synchronization with optimality. Automatica **48**(8), 1598–1611 (2012)
24. Vamvoudakis, K.G., Wan, Y., Lewis, F.L., Cansever, D.: Handbook of Reinforcement Learning and Control. Springer (2021)
25. Liu, D., Li, H., Wang, D.: Online synchronous approximate optimal learning algorithm for multi-player non-zero-sum games with unknown dynamics. IEEE Transactions on Systems, Man, and Cybernetics: Systems **44**(8), 1015–1027 (2014)
26. Song, R., Wei, Q., Zhang, H., Lewis, F.L.: Discrete-time non-zero-sum games with completely unknown dynamics. IEEE Transactions on Cybernetics **51**(6), 2929–2943 (2019)
27. Junkins, J.L., Schaub, H.: Analytical mechanics of space systems. American Institute of Aeronautics and Astronautics (2009)
28. Gao, Y., Li, D., Ge, S.S.: Time-synchronized tracking control for 6-dof spacecraft in rendezvous and docking. IEEE Transactions on Aerospace and Electronic Systems **58**(3), 1676–1691 (2022). 10.1109/TAES.2021.3124865
29. Hughes, P.C.: Spacecraft attitude dynamics. Courier Corporation (2012)
30. Costic, B., Dawson, D., De Queiroz, M., Kapila, V.: Quaternion-based adaptive attitude tracking controller without velocity measurements. Journal of Guidance, Control, and Dynamics **24**(6), 1214–1222 (2001)
31. Bechlioulis, C.P., Rovithakis, G.A.: Robust adaptive control of feedback linearizable mimo nonlinear systems with prescribed performance. IEEE Transactions on Automatic Control **53**(9), 2090–2099 (2008)
32. Bechlioulis, C.P., Rovithakis, G.A.: Adaptive control with guaranteed transient and steady state tracking error bounds for strict feedback systems. Automatica **45**(2), 532–538 (2009)
33. Vamvoudakis, K.G., Lewis, F.L.: Non-zero sum games: Online learning solution of coupled hamilton-jacobi and coupled riccati equations. In: 2011 IEEE International Symposium on Intelligent Control, pp. 171–178 (2011). https://doi.org/10.1109/ISIC.2011.6045401

34. Zhao, D., Zhang, Q., Wang, D., Zhu, Y.: Experience replay for optimal control of nonzero-sum game systems with unknown dynamics. IEEE transactions on cybernetics **46**(3), 854–865 (2015)
35. Başar, T., Olsder, G.J.: Dynamic noncooperative game theory. SIAM (1998)
36. Abu-Khalaf, M., Lewis, F.L.: Nearly optimal control laws for nonlinear systems with saturating actuators using a neural network hjb approach. Automatica **41**(5), 779–791 (2005)
37. Finlayson, B.A.: The method of weighted residuals and variational principles. SIAM (2013)
38. Hornik, K., Stinchcombe, M., White, H.: Universal approximation of an unknown mapping and its derivatives using multilayer feedforward networks. Neural networks **3**(5), 551–560 (1990)
39. Modares, H., Lewis, F.L., Naghibi-Sistani, M.B.: Adaptive optimal control of unknown constrained-input systems using policy iteration and neural networks. IEEE Transactions on Neural Networks and Learning Systems **24**(10), 1513–1525 (2013). https://doi.org/10.1109/TNNLS.2013.2276571
40. Chowdhary, G., Mühlegg, M., Johnson, E.: Exponential parameter and tracking error convergence guarantees for adaptive controllers without persistency of excitation. International Journal of Control **87**(8), 1583–1603 (2014)
41. Dong, H., Zhao, X., Luo, B.: Optimal tracking control for uncertain nonlinear systems with prescribed performance via critic-only adp. IEEE Transactions on Systems, Man, and Cybernetics: Systems (2020)

Chapter 10
Performance Guaranteed Non-Zero-Sum Games for Combined Spacecraft

Abstract Regarding the joint control of the sub-spacecraft in the large-scale combined spacecraft to achieve the mission objective of the combined spacecraft maneuver, this chapter proposes the performance guaranteed non-zero-sum game algorithm for the control inputs of each sub-spacecraft actuator so that these control inputs can be optimized and Nash equilibrium can be achieved. Simultaneously, the mission objectives as joint maneuvering and control fusion of the sub-spacecraft are realized. The performance guaranteed maneuvering and optimization of the combined spacecraft are realized. Theoretical proof, stability analysis, and comparative numerical simulation analysis demonstrate the rationality, effectiveness, and superiority of the algorithm proposed in this chapter.

10.1 Introduction

Since the emergence of the on-orbit service concept, based on the traditional single spacecraft missions, the large-scale combined spacecraft maneuvering control has become a new aerospace challenge for human-crewed spaceflight, space services, and other missions [1, 2].

The on-orbit large spacecraft or facilities are usually composed of multiple small and medium spacecraft. In this book, large spacecraft are referred to as *combined spacecraft*, and small or medium spacecraft assembled into them are collectively referred to as *sub-spacecraft*. These sub-spacecraft generally have their own relatively independent attitude and orbit control systems. However, a single sub-spacecraft's attitude and orbit control system is more challenging to handle the attitude or orbit maneuvering tasks of such a large-scale combined spacecraft. Therefore, it is necessary to consider the cooperative maneuvering of multi-sub-spacecraft. Generally, scholars will design corresponding control allocation algorithms for spacecraft [3]. However, simple spacecraft control allocation design is challenging to achieve optimization or Nash equilibrium for all controllers, resulting in more potential energy loss. On the other hand, different sub-spacecraft may have different actuators, such as momentum wheels, control moment gyros, and

Y. Gao and D. Li, *Spacecraft Maneuver with Performance Guaranteed*,
https://doi.org/10.1007/978-981-99-4653-2_10

jet controllers. However, the traditional control allocation algorithms for individual spacecraft rarely consider the fusion control of different actuators. The work in this chapter expects to design independent optimization and Nash equilibrium control strategies for each sub-spacecraft control system of large combined spacecraft through *performance guaranteed non-zero-sum (NZS) game algorithms*. In the proposed algorithm, the performance guarantees of the combined spacecraft closed-loop system are fully considered, including the expected maximum settling time, maximum overshoot, and maximum steady-state error, to ensure the accuracy, safety, and reliability of the spacecraft closed-loop systems and states. The control strategies with the game solutions can provide specific feasible strategies for the multiple actuators of each sub-spacecraft participating in the maneuvering of the combined spacecraft. Furthermore, these strategies can achieve optimization and Nash equilibrium. Simultaneously, different types of actuators will be uniformly transformed into the coordinate system of the combined spacecraft, and control fusion will be achieved according to each actuator's corresponding positions and optimization indicators.

This chapter introduces the dynamic model of the combined spacecraft and sub-spacecraft. Then, the pre-defined performance guaranteed control and non-zero-sum game methods are introduced. Next, we propose the performance guaranteed game algorithm with experience replay and conduct a comprehensive and detailed theoretical analysis and stability proof. Finally, a comparison with the state-of-the-art algorithms and numerical simulation analysis are given to show the rationality and effectiveness of the algorithm proposed in this chapter, and then the conclusion is summarized.

10.2 Combined Spacecraft Game Model and Preliminaries

10.2.1 Combined Spacecraft Attitude Model with N Sets of Actuators

Although the on-orbit combined spacecraft is composed of multiple sub-spacecraft, it is still a complete spacecraft. Therefore, the attitude dynamics and kinematics of the individual spacecraft are still applied to it. The difference is that the combined spacecraft needs to consider the relationship between the positions of the sub-spacecraft and their actuators and the combined spacecraft. In this chapter, we consider that each sub-spacecraft has a set of actuators, and m sub-spacecraft may have N sets of actuators that can be used for the combined spacecraft maneuvering work. In this section, the relevant coordinate frames are given first, and then the combined spacecraft attitude dynamics and kinematics models are derived from the individual spacecraft model.

The frameworks we adopt as shown in Fig. 10.1 include the Earth-Centered Inertial (ECI) frame $\mathfrak{F}_I$ [4], the local-vertical/local-horizontal (LVLH) reference frame $\mathfrak{F}_l$ [5] centered on the combined spacecraft, the body-fixed Cartesian frame $\mathfrak{F}_{cb}$ of the

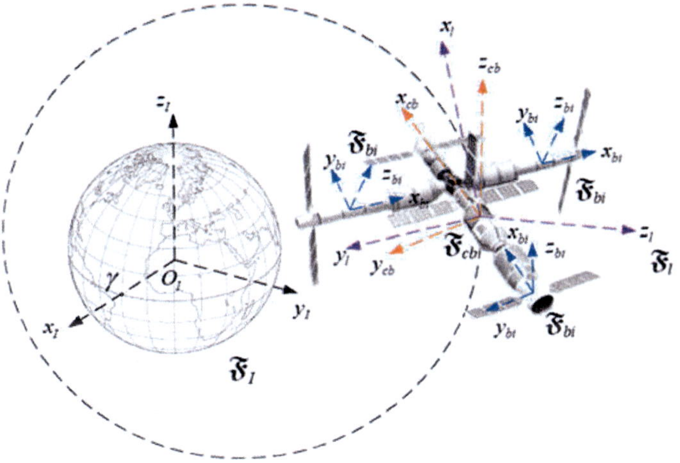

Fig. 10.1 Cartesian coordinate frames

combined spacecraft, and the body-fixed Cartesian frame $\mathfrak{F}_{bi}, i = 1, ..., m$ of the m sub-spacecraft with N sets of actuators. The frames $\mathfrak{F}_{cb}$ and $\mathfrak{F}_{bi}$ are located at centroid on the combined spacecraft and sub-spacecraft, respectively. The attitude movement of the combined spacecraft can be described by the rotation of the body-fixed reference frames $\mathfrak{F}_{cb}$ relative to the $\mathfrak{F}_I$ or other desired reference frames. Matrix R_{cbI} represents the rotation coordinate transformation between $\mathfrak{F}_{cb}$ and $\mathfrak{F}_I$, and R_{cbl} is defined in the same way. For the actuators of the sub-spacecraft in the combined spacecraft, their control inputs need to be transferred from the sub-spacecraft body-frame $\mathfrak{F}_{bi}$ to the combined spacecraft body-frame $\mathfrak{F}_{cb}$. It is even necessary to further transfer the control inputs to the reference coordinate frame expressed in kinematics. In this section, a transition matrix $\Lambda_{ai} \in \mathbb{R}^{3\times3}, i = 1, ..., N$ is abstracted to represent the conversion relationship between the control inputs of different actuators and the kinematics of the system under the reference frames. We assume that Λ_{ai} is reversible. When the reference frame is the combined spacecraft body-frame $\mathfrak{F}_{cb}$, the transition matrix Λ_{ai} can be considered as the installation matrix of the sub-spacecraft actuator on the combined spacecraft. Otherwise, it is necessary to consider the conversion of the actuator control inputs with respect to the attitude movement.

Considering the relative attitude movement of the rigid spacecraft, we can use the unit quaternion $q_a = [q_0 \; q_v]^T \in \mathbb{R}^4$ to describe the rotation of the body-fixed frame $\mathfrak{F}_{cb}$ with respect to the inertial frame $\mathfrak{F}_I$, which is subject to the constraint [6]

$$q_v^T q_v + q_0^2 = 1 \tag{10.1}$$

Then the dynamics and kinematics models of the spacecraft are governed as follows [5, 7]

$$J_c\dot{\omega}_a = -\omega_a^{\times} J_c\omega_a + \tau_a \tag{10.2}$$

$$\dot{q}_v = \frac{1}{2}\left(q_v^{\times}\omega_a + q_0\omega_a\right) \tag{10.3}$$

$$\dot{q}_0 = -\frac{1}{2}q_v^T\omega_a \tag{10.4}$$

where $J_c \in \mathbb{R}^{3\times3}$ is the constant, positive-definite, symmetric inertia matrix of the combined spacecraft, $\omega_a \in \mathbb{R}^3$ is the angular velocity of the body-fixed frame $\mathfrak{F}_{cb}$ with respect to the inertial reference frame $\mathfrak{F}_I$ expressed in $\mathfrak{F}_{cb}$, $\tau_a \in \mathbb{R}^3$ is the vector of control torque of the combined spacecraft in $\mathfrak{F}_{cb}$, and the notation $a^{\times}$, $\forall a = [a_1\ a_2\ a_3]^T$, denotes the following skew-symmetric matrix:

$$a^{\times} \triangleq \begin{bmatrix} 0 & -a_3 & a_2 \\ a_3 & 0 & -a_1 \\ -a_2 & a_1 & 0 \end{bmatrix} \tag{10.5}$$

Considering that the combined spacecraft has multiple sub-spacecraft with N sets of actuators, τ_a is the resultant torque of the multiple sub-spacecraft control input torques $\tau_i \in \mathbb{R}^3$, $i = 1, \ldots, N$. Combined with the transition matrix (installation matrix) Λ_{ai} of each set of actuators, (10.2) can be rewritten as the form as follows that explicitly includes each set of the actuator as the control inputs. In this model, unless otherwise specified, the subscript $i = 1, \ldots, N$ represents the i-th set of actuators belonging to sub-spacecraft in the combined spacecraft.

$$J_c\dot{\omega}_a = -\omega_a^{\times}J_c\omega_a + \sum\nolimits_{i=1}^{N}\Lambda_{ai}\tau_i \tag{10.6}$$

It should be noted that the installation matrix in this chapter is determined and given based on the combined spacecraft and sub-spacecraft. The control allocation is embodied in designing the game optimization algorithm between the control inputs without designing the allocation matrix.

The rotation matrix that brings $\mathfrak{F}_I$ onto $\mathfrak{F}_{bc}$, denoted by $R_a(q_v, q_0) \in SO(3)$, is defined as follows:

$$R_a = (q_0^2 - q_v^T q_v)I_3 + 2q_v q_v^T - 2q_0 q_v^{\times} \tag{10.7}$$

where I_3 is the 3×3 identity matrix, and the angular velocity ω_a can be computed as follows:

$$\omega_a = 2(q_0\dot{q}_v - q_v\dot{q}_0) - 2q_v^{\times}\dot{q}_v \tag{10.8}$$

In order to obtain the attitude error model, we assume that the desired attitude of the combined spacecraft can be described by the desired unit quaternion $q_{ad}(t) = [q_{d0}, q_{dv}]^T \in \mathbb{R}^4$ and define the desired angular velocity $\omega_d \in \mathbb{R}^3$ through the following dynamic equations:

$$\dot{q}_{dv} = \frac{1}{2}\left(q_{dv}^{\times}\omega_d + q_{d0}\omega_d\right) \tag{10.9}$$

$$\dot{q}_{d0} = -\frac{1}{2}q_{dv}^{T}\omega_d \tag{10.10}$$

$$\omega_d = 2(q_{d0}\dot{q}_{dv} - q_{dv}\dot{q}_{d0}) - 2q_{dv}^{\times}\dot{q}_{dv} \tag{10.11}$$

The corresponding rotation matrix $R_d(q_{dv}, q_{d0}) \in SO(3)$ is defined as follows:

$$R_d = (q_{d0}^2 - q_{dv}^T q_{dv})I_3 + 2q_{dv}q_{dv}^T - 2q_{d0}q_{dv}^{\times} \tag{10.12}$$

To quantify the error between the actual and desired spacecraft attitudes, we define the rotation matrix $R_e(q_{ev}, q_{e0}) \in SO(3)$ and quaternion tracking error $q_e(t) = [q_{e0}, q_{ev}]^T \in \mathbb{R}^4$ as follows:

$$R_e = R_a R_d^T = (q_{e0}^2 - q_{ev}^T q_{ev})I_3 + 2q_{ev}q_{ev}^T - 2q_{e0}q_{ev}^{\times} \tag{10.13}$$

$$q_{ev} = q_{d0}q_v - q_0 q_{dv} + q_v^{\times} q_{dv} \tag{10.14}$$

$$q_{e0} = q_0 q_{d0} + q_v^T q_{dv} \tag{10.15}$$

Based on the preceding formulation, the tracking error of the angular velocity $\omega_e \in \mathbb{R}^3$ is defined as follows:

$$\omega_e = \omega_a - R_e\omega_d \tag{10.16}$$

Computing (10.2), (10.3), (10.4), (10.9), (10.10), (10.14), (10.15) and (10.16), we obtain the open-loop tracking error dynamics as follows

$$\begin{aligned} J_c\dot{\omega}_e = &-(\omega_e + R_e\omega_d)^{\times} J_c(\omega_e + R_e\omega_d) \\ &+ J_c(\omega_e^{\times} R_e\omega_d - R_e\dot{\omega}_d) + \sum\nolimits_{i=1}^{N} \Lambda_{ai}\tau_i \end{aligned} \tag{10.17}$$

$$\dot{q}_{ev} = \frac{1}{2}(q_{ev}^{\times} + q_{e0}I_3)\omega_e \tag{10.18}$$

$$\dot{q}_{e0} = -\frac{1}{2}q_{ev}^{T}\omega_e \tag{10.19}$$

where $\dot{R}_e = -\omega_e{}^{\times} R_e$.

To express the open-loop tracking error dynamics, we rewrite (10.18) as follows:

$$\dot{q}_{ev} = \frac{1}{2} T \omega_e \tag{10.20}$$

where $T = q_{ev}^{\times} + q_{e0} I_3$ is a Jacobian-type matrix.

After taking the time derivative of (10.20) and pre-multiplying both sides of the resulting expression by $T^{-T} J_c T^{-1}$, we obtain,

$$M_a \ddot{q}_{ev} = \frac{1}{2} M_a \dot{T} \omega_e + \frac{1}{2} P^T J_c \dot{\omega}_e \tag{10.21}$$

where $M_a = P^T J_c P$ and $P = T^{-1}$.

After substituting (10.17) into the right-hand side of (10.21), we can obtain the following expression for the open-loop tracking error dynamics of the attitude motion with multiple control inputs

$$M_a \ddot{q}_{ev} + C_a \dot{q}_{ev} + G_a = \sum\nolimits_{i=1}^{N} G_i \tau_i \tag{10.22}$$

where

$$G_i = \frac{1}{2} P^T \Lambda_{ai} \tag{10.23}$$

$$C_a = -M_a \dot{P}^{-1} P - 2P^T (J_c P \dot{q}_{ev})^{\times} P \tag{10.24}$$

$$\begin{aligned} G_a = & P^T [(P\dot{q}_{ev})^{\times} J_c R_e \omega_d] + P^T [(R_e \omega_d)^{\times} J_c P \dot{q}_{ev}] \\ & + \frac{1}{2} P^T [(R_e \omega_d)^{\times} J_c R_e \omega_d] - \frac{1}{2} P^T J_c [(2P\dot{q}_{ev})^{\times} R_e \omega_d - R_e \dot{\omega}_d] \end{aligned} \tag{10.25}$$

After obtaining the model we need, we give the control objectives of this chapter as follows. Considering that subscript characters i and j are used to describe sub-spacecraft numbers or actuator numbers in this chapter, therefore, $\iota = 1, 2, 3$ is used for the subscripts of the elements of each sub-spacecraft attitude quaternion vector and the associated performance constraint parameters.

Problem 10.1 Given the initial combined spacecraft system (10.22) and appropriate tracking control indicators include the maximum settling time t_{ps}, the maximum steady-state error ρ_t, and the maximum overshoot χ_t, design the optimal control algorithm τ_i, $i = 1, \ldots, N$, such that the following conditions are satisfied: (1) the closed-loop system is stable, and $t_s \leq t_{ps}$, where t_s is the actual settling time of the closed-loop tracking error; (2) $-\delta\rho_t < q_{ev\,\iota}|_{t \geq t_{ps}} < \rho_t$ when $q_{ev\,\iota}(0) > 0$ or $-\rho_t < q_{ev\,\iota}|_{t \geq t_{ps}} < \delta\rho_t$ when $q_{ev\,\iota}(0) < 0$, where $q_{ev\,\iota}|_{t \geq t_{ps}}$ represents the actual steady-state error when $t \geq t_{ps}$, $\delta \in [0, 1]$ is a constant to be set; (3) the actual maximum overshoot less than χ_t; (4) each τ_i is optimal, and the pair $\{\tau_1, \ldots, \tau_N\}$ reaches the Nash equilibrium.

10.2.2 Performance Guaranteed Technique with Improved Performance Function

In order to ensure the transient performance and steady-state accuracy of the spacecraft closed-loop system, we need to impose strong constraints on the attitude tracking error.

Inspired by [8, 9], we make some improvements to the performance constraint function in the prescribed performance method so that the maximum settling time can be explicitly included in the constraint function to achieve the purpose of defining the settling time in advance. This chapter introduces the so-called *pre-defined maximum settling time constraint function (PMSTF)* that can obviously contain the maximum settling time as follows.

Definition 10.1 A smooth function $\rho(t) : \mathbb{R}^+ \to \mathbb{R}^+ \backslash 0$ will be called a PMSTF if: (1) $\rho(t)$ is positive and decreasing; (2) $\rho(t_{ps}) = \rho_t$, where $t_{ps} > 0$ is the expected upper bound of settling time set in advance according to actual needs, $\rho_t > 0$ is the upper bound of the expected steady-state error specified in advance, and $\rho_0 = \rho(0)$ is the initial value; (3) $\lim_{t\to\infty} \rho(t) = \rho_t - \varepsilon_\rho$, where $0 < \varepsilon_\rho < \rho_t$ is a small constant to provide an asymptote.

According to Definition 10.1, the exponential PMSTF can be designed as follows.

$$\rho(t) = \left(\rho_0 - \rho_t + \varepsilon_\rho\right) \exp\left(\frac{t}{t_{ps}} \ln \frac{\varepsilon_\rho}{\rho_0 - \rho_t + \varepsilon_\rho}\right) + \rho_t - \varepsilon_\rho \tag{10.26}$$

For the combined spacecraft attitude tracking system with multiple inputs and multiple outputs, we apply (10.26) as the constraint functions to force the tracking errors of the closed-loop system to be within the allowable range described by (10.26). This way, the closed-loop system can obtain the desired settling time and transient and steady-state performance. Therefore, we have the following constraints, where $\iota = 1, 2, 3$.

$$\begin{array}{ll} -\delta_\iota \rho_\iota(t) < q_{ev\iota}(t) < \rho_\iota(t) & q_{ev\iota}(0) \geq 0 \\ -\rho_\iota(t) < q_{ev\iota}(t) < \delta_\iota \rho_\iota(t) & q_{ev\iota}(0) < 0 \end{array} \tag{10.27}$$

We take the variable

$$\vartheta_\iota = \begin{cases} \dfrac{2q_{ev\iota}(t)}{(1+\delta_\iota)\,\rho_\iota(t)} + \dfrac{\delta_\iota - 1}{1+\delta_\iota}, & q_{ev\iota}(0) \geq 0 \\ \dfrac{2q_{ev\iota}(t)}{(1+\delta_\iota)\,\rho_\iota(t)} + \dfrac{1-\delta_\iota}{1+\delta_\iota}, & q_{ev\iota}(0) < 0 \end{cases}. \tag{10.28}$$

Then, the constraint inequality (10.27) can be described as $-1 < \vartheta_\iota < 1$. In other words, if ϑ_ι can always remain in $(-1, 1)$, then the tracking error of the closed-loop system will always be constrained in (10.27). If we map $(-1, 1)$ to $(-\infty, +\infty)$, then if the mapped variable associated with ϑ_ι is stable or bounded in $(-\infty, +\infty)$, then

(10.27) holds. This way, we solve the constraint problem by solving the stability problem.

To map ϑ_ι, we define a smooth function $\varphi : \mathbb{R} \to \mathbb{R}$ that satisfies $\lim_{\sigma \to +\infty} \varphi(\sigma_\iota) = 1$, $\lim_{\sigma \to -\infty} \varphi(\sigma_\iota) = -1$, where σ_ι is considered to be the transform parameter. We take a function that satisfies the above conditions, for example, $\varphi(\sigma_\iota) = \tanh(\sigma_\iota)$. We then take $\vartheta_\iota = \varphi(\sigma_\iota)$ and $\sigma_\iota = \varphi^{-1}(\vartheta_\iota)$. If σ_ι is stable or bounded, then $-1 < \vartheta_\iota < 1$ is true.

Considering that the combined spacecraft attitude system is a second-order system, we design a sliding mode surface as follows, where λ_ι is a positive constant.

$$\zeta_\iota = \dot{\sigma}_\iota + \lambda_\iota \sigma_\iota \tag{10.29}$$

Taking the derivative of (10.29) with respect to time and substituting (10.28), we have

$$\dot{\zeta}_\iota = \alpha_\iota + \beta_\iota \ddot{q}_{ev\iota}(t) \tag{10.30}$$

where

$$\begin{aligned} \alpha_\iota = & -\frac{2r_\iota \dot{q}_{ev\iota}(t)\,\dot{\rho}_\iota(t)}{(1+\delta_\iota)\,\rho_\iota{}^2(t)} + \frac{4r_\iota q_{ev\iota}(t)\,\dot{\rho}_\iota^2(t)}{(1+\delta_\iota)\,\rho_\iota{}^3(t)} + \dot{r}_\iota \dot{\vartheta}_\iota \\ & - \frac{2r_\iota \ddot{\rho}_\iota(t)\, q_{ev\iota}(t) + 2r_\iota \dot{\rho}_\iota(t)\,\dot{q}_{ev\iota}(t)}{(1+\delta_\iota)\,\rho_\iota{}^2(t)} + \lambda_\iota r_\iota \dot{\vartheta}_\iota \end{aligned} \tag{10.31}$$

$$\beta_\iota = \frac{2r_\iota}{(1+\delta_\iota)\,\rho_\iota(t)} \tag{10.32}$$

$$r_\iota = \partial\varphi^{-1}(\vartheta_\iota)/\partial\vartheta_\iota \tag{10.33}$$

By expressing (10.30) into a compact form, we have

$$\dot{\zeta} = \ddot{\sigma} + \Gamma\dot{\sigma} = A + B\ddot{q}_{ev}(t) \tag{10.34}$$

where $\zeta = [\zeta_1, \ldots, \zeta_n]^T$, $\sigma = [\sigma_1, \ldots, \sigma_n]^T$, $A = [\alpha_1, \ldots, \alpha_n]^T$, $B = \mathrm{diag}(\beta_1, \ldots, \beta_n)$, $\Gamma = \mathrm{diag}(\lambda_1, \ldots, \lambda_n)$, $n = 3$.

If we can design the controller so that the auxiliary system (10.34) is stable or bounded, then the closed-loop system satisfies the constraint condition (10.27).

10.2.3 Optimal Non-Zero-Sum Games and Nash Equilibrium

The N-actuator Non-Zero-Sum (NZS) differential games is described by [10]

$$\dot{x} = f(x) + \sum\nolimits_{j=1}^{N} g_j(x)u_j(t) \tag{10.35}$$

where $x \in \mathbb{R}^n$ represents state vector, $u_j \in \mathbb{R}^{m_j}$ denotes the j-th control input of each actuator, $f(x) \in \mathbb{R}^n$ and $g_j(x) \in \mathbb{R}^{n \times m_j}$ are smooth nonlinear dynamics, which are assumed to satisfy the Lipschitz condition on a compact set $\Omega \subseteq \mathbb{R}^n$ with $f(0) = 0$.

In order to solve the optimal control problem, the cost functions associated with each actuator are defined as

$$J_i(x_0, u_1, \ldots, u_N) = \int_0^{\infty} \left(Q_i(x) + \sum\nolimits_{j=1}^{N} u_j^T R_{ij} u_j \right) \mathrm{d}t \tag{10.36}$$

where $i, j = 1, \ldots, N$, x_0 is the initial state, $Q_i(x) = x^T Q_i x$ is positive definite and $Q_i > 0$, $R_{ij} > 0$ notes symmetric matrix.

If there is a control policy pair $\mu = \{\mu_1, \ldots, \mu_N\}$, where μ_i is continuous and can stabilize system (10.35) on Ω, and (10.36) is finite $\forall x_0 \in \Omega$, we can call μ as admissible with respect to (10.36) on Ω denoted by $\mu_i \in \Psi(\Omega)$ [10].

So the value functions for any admissible control policy μ_i can be defined as

$$V_i(x, \mu_1, \ldots, \mu_N) = \int_t^{\infty} \left(Q_i(x(\tau)) + \sum\nolimits_{j=1}^{N} \mu_j^T(\tau) R_{ij} \mu_j(\tau) \right) \mathrm{d}\tau \tag{10.37}$$

To solve the NZS game problem, we need to find an optimal admissible control policy pair $\mu^* = \{\mu_1^*, \ldots, \mu_N^*\}$ that minimizes the value functions of each actuator. At the same time, make N controllers reach the Nash equilibrium solution.

Definition 10.2 [11] An N-tuple of policies $\{\mu_1^*, \ldots, \mu_N^*\}$ with $\mu_i^* \in \Psi(\Omega)$ is said to constitute a Nash equilibrium for an N-player games, if the following N inequalities are satisfied for all $\mu_i \in \Psi(\Omega)$:

$$J_i^*(\mu_1^*, \ldots, \mu_i^*, \ldots, \mu_N^*) \le J_i(\mu_1^*, \ldots, \mu_i, \ldots, \mu_N^*), i \in N \tag{10.38}$$

The optimal value function is defined as

$$V_i^*(x) = \min_{\mu_i} \int_t^{\infty} \left(Q_i(x(\tau)) + \sum\nolimits_{j=1}^{N} \mu_j^T(\tau) R_{ij} \mu_j(\tau) \right) \mathrm{d}\tau \tag{10.39}$$

In order to find the optimal control, find the differential equivalent of (10.37), we get

$$0 = Q_i(x) + \sum\nolimits_{j=1}^{N} \mu_j^T R_{ij} \mu_j + (\nabla V_i)^T \left(f + \sum\nolimits_{j=1}^{N} g_j \mu_j \right) \tag{10.40}$$

where $\nabla V_i = \partial V_i / \partial x$, $V_i(0) = 0$. Then we define the Hamiltonian function of system (10.35) and value functions (10.37) as follows

$$\begin{aligned} &H_i(x, \nabla V_i, \mu_1, \ldots, \mu_N) \\ &= Q_i(x) + \sum\nolimits_{j=1}^{N} \mu_j^T R_{ij} \mu_j + (\nabla V_i)^T \left(f + \sum\nolimits_{j=1}^{N} g_j \mu_j \right) \end{aligned} \tag{10.41}$$

With the stationarity conditions $\partial H_i/\partial \mu_i = 0$, we can get each optimal control strategy as follows

$$\mu_i^* = -\frac{1}{2} R_{ii}^{-1} g_i^T \nabla V_i^* \tag{10.42}$$

Substituting optimal control input (10.42) into the Hamiltonian function (10.41), and considering the stationarity conditions, the HJ equation with $V_i^*(0) = 0$ can be described as follows

$$\begin{aligned} 0 = & Q_i(x) + \frac{1}{4} \sum\nolimits_{j=1}^{N} (\nabla V_j^*)^T g_j R_{jj}^{-1} R_{ij} R_{jj}^{-1} g_j^T \nabla V_j^* \\ & + (\nabla V_i^*)^T f - \frac{1}{2} (\nabla V_i^*)^T \sum\nolimits_{j=1}^{N} g_j R_{jj}^{-1} g_j^T \nabla V_j^* \end{aligned} \tag{10.43}$$

Because the controllers are coupled, we assume each controller can obtain all the required information. To solve the optimal control problem, we need to obtain V_i^* by solving the coupled HJ equation and obtaining the optimal control policy pair that satisfies the Nash equilibrium solution. Next section, we will solve V_i^* through a single critic network.

10.3 Performance Guaranteed NZS Games with Experience Replay

10.3.1 Algorithm Design

Considering the requirements of the pre-defined settling time and pre-defined performance, and at the same time, in order to facilitate the design of the optimized control algorithm, we design the auxiliary system as (10.34). Substituting (10.22) into (10.34), we have

$$\dot{\varsigma} = f + \sum\nolimits_{i=1}^{N} g_i \tau_{id} \tag{10.44}$$

where $i = 1, ..., N$, $f = A + BM_a^{-1}(-C_a \dot{q}_{ev} - G_a)$, $g_i = M_a^{-1} G_i \kappa_g^{-1}$, $\tau_{id} = \kappa_g G_i^{-1} M_a \cdot BM_a^{-1} G_i \tau_i$, $\kappa_g = \text{diag}(\frac{2\delta_1 \rho_1(t)}{1+\delta_1}, ..., \frac{2\delta_n \rho_n(t)}{1+\delta_n})$, $n = 3$, $\delta_1, ..., \delta_n$ are the overshoot adjustment parameter in the performance constraint. When the system error converges, that is, when r_i converges, there is $\kappa_g G_i^{-1} M_a B M_a^{-1} G_i \to 1$, so when τ_{id} is the optimal control, τ_i can converge to the optimal control. At the same time, it can be seen that g_i is bounded. Design the optimal control allocation algorithm to make the closed-loop system state ς converge or UUB and the two objectives that pre-defined performance and optimal control allocation with game solutions can be achieved.

Now, we begin to find the optimal control strategy through an adaptive method. We can find the independent basis sets such that V_i and ∇V_i are uniformly approximated by the Weierstrass higher-order approximation theorem [12–14]. Therefore, we define constant neural network weights W_i as follows, which make V_i be approximated on the compact set Ω,

$$V_i(\zeta) = W_i^T \phi_i(\zeta) + \varepsilon_i(\zeta) \tag{10.45}$$

where $\phi_i(\zeta) : \mathbb{R}^n \to \mathbb{R}^K$ is the neural network activation function basis set vectors, K represents the number of neurons in the hidden layer, and $\varepsilon_i(\zeta)$ denotes the neural network approximation errors.

The derivative of the value function with respect to ζ shows

$$\nabla V_i = \nabla \phi_i^T(\zeta) W_i + \nabla \varepsilon_i(\zeta) \tag{10.46}$$

It is worth noting that when $K \to \infty$, the approximation errors $\varepsilon_i(\zeta) \to 0, \nabla \varepsilon_i(\zeta) \to 0$ are uniformly, when K is fixed, the errors $\varepsilon_i(\zeta)$ and $\nabla \varepsilon_i(\zeta)$ are bounded on the compact set Ω.

Utilizing the estimated value of the neural network and the control strategy (10.42), we obtain the Hamiltonians as follows

$$\begin{aligned} H_i =& \zeta^T Q_i \zeta + \frac{1}{4} \sum\nolimits_{j=1}^{N} W_j^T \nabla \phi_j g_j R_{jj}^{-T} R_{ij} R_{jj}^{-1} g_j^T \nabla \phi_j^T W_j \\ &+ W_i^T \nabla \phi_i f - \frac{1}{2} W_i^T \nabla \phi_i \sum\nolimits_{j=1}^{N} g_j R_{jj}^{-1} g_j^T \nabla \phi_j^T W_j \triangleq e_{HJi} \end{aligned} \tag{10.47}$$

where $e_{HJi} = -(\nabla \varepsilon_i)^T \left(f(\zeta) + \sum_{j=1}^{N} g_j(\zeta) \tau_{jd}(t) \right)$. For unknown W_i, we define $\hat{W}_i$ as their estimate values, $\tilde{W}_i = W_i - \hat{W}_i$ as the weight estimate errors, and then the neural network of the value function becomes

$$\hat{V}_i(\zeta) = \hat{W}_i^T \phi_i(\zeta) \tag{10.48}$$

Then we obtain the approximation value of the Hamilton function as

$$\begin{aligned} H_i =& \zeta^T Q_i \zeta + \frac{1}{4} \sum\nolimits_{j=1}^{N} \hat{W}_j^T \nabla \phi_j g_j R_{jj}^{-T} R_{ij} R_{jj}^{-1} g_j^T \nabla \phi_j^T \hat{W}_j \\ &+ \hat{W}_i^T \nabla \phi_i f - \frac{1}{2} \hat{W}_i^T \nabla \phi_i \sum\nolimits_{j=1}^{N} g_j R_{jj}^{-1} g_j^T \nabla \phi_j^T \hat{W}_j \triangleq e_i \end{aligned} \tag{10.49}$$

In order to obtain more accurate estimates, that is to say, $\hat{W}_i \to W_i$ and $e_i \to e_{HJi}$, we need to minimize the square residual error function as follows.

$$E = \sum\nolimits_{i=1}^{N} E_i = \frac{1}{2} \sum\nolimits_{i=1}^{N} e_i^T e_i \tag{10.50}$$

Then we utilize a gradient descent algorithm for the critic network weights [10, 15]

$$\dot{\hat{W}}_i = -\eta_i \frac{\partial E_i}{\partial \hat{W}_i} = -\eta_i \frac{\theta_i}{\left(1+\theta_i^T\theta_i\right)^2} e_i \tag{10.51}$$

where $\theta_i = \nabla\phi_i \left(f - \frac{1}{2}\sum_{j=1}^{N} g_j R_{jj}^{-1} g_j^T \nabla\phi_j^T \hat{W}_j\right)$, and $\eta_i > 0$ is the adaptive rate of critic networks. There exist a positive constant θ_{iM} such that $\|\theta_i\| \le \theta_{iM}$ [15].

In traditional approximate/adaptive dynamic programming (ADP) algorithm, θ_i needs to satisfy the Persistence of Excitation (PE) conditions in the update law (10.51) and ensure $\|\theta_i\| \ge \theta_{im}$, where θ_{im} are positive constants, especially in online situations. In order to relax this condition, inspired by the literature [16], we design the critic neural networks update law based on experience replay technique as follows. Among them, e_{ik} and θ_{ik} are the stored data in the past, and l is the amount of data stored.

$$\dot{\hat{W}}_i = -a_i \frac{\theta_i}{\left(1+\theta_i^T\theta_i\right)^2} e_i - a_i \sum_{k=1}^{l} \frac{\theta_{ik}}{\left(1+\theta_{ik}^T\theta_{ik}\right)^2} e_{ik} \tag{10.52}$$

Because our model is known and the optimal control strategies in the form of a closed-loop system are known, we only need to train the critic network instead of the critic-action network. By the approximated value function, the optimal control policies can be written as

$$\tau_{id} = -\frac{1}{2} R_{ii}^{-1} g_i^T \left(\nabla\phi_i^T(\zeta) W_i + \nabla\varepsilon_i(\zeta)\right) \tag{10.53}$$

Then the approximate control policies can be expressed as

$$\hat{\tau}_{id} = -\frac{1}{2} R_{ii}^{-1} g_i^T \nabla\phi_i^T(\zeta) \hat{W}_i \tag{10.54}$$

10.3.2 Stability Analysis

Theorem 10.1 *Considering the combined spacecraft attitude system (10.22) with multiple control inputs, if the admissible performance guaranteed control policies (10.54) with the experience-replay-based weight updating laws (10.52) of the critic neural networks are chosen. Then, the weight estimate errors of the critic neural networks are uniformly ultimately bounded (UUB).*

Proof Select candidate Lyapunov function as

$$L = \sum_{i=1}^{N} L_i = \frac{1}{2} \sum_{i=1}^{N} a_i^{-1} \tilde{W}_i^T \tilde{W}_i \tag{10.55}$$

Find the derivative of L_i with respect to time as

$$
\begin{aligned}
\dot{L}_i &= a_i^{-1}\tilde{W}_i^T\dot{\tilde{W}}_i = -a_i^{-1}\tilde{W}_i^T\dot{\hat{W}}_i \\
&= +\tilde{W}_i^T\frac{\theta_i}{\left(1+\theta_i^T\theta_i\right)^2}e_i + \tilde{W}_i^T\sum\nolimits_{k=1}^{l}\frac{\theta_{ik}}{\left(1+\theta_{ik}^T\theta_{ik}\right)^2}e_{ik}
\end{aligned}
\tag{10.56}
$$

Base on (10.47) and (10.49), we obtain

$$
\begin{aligned}
e_i =& e_{HJi} - W_i^T\nabla\phi_i f + \hat{W}_i^T\nabla\phi_i f \\
&-\frac{1}{2}\hat{W}_i^T\sum\nolimits_{j=1}^{N}D_{ij}\hat{W}_j + \frac{1}{2}W_i^T\sum\nolimits_{j=1}^{N}D_{ij}W_j \\
&-\frac{1}{4}\sum\nolimits_{j=1}^{N}W_j^TE_{ij}W_j + \frac{1}{4}\sum\nolimits_{j=1}^{N}\hat{W}_j^TE_{ij}\hat{W}_j \\
=& e_{HJi} - \tilde{W}_i^T\theta_i + \frac{1}{2}\sum\nolimits_{j=1}^{N}W_i^TD_{ij}\tilde{W}_j \\
&-\frac{1}{2}\sum\nolimits_{j=1}^{N}W_j^TE_{ij}\tilde{W}_j + \frac{1}{4}\sum\nolimits_{j=1}^{N}\tilde{W}_j^TE_{ij}\tilde{W}_j
\end{aligned}
\tag{10.57}
$$

where $D_{ij} = \nabla\phi_i g_j R_{jj}^{-1} g_j^T \nabla\phi_j^T$, $E_{ij} = \nabla\phi_j g_j R_{jj}^{-T} R_{ij} R_{jj}^{-1} g_j^T \nabla\phi_j^T$. We simply write the above formula as

$$
e_i(t) = -\tilde{W}_i^T\theta_i + \Theta_i \tag{10.58}
$$

where

$$
\begin{aligned}
\Theta_i =& +\frac{1}{2}W_i^T\sum\nolimits_{j=1}^{N}D_{ij}\tilde{W}_j - \frac{1}{2}\sum\nolimits_{j=1}^{N}W_j^TE_{ij}\tilde{W}_j \\
&+\frac{1}{4}\sum\nolimits_{j=1}^{N}\tilde{W}_j^TE_{ij}\tilde{W}_j + e_{HJi}
\end{aligned}
\tag{10.59}
$$

Then, we can get similar items related to experience replay as follows

$$
e_{ik}(t) = -\tilde{W}_i^T\theta_{ik} + \Theta_{ik} \tag{10.60}
$$

Then, we have

$$
\begin{aligned}
\dot{L}_i =& -\psi_i^{-2}\left(\tilde{W}_i^T\theta_i\right)\left(\tilde{W}_i^T\theta_i\right) \\
&+\frac{1}{2}\psi_i^{-2}\tilde{W}_i^T\theta_i\sum\nolimits_{j=1}^{N}W_i^TD_{ij}\tilde{W}_j \\
&-\frac{1}{2}\psi_i^{-2}\tilde{W}_i^T\theta_i\sum\nolimits_{j=1}^{N}W_j^TE_{ij}\tilde{W}_j \\
&+\frac{1}{4}\psi_i^{-2}\tilde{W}_i^T\theta_i\sum\nolimits_{j=1}^{N}\tilde{W}_j^TE_{ij}\tilde{W}_j \\
&+\psi_i^{-2}\tilde{W}_i^T\theta_i e_{HJi} \\
&-\sum\nolimits_{k=1}^{l}\psi_{ik}^{-2}\left(\tilde{W}_i^T\theta_{ik}\right)\left(\tilde{W}_i^T\theta_{ik}\right) \\
&+\sum\nolimits_{k=1}^{l}\psi_{ik}^{-2}\tilde{W}_i^T\theta_{ik}\Theta_{ik}
\end{aligned}
\tag{10.61}
$$

where $\psi_i = 1 + \theta_i^T \theta_i$, $\psi_{ik} = 1 + \theta_{ik}^T \theta_{ik}$, ψ_i^{-2} and ψ_{ik}^{-2} are positive definite.
Considering $\theta_i = \theta_i{}^* + \frac{1}{2}\nabla\phi_i \sum_{j=1}^N g_j R_{jj}^{-1} g_j^T \nabla\phi_j^T \tilde{W}_j$,
$\theta_i{}^* = \nabla\phi_i \left(f - \frac{1}{2}\sum_{j=1}^N g_j R_{jj}^{-1} g_j^T \nabla\phi_j^T W_j\right)$, we have

$$\begin{aligned}
\dot{L}_i =& -\psi_i^{-2}\left(\tilde{W}_i^T \theta_i{}^*\right)\left(\tilde{W}_i^T \theta_i{}^*\right) \\
& -\psi_i^{-2}\tilde{W}_i^T \theta_i{}^* \sum_{j=1}^N \tilde{W}_i^T D_{ij} \tilde{W}_j \\
& -\psi_i^{-2}\left(\frac{1}{2}\sum_{j=1}^N \tilde{W}_i^T D_{ij}\tilde{W}_j\right)\left(\frac{1}{2}\sum_{j=1}^N \tilde{W}_i^T D_{ij}\tilde{W}_j\right) \\
& +\frac{1}{2}\psi_i^{-2}\tilde{W}_i^T \theta_i \sum_{j=1}^N W_i^T D_{ij}\tilde{W}_j \\
& -\frac{1}{2}\psi_i^{-2}\tilde{W}_i^T \theta_i \sum_{j=1}^N W_j^T E_{ij}\tilde{W}_j \\
& +\frac{1}{4}\psi_i^{-2}\tilde{W}_i^T \theta_i \sum_{j=1}^N \tilde{W}_j^T E_{ij}\tilde{W}_j \\
& +\psi_i^{-2}\tilde{W}_i^T \theta_i e_{HJi} \\
& -\sum_{k=1}^l \psi_{ik}^{-2}\left(\tilde{W}_i^T \theta_{ik}\right)\left(\tilde{W}_i^T \theta_{ik}\right) \\
& +\sum_{k=1}^l \psi_{ik}^{-2}\tilde{W}_i^T \theta_{ik}\Theta_{ik}
\end{aligned} \tag{10.62}$$

For the 1st item $-\psi_i^{-2}\left(\tilde{W}_i^T \theta_i{}^*\right)\left(\tilde{W}_i^T \theta_i{}^*\right)$ in (10.62), we expand it as

$$-\psi_i^{-2}\left(\tilde{W}_i^T \theta_i{}^*\right)\left(\tilde{W}_i^T \theta_i{}^*\right) \le -\psi_i^{-2}\theta_{im}^{*2}\left\|\tilde{W}_i\right\|^2 \tag{10.63}$$

where $\theta_{im}^* \le \|\theta_i{}^*\| \le \theta_{iM}^*$, θ_{im}^* are θ_{iM}^* positive constants.
Considering the variables a and b, and the non-zero coefficient κ, we have [17]

$$ab = \frac{1}{2}\left(-\left(\kappa a - \kappa^{-1}b\right)^2 + \kappa^2 a^2 + \kappa^{-2}b^2\right) \tag{10.64}$$

$$-ab = -\frac{1}{2}\left(\left(\kappa a + \kappa^{-1}b\right)^2 - \kappa^2 a^2 - \kappa^{-2}b^2\right) \tag{10.65}$$

According to the product relationship as (10.64) and (10.65), we deal with the remaining function terms.
For the 2nd item $-\psi_i^{-2}\tilde{W}_i^T \theta_i{}^* \sum_{j=1}^N \tilde{W}_i^T D_{ij}\tilde{W}_j$ in (10.62), we expand it as

$$\begin{aligned}
&-\psi_i^{-2}\tilde{W}_i^T \theta_i{}^* \sum_{j=1}^N \tilde{W}_i^T D_{ij}\tilde{W}_j \\
&\quad = -\psi_i^{-2}\tilde{W}_i^T \theta_i{}^* \left(\tilde{W}_i^T D_{i1}\tilde{W}_1 + \ldots + \tilde{W}_i^T D_{iN}\tilde{W}_N\right)
\end{aligned} \tag{10.66}$$

Considering (10.65), we have

$$
\begin{aligned}
&-\psi_i^{-2}\left(\tilde{W}_i^T\theta_i{}^*\right)\left(\tilde{W}_i^T D_{ij}\tilde{W}_j\right)\\
&\quad=-\frac{1}{2}\psi_i^{-2}\left(\left(\kappa_{2ij}\left(\tilde{W}_i^T\theta_i{}^*\right)+\kappa_{2ij}^{-1}\left(\tilde{W}_i^T D_{ij}\tilde{W}_j\right)\right)^2\right.\\
&\qquad\left.-\psi_i^{-2}\kappa_{2ij}^2\left(\tilde{W}_i^T\theta_i{}^*\right)^2-\psi_i^{-2}\kappa_{2ij}^{-2}\left(\tilde{W}_i^T D_{ij}\tilde{W}_j\right)^2\right)\\
&\quad\le\frac{1}{2}\psi_i^{-2}\kappa_{2ij}^2\left(\tilde{W}_i^T\theta_i{}^*\right)^2+\frac{1}{2}\psi_i^{-2}\kappa_{2ij}^{-2}\left(\tilde{W}_i^T D_{ij}\tilde{W}_j\right)^2\\
&\quad\le\frac{1}{2}\psi_i^{-2}\kappa_{2ij}^2\theta_{iM}^{*2}\left\|\tilde{W}_i\right\|^2+\frac{1}{2}\psi_i^{-2}\kappa_{2ij}^{-2}\left\|D_{ij}\right\|^2\left\|\tilde{W}_i\right\|^2\left\|\tilde{W}_j\right\|^2
\end{aligned}
\tag{10.67}
$$

and

$$
\begin{aligned}
&-\psi_i^{-2}\tilde{W}_i^T\theta_i{}^*\sum_{j=1}^N\tilde{W}_i^T D_{ij}\tilde{W}_j\\
&\quad\le+\frac{1}{2}\sum_{j=1}^N\psi_i^{-2}\kappa_{2ij}^2\theta_{iM}^{*2}\left\|\tilde{W}_i\right\|^2\\
&\qquad+\frac{1}{2}\sum_{j=1}^N\psi_i^{-2}\kappa_{2ij}^{-2}\left\|D_{ij}\right\|^2\left\|\tilde{W}_i\right\|^2\left\|\tilde{W}_j\right\|^2
\end{aligned}
\tag{10.68}
$$

For the 3rd item $-\frac{\psi_i^{-2}}{4}\left(\sum_{j=1}^N\tilde{W}_i^T D_{ij}\tilde{W}_j\right)\left(\sum_{j=1}^N\tilde{W}_i^T D_{ij}\tilde{W}_j\right)<0$ in (10.62), there exists a positive definite matrix M_1 so that

$$
-\frac{1}{4}\sum_{i=1}^N\psi_i^{-2}\left(\sum_{j=1}^N\tilde{W}_i^T D_{ij}\tilde{W}_j\right)\left(\sum_{j=1}^N\tilde{W}_i^T D_{ij}\tilde{W}_j\right)\le -Z^T M_1 Z \tag{10.69}
$$

where we define $Z=\left[\left\|\tilde{W}_1\right\|^2,\ldots,\left\|\tilde{W}_N\right\|^2\right]^T\in\mathbb{R}^N$.

For the 4th item $\frac{1}{2}\psi_i^{-2}\tilde{W}_i^T\theta_i\sum_{j=1}^N W_i{}^T D_{ij}\tilde{W}_j$ in (10.62), we expand it as

$$
\begin{aligned}
&\frac{1}{2}\left(\tilde{W}_i^T\theta_i\right)\left(W_i{}^T D_{ij}\tilde{W}_j\right)\\
&\quad=-\frac{1}{4}\left(\kappa_{4ij}\left(\tilde{W}_i^T\theta_i\right)-\kappa_{4ij}^{-1}\left(W_i{}^T D_{ij}\tilde{W}_j\right)\right)^2\\
&\qquad+\frac{1}{4}\kappa_{4ij}^2\left(\tilde{W}_i^T\theta_i\right)^2+\frac{1}{4}\kappa_{4ij}^{-2}\left(W_i{}^T D_{ij}\tilde{W}_j\right)^2\\
&\quad\le\frac{1}{4}\kappa_{4ij}^2\theta_{iM}^2\left\|\tilde{W}_i\right\|^2+\frac{1}{4}\kappa_{4ij}^{-2}\left\|D_{ij}\right\|^2\left\|W_i\right\|^2\left\|\tilde{W}_j\right\|^2
\end{aligned}
\tag{10.70}
$$

where $\theta_{im}\le\|\theta_i\|\le\theta_{iM}$, θ_{im} are θ_{iM} positive constants. So that

$$
\begin{aligned}
&\frac{1}{2}\psi_i^{-2}\tilde{W}_i^T\theta_i\sum_{j=1}^{N}W_i{}^T D_{ij}\tilde{W}_j\\
&\quad\le+\frac{1}{4}\psi_i^{-2}\sum_{j=1}^{N}\kappa_{4ij}^{-2}\left\|D_{ij}\right\|^2\|W_i\|^2\left\|\tilde{W}_j\right\|^2\\
&\quad+\frac{1}{4}\psi_i^{-2}\sum_{j=1}^{N}\kappa_{4ij}^2\theta_{iM}^2\left\|\tilde{W}_i\right\|^2
\end{aligned}
\tag{10.71}
$$

For the 5th item $-\frac{1}{2}\psi_i^{-2}\tilde{W}_i^T\theta_i\sum_{j=1}^{N}W_j{}^T E_{ij}\tilde{W}_j$ in (10.62), we expand it as

$$
\begin{aligned}
&-\frac{1}{2}\left(\tilde{W}_i^T\theta_i\right)\left(W_j{}^T E_{ij}\tilde{W}_j\right)\\
&\quad=-\frac{1}{4}\left(\kappa_{5ij}\left(\tilde{W}_i^T\theta_i\right)+\kappa_{5ij}^{-1}\left(W_j{}^T E_{ij}\tilde{W}_j\right)\right)^2\\
&\quad+\frac{1}{4}\kappa_{5ij}^2\left(\tilde{W}_i^T\theta_i\right)^2+\frac{1}{4}\kappa_{5ij}^{-2}\left(W_j{}^T E_{ij}\tilde{W}_j\right)^2\\
&\le\frac{1}{4}\kappa_{5ij}^2\theta_{iM}^2\left\|\tilde{W}_i\right\|^2+\frac{1}{4}\kappa_{5ij}^{-2}E_{ij}^2\left\|W_j\right\|^2\left\|\tilde{W}_j\right\|^2
\end{aligned}
\tag{10.72}
$$

Then, we have

$$
\begin{aligned}
&-\frac{1}{2}\psi_i^{-2}\tilde{W}_i^T\theta_i\sum_{j=1}^{N}W_j{}^T E_{ij}\tilde{W}_j\\
&\le\frac{1}{4}\psi_i^{-2}\sum_{j=1}^{N}\kappa_{5ij}^2\theta_{iM}^2\left\|\tilde{W}_i\right\|^2\\
&\quad+\frac{1}{4}\psi_i^{-2}\sum_{j=1}^{N}\kappa_{5ij}^{-2}\left\|E_{ij}\right\|^2\left\|W_j\right\|^2\left\|\tilde{W}_j\right\|^2
\end{aligned}
\tag{10.73}
$$

For the 6th item $+\frac{1}{4}\psi_i^{-2}\tilde{W}_i^T\theta_i\sum_{j=1}^{N}\tilde{W}_j^T E_{ij}\tilde{W}_j$ in (10.62), we expand it as

$$
\begin{aligned}
&\frac{1}{4}\left(\tilde{W}_i^T\theta_i\right)\left(\tilde{W}_j^T E_{ij}\tilde{W}_j\right)\\
&\quad=-\frac{1}{8}\left(\kappa_{6ij}\left(\tilde{W}_i^T\theta_i\right)-\kappa_{6ij}^{-1}\left(\tilde{W}_j^T E_{ij}\tilde{W}_j\right)\right)^2\\
&\quad+\frac{1}{8}\kappa_{6ij}^2\left(\tilde{W}_i^T\theta_i\right)^2+\frac{1}{8}\kappa_{6ij}^{-2}\left(\tilde{W}_j^T E_{ij}\tilde{W}_j\right)^2\\
&\quad\le\frac{1}{8}\kappa_{6ij}^2\theta_{iM}^2\left\|\tilde{W}_i\right\|^2+\frac{1}{8}\kappa_{6ij}^{-2}\left\|E_{ij}\right\|^2\left\|\tilde{W}_j\right\|^2\left\|\tilde{W}_j\right\|^2
\end{aligned}
\tag{10.74}
$$

Then, we have

$$
\begin{aligned}
\frac{1}{4}\psi_i^{-2}\tilde{W}_i^T\theta_i\sum_{j=1}^{N}\tilde{W}_j^T E_{ij}\tilde{W}_j&\le\frac{1}{8}\psi_i^{-2}\sum_{j=1}^{N}\kappa_{6ij}^2\theta_{iM}^2\left\|\tilde{W}_i\right\|^2\\
&+\frac{1}{8}\psi_i^{-2}\sum_{j=1}^{N}\kappa_{6ij}^{-2}\left\|E_{ij}\right\|^2\left\|\tilde{W}_j\right\|^2\left\|\tilde{W}_j\right\|^2
\end{aligned}
\tag{10.75}
$$

For the 7th item $\psi_i^{-2}\tilde{W}_i^T\theta_i e_{HJi}$ in (10.62), we expand it as

$$\psi_i^{-2}\tilde{W}_i^T\theta_i e_{HJi} \le \frac{1}{2}\psi_i^{-2}\kappa_{7i}^2\left\|\tilde{W}_i\right\|^2 + \frac{1}{2}\psi_i^{-2}\kappa_{7i}^{-2}\|\theta_i\|^2\|e_{HJi}\|^2 \tag{10.76}$$

For the 8th item $-\sum_{k=1}^{l}\psi_{ik}^{-2}\left(\tilde{W}_i^T\theta_{ik}\right)\left(\tilde{W}_i^T\theta_{ik}\right) \le 0$ in (10.62), we expand it as

$$-\sum_{k=1}^{l}\psi_{ik}^{-2}\left(\tilde{W}_i^T\theta_{ik}\right)\left(\tilde{W}_i^T\theta_{ik}\right) \le -\left(\sum_{k=1}^{l}\psi_{ik}^{-2}\theta_{ikm}^2\right)\left\|\tilde{W}_i\right\|^2 \tag{10.77}$$

where $\theta_{ikm} \le \|\theta_{ik}\| \le \theta_{ikM}$, θ_{ikm} are θ_{ikM} positive constants.

For the 9th item $+\sum_{k=1}^{l}\psi_{ik}^{-2}\tilde{W}_i^T\theta_{ik}\Theta_{ik}$ in (10.62), we expand it as

$$\begin{aligned}\sum_{k=1}^{l}\psi_{ik}^{-2}\tilde{W}_i^T\theta_{ik}\Theta_{ik} \le{}& \frac{1}{2}\left(\sum_{k=1}^{l}\psi_{ik}^{-2}\kappa_{9i}^2\right)\left\|\tilde{W}_i\right\|^2 \\ &+\frac{1}{2}\sum_{k=1}^{l}\psi_{ik}^{-2}\kappa_{9i}^{-2}\|\theta_{ik}\|^2\|\Theta_{ik}\|^2\end{aligned} \tag{10.78}$$

In the above process, κ_{2ij}, κ_{4ij}, κ_{5ij}, κ_{6ij}, κ_{7i}, and κ_{9i} are all adjustable constants. Organizing the above formula, we have

$$\begin{aligned}\dot{L}_i \le{}& -\psi_i^{-2}\theta_{im}^{*2}\left\|\tilde{W}_i\right\|^2 + \frac{1}{2}\sum_{j=1}^{N}\psi_i^{-2}\kappa_{2ij}^2\theta_{iM}^{*2}\left\|\tilde{W}_i\right\|^2 \\ &+\frac{1}{2}\sum_{j=1}^{N}\psi_i^{-2}\kappa_{2ij}^{-2}\|D_{ij}\|^2\left\|\tilde{W}_i\right\|^2\left\|\tilde{W}_j\right\|^2 \\ &-\psi_i^{-2}\left(\frac{1}{2}\sum_{j=1}^{N}\tilde{W}_i^T D_{ij}\tilde{W}_j\right)\left(\frac{1}{2}\sum_{j=1}^{N}\tilde{W}_i^T D_{ij}\tilde{W}_j\right) \\ &+\frac{1}{4}\psi_i^{-2}\sum_{j=1}^{N}\kappa_{4ij}^2\theta_{iM}^2\left\|\tilde{W}_i\right\|^2 \\ &+\frac{1}{4}\psi_i^{-2}\sum_{j=1}^{N}\kappa_{4ij}^{-2}\|D_{ij}\|^2\|W_i\|^2\left\|\tilde{W}_j\right\|^2 \\ &+\frac{1}{4}\psi_i^{-2}\sum_{j=1}^{N}\kappa_{5ij}^2\theta_{iM}^2\left\|\tilde{W}_i\right\|^2 \\ &+\frac{1}{4}\psi_i^{-2}\sum_{j=1}^{N}\kappa_{5ij}^{-2}\|E_{ij}\|^2\|W_j\|^2\left\|\tilde{W}_j\right\|^2 \\ &+\frac{1}{8}\psi_i^{-2}\sum_{j=1}^{N}\kappa_{6ij}^2\theta_{iM}^2\left\|\tilde{W}_i\right\|^2 \\ &+\frac{1}{8}\psi_i^{-2}\sum_{j=1}^{N}\kappa_{6ij}^{-2}\|E_{ij}\|^2\left\|\tilde{W}_j\right\|^2\left\|\tilde{W}_j\right\|^2 \\ &+\frac{1}{2}\psi_i^{-2}\kappa_{7i}^2\left\|\tilde{W}_i\right\|^2 + \frac{1}{2}\psi_i^{-2}\kappa_{7i}^{-2}\|\theta_i\|^2\|e_{HJi}\|^2 \\ &+\left(\frac{1}{2}\sum_{k=1}^{l}\psi_{ik}^{-2}\kappa_{9i}^2 - \sum_{k=1}^{l}\psi_{ik}^{-2}\theta_{ikm}^2\right)\left\|\tilde{W}_i\right\|^2 \\ &+\frac{1}{2}\sum_{k=1}^{l}\psi_{ik}^{-2}\kappa_{9i}^{-2}\|\theta_{ik}\|^2\|\Theta_{ik}\|^2\end{aligned} \tag{10.79}$$

Based on the existing research results [10, 15], we can assume that $\|e_{HJi}\| \leq \kappa_{e_{HJi}}$, $\|\Theta_{ik}\| \leq \kappa_{\Theta_{ik}}$, $\|\theta_i\| \leq \kappa_{\theta_{iM}}$, and $\|\theta_{ik}\| \leq \kappa_{\theta_{ikM}}$.

According to $\dot{L}_i$, we can obtain the inequality about $\dot{L}$ as follows

$$\dot{L} = \sum_{i=1}^{N} \dot{L}_i \leq -Z^T M Z + \Psi Z + K_\kappa \tag{10.80}$$

where

$$M = M_1 - M_2 - M_3 \tag{10.81}$$

$$M_2 = \begin{bmatrix} \frac{1}{2}\psi_1^{-2}\kappa_{211}^{-2}\|D_{11}\|^2 & \cdots & \frac{1}{2}\psi_1^{-2}\kappa_{21N}^{-2}\|D_{1N}\|^2 \\ \vdots & \frac{1}{2}\psi_i^{-2}\kappa_{2ij}^{-2}\|D_{ij}\|^2 & \vdots \\ \frac{1}{2}\psi_N^{-2}\kappa_{2N1}^{-2}\|D_{N1}\|^2 & \cdots & \frac{1}{2}\psi_N^{-2}\kappa_{2NN}^{-2}\|D_{NN}\|^2 \end{bmatrix} \tag{10.82}$$

$$M_3 = \begin{bmatrix} \sum_{i=1}^{N} \frac{\|E_{i1}\|^2}{8\psi_i^2\kappa_{6i1}^2} & & \\ & \sum_{i=1}^{N} \frac{\|E_{ij}\|^2}{8\psi_i^2\kappa_{6ij}^2} & \\ & & \sum_{i=1}^{N} \frac{\|E_{iN}\|^2}{8\psi_i^2\kappa_{6iN}^2} \end{bmatrix} \tag{10.83}$$

$$\begin{aligned} K_\kappa = &+ \sum_{i=1}^{N} \frac{1}{2}\psi_i^{-2}\kappa_{7i}^{-2}\kappa_{e_{HJi}}^2\kappa_{\theta_{iM}}^2 \\ &+ \sum_{i=1}^{N}\sum_{k=1}^{l} \frac{1}{2}\psi_{ik}^{-2}\kappa_{9i}^{-2}\kappa_{\Theta_{ik}}^2\kappa_{\theta_{ikM}}^2 \end{aligned} \tag{10.84}$$

$$\Psi = \Psi_1 + \Psi_2 \tag{10.85}$$

$$\Psi_1 = [\Psi_{11}, \ldots, \Psi_{1i}, \ldots, \Psi_{1N}]^T \tag{10.86}$$

$$\Psi_2 = [\Psi_{21}, \ldots, \Psi_{2i}, \ldots, \Psi_{2N}]^T \tag{10.87}$$

$$\begin{aligned} \Psi_{1i} = &- \psi_i^{-2}\theta_{im}^{*2} + \frac{1}{2}\sum_{j=1}^{N} \psi_i^{-2}\kappa_{2ij}^2\theta_{iM}^{*2} \\ &+ \frac{1}{4}\sum_{j=1}^{N} \psi_i^{-2}\kappa_{4ij}^2\theta_{iM}^2 + \frac{1}{4}\sum_{j=1}^{N} \psi_i^{-2}\kappa_{5ij}^2\theta_{iM}^2 \\ &+ \frac{1}{8}\sum_{j=1}^{N} \psi_i^{-2}\kappa_{6ij}^2\theta_{iM}^2 + \frac{1}{2}\psi_i^{-2}\kappa_{7i}^2 \\ &- \sum_{k=1}^{l} \psi_{ik}^{-2}\theta_{ikm}^2 + \frac{1}{2}\sum_{k=1}^{l} \psi_{ik}^{-2}\kappa_{9i}^2 \end{aligned} \tag{10.88}$$

$$\begin{aligned} \Psi_{2i} = &+ \sum_{j=1}^{N} \left(\frac{1}{4}\psi_i^{-2}\kappa_{5ij}^{-2}\|E_{ij}\|^2\|W_j\|^2\right) \\ &+ \sum_{j=1}^{N} \left(\frac{1}{4}\psi_i^{-2}\kappa_{4ij}^{-2}\|D_{ij}\|^2\|W_i\|^2\right) \end{aligned} \tag{10.89}$$

By selecting the parameters $\kappa_{(\cdot)}$, M can be guaranteed to be positive [10, 17]. Then $\dot{L} \leq 0$, when

$$\|Z\| > \frac{\Psi_M + \sqrt{\Psi_M^2 + 4\kappa_M \kappa_K}}{2\kappa_m} \tag{10.90}$$

where $\kappa_m \leq ||M|| \leq \kappa_M$, $||\Psi|| \leq \Psi_M$, and $||K_\kappa|| \leq \kappa_K$. So $||Z||$ is UUB, and $||\tilde{W}_i||$ is UUB. □

Theorem 10.2 *Considering the combined spacecraft attitude system (10.22) with multiple control inputs, if the admissible performance guaranteed control policies (10.54) with the experience-replay-based weight updating laws (10.52) of the critic neural networks are chosen. For initial conditions t_0 and ζ_0, There exists the time $T(\zeta_0, \kappa_B)$ such that $\zeta(t)$ is UUB with the bound κ_B described as*

$$\|\zeta(t)\| \leq \kappa_B, t \geq t_0 + T \tag{10.91}$$

where $\kappa_B \triangleq \max\left\{\sqrt{\kappa_{\varpi 1}\lambda_{\min}^{-1}(Q_1)}, \ldots, \sqrt{\kappa_{\varpi N}\lambda_{\min}^{-1}(Q_N)}\right\}$, $\kappa_{\varpi i} \in \mathbb{R}^+, i = 1, \ldots, N$.

Proof We take the derivative of the value function and substitute the optimal control policies to get

$$\dot{V}_i(t) = (\nabla V_i(\zeta))^T \left(f + \sum_{j=1}^{N} g_j \hat{\tau}_{jd}\right) \tag{10.92}$$

Consider the HJ equation (10.43), we have

$$\begin{aligned}(\nabla V_i)^T f = &-\frac{1}{4}\sum_{j=1}^{N} (\nabla V_j)^T g_j R_{jj}^{-1} R_{ij} R_{jj}^{-1} g_j^T \nabla V_j \\ &- \zeta^T Q_i \zeta + \frac{1}{2}(\nabla V_i)^T \sum_{j=1}^{N} g_j R_{jj}^{-1} g_j^T \nabla V_j\end{aligned} \tag{10.93}$$

Substituting (10.93) into (10.92) yields

$$\begin{aligned}\dot{V}_i(t) = &- \zeta^T Q_i \zeta + (\nabla V_i)^T \sum_{j=1}^{N} g_j \hat{\tau}_{jd} \\ &+ \frac{1}{2}(\nabla V_i)^T \sum_{j=1}^{N} g_j R_{jj}^{-1} g_j^T \nabla V_j \\ &- \frac{1}{4}\sum_{j=1}^{N} (\nabla V_j)^T g_j R_{jj}^{-1} R_{ij} R_{jj}^{-1} g_j^T \nabla V_j\end{aligned} \tag{10.94}$$

and

$$\begin{aligned}\dot{V}_i(t) = &- \zeta^T Q_i \zeta - \frac{1}{4}\sum_{j=1}^{N} (\nabla V_j)^T g_j R_{jj}^{-1} R_{ij} R_{jj}^{-1} g_j^T \nabla V_j \\ &- (\nabla V_i)^T \sum_{j=1}^{N} g_j \left(\tau_{jd} - \hat{\tau}_{jd}\right)\end{aligned} \tag{10.95}$$

Substituting (10.46), (10.54) into (10.95), we have

$$\dot{V}_i(t) = -\zeta^T Q_i \zeta - \frac{1}{4}\sum\nolimits_{j=1}^{N} (\nabla V_j)^T g_j R_{jj}^{-1} R_{ij} R_{jj}^{-1} g_j^T \nabla V_j + \varpi_i \tag{10.96}$$

where $\varpi_i = \frac{1}{2}(\nabla\phi_i^T W_i + \nabla\varepsilon_i)^T \sum_{j=1}^{N} g_j R_{jj}^{-1} g_j^T (\nabla\phi_i^T(\zeta)\tilde{W}_i + \nabla\varepsilon_i(\zeta))$. Considering that the items in ϖ_i are all UUB, then ϖ_i has a boundary as follows

$$\begin{aligned} \varpi_i \le & \left\| \frac{1}{2}(\nabla\phi_i^T W_i + \nabla\varepsilon_i)^T \right. \\ & \left. \sum\nolimits_{j=1}^{N} g_j R_{jj}^{-1} g_j^T \left(\nabla\phi_i^T(\zeta)\tilde{W}_i + \nabla\varepsilon_i(\zeta)\right) \right\| \le \kappa_{\varpi i} \end{aligned} \tag{10.97}$$

where $\kappa_{\varpi i}$ are constant. Considering that R_{ij} is a positive definite symmetric matrix, then the following term is also a positive definite

$$\frac{1}{4}\sum\nolimits_{j=1}^{N} (\nabla V_j)^T g_j R_{jj}^{-1} R_{ij} R_{jj}^{-1} g_j^T \nabla V_j > 0 \tag{10.98}$$

Then we have

$$\dot{V}_i(t) \le -\zeta^T Q_i \zeta + \kappa_{\varpi i} \le -\lambda_{\min}(Q_i)||\zeta||^2 + \kappa_{\varpi i} \tag{10.99}$$

where $\lambda_{\min}(Q_i)$ represents the least eigenvalue of the matrix Q_i. Define the compact set

$$\begin{aligned} \Omega_\zeta \triangleq \{\zeta : ||\zeta|| \le \max\{&\sqrt{\kappa_{\varpi 1}\lambda_{\min}^{-1}(Q_1)}, \ldots, \\ &\sqrt{\kappa_{\varpi N}\lambda_{\min}^{-1}(Q_N)}\}\} \end{aligned} \tag{10.100}$$

If ζ is inside Ω_ζ, ζ is bounded. If ζ is outside Ω_ζ, then all the derivatives $\dot{V}_i(t)$ are negative. So $||\zeta||$ is UUB. □

10.3.3 Nash Equilibrium Solution

Theorem 10.3 (Nash Solution of the Game) *Considering the combined spacecraft attitude system (10.22) with multiple control inputs, if the admissible performance guaranteed bounded control policies (10.54) with the experience-replay-based weight updating laws (10.52) of the critic neural networks are chosen. Theorems 10.1 and 10.2 hold, then H_i are UUB, i.e., $\hat{W}_i$ converge to the approximate coupled HJ solution, and the control policy pair $\hat{\tau}_d = \{\hat{\tau}_{1d}, \ldots, \hat{\tau}_{Nd}\}$ converge to the approximate Nash solution of the game.*

Proof According to the proof of Theorem 10.1, $\tilde{W}_i$ are UUB. We write the approximate coupled HJ equations as

$$
\begin{aligned}
H_i = & -\tilde{W}_i^T \nabla\phi_i f - e_{HJi} \\
& -\frac{1}{4}\sum_{j=1}^{N} W_j^T E_{ij} W_j + \frac{1}{4}\sum_{j=1}^{N} \hat{W}_j^T E_{ij} \hat{W}_j \\
& +\frac{1}{2} W_i^T \sum_{j=1}^{N} D_{ij} W_j - \frac{1}{2}\hat{W}_i^T \sum_{j=1}^{N} D_{ij} \hat{W}_j
\end{aligned} \tag{10.101}
$$

Considering $\hat{W}_i = -\tilde{W}_i + W_i$ and $||W_i|| \le W_{i\max}$, we have $||\hat{W}_i|| = || - \tilde{W}_i + W_i|| \le ||\tilde{W}_i|| + ||W_i|| \le ||\tilde{W}_i|| + W_{i\max}$. So, (10.101) can be written as

$$
\begin{aligned}
||H_i|| \le & ||\tilde{W}_i^T|| ||\nabla\phi_i|| ||f|| + ||e_{HJi}|| \\
& + \frac{1}{4}\sum_{j=1}^{N} W_{j\max}^2 ||E_{ij}|| \\
& + \frac{1}{4}\sum_{j=1}^{N} ||E_{ij}|| (||\tilde{W}_j|| + W_{j\max})^2 \\
& + \frac{1}{2}\sum_{j=1}^{N} W_{i\max} ||D_{ij}|| ||W_j|| \\
& + \frac{1}{2}\sum_{j=1}^{N} ||D_{ij}|| (||\tilde{W}_i|| + W_{i\max})(||\tilde{W}_j|| + W_{j\max})
\end{aligned} \tag{10.102}
$$

According to the proof of Theorem 10.1, all terms on the right side of the inequality (10.102) are UUB, so H_i is UUB.

From the proofs of Theorems 10.1 and 10.2, we know that $||\tau_{id} - \hat{\tau}_{id}||$ is bounded because $||\tilde{W}_i||$ is bounded. So the control strategy set $\hat{\tau}_d = \{\hat{\tau}_{1d}, \ldots, \hat{\tau}_{Nd}\}$ is a set of the approximate Nash equilibrium solution of the game. □

10.4 Simulation Analysis

The numerical example is shown in this section to demonstrate the performance of the developed performance guaranteed non-zero-sum game attitude control strategy. For the convenience of distinguishing from the comparison controllers, we label it as 'A' for (10.54). If the performance constraint part of the controller (10.54) is not considered, then the controller to be compared marked as 'B' can be obtained to illustrate the function of the performance constraint mechanism in 'A'. Similarly, to illustrate the critical role of performance constraints, the online adaptive learning solution for multi-player non-zero-sum games in [10] is used for comparison and labeled as 'C'. In order to illustrate the optimization effect of the proposed algorithm, the traditional prescribed performance controller in [8] is introduced for comparison and marked as 'D'. The finite-time controller in [18] is used for comparison and marked as 'E' to show that the controller proposed in this chapter has the ability to describe the settling time and has the optimization effect. In the simulation diagram and analysis, the symbols or subscripts 'A', 'B', 'C', 'D', and 'E' represent the above controllers.

10.4.1 Simulation Parameters

For the combined spacecraft, the orbital parameters are given in Table 10.1. Without loss of generality, we directly give the initial parameters of the combined spacecraft simulation model here, $q_{v0} = [0.6, 0.3, -0.3]^T$, $q_0(0) = \sqrt{1 - q_{v0}^T q_{v0}}$, $\omega_{a0} = [0, 0, 0]^T$ rad/s,

$$J_c = \begin{bmatrix} 100 & -10 & 5 \\ -10 & 120 & -3 \\ 5 & -3 & 100 \end{bmatrix} \text{kg} \cdot \text{m}^2.$$

The parameters used for the simulation are all replaceable. The desired signal is given by the following scenario. In this simulation, the combined spacecraft tracks the attitude of the expected $\mathfrak{F}_{cbd}$. We assume that the $\mathfrak{F}_{cbd}$ coincides with the $\mathfrak{F}_I$, i.e., it is inertially oriented.

Next, the parameters selected in the simulation of the algorithm proposed in this chapter are given, as shown in the following table (Table 10.2). In the simulation, the combined spacecraft is considered to have 3 sub-spacecraft with actuators 1, 2, and 3. Where ρ_0 needs to be greater than the initial quaternion errors, we select it as 1 so that it can be greater than any quaternion element; ρ_t and ε_ρ provide the steady-state error boundary; δ_ι is about whether to suppress overshoot; λ_ι is the proportional coefficient of the sliding-mode surface in the controller; a_1, a_2, and a_3 are the ratio coefficients of each sub-spacecraft actuator control input associated neural network weight adaptive law. The position transformation matrixes of the multiple sub-spacecraft actuators are temporarily taken as $\Lambda_{a1} = \text{diag}(1, 0.5, 0.5)$, $\Lambda_{a2} = \text{diag}(0.5, 1, 0.5)$, $\Lambda_{a3} = \text{diag}(0.5, 0.5, 1)$. The selection of the position transformation matrixes is general,

Table 10.1 The orbital parameters of the combined spacecraft

Parameters	Values
Semimajor axis	7.35×10^6m
Eccentricity	0.0068
Inclination	60°
Argument of periapsis	0°
Longitude of the ascending node	0°
Mean anomaly	0°

Table 10.2 The initial parameters of the proposed control policies (10.54)

Param	Value	Param	Value	Param	Value
t_{ps}	110s	ε_ρ	0.01	a_1	0.1
ρ_0	1	δ_ι	1	a_2	0.1
ρ_t	0.1	λ_ι	1	a_3	0.1

and only some parameters are assigned here for simulation verification. For practical applications, actual parameters need to be assigned. The matrix in the value function is temporarily taken as $R_{11} = R_{22} = R_{33} = 0.05I_{3\times3}$, $R_{12} = R_{21} = R_{13} = R_{31} = R_{23} = R_{32} = 0.05I_{3\times3}$, $Q_1 = Q_2 = Q_3 = 0.01I_{3\times3}$, where $I_{3\times3}$ is the unit matrix of 3×3. In this simulation example, the parameter values are general values, which can be adjusted according to specific conditions. In addition, the neural network basis set used in this chapter is $\phi = [\zeta_1^2, \zeta_2^2, \zeta_3^2, \zeta_1\zeta_2, \zeta_2\zeta_3, \zeta_3\zeta_1]^T$. The controllers in this chapter do not consider external disturbances for the time being.

To show the error in the whole simulation process, we define the error-index for the controllers as $I_a = ||q_{ev}|| + ||\omega_e||$. For illustrating the energy consumption of the proposed controllers, we use the energy index [19], which is related to the amount of energy (chemical or electrical energy) that actuators (thrusters or reaction flywheels) consume during closed-loop system stabilization, as $E_a = \sum_{i=1}^{N} \int_0^t ||\tau_i(t)||^2 dt$.

10.4.2 Performance and Comparison

Figures 10.2, 10.3, 10.4, 10.5, 10.6, 10.7, 10.8, 10.9, 10.10, 10.11, 10.12, 10.13, 10.14 and 10.15 show the numerical simulation results of the proposed control algorithm and the controllers to be compared. Figures 10.2, 10.3 and 10.4 are the tracking error curves of the combined spacecraft attitude closed-loop systems. The error elements of each dimension are plotted separately to compare the effects of different controllers on the same state variable. The labels 'A'–'E' correspond to the different controllers to be compared in the simulation diagram. It can be seen from these three graphs that under the action of the controller 'A' proposed in this chapter, the tracking errors converge and satisfy the performance constraints, including faster convergence to the predefined steady-state error area within the predefined maximum settling time with better transient performance and predefined performance guarantees. The control algorithm 'B' is where the performance constraints are not considered in the proposed algorithm. Under its effect, the performance of the tracking error convergence of the closed-loop system is similar to the effect under the traditional game strategy 'C'. Neither 'B' nor 'C' can make the closed-loop system better meet the performance constraints under the same energy consumption as 'A'. For precise aerospace missions, the optimal strategies 'B' and 'C' require many parameter adjustments to achieve precise control objectives. The proposed control algorithm can meet the performance constraints in principle and, at the same time, can achieve optimal control under the performance constraints. It can reveal the advantages of the proposed control algorithms in the traditional game strategies and the performance-constrained control algorithms. Compared with the traditional prescribed performance controller 'D' and the finite-time controller 'E', the control algorithm proposed in this chapter still has better transient and steady-state performance under the condition of less energy consumption and finally achieves optimal solution and Nash equilibrium.

The tracking error curves of the combined spacecraft attitude angular velocity are shown in the Figs. 10.5, 10.6 and 10.7. Under the proposed algorithm, the combined

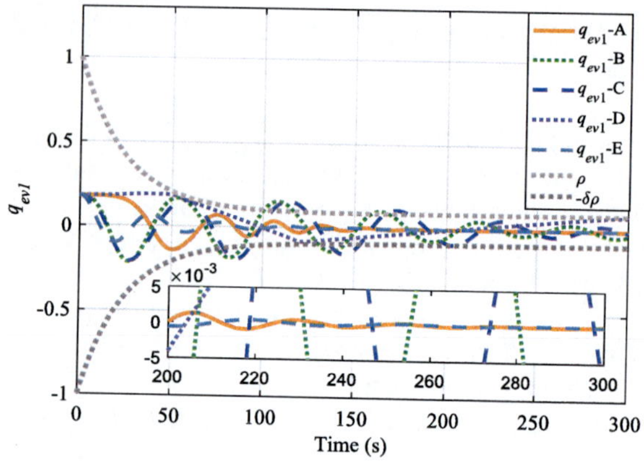

Fig. 10.2 The spacecraft attitude tracking error q_{ev1}

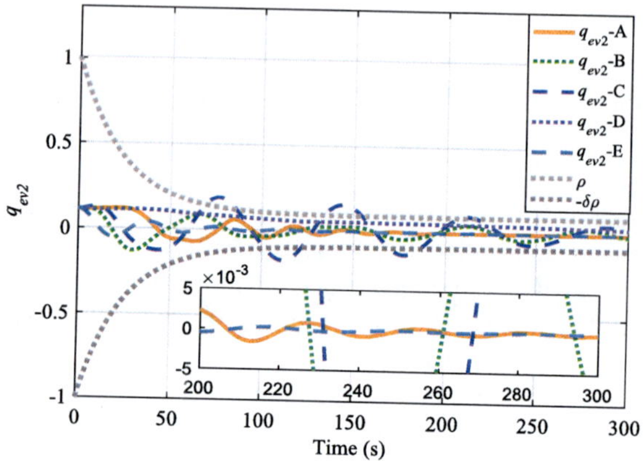

Fig. 10.3 The spacecraft attitude tracking error q_{ev2}

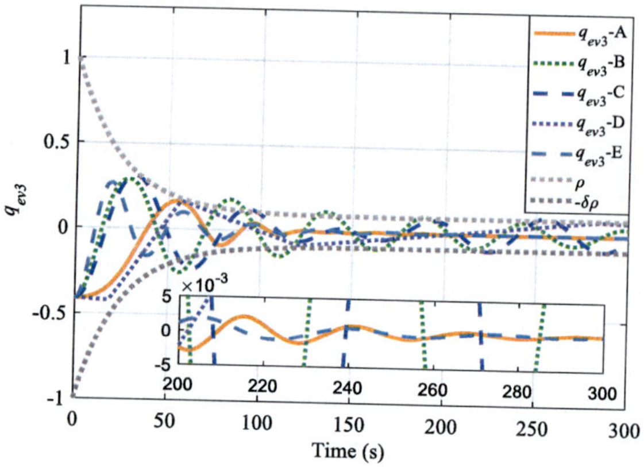

Fig. 10.4 The spacecraft attitude tracking error q_{ev3}

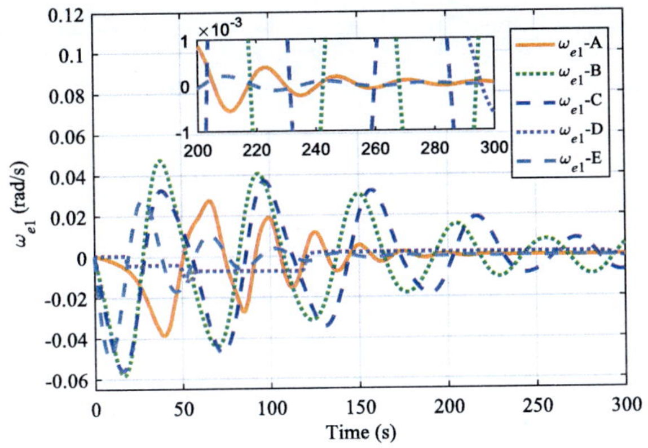

Fig. 10.5 The angular velocity tracking error ω_{e1}

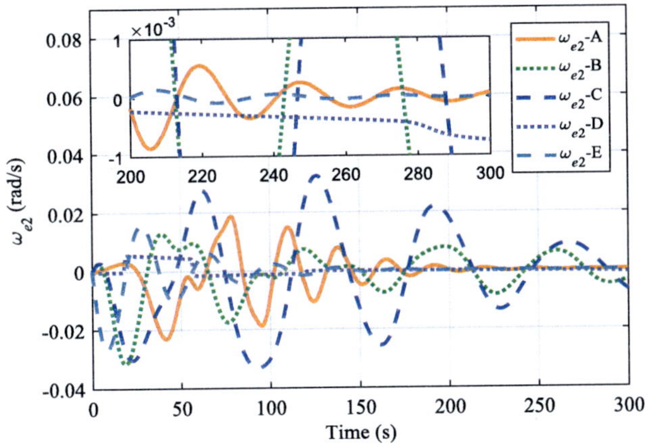

Fig. 10.6 The angular velocity tracking error ω_{e2}

Fig. 10.7 The angular velocity tracking error ω_{e3}

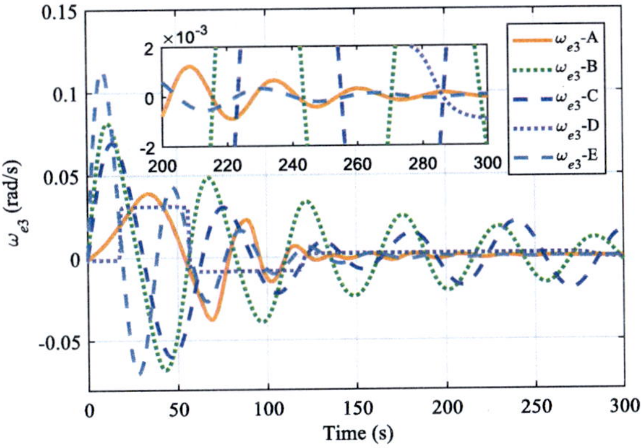

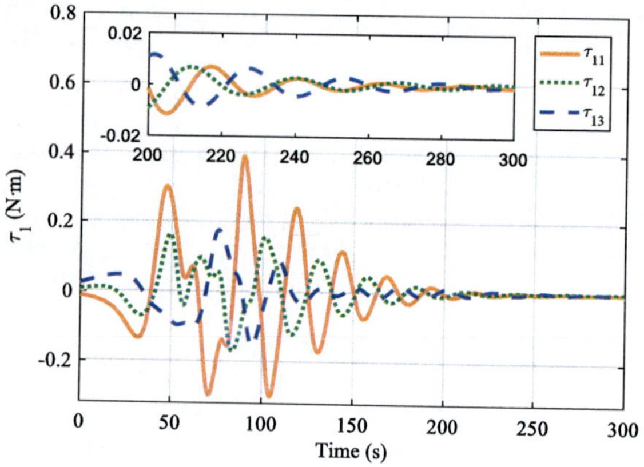

Fig. 10.8 The control input of actuator 1

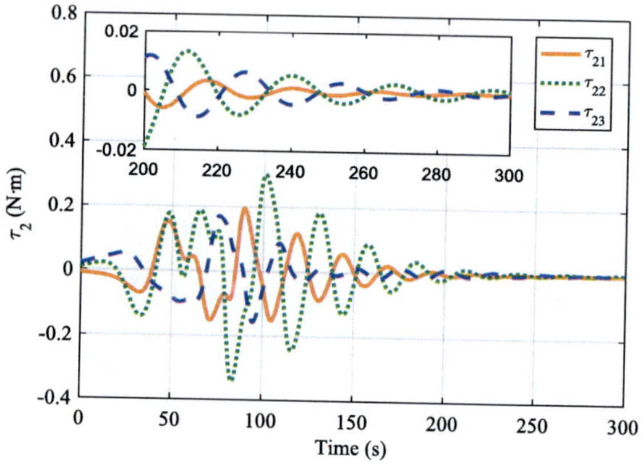

Fig. 10.9 The control input of actuator 2

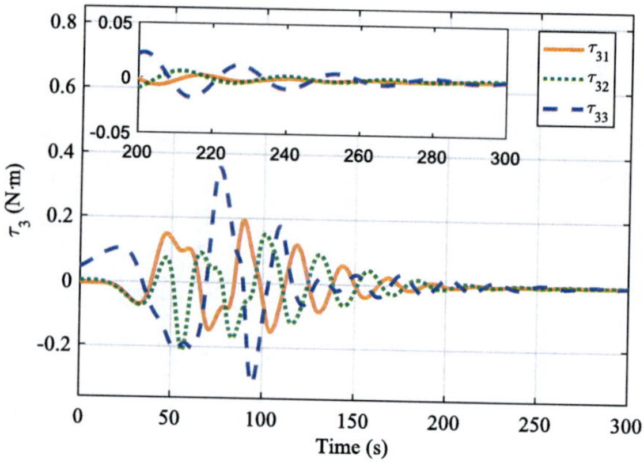

Fig. 10.10 The control input of actuator 3

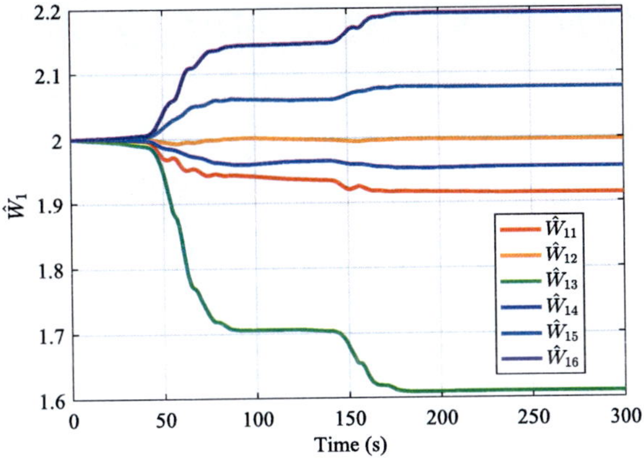

Fig. 10.11 The critic network adaptive weights of the actuator 1

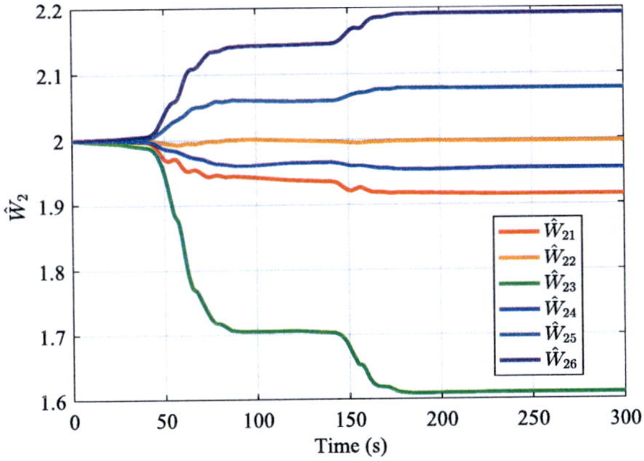

Fig. 10.12 The critic network adaptive weights of the actuator 2

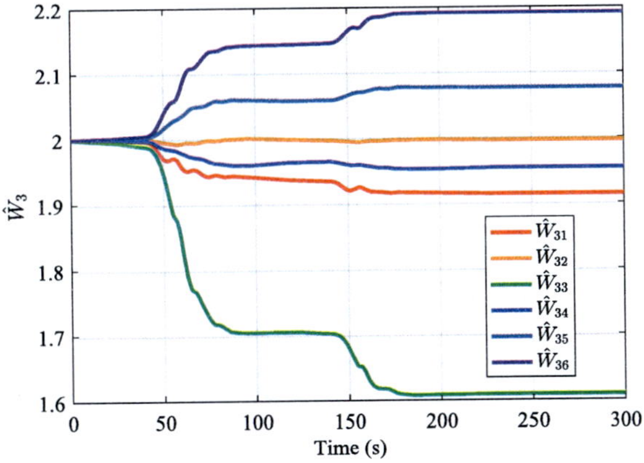

Fig. 10.13 The critic network adaptive weights of the actuator 3

Fig. 10.14 The error index I_a

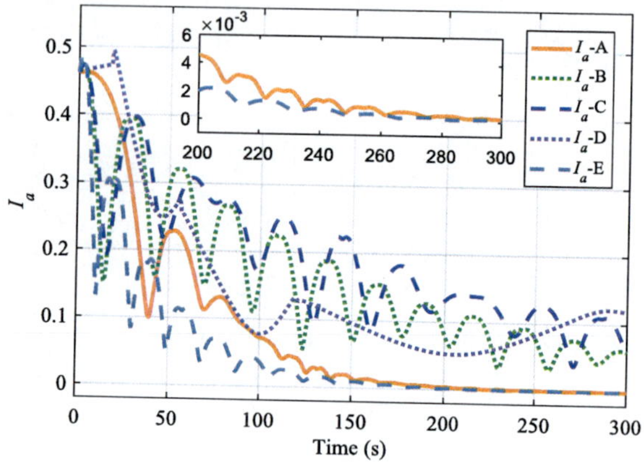

Fig. 10.15 The energy consumption

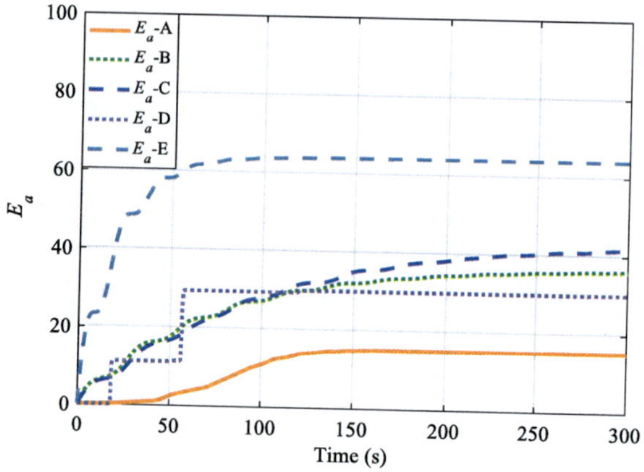

spacecraft attitude angular velocity tracking error can converge faster, and the steady-state error is small. Better transient and steady-state performance can be obtained. Figures 10.8, 10.9 and 10.10 are the control inputs of the three sub-spacecraft actuators of the combined spacecraft under the control strategy 'A'. Figures 10.11, 10.12 and 10.13 respectively show the adaptive weight parameters of the critic neural network of the corresponding sub-spacecraft actuator control inputs. It can be seen from the figures that the adaptive weights can stabilize around the constants after a period of time. Figures 10.14 and 10.15 show the attitude error index $I_a = ||q_{ev}|| + ||\omega_e||$ of the combined spacecraft and the energy consumption E_a under the action of the controllers. By comparison, under the action of the proposed control algorithm, more significant convergence rates and more minor steady-state errors can be obtained, and less energy is consumed. Compared with the traditional game strategies 'B' and 'C', performance constraints make the tracking error converge faster and reduce energy consumption due to the reduction of redundant motion. Compared with the traditional prescribed performance controller 'D' and the finite-time controller 'E', due to the optimization strategy considered under the proposed algorithm, the energy consumption is also smaller in the process of the control inputs fitting to the optimal solution. This reflects the proposed algorithm's advantages in energy consumption compared to traditional controllers.

Through comparison and simulation, it is fully illustrated that under the control algorithm proposed in this chapter, better transient and steady-state performance can be achieved by consuming less energy. The performance is guaranteed. That is, the practical maximum settling time of the combined spacecraft tracking errors is less than the pre-defined maximum settling time, the steady-state errors are less than the pre-defined values, and the overshoot meets the constraint condition is fully guaranteed. Furthermore, the coupled control inputs are optimal for the actuators of all sub-spacecraft. Since the optimal solution of each sub-spacecraft actuator is considered, all control inputs involved in attitude control constitute a Nash equilibrium solution.

10.5 Conclusion

Considering the large combined spacecraft composed of multiple sub-spacecraft with multiple actuators, the actuators of an individual spacecraft often find it challenging to handle the overall maneuvering task of the entire combined spacecraft. Therefore, it is necessary to consider the joint action of multiple sub-spacecraft actuators. Traditional and straightforward control allocation algorithms can solve this difficulty, but they may not be able to obtain the optimal solution for each actuator. Simultaneously considering the performance constraints of the combined spacecraft maneuver and the joint control of multiple cooperative sub-spacecraft, this chapter proposes the performance guaranteed non-zero-sum game algorithm to solve the problem of the combined spacecraft optimized maneuver with the multiple sub-spacecraft inputs. The optimized control design is performed on the sub-spacecraft control inputs of all actuators with fixed installation positions relative to the combined spacecraft. The various controllers form a cooperative game relationship, reach the Nash equilibrium, and jointly complete the performance guaranteed attitude control of the combined spacecraft. Finally, through many comparisons and numerical simulation experiments with the existing advanced controllers, the rationality, effectiveness, applicability, and advantages of the control algorithm proposed in this chapter are proved.

References

1. Waltz, D.M.: On-orbit servicing of space systems. Krieger Publishing Company (1993)
2. Li, W., Cheng, D., Liu, X., Wang, Y., Shi, W., Tang, Z., Gao, F., Zeng, F., Chai, H., Luo, W., et al.: On-orbit service (oos) of spacecraft: A review of engineering developments. Progress in Aerospace Sciences **108**, 32–120 (2019)
3. Hu, Q., Li, B., Xiao, B., Zhang, Y.: Control Allocation for Spacecraft Under Actuator Faults. Springer (2021)
4. Junkins, J.L., Schaub, H.: Analytical mechanics of space systems. American Institute of Aeronautics and Astronautics (2009)
5. Gao, Y., Li, D., Ge, S.S.: Time-synchronized tracking control for 6-dof spacecraft in rendezvous and docking. IEEE Transactions on Aerospace and Electronic Systems **58**(3), 1676–1691 (2022). https://doi.org/10.1109/TAES.2021.3124865
6. Hughes, P.C.: Spacecraft attitude dynamics. Courier Corporation (2012)
7. Costic, B., Dawson, D., De Queiroz, M., Kapila, V.: Quaternion-based adaptive attitude tracking controller without velocity measurements. Journal of Guidance, Control, and Dynamics **24**(6), 1214–1222 (2001)
8. Bechlioulis, C.P., Rovithakis, G.A.: Robust adaptive control of feedback linearizable mimo nonlinear systems with prescribed performance. IEEE Transactions on Automatic Control **53**(9), 2090–2099 (2008)
9. Bechlioulis, C.P., Rovithakis, G.A.: Adaptive control with guaranteed transient and steady state tracking error bounds for strict feedback systems. Automatica **45**(2), 532–538 (2009)
10. Vamvoudakis, K.G., Lewis, F.L.: Multi-player non-zero-sum games: Online adaptive learning solution of coupled hamilton–jacobi equations. Automatica **47**(8), 1556–1569 (2011)
11. Başar, T., Olsder, G.J.: Dynamic noncooperative game theory. SIAM (1998)
12. Abu-Khalaf, M., Lewis, F.L.: Nearly optimal control laws for nonlinear systems with saturating actuators using a neural network hjb approach. Automatica **41**(5), 779–791 (2005)
13. Finlayson, B.A.: The method of weighted residuals and variational principles. SIAM (2013)
14. Hornik, K., Stinchcombe, M., White, H.: Universal approximation of an unknown mapping and its derivatives using multilayer feedforward networks. Neural networks **3**(5), 551–560 (1990)
15. Liu, D., Li, H., Wang, D.: Online synchronous approximate optimal learning algorithm for multi-player non-zero-sum games with unknown dynamics. IEEE Transactions on Systems, Man, and Cybernetics: Systems **44**(8), 1015–1027 (2014)
16. Modares, H., Lewis, F.L., Naghibi-Sistani, M.B.: Integral reinforcement learning and experience replay for adaptive optimal control of partially-unknown constrained-input continuous-time systems. Automatica **50**(1), 193–202 (2014)
17. Han, N., Luo, J., Chai, Y.: Differential game learning approach for multiple microsatellites takeover of the attitude movement of failed spacecraft. SCIENTIA SINICA Informationis **50**(4), 588–602 (2020)
18. Du, H., Li, S., Qian, C.: Finite-time attitude tracking control of spacecraft with application to attitude synchronization. IEEE Transactions on Automatic Control **56**(11), 2711–2717 (2011)
19. Jiang, B., Li, C., Ma, G.: Finite-time output feedback attitude control for spacecraft using "adding a power integrator" technique. Aerospace Science and Technology **66**, 342–354 (2017)

Part III
Performance Guaranteed Cooperation for Spacecraft Formation

In this part, the performance guaranteed focuses on the systems' state constraints and formation-containment control for spacecraft/aerospace formation behaviors. The state constraints are the guarantees of the time-domain response performance of the system states. The formation-containment control can be considered as a control technique by considering the formation configuration constraints that guarantee formation performance in which some spacecraft serve as the movement constraints of another part of the spacecraft. Such a design can guarantee the formation flight configuration and the motion performance of all spacecraft. By changing the configuration, other spacecraft can be kept within the constraints of the configuration, which also constitutes the performance guaranteed of the spacecraft formation. Control of this behavior can be referred to as *multi-spacecraft formation-containment control.* Considering that in this chapter, multi-spacecraft systems are often networked, this behavior can also be referred to as *networked spacecraft formation-containment control.*

With the primary objective of performance guaranteed cooperation and consistency, a series of cooperative control methods are introduced for the multi-spacecraft systems or network Euler-Lagrange systems. Considering the related methods of performance guaranteed as the barrier Lyapunov function and formation-containment control, this part introduces the control algorithms with performance and full-state constraints. Inclusively, the output feedback problems with state constraints are also solved. This part addresses the constrained multi-spacecraft formation-containment control problems with output feedback and with collision avoidance. Further, the layered multi-spacecraft formation-containment control problem in the constrained space is also solved.

Chapter 11
Distributed Coordinated Control for Network Systems with Full-State Constraints

Abstract This chapter addresses the distributed coordinated tracking problem for network systems modeled by Euler-Lagrange systems (ELSs), where full-state constraints (FSCs) and model uncertainties are considered. We first introduce a distributed estimator to obtain accurate estimations of the target trajectory. To guarantee the performance or FSCs of ELSs, we next use the barrier Lyapunov function (BLF) technique to design a distributed coordinated tracking control law combined with a Moore-Penrose inverse term. To tackle the model uncertainties, the adaptive neural networks (NNs) are then integrated into the proposed control law. By using the Lyapunov stability theory, the semi-global uniform ultimate boundedness (SGUUB) of state errors can be guaranteed. Simulation results are finally given to show the feasibility of the proposed approaches.

11.1 Introduction

Different from the previous Parts to abstract more spacecraft formation attitude motions or relative orbital motions, multiple spacecraft will be constructed as networked systems starting from this chapter. On the one hand, graph theory-based control design and networked system modeling can describe larger-scale and larger-number spacecraft formation behaviors. On the other hand, the networked system design also simplifies and unifies the specific model of each member spacecraft. These models, which the Euler-Lagrange systems (ELSs) can describe, can represent not only the attitude motions of the spacecraft but also the relative orbital motions and even the motions of the aerospace manipulators. In this chapter, based on the basic networked systems, the cooperative tracking control considering the full-state constraints is designed.

Distributed tracking problems with all members of the network systems tracking a moving leader have been an active area due to its extensive utilization and advantages, which include less communication requirement, stronger robustness and higher efficiency. At first, cooperative tracking for network systems described by linear systems are studied [1–4]. And ever since then, this cooperative problem has been

Y. Gao and D. Li, *Spacecraft Maneuver with Performance Guaranteed*,
https://doi.org/10.1007/978-981-99-4653-2_11

investigated from different interesting aspects, e.g., without using velocity measurements, the synchronization control laws of networked systems are proposed by output feedback in [5–10].

As ELSs can characterize a broad class of nonlinear systems in line with reality, such as spacecraft formation, robot manipulators, and autonomous vehicles, the tracking problems are also considered with ELSs [11–16]. In practice, a variety of constraints should be met for ELSs since systems in the real world are unavoidably subjected to various constraints, e.g., task performance guaranteed requirements, safety specifications, or limits of their own physical structures. Without taking these constraints into consideration, the performance and stability of the closed-loop system will be influenced. For example, it is essential for a spacecraft to guarantee attitude performance constraints in missions to point along a prescribed direction [17]. Any violation of the attitude constraints will lead to system failure or damage of payloads, because some equipped payloads (such as interferometers or infrared telescopes) are sensitive which should be kept a proper distance away from any bright energy sources or undesired celestial objects [18].

In this chapter, full-state constraints (FSCs) are considered, where the states of network systems are supposed to stay within a given compact set since any violation may induce severe performance degradation [19]. In addition, despite the practical importance, there are few studies on distributed tracking control of network systems modeled by ELSs with FSCs, where the FSCs of each member of the network systems should be guaranteed by properly designed control law. Similar yet different problems have been discussed in works on other types of systems. With state constraints, network system consensus problem is studied in [20]. For a class of nonlinear network systems, an adaptive cooperative control law is proposed in [21] with partial tracking errors constrained. For a piecewise constant value, the distributed tracking problem is investigated with constrained linear systems in [22]. Considering communication delays in an unbalanced network, a constrained consensus control law is introduced in [23]. A consensus control scheme is developed in [24], while ensuring the output constraints. Unfortunately, only output constraint problem are solved in [20–25]. In [26], the robust constrained formation control is investigated with external disturbances and model uncertainties, while only the input constraint is considered and the linearized model is required. Compared with output constraints, full-state constraints are more general in actual systems, and control problems with full-state constraints are more challenging. For example, mobile robots are usually required to limit their positions and velocities for safety demands, which involves not only the output constraints but full-state constraints. Therefore, the research of full-state constraints in uncertain EL-systems is necessary for practical network systems applications.

Compared with the consensus problem of network systems investigated in [20, 23, 24], the distributed coordinated tracking problem studied in this chapter is nontrivial as the target trajectory is a time-varying value. On the other hand, the controllers in [20–26] only deal with the output constraints, partial state constraints or input constraints, and especially some references consider linear systems or SISO systems.

In this chapter, the coordinated tracking for networked ELSs with FSCs in the presence of nonlinear uncertainties is addressed. The main contributions of our study are threefold:

(1) A distributed control law is designed with a Moore-Penrose inverse term and the barrier Lyapunov function (BLF) technique which can guarantee the FSCs of network systems. In comparison, only the output constraints are considered in [20–25]. Different from the control for a single system with constraints [27, 28], the approach in this chapter tackles the constraints of network systems, which is rather challenging.

(2) A distributed estimator is introduced to obtain a precise estimation of the leader's information in a finite time. Compare with the distributed finite-time sliding-mode estimator (DFSE) in [29], the scheme in this chapter can estimate the leader's position, while the acceleration information is no longer required. It is noteworthy that this approach is fully distributed without using any global information.

(3) Compared with most existing results dealing with parametric uncertainties requiring the 'linearity-in-parameters' assumption [30, 31], a coordinated tracking control law integrated with the neural networks (NNs) is proposed, which can approximate nonlinear uncertainties. By utilizing the Lyapunov direct method, the semi-global uniform ultimate boundedness (SGUUB) is proved.

Notations: Let $\mathbb{R}$ denote the set of real numbers. Let $I_n \in \mathbb{R}^{n\times n}$ denote the n-dimensional identity matrix. Let $\lambda_{\min}(\bullet)$ and $\lambda_{\max}(\bullet)$ represent, respectively, the minimum and maximum eigenvalues of the matrix $\bullet$. Let $\|\cdot\|$ be the Euclidean norm of a vector. Let $\mathrm{diag}(z_1, \ldots, z_n) \in \mathbb{R}^{n\times n}$ be the diagonal matrix, where $z_1, \ldots, z_n$ are the diagonal entries. Let $\mathrm{sgn}(\bullet)$ stand for the signum function.

11.2 Preliminaries and Problem Description

We utilize a communication graph $\mathcal{G} = (V, E)$ to characterize the communication topology among the network systems, where $E \subseteq V \times V$ and $V = \{1, \ldots, k\}$ denote the arc set and the vertex set, respectively. The adjacency matrix $\mathcal{A} = [a_{ij}] \in \mathbb{R}^{k\times k}$ is a nonnegative matrix defined as $a_{ij} > 0$ if $(j, i) \in E$, and $a_{ij} = 0$ otherwise. We define $a_{ii} = 0$, and the Laplacian matrix $\mathcal{A}$ is defined as $\mathcal{L} = [l_{ij}] \in \mathbb{R}^{k\times k}$, where $l_{ii} = \sum_{j=1, j\neq i}^{k} a_{ij}$ and $l_{ij} = -a_{ij}$, $i \neq j$. Assume that the information flow among followers is bidirectional. Let $\bar{\mathcal{G}} \in \mathbb{R}^{(N+1)\times(N+1)}$ denote the topology among N followers (labeled as $1, 2, \ldots, N$) and 1 leader (labeled as $N+1$), with corresponding Laplacian matrix $\bar{\mathcal{L}} \in \mathbb{R}^{(N+1)\times(N+1)}$ taking the following form

$$\bar{\mathcal{L}} = \begin{bmatrix} \mathcal{L}_1 & \mathcal{L}_2 \\ 0_{1\times N} & 0 \end{bmatrix}, \tag{11.1}$$

where $\mathcal{L}_1 \in \mathbb{R}^{N\times N}$ is symmetric positive definite.

Assumption 11.1 Supposing that in the communication topology $\bar{\mathcal{G}}$, the leader has paths to all followers.

Lemma 11.1 [12] *Each entry of* $-\mathcal{L}_1^{-1}\mathcal{L}_2$ *is nonnegative and each row sum of* $-\mathcal{L}_1^{-1}\mathcal{L}_2$ *is equal to one.*

In this chapter, the Euler-Lagrange systems are utilized to describe the network systems because of their abstraction of different control objects and the transferability of control algorithms. Then, each follower is modeled by Euler-Lagrange system equation in $\mathbb{R}^p$ as

$$M_i(q_i)\ddot{q}_i + C_i(q_i, \dot{q}_i)\dot{q}_i + g_i(q_i) = \tau_i(t), \tag{11.2}$$

where $q_i \in \mathbb{R}^p$ denotes the position vector of member i of the network systems, $M_i(q_i) \in \mathbb{R}^{p\times p}$ the symmetric positive definite inertia matrix, $C_i(q_i, \dot{q}_i) \in \mathbb{R}^{p\times p}$ the vector of Coriolis and centrifugal matrix, $g_i(q_i) \in \mathbb{R}^p$ the gravitational force, and $\tau_i \in \mathbb{R}^p$ the control input. The following properties hold for (11.2):

Property 11.1 [32] *For any* i, $M_i(q_i)$ *is bounded, which is bounded as*

$$0 < k_{\underline{m}}\|q_i\|^2 \le q_i^T M_i(q_i) q_i \le k_{\overline{m}}\|q_i\|^2, \ \forall q_i \in \mathbb{R}^p, \tag{11.3}$$

where $k_{\underline{m}}$ *and* $k_{\overline{m}}$ *are positive constants.*

Property 11.2 [33] *The matrix* $\dot{M}_i(q_i) - 2C_i(q_i, \dot{q}_i)$ *is skew-symmetric.*

Assumption 11.2 The leader is independent of the followers. The acceleration of the leader is bounded by a nonnegative constant δ_L as

$$\|\ddot{q}_{N+1}\| \le \delta_L < \infty. \tag{11.4}$$

11.3 Control Design

First, in an undirected graph, we introduce a distributed estimator with finite-time convergence property. Next, applying a Moore-Penrose inverse term and a BLF, the distributed tracking control law is designed for ELSs with FSCs, and the NNs are integrated to compensate for the nonlinear nonlinearities of the system model.

11.3.1 Distributed Finite-Time Sliding-Mode Estimator

Inspired by [29], we propose the following distributed finite-time sliding-mode estimator (DFSE) protocol as

$$\dot{\hat{q}}_i = \hat{v}_i - \beta_1 \text{sgn} \sum_{j=1}^{N+1} a_{ij} \left(\hat{q}_i - \hat{q}_j\right), \tag{11.5}$$

$$\dot{\hat{v}}_i = -\beta_2 \text{sgn} \sum_{j=1}^{N+1} a_{ij} \left(\hat{v}_i - \hat{v}_j\right), \ i = 1, 2, \ldots, N, \tag{11.6}$$

where β_1 and β_2 are positive constants, $\hat{q}_k$ and $\hat{v}_k$, $k = 1, \ldots, N+1$ are the estimations of desired position and velocity for each follower, respectively, and $\hat{q}_{N+1} = q_{N+1}$, $\hat{v}_{N+1} = \dot{q}_{N+1}$. Define

$$Q_F = \left[q_1^T, q_2^T, \ldots, q_N^T\right]^T, \tag{11.7}$$

$$Q_L = \left[q_{N+1}^T, q_{N+1}^T, \ldots, q_{N+1}^T\right]^T. \tag{11.8}$$

Theorem 11.1 *Let $\beta_1 > 0$ and $\beta_2 > \delta_L$. If Assumptions 11.1 and 11.2 hold, under the DFSEs (11.5) and (11.6), the followers can acquire the precise estimations of the leader's position and velocity in finite time, respectively, which means*

$$\hat{q}_i \to q_{N+1}, \tag{11.9}$$

$$\hat{v}_i \to \dot{q}_{N+1}, \ i = 1, 2, \ldots, N. \tag{11.10}$$

Proof Let $\tilde{v}_i = \hat{v}_i - \dot{q}_{N+1}$ and $\tilde{q}_i = \hat{q}_i - q_{N+1}$, $i = 1, \ldots, N$. Define the vectors $\tilde{v}_L = \left[\tilde{v}_1, \tilde{v}_2, \ldots, \tilde{v}_N\right]^T$ and $\tilde{q}_L = \left[\tilde{q}_1, \tilde{q}_2, \ldots, \tilde{q}_N\right]^T$. Consider the following Lyapunov function according to (11.6).

$$V_1 = \frac{1}{2} \tilde{v}_L^T \left(\mathcal{L}_1 \otimes I_p\right)^T \tilde{v}_L. \tag{11.11}$$

According to (11.5) and (11.6), taking the derivative of V_1 yields

$$\begin{aligned}
\dot{V}_1 =& \tilde{v}_L^T (\mathcal{L}_1 \otimes I_p)^T \dot{\tilde{v}}_L \\
=& \tilde{v}_L^T \left(\mathcal{L}_1 \otimes I_p\right)^T \left[-\beta_2 \text{sgn}\left(\left(\mathcal{L}_1 \otimes I_p\right) \tilde{v}_L\right) - \ddot{Q}_L\right] \\
=& \left(\left(\mathcal{L}_1 \otimes I_p\right) \tilde{v}_L\right)^T \left[-\beta_2 \text{sgn}\left(\left(\mathcal{L}_1 \otimes I_p\right) \tilde{v}_L\right)\right. \\
& \left. + \left(\mathcal{L}_1^{-1} \mathcal{L}_2 \otimes I_p\right) \ddot{Q}_L\right].
\end{aligned} \tag{11.12}$$

According to [34], to any $x \subseteq \mathbb{R}^p$, it holds

$$\|x\|_1 \geq \|x\|_2, \tag{11.13}$$

where $\|x\|_2 = \sqrt{x^T x}$, $\|x\|_1 = \sum_{i=1}^p |x_i|$, x_i is ith element of vector x. According to (11.13) and Lemma 11.1, we can get

$$\begin{aligned}
\dot{V}_1 \leq & -\left(\beta_2 - \left\|\mathcal{L}_1^{-1}\mathcal{L}_2\right\| \delta_L\right) \left\|\left(\mathcal{L}_1 \otimes I_p\right) \tilde{v}_L\right\|_1 \\
\leq & -\left(\beta_2 - \delta_L\right) \left\|\left(\mathcal{L}_1 \otimes I_p\right) \tilde{v}_L\right\|_2 \\
\leq & -\left(\beta_2 - \delta_L\right) \sqrt{\tilde{v}_L^T \left(\mathcal{L}_1 \otimes I_p\right)^2 \tilde{v}_L} \\
\leq & -\left(\beta_2 - \delta_L\right) \lambda_{\min}\left(\mathcal{L}_1\right) \|\tilde{v}_L\|_2 \\
\leq & -\sqrt{2}\left(\beta_2 - \delta_L\right) \frac{\lambda_{\min}\left(\mathcal{L}_1\right)}{\sqrt{\lambda_{\max}\left(\mathcal{L}_1\right)}} V_1^{\frac{1}{2}}.
\end{aligned} \tag{11.14}$$

Thus, it follows from Lemma 2.3 that $\hat{v}_i \to \dot{q}_{N+1}$, $i = 1, 2, \ldots, N$ in a settling time

$$T_1 = \frac{\sqrt{2}\sqrt{\lambda_{\max}\left(\mathcal{L}_1\right)}}{\left(\beta_2 - \delta_L\right) \lambda_{\min}\left(\mathcal{L}_1\right)} V_1^{\frac{1}{2}}(0), \tag{11.15}$$

and $\dot{q}_{N+1}$ can be utilized to replace $\hat{v}_i$, when $t \geq T_1$.

When $t \geq T_1$, construct the Lyapunov function $V_2 = \frac{1}{2}\tilde{q}_L^T \left(\mathcal{L}_1 \otimes I_p\right)^T \tilde{q}_L$, and take the derivative of V_2 yields,

$$\begin{aligned}
\dot{V}_2 = & \tilde{q}_L^T \left(\mathcal{L}_1 \otimes I_p\right)^T \dot{\tilde{q}}_L \\
= & \tilde{q}_L^T \left(\mathcal{L}_1 \otimes I_p\right)^T \left[-\beta_1 \mathrm{sgn}\left(\left(\mathcal{L}_1 \otimes I_p\right) \tilde{q}_L\right)\right] \\
\leq & -\beta_1 \left\|\left(\mathcal{L}_1 \otimes I_p\right) \tilde{q}_L\right\|_1 \\
\leq & -\beta_1 \left\|\left(\mathcal{L}_1 \otimes I_p\right) \tilde{q}_L\right\|_2 \\
\leq & -\beta_1 \sqrt{\tilde{q}_L^T \left(\mathcal{L}_1 \otimes I_p\right)^2 \tilde{q}_L} \\
\leq & -\beta_1 \lambda_{\min}\left(\mathcal{L}_1\right) \|\tilde{q}_L\|_2 \\
\leq & -\sqrt{2}\beta_1 \frac{\lambda_{\min}\left(\mathcal{L}_1\right)}{\sqrt{\lambda_{\max}\left(\mathcal{L}_1\right)}} V_2^{\frac{1}{2}}.
\end{aligned} \tag{11.16}$$

Next, from Lemma 2.3, we have $\hat{q}_i \to q_{N+1}$, $i = 1, 2, \ldots, N$ in finite time with a settling time

$$T_2 = T_1 + \frac{\sqrt{2}\sqrt{\lambda_{\max}\left(\mathcal{L}_1\right)}}{\beta_1 \lambda_{\min}\left(\mathcal{L}_1\right)} V_2^{\frac{1}{2}}\left(T_1\right), \tag{11.17}$$

and q_{N+1} can be used to replace $\hat{q}_i$, when $t \geq T_2$.

Therefore, $\hat{q}_i$ and $\hat{v}_i$, $i = 1, 2, \ldots, N$ can be used to replace the position and velocity of the leader when $t \geq T_2$. □

Remark 11.1 The solutions (11.5) and (11.6) can be investigated according to differential inclusions utilizing nonsmooth analysis [35], since the right-hand sides of (11.5) and (11.6) are discontinuous. Note that the signum functions are locally essentially bounded and measurable. Therefore, the Filippov solutions for the closed-loop systems (11.2) using (11.5) and (11.6) exist and are unique [35].

Remark 11.2 Although $\dot{\hat{q}}_i$ in (11.5) is discontinuous and $\dot{q}_0$ might be continuous, $\dot{\hat{q}}_i$ can still be used to replace $\dot{q}_0$ or $\hat{v}_i$ when $t \geq T_2$. $\dot{\hat{q}}_i$ can be written as

$$\dot{\hat{q}}_i(t) = \hat{v}_i(t) + \upsilon_i(t), \tag{11.18}$$

where $\upsilon_i(t)$ satisfying $\int_{t_1}^{t_2} \upsilon_i(t)\,\mathrm{d}t = 0, \forall t_2 > t_1 > T_2$ with infinitely large frequency of $\upsilon_i(t)$. Since integration-based systems can be seen as ideal low-pass filters, $\upsilon_i(t)$ will be filtered out by these integration-based systems.

Lemma 11.2 *The estimation error $\tilde{q}_i$, $i = 1, 2, \ldots, N$ is bounded for $0 \leq t \leq T_2$.*

Proof Let $\tilde{v}_i = \hat{v}_i - \dot{q}_{N+1}$ and $\tilde{q}_i = \hat{q}_i - q_{N+1}$, $i = 1, \ldots, N$. It follows from (11.6) that

$$|\tilde{v}_i(t)| \leq |\tilde{v}_i(0)| + (\delta_L + \beta_2)\, t \tag{11.19}$$

for $0 \leq t \leq T_2$. From (11.5), we know

$$|\tilde{q}_i(t)| \leq |\tilde{q}_i(0)| + (|\tilde{v}_i(0)| + \beta_1)\, t + \frac{(\delta_L + \beta_2)\, t^2}{2} \tag{11.20}$$

for $0 \leq t \leq T_2$. Thus, $\tilde{q}_i$ is bounded for $0 \leq t \leq T_2$. □

11.3.2 Distributed Coordinated Control with Full-State Constraints

Considering a dynamic leader, the objective is to design a distributed controller to drive all the followers to track the leader while all the closed-loop errors are SGUUB and the FSCs are never violated, i.e., for $\forall i = 1, 2, \ldots, N$, we have $|q_i| \leq |k_{c1i}|$, $|\dot{q}_i| \leq |k_{c2i}|$, $\forall t \geq 0$, where k_{c1i}, $k_{c2i} \in \mathbb{R}^p$ are positive constant vectors. Firts, in the case of $t \geq T_2$, we define the following auxiliary variables

$$z_{1i} = q_i - \hat{q}_i, \tag{11.21}$$

$$z_{2i} = \dot{q}_i - \alpha_{1i}, \tag{11.22}$$

$$\alpha_{1i} = \hat{v}_i - A_i. \tag{11.23}$$

Vector A_i is designed as

$$A_i = \begin{bmatrix} \left(k_{ai,1}^2 - z_{1i,1}^2\right) k_{1i,1} z_{1i,1} \\ \left(k_{ai,2}^2 - z_{1i,2}^2\right) k_{1i,2} z_{1i,2} \\ \cdots \\ \left(k_{ai,p}^2 - z_{1i,p}^2\right) k_{1i,p} z_{1i,p} \end{bmatrix}, \tag{11.24}$$

where $k_{1i,j}$, $j = 1, 2, \ldots, p$ is a positive constant, $k_{ai} = k_{c1i} - X_0 = \left[k_{ai,1}, k_{ai,2}, \ldots, k_{ai,p}\right]^T$, and X_0 is a positive constant vector satisfying $|q_{N+1}| \leq X_0 < k_{c1i}$. In fact, A_i is designed to cancel the coupling term in the derivative of BLF candidate in the later analysis, which is essential to get the uniformly ultimately bounded results in Theorem 11.2.

Differentiating z_{2i} with respect to time, we can get

$$\dot{z}_{2i} = M_i^{-1}(q_i)\left[\tau_i - C_i(q_i, \dot{q}_i)\dot{q}_i - g_i(q_i)\right] - \dot{\alpha}_{1i}. \tag{11.25}$$

In contrast to the general network systems modelled by linear dynamics in most research works, ELSs in this chapter can characterize a broad class of systems, which exhibit inherent nonlinearity. Also, the results for general network systems cannot be directly applied to ELSs due to their the high nonlinearities in $M_i(q_i)$, $C_i\left(q_{i,}\dot{q}_i\right)$ and $g_i(q_i)$. Therefore, these terms cannot by directly utilized in a control law. We next introduce NNs to approximate the model uncertainties f_i as

$$f_i(q_i, \dot{q}_i, \alpha_{1i}, \dot{\alpha}_{1i}) = W_i^{*T}\phi_i(q_i, \dot{q}_i, \alpha_{1i}, \dot{\alpha}_{1i}) + \epsilon_i, \tag{11.26}$$

where $W_i^* \in \mathbb{R}^{l\times p}$ is the ideal constant weight matrix, $l > 1$ is the NN node number, $f_i(q_i, \dot{q}_i, \alpha_{1i}, \dot{\alpha}_{1i}) = -g_i(q_i) - C_i(q_i, \dot{q}_i)\alpha_{1i} - M_i(q_i)\dot{\alpha}_{1i}(t)$ represents the nonlinear uncertainties, and $\phi_i = [\varphi_{i1}, \varphi_{i2}, \ldots, \varphi_{il}]^T \in \mathbb{R}^l$ is the activation function. Within a compact set, the approximation error ϵ_i is bounded, i.e., $||\epsilon_i(Z_i)|| \leq \bar{\epsilon}_i$, where $\bar{\epsilon}_i > 0$ and $Z_i = \left[q_i^T, \dot{q}_i^T, \alpha_{1i}^T, \dot{\alpha}_{1i}^T\right]^T$ [36, 37]. The activation function is chosen as

$$\varphi_{ir}(Z_i) = \exp\left[\frac{-(Z_i - \mu_r)^T(Z_i - \mu_r)}{{\eta_r}^2}\right], \; r = 1, \ldots, l, \tag{11.27}$$

where η_r and μ_r denote the Gaussian function width and the receptive field center, respectively [38].

Remark 11.3 In terms of selecting the parameters of neural networks, there are theoretical guidelines in the literature, e.g., [39], while in this approach, the trial and error method is applied as the neural structure is relatively simple.

Let $\hat{W}_i$ denote the estimate of the ideal weight matrix W_i^* and $\hat{W}_{i,k}$ the kth column vector of the matrix $\hat{W}_i$. We next present the control algorithms as

$$
\begin{aligned}
\tau_i =& -\left(z_{2i}^T\right)^{\dagger} \sum_{k=1}^{p} \left[\frac{k_{1i,k} z_{1i,k}^2}{k_{ai,k}^2 - z_{1i,k}^2} + \frac{z_{2i,k}\left(\ddot{q}_{i,k} - \dot{\alpha}_{i,k}\right)}{k_{bi,k}^2 - z_{2i,k}^2} \right. \\
& \left. + \frac{k_{2i,k} z_{2i,k}^2}{k_{bi,k}^2 - z_{2i,k}^2} + \frac{z_{1i,k} z_{2i,k}}{k_{ai,k}^2 - z_{1i,k}^2} \right] \\
& - \hat{W}_i^T \phi_i\left(Z_i\right) - K_{3i} z_{2i},
\end{aligned} \tag{11.28}
$$

$$
\dot{\hat{W}}_{i,k} = \gamma_i \left(\phi_i\left(Z_i\right) z_{2i,k} - \sigma_i \left| z_{2i,k} \right| \hat{W}_{i,k} \right), \tag{11.29}
$$

where γ_i, σ_i and $k_{2i,j}$, $j = 1, 2, \ldots, p$, are positive constants, K_{3i} is a positive definite matrix, $k_{bi} = k_{c2i} - Y_0 = \left[k_{bi,1}, k_{bi,2}, \ldots, k_{bi,p}\right]^T$, Y_0 is a positive constant vector satisfying $|\dot{q}_{N+1}| \leq Y_0 < k_{c2i}$, and $\left(z_{2i}^T\right)^{\dagger}$ is the Moore-Penrose inverse of z_{2i}^T defined as

$$
z_{2i}^T \left(z_{2i}^T\right)^{\dagger} = \begin{cases} 0, & z_{2i} = [0, 0, \ldots, 0]^T, \\ 1, & \text{otherwise.} \end{cases} \tag{11.30}
$$

Before presenting the main result, we introduce a useful lemma for the adaptive law (11.29).

Lemma 11.3 [40] *For adaptive law (11.29), there exits a compact set*

$$
\Omega_{\omega 1i} = \left\{ \hat{W}_{i,k} \middle| \left\| \hat{W}_{i,k} \right\| \leq \frac{s_i}{\sigma_i} \right\}, \tag{11.31}
$$

where $\hat{W}_{i,k}$ is the kth column vector of the matrix $\hat{W}_i$, s_i is a positive constant which satisfies $s_i \geq ||\phi_i\left(Z_i\right)||$, and each element of $\phi_i\left(Z_i\right)$ is positive , such that $\hat{W}_{i,k}(t) \in \Omega_{\omega 1i}$, $\forall t \geq 0$, provided that $\hat{W}_{i,k}(0) \in \Omega_{\omega 1i}$.

Remark 11.4 Note that the acceleration information is required in the controller (11.28). In practice, in the case without acceleration sensors, this acceleration can be indirectly calculated by numerical differentiation of the velocities [41]. In terms of the algebraic loops induced by the usage of acceleration information, we can analyze the stability with delayed acceleration information [42] to solve the problem of algebraic loops.

Theorem 11.2 *Under Assumptions 11.1 and 11.2, given that the initial conditions of the networked ELSs satisfy $z_{1i}(0) \in \Omega_{0_{z1}} := \{z_{1i} \in R^p : -k_{ai} < z_{1i} < k_{ai}\}$, $z_{2i}(0) \in \Omega_{0_{z2}} := \{z_{2i} \in R^p : -k_{bi} < z_{2i} < k_{bi}\}$, the closed-loop error signals are SGUUB under the control laws (11.28) and (11.29), while the FSCs are never violated, i.e., $|q_i| < k_{c1i}$, $|\dot{q}_i| < k_{c2i}$, $\forall t > 0$. Error signals z_{1i}, z_{2i} and $\tilde{W}_i$ will stay within the following compact sets $\Omega_{z_{1i}}$, $\Omega_{z_{2i}}$ and $\Omega_{\tilde{W}_i}$*

$$\Omega_{z_{1i}} := \left\{ z_{1i} \in \mathbb{R}^{p\times 1} | \left\| z_{1i,k} \right\| \leq \sqrt{k_{ai,k}^2 \left(1 - e^{-D_i}\right)} \right\}, \tag{11.32}$$

$$\Omega_{z_{2i}} := \left\{ z_{2i} \in \mathbb{R}^{p\times 1} | \left\| z_{2i,k} \right\| \leq \sqrt{k_{bi,k}^2 \left(1 - e^{-D_i}\right)} \right\} \cap \left\{ z_{2i} \in \mathbb{R}^{p\times 1} | \left\| z_{2i} \right\| \leq \sqrt{D_i / \lambda_{\min}(M_i)} \right\}, \tag{11.33}$$

$$\Omega_{\tilde{W}_i} := \left\{ \tilde{W}_i \in \mathbb{R}^{l\times p} | \left\| \tilde{W}_i \right\| \leq \sqrt{\gamma_i D_i} \right\}, \tag{11.34}$$

where $\tilde{W}_i$ is defined as $\tilde{W}_i = \hat{W}_i - W_i^*$, and D_i is detailed in (11.54) at the end of the proof.

Proof Consider the BLF candidate

$$V_{1i} = \frac{1}{2} \sum_{k=1}^{p} \log \frac{k_{ai,k}^2}{k_{ai,k}^2 - z_{1i,k}^2}. \tag{11.35}$$

Taking the derivative of V_{1i} yields

$$\dot{V}_{1i} = \sum_{k=1}^{p} \frac{z_{1i,k} \dot{z}_{1i,k}}{k_{ai,k}^2 - z_{1i,k}^2}. \tag{11.36}$$

Substituting (11.21)–(11.24) into (11.36), we have

$$\dot{V}_{1i} = -\sum_{k=1}^{p} k_{1i} z_{1i,k}^2 + \sum_{k=1}^{p} \frac{z_{1i,k} z_{2i,k}}{k_{ai,k}^2 - z_{1i,k}^2}. \tag{11.37}$$

We then consider the BLF candidate as

$$V_{2i} = V_{1i} + \frac{1}{2} z_{2i}^T M_i z_{2i} + \frac{1}{2} \sum_{k=1}^{p} \log \frac{k_{bi,k}^2}{k_{bi,k}^2 - z_{2i,k}^2}. \tag{11.38}$$

The derivative of (11.38) takes the form

$$\dot{V}_{2i} = -\sum_{k=1}^{p} k_{1i} z_{1i,k}^2 + z_{2i}^T M_i \dot{z}_{2i} + \frac{1}{2} z_{2i}^T \dot{M}_i z_{2i} + \sum_{k=1}^{p} \left(\frac{z_{1i,k} z_{2i,k}}{k_{ai,k}^2 - z_{1i,k}^2} + \frac{z_{2i,k} \dot{z}_{2i,k}}{k_{bi,k}^2 - z_{2i,k}^2} \right). \tag{11.39}$$

Consider the Lyapunov function

$$V_{3i} = V_{2i} + \frac{1}{2\gamma_i}\sum_{k=1}^{p} \tilde{W}_{i,k}^T \tilde{W}_{i,k}. \tag{11.40}$$

Taking the derivative of V_{3i} yields

$$\begin{aligned}\dot{V}_{3i} =& -\sum_{k=1}^{p} k_{1i} z_{1i,k}^2 + \frac{1}{\gamma_i}\sum_{k=1}^{p} \tilde{W}_{i,k}^T \dot{\tilde{W}}_{i,k} + z_{2i}^T M_i \dot{z}_{2i} \\ &+ \sum_{k=1}^{p}\left(\frac{z_{1i,k} z_{2i,k}}{k_{ai,k}^2 - z_{1i,k}^2} + \frac{z_{2i,k}\dot{z}_{2i,k}}{k_{bi,k}^2 - z_{2i,k}^2}\right) + \frac{1}{2} z_{2i}^T \dot{M}_i z_{2i}.\end{aligned} \tag{11.41}$$

Substituting (11.25) into (11.41), we can get

$$\begin{aligned}\dot{V}_{3i} =& -\sum_{k=1}^{p} k_{1i} z_{1i,k}^2 + \frac{1}{\gamma_i}\sum_{k=1}^{p} \tilde{W}_{i,k}^T \dot{\tilde{W}}_{i,k} + \sum_{k=1}^{p}\frac{z_{1i,k} z_{2i,k}}{k_{ai,k}^2 - z_{1i,k}^2} \\ &+ z_{2i}^T\left[\tau_i - C_i\left(q_i, \dot{q}_i\right)\alpha_{1i} - g_i\left(q_i\right)\right] - z_{2i}^T M_i \dot{\alpha}_{1i} \\ &+ \sum_{k=1}^{p}\frac{z_{2i,k}\left(\ddot{q}_{i,k} - \dot{\alpha}_{1i,k}\right)}{k_{bi,k}^2 - z_{2i,k}^2} + \frac{1}{2} z_{2i}^T \dot{M}_i z_{2i}.\end{aligned} \tag{11.42}$$

Based on Property 11.2, from (11.28) and (11.29), we can obtain

$$\begin{aligned}\dot{V}_{3i} =& -\sum_{k=1}^{p} k_{1i,k} z_{1i,k}^2 - \sum_{k=1}^{p} \tilde{W}_{i,k}^T \phi_i\left(Z_i\right) z_{2i,k} - z_{2i}^T K_{3i} z_{2i} \\ &+ \sum_{k=1}^{p}\frac{z_{2i,k}\left(\ddot{q}_{i,k} - \dot{\alpha}_{1i,k}\right) - z_{2i}^T\left(z_{2i}^T\right)^{\dagger} z_{2i,k}\left(\ddot{q}_{i,k} - \dot{\alpha}_{1i,k}\right)}{k_{bi,k}^2 - z_{2i,k}^2} \\ &- \sum_{k=1}^{p}\frac{z_{2i}^T\left(z_{2i}^T\right)^{\dagger} k_{1i,k} z_{1i,k}^2}{k_{ai,k}^2 - z_{1i,k}^2} - \sum_{k=1}^{p}\frac{z_{2i}^T\left(z_{2i}^T\right)^{\dagger} k_{2i,k} z_{2i,k}^2}{k_{bi,k}^2 - z_{2i,k}^2} \\ &+ z_{2i}^T\left[\left(W_i^{*T} - \hat{W}_i^T\right)\phi_i\left(Z_i\right) + \epsilon_i\right] \\ &- \sum_{k=1}^{p}\sigma_i\left|z_{2i,k}\right| \tilde{W}_{i,k}^T \hat{W}_{i,k}.\end{aligned} \tag{11.43}$$

When $z_{2i} = [0, 0, \ldots, 0]^T$, $\dot{V}_{3i} = -\sum_{k=1}^{p} k_{1i,k} z_{1i,k}^2$. The asymptotic stability can be proved with Barbalat Lemma in [43]. Otherwise, if $z_{2i} \neq [0, 0, \ldots, 0]^T$, we can get

$$\begin{aligned}\dot{V}_{3i} \le &-\sum_{k=1}^{p} k_{1i,k} z_{1i,k}^2 - z_{2i}^T K_{3i} z_{2i} + \frac{1}{2}\left(z_{2i}^T z_{2i} + \|\bar{\epsilon}_i\|^2\right)\\ &-\sum_{k=1}^{p} \frac{k_{1i,k} z_{1i,k}^2}{k_{ai,k}^2 - z_{1i,k}^2} - \sum_{k=1}^{p} \frac{k_{2i,k} z_{2i,k}^2}{k_{bi,k}^2 - z_{2i,k}^2}\\ &-\sum_{k=1}^{p} \sigma_i \left|z_{2i,k}\right| \tilde{W}_{i,k}^T \hat{W}_{i,k}\\ \le &\sum_{k=1}^{p} \frac{\sigma_i^2}{4}\left(\|W_{i,k}^*\|^4 + \left\|\tilde{W}_{i,k}\right\|^4 - 2\|W_{i,k}^*\|^2\left\|\tilde{W}_{i,k}\right\|^2\right)\\ &-\sum_{k=1}^{p} \frac{k_{1i,k} z_{1i,k}^2}{k_{ai,k}^2 - z_{1i,k}^2} - \sum_{k=1}^{p} \frac{k_{2i,k} z_{2i,k}^2}{k_{bi,k}^2 - z_{2i,k}^2}\\ &- z_{2i}^T \left(K_{3i} - I_p\right) z_{2i} + \frac{1}{2}\|\bar{\epsilon}_i\|^2. \end{aligned} \tag{11.44}$$

According to Lemma 11.3, we have

$$\begin{aligned}\left\|\tilde{W}_{i,k}\right\| = \left\|\hat{W}_{i,k} - W_{i,k}^*\right\| \le \left\|\hat{W}_{i,k}\right\| + \|W_{i,k}^*\|\\ \le \frac{s_i}{\sigma_i} + \|W_{i,k}^*\|. \end{aligned} \tag{11.45}$$

From Lemma 4 in [28], we know

$$-\sum_{k=1}^{p} \frac{k_{1i,k} z_{1i,k}^2}{k_{ai,k}^2 - z_{1i,k}^2} \le -\sum_{k=1}^{p} \log \frac{k_{1i,k} k_{ai,k}^2}{k_{ai,k}^2 - z_{1i,k}^2}, \tag{11.46}$$

$$-\sum_{k=1}^{p} \frac{k_{2i,k} z_{2i,k}^2}{k_{bi,k}^2 - z_{2i,k}^2} \le -\sum_{k=1}^{p} \log \frac{k_{2i,k} k_{bi,k}^2}{k_{bi,k}^2 - z_{2i,k}^2}. \tag{11.47}$$

We thus obtain

$$\dot{V}_{3i} \le -\rho_i V_{3i} + C_i, \tag{11.48}$$

where

$$\begin{aligned}\rho_i =&\min\left(\min\left(\frac{\sigma_i^2\|W_i^*\|^2}{\gamma_i}\right), \min(2k_{1i}),\right.\\&\left.\frac{2\lambda_{\min}(K_{3i}-I_p)}{\lambda_{\max}(M_i)}, \min(2k_{2i})\right),\end{aligned} \tag{11.49}$$

$$\begin{aligned}C_i =&\sum_{k=1}^{p}\frac{\sigma_i^2}{4}\left[\|W_{i,k}^*\|^2+\left(\frac{s_i}{\sigma_i}+\|W_{i,k}^*\|\right)^4\right]\\&+\frac{1}{2}\|\bar{\epsilon}_i\|^2,\end{aligned} \tag{11.50}$$

To guarantee $\rho_i \geq 0$, we choose the control gain K_{3i} as

$$\lambda_{\min}(K_{3i}-I_p) > 0. \tag{11.51}$$

Multiply (11.48) by $e^{-\rho_i t}$ and integrate the both sides yielding

$$\begin{aligned}V_{3i}(t) \leq& V_{3i}(0)\, e^{-\rho_i t}+\frac{C_i}{\rho_i}\left(1-e^{-\rho_i t}\right)\\ \leq& V_{3i}(0)+\frac{C_i}{\rho_i}.\end{aligned} \tag{11.52}$$

Thus, we have

$$\|z_{1i,k}\| \leq \sqrt{k_{ai,k}^2\left(1-e^{-D_i}\right)}, \tag{11.53}$$

where

$$D_i = 2\left(V_{3i}(0)+C_i/\rho_i\right). \tag{11.54}$$

Therefore, z_{1i} remains within $\Omega_{z_{1i}}$. Similarly, we can prove z_{2i} and $\tilde{W}_i$ are SGUUB. According to Theorem 11.1, we have $\hat{q}_i(t) = q_{N+1}(t)$ and $\hat{v}_i(t) = \dot{q}_{N+1}(t)$, $i = 1, 2, \ldots, N$, $t \geq T_2$. Then, given that $|z_{1i}(0)| < k_{ai}$ and $|z_{2i}(0)| < k_{bi}$, using Lemma 1 in [27], we have $|z_{1i}| < k_{ai}$, $|z_{2i}| < k_{bi}$, $\forall t > 0$. From $q_i(t) = z_{1i}(t) + q_{N+1}(t)$, $\dot{q}_i(t) = z_{2i}(t) + \alpha_{i1}(t)$, $|q_{N+1}(t)| \leq X_0$, and $|\alpha_{i1}(t)| \leq Y_0$, we have $|q_i(t)| \leq k_{ai} + X_0 = k_{c1i}$ and $|\dot{q}_i(t)| \leq k_{bi} + Y_0 = k_{c2i}$, which concludes the proof and directly presents that FSCs are guaranteed. □

Lemma 11.4 *Under the control laws (11.28) and (11.29), when $t \leq T_M$, the error signals z_{1i}, z_{2i}, and $\tilde{W}_i$ are also bounded, and the FSCs are never violated.*

Proof According to Lemma 11.2, the estimation error $\tilde{q}_i$ is bounded for $0 \leq t \leq T_2$. Utilizing manipulations similar to (11.19)–(11.20), we can get $\tilde{v}_i$ is bounded for $0 \leq t \leq T_2$. Therefore, according to Theorem 11.2 and (11.33)–(11.34), we know the error signals z_{1i}, z_{2i}, and $\tilde{W}_i$ are bounded when $t \leq T_2$. Then, we can always meet the state constraints by properly choosing the parameters of the distributed estimator (11.5)–(11.6) and the control law (11.28)–(11.29). □

11.4 Simulation

Simulation examples been conducted to illustrate the full power of the proposed algorithm. For convenience, we label the leader as 0 in this simulation. To demonstrate the generality of the algorithm proposed in this chapter for the network systems and the future application of the manipulator payloads on the on-orbit service spacecraft, the simulation analysis in this chapter takes the manipulators in the form of EL systems as an example. Consider 4 manipulators with 2 joints modeled by the EL equation (11.2), where (see (11.55)–(11.57)), in which $q_i = [q_i^{(1)}, q_i^{(2)}]^T$,

$$M_i(q_i) = \begin{bmatrix} \Upsilon_{i1} + \Upsilon_{i2} + 2\Upsilon_{i3}\cos q_i^{(2)} & \Upsilon_{i2} + \Upsilon_{i3}\cos q_i^{(2)} \\ \Upsilon_{i2} + \Upsilon_{i3}\cos q_i^{(2)} & \Upsilon_{i2} \end{bmatrix}, \tag{11.55}$$

$$C_i(q_i, \dot{q}_i) = \begin{bmatrix} -\Upsilon_{i3}\left(\sin q_i^{(2)}\right)\dot{q}_i^{(2)} & -\Upsilon_{i3}\left(\sin q_i^{(2)}\right)\left(\dot{q}_i^{(1)} + \dot{q}_i^{(2)}\right) \\ \Upsilon_{i3}\left(\sin q_i^{(2)}\right)\dot{q}_i^{(1)} & 0 \end{bmatrix}, \tag{11.56}$$

$$g_i(q_i) = \begin{bmatrix} \Upsilon_{i4} g\cos q_i^{(1)} + \Upsilon_{i5} g\cos\left(q_i^{(1)} + q_i^{(2)}\right) \\ \Upsilon_{i5} g\cos\left(q_i^{(1)} + q_i^{(2)}\right) \end{bmatrix}, \tag{11.57}$$

$\Upsilon_{i1} = J_{i1} + m_{i2}l_{i1}^2 + 0.25m_{i1}l_{i1}^2$, $\Upsilon_{i2} = 0.25m_{i2}l_{i2}^2 + J_{i2}$, $\Upsilon_{i3} = 0.5m_{i2}l_{i1}l_{i2}$, $\Upsilon_{i4} = (0.5m_{i1} + m_{i2})l_{i1}$, and $\Upsilon_{i5} = 0.5m_{i2}l_{i2}$ are unknown constants. m_{i1}, m_{i2}, J_{i1} and J_{i2} given in Table 11.1, respectively, denote the masses and the inertia moments of joint 1 and 2 for 4 manipulators. The lengths of link 1 and 2 for four followers are chosen as $l_{i1} = 1$ m and $l_{i2} = 1$ m. $g = 9.8\ \mathrm{m/s^2}$ is the gravity acceleration. The communication topology is given in Fig. 11.1.

Table 11.1 The parameters of the 2-link manipulators

Parameter	Follower 1	Follower 2	Follower 3	Follower 4
m_{i1}/kg	0.95	1.06	1.02	1.04
m_{i2}/kg	0.98	1.05	1.15	1.08
$J_{i1}/\mathrm{kg\cdot m^2}$	0.23	0.18	0.19	0.22
$J_{i2}/\mathrm{kg\cdot m^2}$	0.24	0.29	0.18	0.25

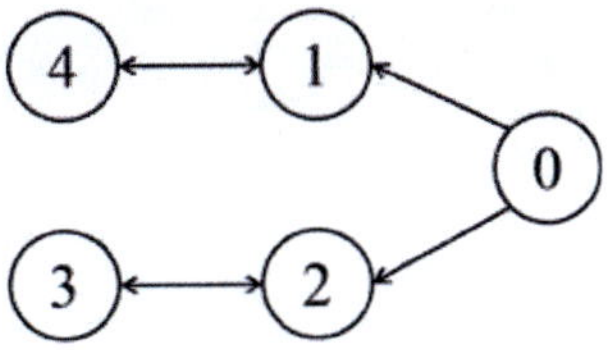

Fig. 11.1 The communication topology between a leader and four followers

The initial conditions and the leader trajectory are given as

$$q_0^{(1)} = 0.6 \sin(0.3t) \text{ rad},$$
$$q_0^{(2)} = 0.4 \sin(0.4t) \text{ rad},$$
$$q_i^{(1)}(0) = q_i^{(2)}(0) = [0, 0]^T \text{ rad/s},$$
$$\dot{q}_i^{(1)}(0) = \dot{q}_i^{(2)}(0) = [0, 0]^T \text{ rad/s},$$

where $t \in [0, t_f]$, and $t_f = 100$ s.

Here, we consider two different cases. First, considering a tracking case with a dynamic leader, we evaluate the proposed control algorithms (11.28) and (11.29) for networked ELSs with FSCs. Second, we apply our control strategy to a special case of synchronization with the leader state $q_0(t) = 0, \ \forall t \geq 0$.

For the proposed control strategy (11.28) and (11.29), to guarantee the FSCs $|q_i| \leq k_{c1i} = [0.85, \ 0.65]^T$, and $|\dot{q}_i| \leq k_{c2i} = [0.43, \ 0.41]^T$, $i = 1, 2, \ldots, N$, we select $X_0 = [0.6, \ 0.4]^T$ and $Y_0 = [0.18, \ 0.16]^T$, which satisfy $-X_0 \leq q_0 \leq X_0$ and $-Y_0 \leq \dot{q}_0 \leq Y_0$. Hence, the constraints of z_{1i} and z_{2i} are $k_{ai} = k_{c1i} - X_0 = [0.25, \ 0.25]^T$ and $k_{bi} = k_{c2i} - Y_0 = [0.25, \ 0.25]^T$, respectively. Control parameters are chosen as $K_{1i} = \text{diag}(k_{1i,1}, k_{1i,2}) = \text{diag}(20, 20)$, $K_{2i} = \text{diag}(k_{2i,1}, k_{2i,2}) = \text{diag}(50, 50)$, and $K_{3i} = \text{diag}(100, 100)$, which satisfies the stable condition in (11.51). For the NN technique, we choose 256 NN nodes for every $\phi_i(Z_i)$, where the receptive field center is set in an area of $[-1, 1]^8$. Variance of centers is given as $\eta_r = 10$. Let $\gamma_i = 4$, $\sigma_i = 5$, and the initial weights $\hat{W}_i = 0$.

In Fig. 11.2a and b, the tracking performances are given, where the angle constraints $|q_i| \leq k_{c1i} = [0.85, \ 0.65]^T$ are never violated. Figures 11.3a and b demonstrate that tracking errors for the followers are constrained by $z_{1i} \leq k_{ai}$, $i = 1, 2, \ldots, N$. An interesting thing is that the errors are reducing as time goes by, which will be explained later. Figures 11.4 and 11.5 show the angular velocities and the angular velocity errors of four followers, respectively. From Figs. 11.5, 11.6a and b, it is found that the errors of angular velocity are constrained by $z_{2i} \leq k_{bi}$, $i = 1, 2, \ldots, N$.

Fig. 11.2 The differences between angles of four followers and the leader

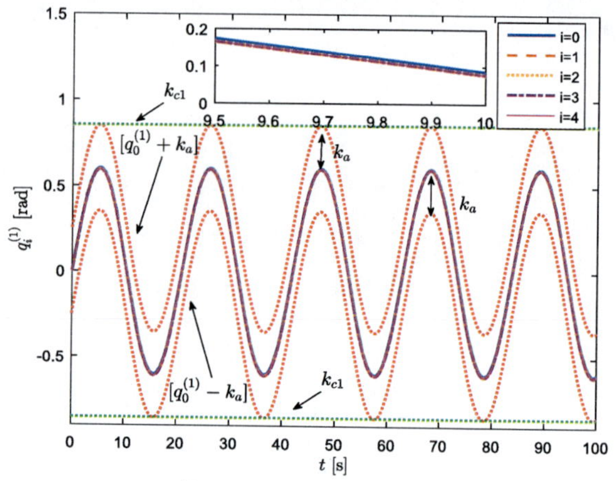

(a) $q_i^{(1)}, i = 0, 1, \ldots, 4$, for the first joint.

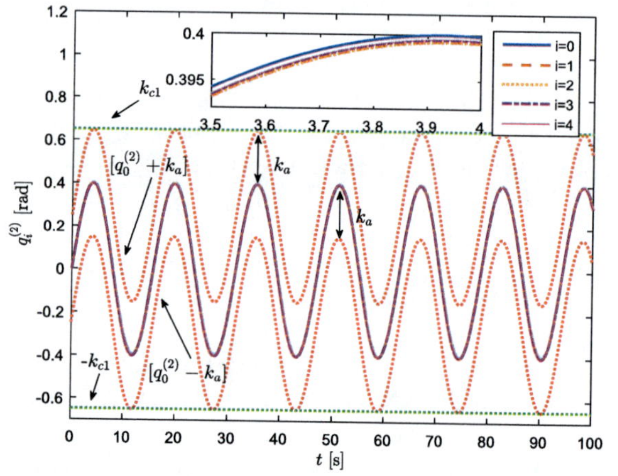

(b) $q_i^{(2)}, i = 0, 1, \ldots, 4$, for the second joint.

From Fig. 11.6, we know the adaptive NNs can approximate the nonlinear uncertainties [44]. In particular, Fig. 11.3a and b show that the tracking errors reduces along with the updating of the adaptation weights, which is the reason why errors are smaller as time pass. A proper increase of the adaptive gain γ_i can yield a better system performance. The control inputs are shown in Fig. 11.7a and b, which is acceptable to practical robotic manipulators. Besides, the errors between the leader states and their estimations obtained by the DFSEs are shown in Figs. 11.8 and 11.9, which imply that the proposed DFSEs yield accurate estimations in finite time.

Fig. 11.3 The angle tracking error of each follower

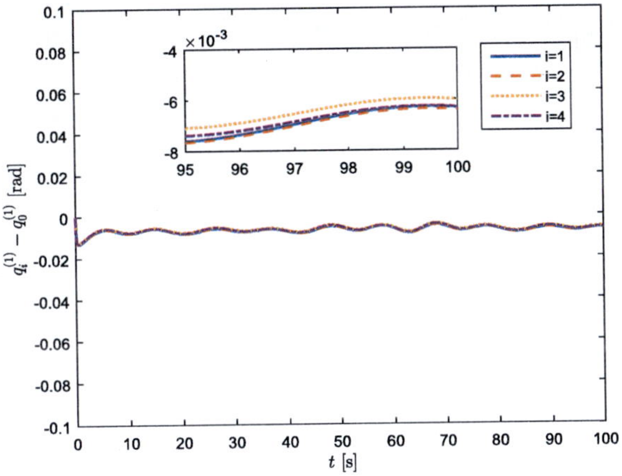

(a) $q_i^{(1)} - q_0^{(1)}, i = 1, 2, \ldots, 4$ for the first joint.

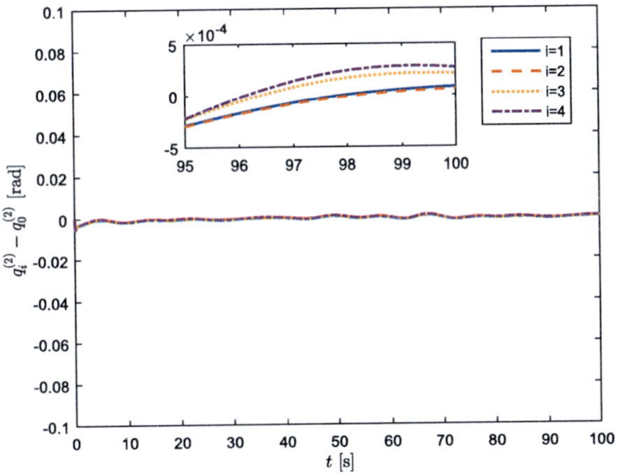

(b) $q_i^{(2)} - q_0^{(2)}, i = 1, 2, \ldots, 4$ for the second joint.

We next apply control strategies (11.28) and (11.29) to the distributed synchronization problem (with the leader state $q_0(t) = 0, \ \forall t \geq 0$) of four networked manipulators with two revolute joints. The initial positions of four manipulators are given as

$$q_1(0) = [0.60, 0.40]^T, \ q_2(0) = [0.40, 0.20]^T,$$
$$q_3(0) = [-0.30, -0.50]^T, \ q_4(0) = [-0.60, -0.35]^T,$$
$$\dot{q}_i(0) = [0, 0]^T, \ i = 1, \ldots, 4.$$

Fig. 11.4 The velocity tracking performance of each follower

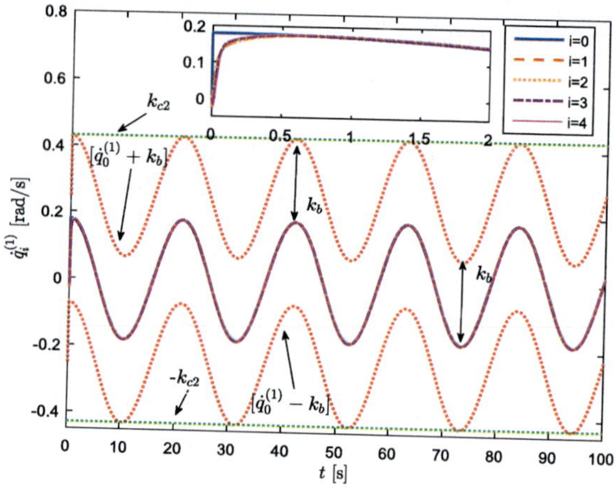

(a) $\dot{q}_i^{(1)}, i = 0, 1, \ldots, 4$, for the first joint.

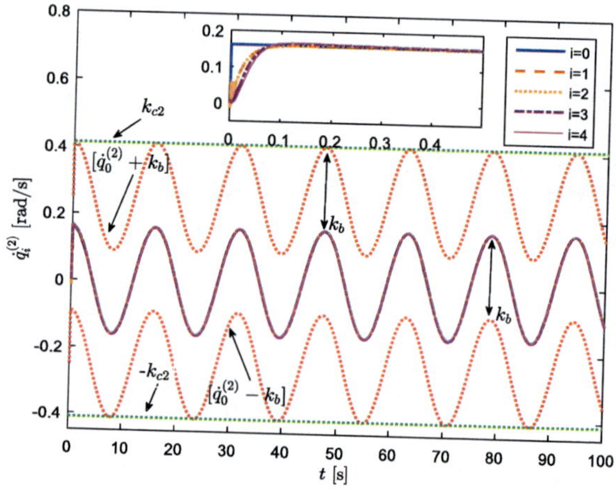

(b) $\dot{q}_i^{(2)}, i = 0, 1, \ldots, 4$, for the second joint.

And the constraint parameters are selected as $k_{ai} = [1,\ 1]^T$ and $k_{bi} = [1,\ 1]^T$, respectively, while other parameters of the controller and the adaptive law are chosen the same as those in the previous coordinated tracking case. Figure 11.10a and b are given to present the movements of the four followers, and it can be observed that all followers approach a common value within 0.3 s. Norms of the adaptation weights in this special case are given in Fig. 11.11.

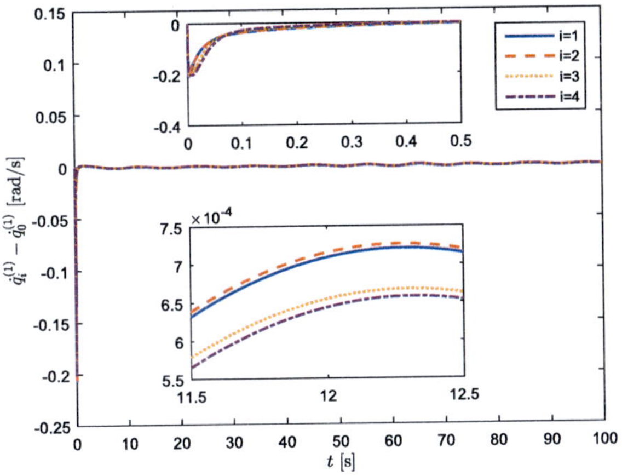

(a) $\dot{q}_i^{(1)} - \dot{q}_0^{(1)}, i = 1, \ldots, 4$, for the first joint.

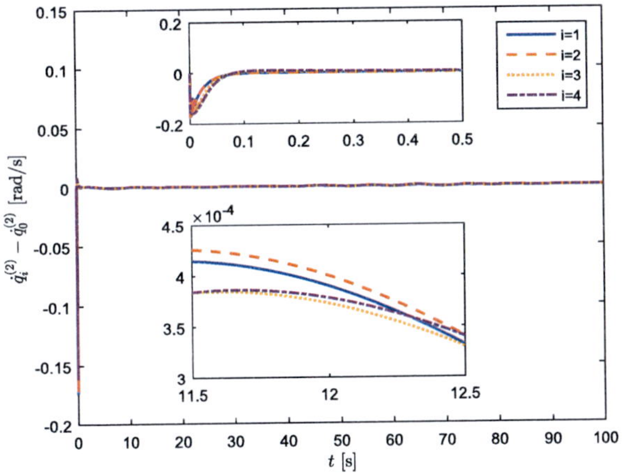

(b) $\dot{q}_i^{(2)} - \dot{q}_0^{(2)}, i = 1, \ldots, 4$, for the second joint.

Fig. 11.5 The angular velocity error of each follower

11.5 Conclusion

In this chapter, for networked ELSs with FSCs and model uncertainties, the coordinated tracking control algorithm has been proposed. A distributed estimator has been introduced to obtain the information of a dynamic target. A BLF based controller and adaptive law are properly designed to provide performance guaranteed and avoid any violation of the FSCs. To approximate uncertainties of ELSs, the NNs

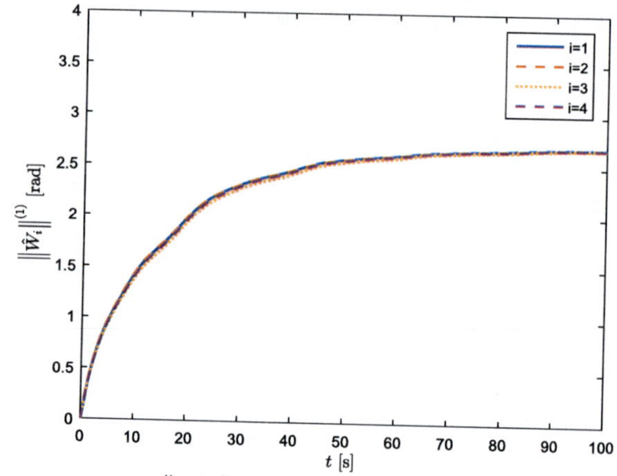

(a) $\left\|\hat{W}_i^{(1)}\right\|$, $i = 1, \ldots, 4$, for the first joint.

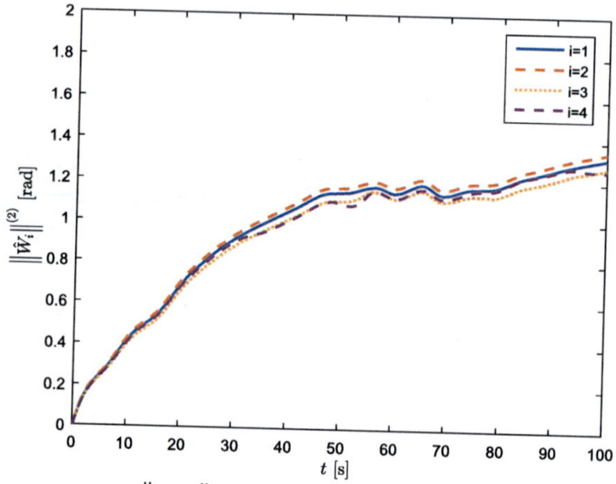

(b) $\left\|\hat{W}_i^{(2)}\right\|$, $i = 1, \ldots, 4$, for the second joint.

Fig. 11.6 The norms of the adaptation weights

have been integrated into the proposed control law. With proper control parameters, the proposed approaches can drive all followers to track a dynamic leader while guaranteeing the FSCs. Some practical issues, such as communication constraints and delays, will be considered in future work.

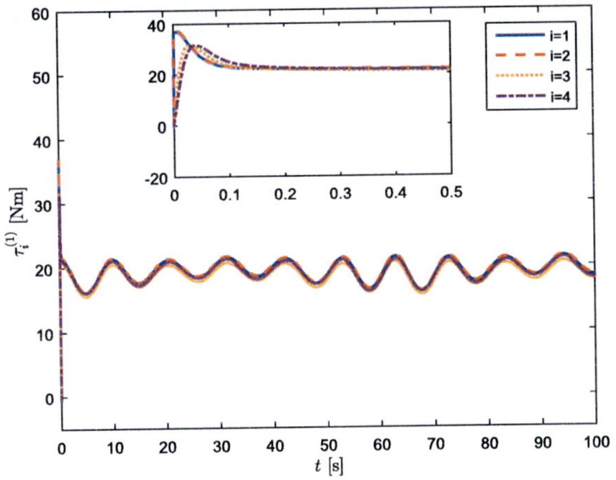

(a) $\tau_i^{(1)}, i = 1, \ldots, 4$, for the first joint.

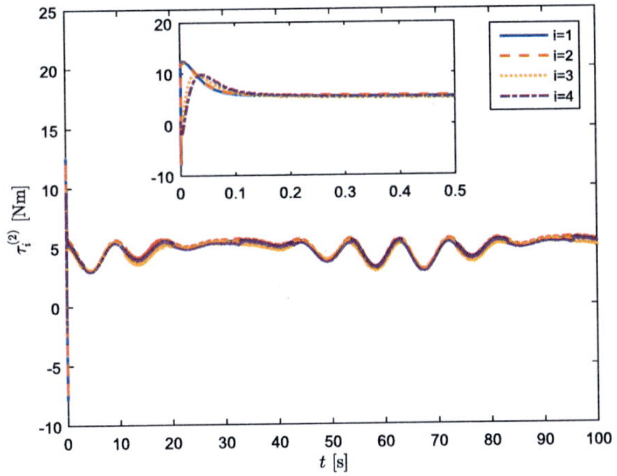

(b) $\tau_i^{(2)}, i = 1, \ldots, 4$, for the second joint.

Fig. 11.7 The control inputs of four followers

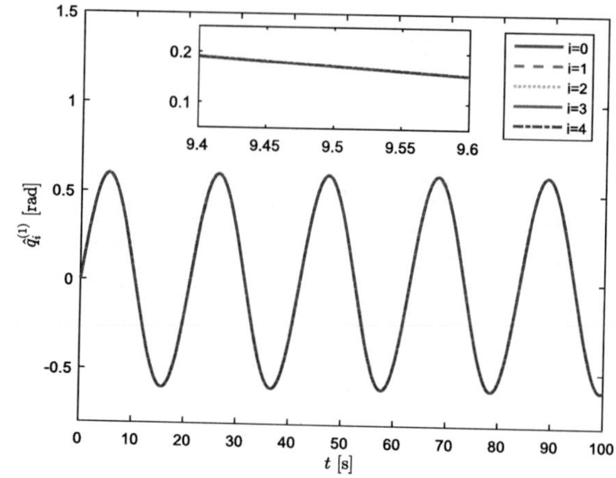

(a) $\hat{x}_i^{(1)}, i = 1, \ldots, 4$, for the first joint.

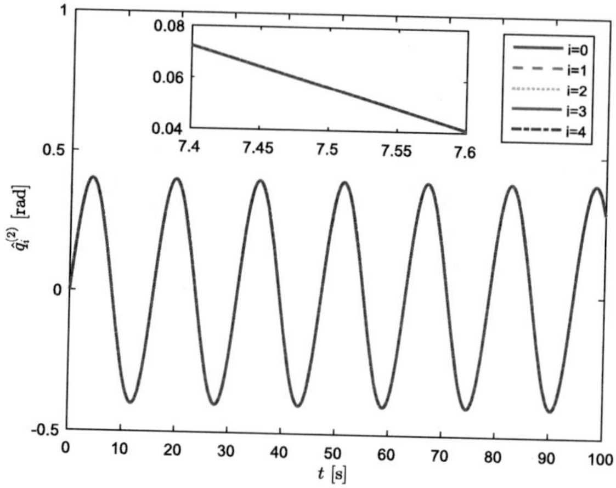

(b) $\hat{x}_i^{(2)}, i = 1, \ldots, 4$, for the second joint.

Fig. 11.8 The differences between the leader angle and the estimations by (11.5)

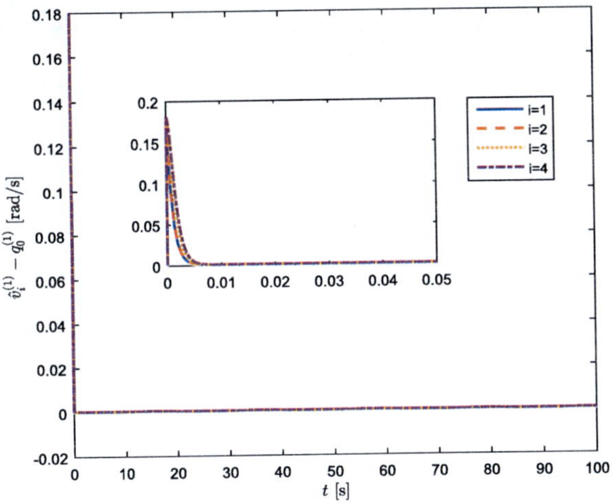

(a) $\hat{v}_i^{(1)}, i = 1, \ldots, 4$, for the first joint.

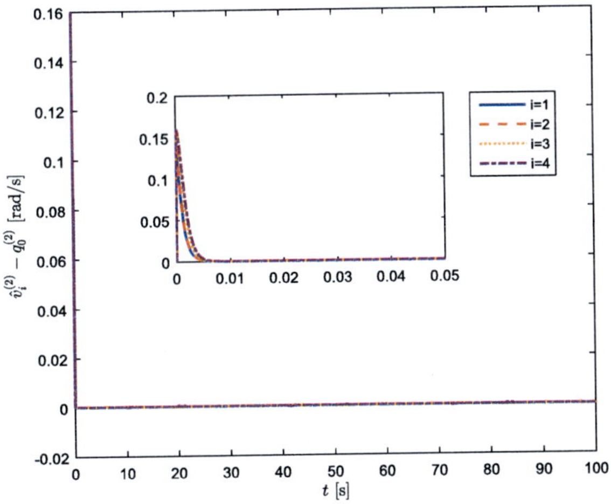

(b) $\hat{v}_i^{(2)}, i = 1, \ldots, 4$, for the second joint.

Fig. 11.9 The angular velocity estimation error of each follower by (11.6)

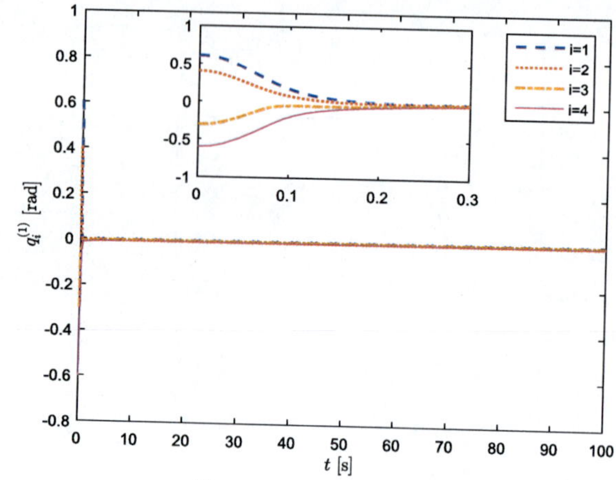

(a) $q_i^{(1)}, i = 1, \ldots, 4$, for the first joint.

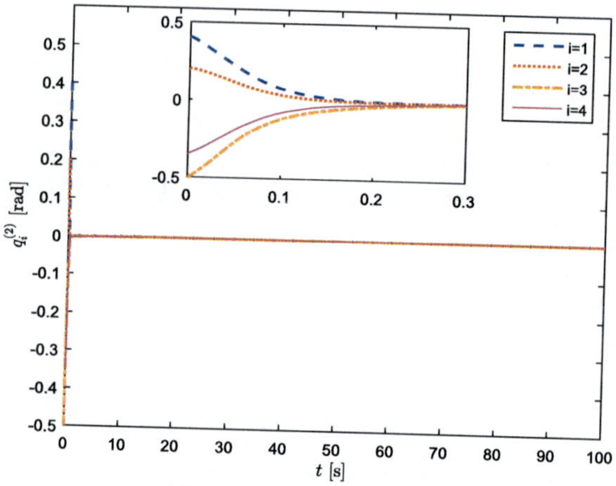

(b) $q_i^{(2)}, i = 1, \ldots, 4$, for the second joint.

Fig. 11.10 The joint angles of four manipulators in the special case with $q_0(t) = 0$

Fig. 11.11 The norms of the adaptation weights in the special case with $q_0(t) = 0$

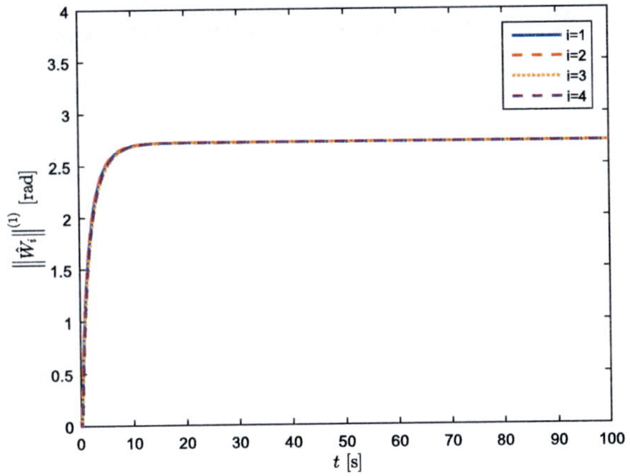

(a) $\left\|\hat{W}_i^{(1)}\right\|, i = 1, \ldots, 4$, for the first joint.

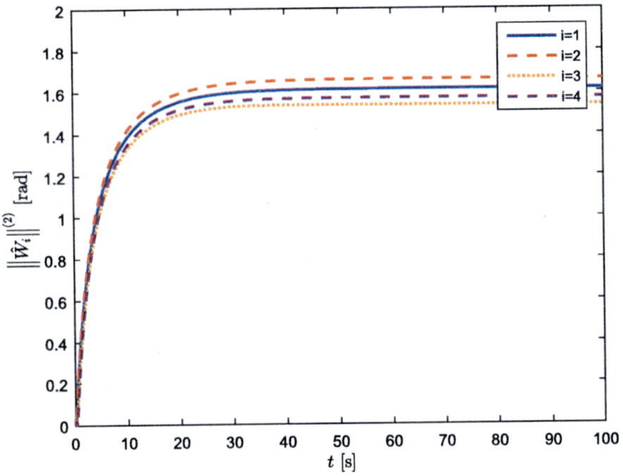

(b) $\left\|\hat{W}_i^{(2)}\right\|, i = 1, \ldots, 4$, for the second joint.

References

1. Cao, Y., Ren, W., Egerstedt, M.: Distributed containment control with multiple stationary or dynamic leaders in fixed and switching directed networks. Automatica **48**(8), 1586–1597 (2012)
2. Peng, K., Yang, Y.: Leader-following consensus problem with a varying-velocity leader and time-varying delays. Physica A Statistical Mechanics & Its Applications **388**(388), 193–208 (2009)
3. Hong, Y., Chen, G., Bushnell, L.: Distributed observers design for leader-following control of multi-agent networks. Automatica **44**(3), 846–850 (2008)

4. Meng, Z., Yang, T., Dimarogonas, D.V., Johansson, K.H.: Coordinated output regulation of heterogeneous linear systems under switching topologies. Automatica **53**, 362–368 (2015). https://doi.org/10.1016/j.automatica.2015.01.009
5. Cicek, E., Dasdemir, J.: Output feedback synchronization of multiple robot systems under parametric uncertainties. Transactions of the Institute of Measurement and Control **39**(3), 277–287 (2017)
6. Cruz-Zavala, E., Nuño, E., Moreno, J.A.: Leaderless and Leader-Follower Consensus of Euler-Lagrange Agents: Finite-Time Convergence Without Velocity Measurements. In: 2018 IEEE Conference on Decision and Control (CDC), pp. 6139–6144. IEEE (2018)
7. Zhang, B., Jia, Y., Du, J.: Adaptive synchronization control of networked robot systems without velocity measurements. International Journal of Robust and Nonlinear Control **28**(11), 3606–3622 (2018)
8. Nuno, E.: Consensus of Euler-Lagrange systems using only position measurements. IEEE Transactions on Control of Network Systems **5**(1), 489–498 (2016)
9. Xu, T., Duan, Z., Sun, Z.: Event-based distributed robust synchronization control for multiple Euler-Lagrange systems without relative velocity measurements. International Journal of Robust and Nonlinear Control **29**(11), 3684–3700 (2019)
10. Nuño, E., Ortega, R.: Achieving consensus of Euler–Lagrange agents with interconnecting delays and without velocity measurements via passivity-based control. IEEE Transactions on Control Systems Technology **26**(1), 222–232 (2017)
11. Chung, S.J., Slotine, J.J.E.: Cooperative robot control and concurrent synchronization of Lagrangian systems. IEEE Transactions on Robotics **25**(3), 686–700 (2009)
12. Meng, Z., Ren, W., You, Z.: Distributed finite-time attitude containment control for multiple rigid bodies. Automatica **46**(12), 2092–2099 (2010). https://doi.org/10.1016/j.automatica.2010.09.005
13. Dong, W.: On consensus algorithms of multiple uncertain mechanical systems with a reference trajectory. Automatica **47**(9), 2023–2028 (2011)
14. Mei, J., Ren, W., Ma, G.: Distributed coordinated tracking with a dynamic leader for multiple Euler-Lagrange systems. IEEE Transactions on Automatic Control **56**(6), 1415–1421 (2011)
15. Cheah, C.C., Hou, S.P., Slotine, J.J.E.: Region-based shape control for a swarm of robots. Automatica **45**(10), 2406–2411 (2009)
16. Meng, Z., Dimarogonas, D.V., Johansson, K.H.: Leader–follower coordinated tracking of multiple heterogeneous lagrange systems using continuous control. IEEE Transactions on Robotics **30**(3), 739–745 (2013)
17. McInnes, C.R.: Large angle slew maneuvers with autonomous sun vector avoidance. Journal of guidance, control, and dynamics **17**(4), 875–877 (1994)
18. Shen, Q., Yue, C., Goh, C.H.: Velocity-Free Attitude Reorientation of a Flexible Spacecraft with Attitude Constraints. Journal of Guidance Control, and Dynamics **40**(5), 1–7 (2017). 10.2514/1.G002129
19. Chen, M., Ge, S.S., Ren, B.: Adaptive tracking control of uncertain MIMO nonlinear systems with input constraints. Automatica **47**(3), 452–465 (2011). https://doi.org/10.1016/j.automatica.2011.01.025
20. Liu, Z.X., Chen, Z.Q.: Discarded consensus of network of agents with state constraint. IEEE Transactions on Automatic Control **57**(11), 2869–2872 (2012). https://doi.org/10.1109/TAC.2012.2190199
21. Shahvali, M., Askari, J.: Cooperative adaptive neural partial tracking errors constrained control for nonlinear multi-agent systems. International Journal of Adaptive Control & Signal Processing **30**(7), 1019–1042 (2016)
22. Conte, C., Zeilinger, M.N., Morari, M., Jones, C.N.: Cooperative distributed tracking MPC for constrained linear systems: Theory and synthesis. Proceedings of the IEEE Conference on Decision and Control pp. 3812–3817 (2013). https://doi.org/10.1109/CDC.2013.6760471
23. Lin, P., Ren, W.: Constrained consensus in unbalanced networks with communication delays. IEEE Transactions on Automatic Control **59**(3), 775–781 (2014). https://doi.org/10.1109/TAC.2013.2275671

24. Nguyen, D.H., Narikiyo, T., Kawanishi, M.: On Consensus of Nonlinear Multi-Agent Systems with Output Constraints. IFAC-PapersOnLine **49**(22), 103–108 (2016). https://doi.org/10.1016/j.ifacol.2016.10.380
25. Meng, W., Yang, Q., Pan, D., Zheng, H., Wang, G., Sun, Y.: NN-based asymptotic tracking control for a class of strict-feedback uncertain nonlinear systems with output constraints. Proceedings of the IEEE Conference on Decision and Control pp. 5410–5415 (2012). https://doi.org/10.1109/CDC.2012.6426120
26. Shahbazi, B., Malekzadeh, M., Koofigar, H.R.: Robust constrained control of spacecraft formation flying. IEEE Transactions on Aerospace and Electronic Systems **9251**(c), 1–1 (2017). https://doi.org/10.1109/TAES.2017.2704160
27. Tee, K.P., Ge, S.S., Tay, E.H.: Automatica Barrier Lyapunov Functions for the control of output-constrained. Automatica **45**(4), 918–927 (2009). https://doi.org/10.1016/j.automatica.2008.11.017
28. He, W., David, A.O., Yin, Z., Sun, C.: Neural Network Control of a Robotic Manipulator With Input Deadzone and Output Constraint. IEEE Transactions on Systems, Man, and Cybernetics: Systems **46**(6), 759–770 (2016). https://doi.org/10.1109/TSMC.2015.2466194
29. Mei, J., Ren, W., Ma, G.: Distributed containment control for Lagrangian networks with parametric uncertainties under a directed graph. Automatica **48**(4), 653–659 (2012). https://doi.org/10.1016/j.automatica.2012.01.020
30. Ghapani, S., Mei, J., Ren, W., Song, Y.: Fully distributed flocking with a moving leader for Lagrange networks with parametric uncertainties. Automatica **67**, 67–76 (2016). https://doi.org/10.1016/j.automatica.2016.01.004
31. Mehrabian, A.R., Khorasani, K.: Constrained distributed cooperative synchronization and reconfigurable control of heterogeneous networked Euler-Lagrange multi-agent systems. Information Sciences **370-371**(5), 578–597 (2016). https://doi.org/10.1016/j.ins.2015.09.032
32. Ge, S.S., Lee, T.H., Harris, C.J.: Adaptive Neural Network Control of Robotic Manipulators. Proceedings of SPIE - The International Society for Optical Engineering **4390**, 236–242 (2001)
33. Li, D., Ma, G., He, W., Zhang, W., Li, C., Ge, S.S.: Distributed coordinated tracking control of multiple Euler-Lagrange systems by state and output feedback. IET Control Theory & Applications **11**(14), 2213–2221 (2017)
34. Horn, R., Johnson, C.: Matrix analysis. Cambridge University Press, Cambridge UK (1990)
35. Filippov, A.F.: Differential equations with discontinuous right-hand side. Matematicheskii sbornik **93**(1), 99–128 (1960)
36. Kubat, M.: Neural networks: a comprehensive foundation by Simon Haykin, Macmillan, 1994, ISBN 0-02-352781-7. Knowledge Engineering Review **13**(4), 409–412 (1999)
37. Xu, B., Sun, F.: Composite intelligent learning control of strict-feedback systems with disturbance. IEEE Transactions on Cybernetics **48**(2), 730–741 (2017)
38. Ge, S.S., Wang, C.: Adaptive neural control of uncertain mimo nonlinear systems. IEEE Transactions on Neural Networks **15**(3), 674–692 (2004)
39. Wu, Y., Wang, H., Zhang, B., Du, K.L.: Using radial basis function networks for function approximation and classification. ISRN Applied Mathematics **2012** (2012)
40. Huang, S.N., Tan, K.K., Lee, T.H.: Further results on adaptive control for a class of nonlinear systems using neural networks. IEEE Transactions on Neural Networks **14**(3), 719–722 (2003)
41. Chen, G., Lewis, F.L.: Distributed adaptive controller design for the unknown networked Lagrangian systems. In: IEEE Conference on Decision and Control (CDC), pp. 6698–6703 (2010). https://doi.org/10.1109/CDC.2010.5717728
42. Liu, S., Xie, L., Lewis, F.L.: Synchronization of multi-agent systems with delayed control input information from neighbors. Automatica **47**(10), 2152–2164 (2011)
43. Slotine, J.J.E., Li, W., Others: Applied nonlinear control. prentice-Hall Englewood Cliffs, NJ (1991)
44. Liu, Z., Chen, C., Zhang, Y., Chen, C.L.P.: Adaptive Neural Control for Dual-Arm Coordination of Humanoid Robot With Unknown Nonlinearities in Output Mechanism. IEEE Transactions on Cybernetics **45**(3), 507–518 (2015). https://doi.org/10.1109/TCYB.2014.2329931

Chapter 12
Distributed Coordinated Control for Networked Spacecraft with Output Constraints

Abstract This chapter addresses the distributed attitude coordinated tracking problem for multiple constrained spacecraft by state and output feedback. Since only parts of the followers can acquire the states of the dynamic leader, a distributed finite-time sliding-mode estimator is firstly proposed to reconstruct the information of the leader's attitude and angular velocity accurately. Then, to guarantee the attitude constraints of spacecraft systems, we use the integral barrier Lyapunov function (BLF) technique to propose a distributed state feedback attitude tracking control algorithm. Next, in the case without angular velocity sensors, we design a distributed attitude tracking control law via output feedback using high-gain observers. In the two proposed control schemes, the adaptive neural networks (NNs) are introduced to tackle the model uncertainties. The uniform ultimate boundedness of all the state errors can be guaranteed by Lyapunov stability theory. Finally, numerical results are given to show the feasibility of the proposed schemes.

12.1 Introduction

Distributed attitude coordinated control for multiple spacecraft has been an active area due to its extensive utilizations, such as on-orbit services and deep space explorations with stronger robustness, less communication requirement, and higher efficiency.

Currently, approaches to attitude coordination control mainly focus on the consensus problem and the tracking problem. As for the consensus problem, all the spacecraft attitudes reach a common value by updating the information from neighbors [1, 2]. With a moving leader (or a prescribed trajectory), the coordinated attitude tracking problems have also been intensively studied for multiple spacecraft. In [3], the attitude coordination problem is investigated for multiple rigid bodies. A coordinated estimator is introduced to estimate the acceleration and the velocity of the desired signal, where only relative attitude and relative angular velocity information are used via information exchanges. The authors in [4] study the adaptive attitude tracking control problem, and the approach is robust to disturbances. In [5], the finite-

Y. Gao and D. Li, *Spacecraft Maneuver with Performance Guaranteed*,
https://doi.org/10.1007/978-981-99-4653-2_12

time attitude coordination problem is considered for multiple spacecraft, while no singularity problem exists in the proposed algorithm. A tracking algorithm is studied with time-varying delays in [6], which can also be applied to the formation problem. The authors in [7] address a finite-time decentralized control approach of multiple autonomous vehicles, where both first-order and second-order estimators are introduced. As a main interest, the attitude coordinated tracking case is more challenging because it can fully command the behavior of a group of spacecraft systems, which can be regarded as an extension of the leaderless consensus problem. In [8, 9], the study on the orbit and attitude control of spacecraft formation flying is addressed.

It should be noted that the control schemes in references mentioned above need the angular velocity measurements to be implemented. In practice, angular velocity sensors might not be equipped on the spacecraft to save cost and space. The coordination problem is investigated in [10] without velocity measurements. The authors in [11] study the distributed control problem for multi-agent systems to achieve consensus by output feedback. Under a condition that all followers can only receive information from neighbors via an undirected topology, the nonsmooth control algorithms are designed in [12]. Unfortunately, only linear systems are considered in [10–12]. The multi-agent coordination problem for multiple Euler-Lagrange systems is studied in [13] without velocity measurements. Under a ring communication topology, an attitude control law for multiple spacecraft is presented without using velocity measurements in [1], while the final angular velocity is constrained to zero. In [14], based on a novel velocity observer, the distributed attitude control problem for multiple spacecraft is considered via output feedback. Unfortunately, the assumption that each follower has access to the leader's information is required in [14]. In [15], the coordinated tracking control of multiple Euler-lagrange systems is studied using output feedback. Unfortunately, the state constraints are not considered in the proposed controller [15], which may result in system failure in practice.

For spacecraft, attitude constraints require the attitude of the spacecraft to stay within a desired range. It is essential for a spacecraft to guarantee attitude constraints in missions when the spacecraft is required to point along a prescribed direction [16]. Any violation of attitude constraints will lead to severe performance degradation and damage of payloads, because some equipped payloads (such as interferometers or infrared telescopes) are sensitive which should be kept a proper distance away from any bright energy sources or undesired celestial objects [17]. However, most of the papers mentioned above considered the attitude control without constraints, which may lead to performance degradation in practical missions. Presently, it is a challenging issue to design an effective controller to keep the spacecraft attitude within the constraints. Some approaches utilize the path planning schemes to tackle the constrained attitude tracking problems [18–20]. In the presence of bright light sources, the attitude path planning is studied while a communication link between the spacecraft and a ground station is guaranteed in [18]. In [19], by employing an icosahedron-based pixelization subroutine, the unit celestial sphere is converted into a graph, and the A^* path-finding method is utilized to yield a path with minimum cost. In [20], the time-optimal attitude tracking problem is investigated by transforming the task into a path planning. However, it is hard for the above mentioned path-planning-based schemes to be implemented in a more complex scenario, as these schemes are often realized by computationally demanding search methods which can hardly guarantee the system stability and the computational tractability [21]. Based on a logarithmic potential function, two constrained attitude control algorithms are proposed based on the backstepping method in [21].

In this chapter, a barrier Lyapunov function is applied to tackle the control with attitude constraints, where a barrier Lyapunov function (BLF) approaches to infinity as long as the states are reaching the constraints [22, 23]. We consider the distributed attitude tracking problem for multiple spacecraft with attitude constraints and model uncertainties via state and output feedback, of which, to the best of our knowledge, no results have been published hitherto. This chapter contains the following contributions:

(1) By employing the integral BLF technique in our distributed attitude control algorithms, the attitude constraints are never violated. As a granted property of the integral BLF, the constraints are incorporated with the tracking errors, while the typical logarithmic BLF in [22–24] composed of only errors contains some degree of conservatism. By using the Lyapunov stability theory, the closed-loop systems are uniformly ultimately bounded (UUB) and the tracking errors can converge to the origin within a small scale.

(2) Compared with the control algorithms [18–25] using full-state information, the proposed distributed attitude control algorithm is feasible using output feedback when the angular velocity measurements are no longer accessible in practice.

(3) A distributed finite-time sliding-mode estimator (DFSE) algorithm is designed to access accurate estimations of the leader's attitude and angular velocity in finite time. In addition, the control law proposed in this chapter is fully distributed, while the Laplacian matrix or some other global information is required in [26].

(4) Two proposed control algorithms are combined with the neural network (NN) technique, which will compensate for the unknown model uncertainties. In comparison, the references above consider the model-based control [27, 28], or Euler-Lagrange systems with the parametric uncertainties, which require the 'linearity-in-parameter' assumption [13, 29].

The rest of this chapter is organized as follows. In Sect. 12.2, the graph theory, the spacecraft attitude kinematics and dynamics, and other technical Lemmas are given. Section 12.3 covers the main results of this study, where the distributed attitude coordinated tracking control algorithms are proposed combined with the NNs and the integral BLF technique by state and output feedback, respectively. In Sect. 12.4, simulations are given, and a conclusion is drawn in Sect. 12.5.

Notations: Let $I_n \in \mathbb{R}^{n\times n}$ be the n-dimensional identity matrix. Let $\|\bullet\|$ denote the Euclidean norm of a vector. Let $\lambda_{\max}(\bullet)$ and $\lambda_{\min}(\bullet)$ represent, respectively, the maximal and minimum eigenvalues of the matrix $\bullet$. Let $\mathrm{diag}(z_1, \ldots, z_n) \in \mathbb{R}^{n\times n}$ denote a diagonal matrix, where $z_1, \ldots, z_n$ are the diagonal entries. Let $\mathrm{sgn}(\bullet)$ stand for the signum function.

Before further proceeding, a list of abbreviations is given in Table 12.1.

Table 12.1 Abbreviation list

Abbreviations	Meanings
BLF	Barrier Lyapunov Function
NN	Neural Network
UUB	Uniformly Ultimately Bounded
DFSE	Distributed Finite-Time Sliding-Mode Estimator

12.2 Preliminaries and Problem Description

We utilize a communication graph $\mathcal{G} = (V, E)$ to characterize the communication topology among the spacecraft, where $E \subseteq V \times V$ and $V = \{1, \ldots, k\}$ denote the arc set and the vertex set, respectively. An arc $(i, j) \in E$ means that the spacecraft j can access the information from spacecraft i. A path from the vertex i to the vertex j is a sequence of arcs of the form $(i_1, i_2), (i_2, i_3), \ldots$. The adjacency matrix $\mathcal{A} = [a_{ij}] \in \mathbb{R}^{k\times k}$ is a nonnegative matrix defined as $a_{ij} > 0$ if $(j, i) \in E$, and $a_{ij} = 0$ otherwise. We define $a_{ii} = 0$, and the Laplacian matrix $\mathcal{A}$ is defined as $\mathcal{L} = [l_{ij}] \in \mathbb{R}^{k\times k}$, where $l_{ii} = \sum_{j=1, j\neq i}^{k} a_{ij}$ and $l_{ij} = -a_{ij}, i \neq j$.

We consider the attitude coordination problem with a dynamic leader labeled as 0, and N followers labeled as $1, \ldots, N$, where the communication topology between the followers is bidirectional. Let $\mathcal{G} \in \mathbb{R}^{(N+1)\times(N+1)}$ be the graph composed of the leader and N followers, while the corresponding Laplacian matrix $\mathcal{L} \in \mathbb{R}^{(N+1)\times(N+1)}$ is given as

$$\mathcal{L} = \begin{bmatrix} 0 & 0_{1\times N} \\ \mathcal{L}_2 & \mathcal{L}_1 \end{bmatrix}, \tag{12.1}$$

where $\mathcal{L}_1 \in \mathbb{R}^{N\times N}$ is a nonsingular M-matrix which is symmetric positive definite and $\mathcal{L}_2 \in \mathbb{R}^{N\times 1}$ has no positive entry.

Assumption 12.1 In the communication topology $\mathcal{G}$, suppose that the leader has a path to each follower.

Lemma 12.1 [30] *Each entry of $-\mathcal{L}_1^{-1}\mathcal{L}_2$ is nonnegative and each row sum of $-\mathcal{L}_1^{-1}\mathcal{L}_2$ is equal to one.*

With respect to the inertia frame, the attitude kinematic and dynamic of the spacecraft using modified Rodrigues parameters is given as [31]

$$\dot{\sigma}_i = G_i(\sigma_i)\,\omega_i, \tag{12.2}$$

$$J_i\dot{\omega}_i = -\omega_i^{\times} J_i\omega_i + \tau_{ai}, \tag{12.3}$$

$\sigma_i = e_i \tan(\Phi_i/4)$, Φ_i is the Euler angle, e_i is the Euler axis, $\omega_i \in \mathbb{R}^3$ denotes the angular velocity of spacecraft i . $J_i \in \mathbb{R}^{3\times 3}$ is the inertia matrix, $\tau_{ai} \in \mathbb{R}^3$ denotes the control torque, and the matrix $G_i(\sigma_i)$ is given by

$$G_i(\sigma_i) = \frac{1}{2}\left[\frac{1-\sigma_i^T\sigma_i}{2} I_3 + \sigma_i^\times + \sigma_i\sigma_i^T\right]. \tag{12.4}$$

Except for the singularity at $\Phi_i = \pm 360°$, the stability analysis is valid for any initial attitude. (12.2) and (12.3) can be rewritten as

$$M_i(\sigma_i)\ddot{\sigma}_i + C_i(\sigma_i, \dot{\sigma}_i)\dot{\sigma}_i = \tau_i(t), \tag{12.5}$$

where

$$M_i(\sigma_i) = G_i^{-T}(\sigma_i) J_i G_i^{-1}(\sigma_i), \tag{12.6}$$

$$C_i(\sigma_i, \dot{\sigma}_i) = -G_i^{-T} J_i G_i^{-1}\dot{G}_i G_i^{-1} - G_i^{-T}(J_i\omega_i)^\times G_i^{-1}, \tag{12.7}$$

$$\tau_i = G_i^{-T}(\sigma_i)\tau_{ai}, \tag{12.8}$$

which has following properties:

Property 12.1 [32] *$M_i(\sigma_i)$ is bounded as*

$$0 < k_{\underline{m}}\|\sigma_i\|^2 \le \sigma_i^T M_i(\sigma_i)\sigma_i \le k_{\overline{m}}\|\sigma_i\|^2, \ \forall \sigma_i \in \mathbb{R}^3, \tag{12.9}$$

where $k_{\underline{m}}$ and $k_{\overline{m}}$ are positive constants.

Property 12.2 [32] *$\dot{M}_i(\sigma_i) - 2C_i(\sigma_i, \dot{\sigma}_i)$ is skew-symmetric.*

Assumption 12.2 The leader is independent of the followers. There exist a nonnegative constants δ_L, such that

$$\|\ddot{\sigma}_0\| \le \delta_L < \infty, \tag{12.10}$$

Remark 12.1 Assumption 12.2 is necessary as the NNs can be effective only in a compact set. Compared with the assumptions in [33] that full states of the spacecraft are bounded, the acceleration boundedness assumption in this chapter is less conservative and acceptable in practice.

The lemmas to be considered in this chapter are listed here.

Lemma 12.2 [34] *Considering a system as*

$$\dot{x} = f(x), \ f(0) = 0, \ x \in \mathbb{R}, \tag{12.11}$$

for any real numbers $c > 0$ and $\alpha \in (0, 1)$, the origin of the given system is finite-time stable if there exists a positive definite function $V(x)$ satisfying the following inequality

$$\dot{V}(x) + cV(x)^\alpha \le 0. \tag{12.12}$$

Given the initial state $x(0) = x_0$, *the settling time is estimated by*

$$T(x_0) \leq \frac{V(x_0)^{1-\alpha}}{c(1-\alpha)}. \tag{12.13}$$

12.3 Control Design

In this section, the distributed attitude tracking problem is considered for spacecraft with attitude constraints and model uncertainties. We firstly propose a DFSE to generate the precise estimation of the leader's state in finite time. Then, employing an integral BLF, the distributed attitude coordinated tracking control is proposed for systems with attitude constraints, and the NNs are applied to compensate for the model uncertainties, which can improve the system performance. Next, in the case that the angular velocity measurements are no longer accessible, by employing a high-gain observer, the distributed tracking control by output feedback is proposed, while the attitude constraints will never be violated.

12.3.1 Distributed Finite-Time Sliding-Mode Estimator

Inspired by [7], we propose the following DFSE protocol as

$$\dot{\hat{q}}_i = \hat{v}_i - \beta_1 \mathrm{sgn} \sum_{j=0}^{N} \left(a_{ij} \left(\hat{q}_i - \hat{q}_j \right) \right), \tag{12.14}$$

$$\dot{\hat{v}}_i = -\beta_2 \mathrm{sgn} \sum_{j=0}^{N} \left(a_{ij} \left(\hat{v}_i - \hat{v}_j \right) \right), \ i = 1, 2, \ldots, N, \tag{12.15}$$

where β_1 and β_2 are positive constants, $\hat{q}_k$ and $\hat{v}_k$, $k = 1, \ldots, N$ are the estimations of desired attitude and angular velocity for each follower, respectively, and $\hat{q}_0 = \sigma_0$, $\hat{v}_0 = \dot{\sigma}_0$. Define

$$Q_F = \left[\sigma_1^T, \sigma_2^T, \ldots, \sigma_N^T \right]^T, \tag{12.16}$$

$$Q_L = \left[\sigma_0^T, \sigma_0^T, \ldots, \sigma_0^T \right]^T. \tag{12.17}$$

Theorem 12.1 *Let* $\beta_1 > 0$ *and* $\beta_2 > \delta_L$. *If Assumptions 12.1 and 12.2 hold, under the DFSEs (12.14) and (12.15), the followers can acquire the precise estimations of the leader's attitude and angular velocity in finite time, respectively, which means*

$$\hat{q}_i \to \sigma_0, \tag{12.18}$$

$$\hat{v}_i \to \dot{\sigma}_0,\ i = 1, 2, \ldots, N. \tag{12.19}$$

Proof Let $\tilde{v}_i = \hat{v}_i - \dot{\sigma}_0$ and $\tilde{q}_i = \hat{q}_i - \sigma_0,\ i = 1, \ldots, N$. Define the vectors $\tilde{v} = \left[\tilde{v}_1^T, \tilde{v}_2^T, \ldots, \tilde{v}_N^T\right]^T$ and $\tilde{q} = \left[\tilde{q}_1^T, \tilde{q}_2^T, \ldots, \tilde{q}_N^T\right]^T$. Consider the following Lyapunov function according to (12.15).

$$V_1 = \frac{1}{2}\tilde{v}^T (\mathcal{L}_1 \otimes I_3)^T \tilde{v}. \tag{12.20}$$

Since the topology graph is undirected, we know $\mathcal{L}_1 \otimes I_3$ is symmetric positive definite.

According to (12.14) and (12.15), taking the derivative of V_1 yields

$$\begin{aligned}
\dot{V}_1 &= \tilde{v}^T (\mathcal{L}_1 \otimes I_3)^T \dot{\tilde{v}} \\
&= \tilde{v}^T (\mathcal{L}_1 \otimes I_3)^T \left[-\beta_2 \mathrm{sgn}\left((\mathcal{L}_1 \otimes I_3)\tilde{v}\right) - \ddot{Q}_L\right] \\
&= \left((\mathcal{L}_1 \otimes I_3)\tilde{v}\right)^T \left[-\beta_2 \mathrm{sgn}\left((\mathcal{L}_1 \otimes I_3)\tilde{v}\right)\right. \\
&\quad \left. + \left(\mathcal{L}_1^{-1}\mathcal{L}_2 \otimes I_3\right)\ddot{Q}_L\right].
\end{aligned} \tag{12.21}$$

According to [35], to any $x \subseteq \mathbb{R}^3$, it holds

$$\|x\|_1 \geq \|x\|_2, \tag{12.22}$$

where $\|x\|_2 = \sqrt{x^T x}$, $\|x\|_1 = \sum_{i=1}^{3} |x_i|$, x_i is the ith element of vector x. According to (12.22) and Lemma 12.1, we can get

$$\begin{aligned}
\dot{V}_1 &\leq -\left(\beta_2 - \left\|\mathcal{L}_1^{-1}\mathcal{L}_2\right\| \delta_L\right) \|(\mathcal{L}_1 \otimes I_3)\tilde{v}\|_1 \\
&\leq -(\beta_2 - \delta_L) \|(\mathcal{L}_1 \otimes I_3)\tilde{v}\|_2 \\
&\leq -(\beta_2 - \delta_L)\sqrt{\tilde{v}^T (\mathcal{L}_1 \otimes I_3)^2 \tilde{v}} \\
&\leq -(\beta_2 - \delta_L)\lambda_{\min}(\mathcal{L}_1)\|\tilde{v}\|_2 \\
&\leq -\sqrt{2}(\beta_2 - \delta_L)\frac{\lambda_{\min}(\mathcal{L}_1)}{\sqrt{\lambda_{\max}(\mathcal{L}_1)}} V_1^{\frac{1}{2}}.
\end{aligned} \tag{12.23}$$

Thus, it follows from Lemma 12.2 that $\hat{v}_i \to \dot{\sigma}_0, i = 1, 2, \ldots, N$ in a settling time

$$T_1 = \frac{\sqrt{2}\sqrt{\lambda_{\max}(\mathcal{L}_1)}}{(\beta_2 - \delta_L)\lambda_{\min}(\mathcal{L}_1)} V_1^{\frac{1}{2}}(0), \tag{12.24}$$

where $V_1(0)$ is the value of V_1 at the initial time 0. Then, $\dot{\sigma}_0$ can be utilized to replace $\hat{v}_i$, when $t \geq T_1$.

When $t \geq T_1$, construct the Lyapunov function $V_2 = \frac{1}{2}\tilde{q}^T (\mathcal{L}_1 \otimes I_3)^T \tilde{q}$, and take the derivative of V_2 yields,

$$\begin{aligned}
\dot{V}_2 &= \tilde{q}^T (\mathcal{L}_1 \otimes I_3)^T \dot{\tilde{q}} \\
&= \tilde{q}^T (\mathcal{L}_1 \otimes I_3)^T \left[-\beta_1 \mathrm{sgn}\left((\mathcal{L}_1 \otimes I_3)\,\tilde{q}\right)\right] \\
&\leq -\beta_1 \|(\mathcal{L}_1 \otimes I_3)\,\tilde{q}\|_1 \\
&\leq -\beta_1 \|(\mathcal{L}_1 \otimes I_3)\,\tilde{q}\|_2 \\
&\leq -\beta_1 \sqrt{\tilde{q}^T (\mathcal{L}_1 \otimes I_3)^2 \tilde{q}} \\
&\leq -\beta_1 \lambda_{\min}(\mathcal{L}_1)\,\|\tilde{q}\|_2 \\
&\leq -\sqrt{2}\beta_1 \frac{\lambda_{\min}(\mathcal{L}_1)}{\sqrt{\lambda_{\max}(\mathcal{L}_1)}} V_2^{\frac{1}{2}}.
\end{aligned} \tag{12.25}$$

Next, from Lemma 12.2, we have $\hat{q}_i \to \sigma_0$, $i = 1, 2, \ldots, N$ in finite time with a settling time

$$T_2 = T_1 + \frac{\sqrt{2}\sqrt{\lambda_{\max}(\mathcal{L}_1)}}{\beta_1 \lambda_{\min}(\mathcal{L}_1)} V_2^{\frac{1}{2}}(T_1), \tag{12.26}$$

where $V_1(T_1)$ is the value of V_1 at the time T_1. Then, σ_0 can be used to replace $\hat{q}_i$, when $t \geq T_2$.

Therefore, $\hat{q}_i$ and $\hat{v}_i$, $i = 1, 2, \ldots, N$ can be used to replace the attitude and angular velocity of the leader when $t \geq T_2$. □

Remark 12.2 Compared with the dynamic estimator in [36], the states of followers are not used in the DFSEs (12.14) and (12.15), which provides some convenience for us to design the output-feedback control algorithm in Sect. 12.3.3. More specifically, only the Laplacian matrix is involved in (12.14) and (12.15), which grants the estimators independence from the controller or any system dynamics.

Remark 12.3 Although $\dot{\hat{q}}_i$ in (12.14) is discontinuous and $\dot{q}_0$ might be continuous, $\dot{\hat{q}}_i$ can still be used to replace $\dot{q}_0$ or $\hat{v}_i$ when $t \geq T_2$. $\dot{\hat{q}}_i$ can be written as

$$\dot{\hat{q}}_i(t) = \hat{v}_i(t) + \upsilon_i(t), \tag{12.27}$$

where $\upsilon_i(t)$ satisfying $\int_{t_1}^{t_2} \upsilon_i(t)\,\mathrm{d}t = 0, \forall t_2 > t_1 > T_2$ with infinitely large frequency of $\upsilon_i(t)$. Since integration-based systems can be seen as ideal low-pass filters, $\upsilon_i(t)$ will be filtered out by these integration-based systems.

12.3.2 Full-State Feedback Control with Output Constraints

Considering the case with $t \geq T_2$, the control objective is to let all the followers track a dynamic leader and guarantee that all errors are bounded and the output constraints are never breached, i.e., $|\sigma_i| \leq k_{ci}$, $\forall t \geq 0$, $i = 1, 2, \ldots, N$, where $k_{ci} \in \mathbb{R}^3$ is a positive constant vector. Before moving forward, define the tracking errors as

$$z_{1i} = \sigma_i - \hat{q}_i, \tag{12.28}$$

$$z_{2i} = \dot{\sigma}_i - \alpha_{1i}, \tag{12.29}$$

where the stabilizing function α_{1i} is given as

$$\alpha_{1i} = \begin{bmatrix} -k_{1i,1}z_{1i,1} + \frac{(k_{ci,1}^2 - \sigma_{i,1}^2)\hat{v}_{i,1}\rho_{i,1}}{k_{ci,1}^2} \\ -k_{1i,2}z_{1i,2} + \frac{(k_{ci,2}^2 - \sigma_{i,2}^2)\hat{v}_{i,2}\rho_{i,2}}{k_{ci,2}^2} \\ -k_{1i,3}z_{1i,3} + \frac{(k_{ci,3}^2 - \sigma_{i,3}^2)\hat{v}_{i,3}\rho_{i,3}}{k_{ci,3}^2} \end{bmatrix}, \tag{12.30}$$

in which the control gain $k_{1i,j}$ and the output constraints $k_{ci,j}$, $i = 1, 2, \ldots, N$, $j = 1, 2, 3$ are positive constants, and ρ_i will be defined later. Denote $k_{1i} = \left[k_{1i,1}, k_{1i,2}, k_{1i,3}\right]^T$.

Differentiating z_{2i} with respect to time, we can get

$$\dot{z}_{2i} = M_i^{-1}(\sigma_i)\left[\tau_i - C_i(\sigma_i, \dot{\sigma}_i)\dot{\sigma}_i\right] - \dot{\alpha}_{1i}. \tag{12.31}$$

Due to the model uncertainties in $M_i(\sigma_i)$ and $C_i\left(\sigma_i, \dot{\sigma}_i\right)$, we can not directly use these terms in our controller. Here, we apply the NNs to approximate the model uncertainties [37], where the following function f_i is introduced:

$$f_i(\sigma_i, \dot{\sigma}_i, \alpha_{1i}, \dot{\alpha}_{1i}) = W_i^{*T}\phi_i(\sigma_i, \dot{\sigma}_i, \alpha_{1i}, \dot{\alpha}_{1i}) + \epsilon_i, \tag{12.32}$$

where $W_i^* \in \mathbb{R}^{l\times 3}$ is the ideal weight matrix, $f_i(\sigma_i, \dot{\sigma}_i, \alpha_{1i}, \dot{\alpha}_{1i}) = C_i(\sigma_i, \dot{\sigma}_i)\alpha_{1i} + M_i(\sigma_i)\dot{\alpha}_{1i}$ represents the model uncertainties and $\phi_i = [\varphi_{i1}, \varphi_{i2}, \ldots, \varphi_{il}]^T \in \mathbb{R}^l$ is the activation function. l is the NN node number, and each node involves a Gaussian function taking the following form [38]:

$$\varphi_{ir}(Z_i) = \exp\left[\frac{-(Z_i - \mu_r)^T(Z_i - \mu_r)}{{\eta_r}^2}\right], \quad r = 1, \ldots, l, \tag{12.33}$$

where η_r is the variance, and μ_r is the function center. Within a compact set, the approximation error ϵ_i is bounded, i.e., $\|\epsilon_i(Z_i)\| \leq \bar{\epsilon}_i, \forall Z_i \in \Omega_{z_i}$, where $\bar{\epsilon}_i > 0$ and $Z_i = \left[\sigma_i^T, \dot{\sigma}_i^T, \alpha_{1i}^T, \dot{\alpha}_{1i}^T\right]^T$.

For multiple spacecraft modeled by (12.5), the following distributed attitude control scheme and adaptation law are given as:

$$\tau_i = -K_{2i} z_{2i} + \hat{W}_i^T \phi_i(Z_i) - \begin{bmatrix} \frac{k_{ci,1}^2 z_{1i,1}}{k_{ci,1}^2 - \sigma_{i,1}^2} \\ \frac{k_{ci,2}^2 z_{1i,2}}{k_{ci,2}^2 - \sigma_{i,2}^2} \\ \frac{k_{ci,3}^2 z_{1i,3}}{k_{ci,3}^2 - \sigma_{i,3}^2} \end{bmatrix}, \tag{12.34}$$

$$\dot{\hat{W}}_{i,k} = -\gamma_i \left(\phi_i(Z_i) z_{2i,k} + \chi_i \hat{W}_{i,k} \right), \tag{12.35}$$

where γ_i, χ_i are positive constants, K_{2i} is a positive definite matrix, and $\hat{W}_{i,k}$ is the kth column vector of $\hat{W}_i$, $i = 1, 2, \ldots, N$.

Theorem 12.2 *Under Assumption 12.1, given that the initial conditions of the multiple spacecraft satisfy $q_i(0) \in \Omega_{0i} := \{\sigma_i \in \mathbb{R}^3 : -k_{ci} < \sigma_i < k_{ci}\}$, i.e., the initial conditions are bounded, the closed-loop errors are UUB under the proposed control algorithm (12.34) and the adaptation law (12.35). The distributed tracking is accomplished, i.e., $\sigma_i(t) \to \sigma_0(t)$ and $\dot{\sigma}_i(t) \to \dot{\sigma}_0(t)$ as $t \to \infty$, while the attitude constraints are never violated, i.e., $|\sigma_i| \le k_{ci}$, $\forall t > 0$. The error signals z_{1i}, z_{2i} and $\tilde{W}_i$ will remain within following compact sets $\Omega_{z_{1i}}$, $\Omega_{z_{2i}}$ and $\Omega_{\tilde{W}_i}$, respectively:*

$$\Omega_{z_{1i}} := \left\{ z_{1i} \in \mathbb{R}^3 | \, \|z_{1i}\| \le \sqrt{D_i} \right\}, \tag{12.36}$$

$$\Omega_{z_{2i}} := \left\{ z_{2i} \in \mathbb{R}^3 | \, \|z_{2i}\| \le \sqrt{D_i / \lambda_{\min}(M_i)} \right\}, \tag{12.37}$$

$$\Omega_{\tilde{W}_i} := \left\{ \tilde{W}_i \in \mathbb{R}^{l \times 3} | \, \left\| \tilde{W}_i \right\| \le \sqrt{\gamma_i D_i} \right\}, \tag{12.38}$$

where $i = 1, \ldots, N$, $k = 1, \ldots, p$, $\tilde{W}_i$ is the weight error defined as $\tilde{W}_i = \hat{W}_i - W_i^*$, and D_i is a positive constant, which will be detailed later.

Proof Consider the positive definite integral BLF candidate

$$V_{1i} = \sum_{k=1}^{3} \left(\int_0^{z_{1i,k}} \frac{\mu k_{ci,k}^2}{k_{ci,k}^2 - (\mu + \hat{q}_{i,k})^2} d\mu \right), \tag{12.39}$$

where $k_{ci,k}$ is a positive constant satisfying $k_{ci,k} > |\sigma_{0,k}|$.

According to Theorem 12.1, $\hat{q}_i$ and $\hat{v}_i$, $i = 1, 2, \ldots, N$ can be used to replace the attitude and angular velocity of the leader when $t \ge T_2$. Also, since $\hat{q}_i$ is bounded, we can always find a proper attitude constraint k_{ci} to implement the proposed controller.

From Remark 12.3, note that $\dot{\hat{q}}_i$ can be used to replace $\dot{q}_0$ or $\hat{v}_i$ when $t \ge T_2$, we then taking the derivative of V_{1i} as

$$\dot{V}_{1i} = \sum_{k=1}^{3} \left(\frac{\partial V_{1i}}{\partial z_{1i,k}} \left(z_{2i,k} + \alpha_{1i,k} - \hat{v}_{i,k} \right) + \frac{\partial V_{1i}}{\partial \hat{q}_{i,k}} \hat{v}_{i,k} \right). \tag{12.40}$$

Integrating by parts and substituting $\mu = \beta z_{1i,k}$ [39] into (12.40) (β is defined as a constant used to simplify (12.40)), we have

$$\frac{\partial V_{1i}}{\partial \hat{q}_{i,k}} = z_{1i,k}\left(\frac{k_{ci,k}^2}{k_{ci,k}^2 - \sigma_{i,k}^2} - \rho_{i,k}\right), \tag{12.41}$$

where

$$\rho_{i,k} = \frac{k_{ci,k}}{2z_{1i,k}} \ln \frac{\left(k_{ci,k} + z_{1i,k} + \hat{q}_{i,k}\right)\left(k_{ci,k} - \hat{q}_{i,k}\right)}{\left(k_{ci,k} - z_{1i,k} - \hat{q}_{i,k}\right)\left(k_{ci,k} + \hat{q}_{i,k}\right)}. \tag{12.42}$$

Using L'Hôpital's rule yields

$$\lim_{z_{1i,k}\to 0} \rho_{i,k} = \frac{k_{ci,k}^2}{k_{ci,k}^2 - \hat{q}_{i,k}^2}. \tag{12.43}$$

From (12.42), we know $\rho_{i,k}$ is well-defined in the neighborhood of $z_{1i,k} = 0$.

Then, substituting (12.41) into (12.40), it gives

$$\begin{aligned}\dot{V}_{1i} =& \sum_{k=1}^{3}\left(\frac{z_{1i,k}k_{ci,k}^2}{k_{ci,k}^2 - \sigma_{i,k}^2}\left(z_{2i,k} + \alpha_{1i,k} - \hat{q}_{i,k}\right)\right)\\ &+ \sum_{k=1}^{3}\left(z_{1i,k}\hat{v}_{i,k}\left(\frac{k_{ci,k}^2}{k_{ci,k}^2 - \sigma_{i,k}^2} - \rho_{i,k}\right)\right).\end{aligned} \tag{12.44}$$

Taking the stabilizing function (12.30) into (12.44), we can obtain

$$\dot{V}_{1i} = \sum_{k=1}^{3}\left(-\frac{k_{1i,k}k_{ci,k}^2 z_{1i,k}^2}{k_{ci,k}^2 - \sigma_{i,k}^2} + \frac{k_{ci,k}^2 z_{1i,k} z_{2i,k}}{k_{ci,k}^2 - \sigma_{i,k}^2}\right). \tag{12.45}$$

We then consider the Lyapunov candidate V_{2i} as

$$V_{2i} = V_{1i} + \frac{1}{2}z_{2i}^T M_i z_{2i} + \frac{1}{2\gamma_i}\sum_{k=1}^{3}\tilde{W}_{i,k}^T\tilde{W}_{i,k}. \tag{12.46}$$

Differentiating (12.46) with respect to time and using (12.31) gives

$$\begin{aligned}\dot{V}_{2i} =& \dot{V}_{1i} + z_{2i}^T\left[\tau_i - C_i\left(\sigma_i, \dot{\sigma}_i\right)\dot{\sigma}_i - M_i\dot{\alpha}_{1i}\right]\\ &+ \frac{1}{2}z_{2i}^T\dot{M}_i z_{2i} + \frac{1}{\gamma_i}\sum_{k=1}^{3}\tilde{W}_{i,k}^T\dot{\tilde{W}}_{i,k}.\end{aligned} \tag{12.47}$$

Using Property 12.2 and taking (12.34), (12.35) into (12.47), we can obtain

$$
\begin{aligned}
\dot{V}_{2i} = & -\sum_{k=1}^{3}\left(\frac{k_{1i,k}k_{ci,k}^{2}z_{1i,k}^{2}}{k_{ci,k}^{2}-\sigma_{i,k}^{2}}\right) - z_{2i}^{T}\left[\left(W_i^{*T}-\hat{W}_i^{T}\right)\phi_i\left(Z_i\right)\right] \\
& -\sum_{k=1}^{3}\tilde{W}_{i,k}^{T}\left(\phi_i\left(Z_i\right)z_{2i,k}+\chi_i\hat{W}_{i,k}\right) \\
& - z_{2i}^{T}K_{2i}z_{2i} - z_{2i}^{T}\epsilon_i\left(Z_i\right).
\end{aligned} \tag{12.48}
$$

Then, noticing

$$
z_{2i}^{T}\epsilon_i\left(Z_i\right) \le \frac{1}{2}z_{2i}^{T}z_{2i} + \frac{1}{2}\|\bar{\epsilon}_i\|^2, \tag{12.49}
$$

$$
\begin{aligned}
-\chi_i\tilde{W}_{i,k}^{T}\hat{W}_{i,k} & = -\chi_i\tilde{W}_{i,k}^{T}\left(\tilde{W}_{i,k}+W_{i,k}^{*}\right) \\
& \le -\chi_i\left\|\tilde{W}_{i,k}\right\|^2 + \frac{\chi_i}{2}\left\|\tilde{W}_{i,k}\right\|^2 + \frac{\chi_i}{2}\|W_{i,k}^{*}\|^2 \\
& \le \frac{\chi_i}{2}\left(\|W_{i,k}^{*}\|^2 - \left\|\tilde{W}_{i,k}\right\|^2\right),
\end{aligned} \tag{12.50}
$$

we further have

$$
\begin{aligned}
\dot{V}_{2i} \le & -z_{2i}^{T}\left(K_{2i}-\frac{1}{2}I_3\right)z_{2i} - \sum_{k=1}^{3}\left(\frac{k_{1i,k}k_{ci,k}^{2}z_{1i,k}^{2}}{k_{ci,k}^{2}-\sigma_{i,k}^{2}}\right) \\
& -\frac{\chi_i}{2}\sum_{k=1}^{3}\left\|\tilde{W}_{i,k}\right\|^2 + \frac{\chi_i}{2}\sum_{k=1}^{3}\|W_{i,k}^{*}\|^2 + \frac{1}{2}\|\bar{\epsilon}_i\|^2.
\end{aligned} \tag{12.51}
$$

Define

$$
\omega_{i,k}\left(\mu,\hat{q}_{i,k}\right) = \frac{\mu k_{ci,k}^{2}}{k_{ci,k}^{2}-\left(\mu+\hat{q}_{i,k}\right)^2}, \tag{12.52}
$$

and we have

$$
\frac{\partial\omega_{i,k}\left(\mu,\hat{q}_{i,k}\right)}{\partial\mu} = \frac{k_{ci,k}^{2}+\mu^2-\hat{q}_{i,k}^{2}}{k_{ci,k}^{2}-\left(\mu+\hat{q}_{i,k}\right)^2}, \tag{12.53}
$$

which is positive and $\omega_{i,k}\left(\mu,\hat{q}_{i,k}\right) > \omega_{i,k}\left(0,\hat{q}_{i,k}\right) = 0$ in the set $\left|\mu+\hat{q}_{i,k}\right| < k_{ci,k}$. Then, it can be shown that for $\left|\mu+\hat{q}_{i,k}\right| < k_{ci,k}$,

$$
\int_{0}^{z_{1i,k}}\omega_{i,k}\left(\mu,\hat{q}_{i,k}\right)d\mu \le z_{1i,k}\omega_{i,k}\left(\mu,\hat{q}_{i,k}\right), \tag{12.54}
$$

which leads to

$$\int_0^{z_{1i,k}} \frac{\mu k_{ci,k}^2}{k_{ci,k}^2 - \left(\mu + \hat{q}_{i,k}\right)^2} d\mu \leq \frac{k_{ci,k}^2 z_{1i,k}^2}{k_{ci,k}^2 - \hat{q}_{i,k}^2}. \tag{12.55}$$

Taking (12.55) into (12.51) gives

$$\dot{V}_{2i} \leq -\vartheta_i V_{2i} + C_i, \tag{12.56}$$

where

$$\begin{aligned} \vartheta_i = \min \Big[& \lambda_{\min}\left(\operatorname{diag}\left(k_{1i,1}, k_{1i,2}, k_{1i,3}\right)\right), \\ & \frac{2\lambda_{\min}\left(K_{2i} - \frac{1}{2}I_3\right)}{\lambda_{\max}\left(M_i\right)}, \gamma_i \chi_i \Big], \end{aligned} \tag{12.57}$$

$$C_i = \frac{\chi_i}{2} \sum_{k=1}^{3} \left\| W_{i,k}^* \right\|^2 + \frac{1}{2} \|\bar{\epsilon}_i\|^2. \tag{12.58}$$

To guarantee $\vartheta_i \geq 0$, we choose the control gain $k_{1i,1}, k_{1i,2}, k_{1i,3}$ and K_{2i} as

$$k_{1i,1}, k_{1i,2}, k_{1i,3} > 0, \tag{12.59}$$

$$\lambda_{\min}\left(K_{2i} - \frac{1}{2}I_3\right) > 0. \tag{12.60}$$

Therefore, z_{1i}, z_{2i}, and $\tilde{W}_i$ are UUB. The details of the proof is given here for completeness. Multiplying (12.56) by $e^{-\vartheta_i t}$ and integrating the both sides yields

$$\begin{aligned} V_{2i}(t) \leq & V_{2i}(0) e^{-\vartheta_i t} + \frac{C_i}{\vartheta_i}\left(1 - e^{-\vartheta_i t}\right) \\ \leq & V_{2i}(0) + \frac{C_i}{\vartheta_i}. \end{aligned} \tag{12.61}$$

Using

$$\begin{aligned} V_{1i} = & \sum_{k=1}^{3} \left(\int_0^{z_{1i,k}} \frac{\mu k_{ci,k}^2}{k_{ci,k}^2 - \left(\mu + \hat{q}_{i,k}\right)^2} d\mu \right) \\ \geq & \frac{1}{2} z_{1i}^T z_{1i}, \end{aligned} \tag{12.62}$$

we then have

$$\|z_{1i}\| \leq \sqrt{D_i}, \tag{12.63}$$

where $D_i = 2\left(V_{2i}(0) + C_i/\vartheta_i\right)$. Therefore, z_{1i} remains within $\Omega_{z_{1i}}$, and the bounds for z_{2i} and $\tilde{W}_i$ can be derived by utilizing the similar procedures.

According to Theorem 12.1, the proposed DFSEs can yield accurate desired attitude and angular velocity estimations, respectively, when $t \geq T_2$. As such,

$$\hat{q}_i(t) = \sigma_0(t), \tag{12.64}$$

$$\hat{v}_i(t) = \dot{\sigma}_0(t), \quad i = 1, 2, \ldots, N, \quad t \geq T_2. \tag{12.65}$$

Then, given that $|\sigma_i(0)| < k_{ci}$, using Lemma 1 in [22], we have $|\sigma_i(t)| < k_{ci}, \forall t > 0$. Hence, it is straightforward to present that all the closed-loop error signals z_{1i}, z_{2i}, $\tilde{W}_i$ for each follower are UUB and the attitude constraints will never be breached, which concludes the proof. □

Remark 12.4 An increase in the control gain $k_{1i,1}, k_{1i,2}, k_{1i,3}, K_{2i}$, the adaptive gain γ_i, the NN node number l, and the modification term χ_i can contribute to smaller tracking errors. In addition, the stability of closed-loop systems will be guaranteed for arbitrary-bounded initial conditions if l is selected as a constant large enough.

12.3.3 Output Feedback Control with Output Constraints

Note that the proposed control algorithms (12.34) and (12.35) require the full-sate measurements, which may be not always accessible in practice. When the angular velocity measurements are not available, by using a high-gain observer [40], the unmeasurable state vector z_{2i} is estimated as

$$\hat{z}_{2i} = \frac{\pi_{2i}}{\varepsilon_i} - \alpha_{1i}, \; i = 1, \ldots, N, \tag{12.66}$$

and the dynamics of π_{2i} takes the form

$$\varepsilon_i \dot{\pi}_{1i} = \pi_{2i}, \tag{12.67}$$

$$\varepsilon_i \dot{\pi}_{2i} = -\bar{\lambda}_{1i}\pi_{2i} - \pi_1 i + \sigma_i. \tag{12.68}$$

where ε_i can be any small positive constant and the parameter $\bar{\lambda}_1$ is chosen such that the polynomial $s^2 + \bar{\lambda}_1 is + 1$ is Hurwitz. Then, following property holds [40]:

$$\xi_{2i} = \frac{\pi_{2i}}{\varepsilon_i} - \dot{\sigma}_i = -\varepsilon_i \psi_i^{(2)}, \tag{12.69}$$

$$\psi_i = \pi_{2i} + \bar{\lambda}_{1i}\pi_{1i}, \tag{12.70}$$

where $\psi_i^{(2)}$ denotes the 2th derivative of ψ_i.

Hence, the estimation error $\tilde{z}_{2i}$ takes the form

$$\tilde{z}_{2i} = \hat{z}_{2i} - z_{2i} = \frac{\pi_{2i}}{\varepsilon_i} - \dot{q}_i = \xi_{2i}. \tag{12.71}$$

Moreover, there exist two positive constants t_{li}^*, h_{li} such that $\forall t > t_{li}^*$, we have

$$\|\xi_{2i}\| \le \varepsilon_i h_{li}. \tag{12.72}$$

We propose the following distributed tracking control algorithm (12.73) and the adaptive law (12.74) using only attitude measurements

$$\tau_i = -K_{2i}\hat{z}_{2i} + \hat{W}_i^T \phi_i\left(\hat{Z}_i\right) - \begin{bmatrix} \frac{k_{ci,1}^2 z_{1i,1}}{k_{ci,1}^2 - \sigma_{i,1}^2} \\ \frac{k_{ci,2}^2 z_{1i,2}}{k_{ci,2}^2 - \sigma_{i,2}^2} \\ \frac{k_{ci,3}^2 z_{1i,3}}{k_{ci,3}^2 - \sigma_{i,p}^2} \end{bmatrix}, \tag{12.73}$$

$$\dot{\hat{W}}_{i,k} = -\gamma_i \left(\phi_i\left(\hat{Z}_i\right)\hat{z}_{2i,k} + \chi_i \hat{W}_{i,k}\right), \tag{12.74}$$

where K_{2i} is the control gain, γ_i, χ_i are positive constants, $\hat{W}_{i,k}$ denotes the kth column vector of $\hat{W}_i$, and $\hat{z}_{2i,k}$ stands for the kth element of $\hat{z}_{2i}$, $k = 1, 2, 3$, $i = 1, \ldots, N$,.

Theorem 12.3 *Considering the multiple spacecraft governed by (12.5), under the control scheme (12.73) with the adaptive law (12.74), for each compact set Ω_{i0} where $\sigma_i(0)$, $\dot{\sigma}_i(0)$, $\hat{W}_{i,1}(0), \ldots, \hat{W}_{i,k}(0) \in \Omega_{i0}$, $i = 1, \ldots, N$, $k = 1, 2, 3$, the closed-loop errors are UUB. The attitude constraints are never violated, i.e., $|\sigma_i| < k_{c1i}$, $\forall t > 0$, while the closed-loop error signals z_{1i}, z_{2i} and $\tilde{W}_i$ will remain within the following compact sets $\Omega_{z_{1i}}$, $\Omega_{z_{2i}}$ and $\Omega_{\tilde{W}_i}$, respectively:*

$$\Omega_{z_{1i}} := \left\{z_{1i} \in \mathbb{R}^3 |\ \|z_{1i}\| \le \sqrt{D_i}\right\}, \tag{12.75}$$

$$\Omega_{z_{2i}} := \left\{z_{2i} \in \mathbb{R}^3 |\ \|z_{2i}\| \le \sqrt{D_i/\lambda_{\min}(M_i)}\right\}, \tag{12.76}$$

$$\Omega_{\tilde{W}_i} := \left\{\tilde{W}_i \in \mathbb{R}^{l\times 3} |\ \left\|\tilde{W}_i\right\| \le \sqrt{\gamma_i D_i}\right\}, \tag{12.77}$$

where D_i is a positive constant which will be detailed later.

Proof Consider the following Lyapunov function candidate

$$V_{3i} = \sum_{k=1}^{3}\left(\int_0^{z_{1i,k}} \frac{\mu k_{ci,k}^2}{k_{ci,k}^2 - (\mu + \hat{q}_{i,k})^2} d\mu\right) + \frac{1}{2} z_{2i}^T M_i z_{2i} + \frac{1}{2\gamma_i}\sum_{k=1}^{3} \tilde{W}_{i,k}^T \tilde{W}_{i,k}. \tag{12.78}$$

Substituting (12.31) into the time derivative of V_{3i} yields

$$\dot{V}_{3i} = -\sum_{k=1}^{3}\left(\frac{k_{1i,k}k_{ci,k}^2 z_{1i,k}^2}{k_{ci,k}^2-\sigma_{i,k}^2} - \frac{k_{ci,k}^2 z_{1i,k}z_{2i,k}}{k_{ci,k}^2-\sigma_{i,k}^2}\right) + \frac{1}{2}z_{2i}^T\dot{M}_i z_{2i} + z_{2i}^T\left[\tau_i - C_i\left(\sigma_i,\dot{\sigma}_i\right)\dot{\sigma}_i\right] - z_{2i}^T M_i\dot{\alpha}_{1i} + \frac{1}{\gamma_i}\sum_{k=1}^{3}\tilde{W}_{i,k}^T\dot{\hat{W}}_{i,k}. \tag{12.79}$$

Applying Property 12.1, and substituting (12.73), (12.74) into (12.79), we have

$$\dot{V}_{3i} = -\sum_{k=1}^{3}\frac{k_{1i,k}k_{ci,k}^2 z_{1i,k}^2}{k_{ci,k}^2-\sigma_{i,k}^2} - z_{2i}^T\left(K_{2i}-\frac{1}{2}I_3\right)z_{2i} - z_{2i}^T K_{2i}\tilde{z}_{2i} - \sum_{k=1}^{3}\left[\tilde{W}_{i,k}^T\phi_i\left(\hat{Z}_i\right)\hat{z}_{2i,k} + \chi_i\tilde{W}_{i,k}^T\hat{W}_{i,k}\right] + \frac{1}{2}\|\bar{\epsilon}_i\|^2 - \sum_{k=1}^{3}z_{2i,k}\left[W_{i,k}^{*T}\phi_i\left(Z_i\right) - \hat{W}_{i,k}^T\phi_i\left(\hat{Z}_i\right)\right], \tag{12.80}$$

where $\tilde{z}_{2i} = \hat{z}_{2i} - z_{2i}$. Since $\hat{Z}_i$ denotes the input of Gaussian functions, the following equations hold [41]:

$$\varphi_{ir}(\hat{Z}) = \exp\left[\frac{-\left(\hat{Z}_i-\mu_r\right)^T\left(\hat{Z}_i-\mu_r\right)}{\eta_r{}^2}\right], \quad r = 1,\ldots,l, \tag{12.81}$$

$$\phi_i(\hat{Z}_i) = \phi_i\left(Z_i\right) + \epsilon_i\phi_{ti}, \tag{12.82}$$

where ϵ_i, l are positive constants, and ϕ_{ti} is a bounded vector function.

From (12.81) and (12.82), we can get

$$\left\|\phi_i\left(\hat{Z}_i\right)\right\|^2 \le l, \tag{12.83}$$

$$\hat{W}_{i,k}^T\phi_i\left(\hat{Z}_i\right) = \tilde{W}_i^T\phi_i\left(\hat{Z}_i\right) + W_{i,k}^{*T}\left(\phi_i\left(Z_i\right)+\epsilon_i\phi_{ti}\right) = \tilde{W}_i^T\phi_i\left(\hat{Z}_i\right) + W_{i,k}^{*T}\phi_i\left(Z_i\right) + \epsilon_i W_{i,k}^{*T}\phi_{ti}. \tag{12.84}$$

Substituting (12.50) and (12.84) to (12.80) yields

$$\begin{aligned}\dot{V}_{3i} =& -\sum_{k=1}^{3}\frac{k_{1i,k}k_{ci,k}^{2}z_{1i,k}^{2}}{k_{ci,k}^{2}-\sigma_{i,k}^{2}} - z_{2i}^{T}\left(K_{2i}-\frac{1}{2}I_{3}\right)z_{2i} - z_{2i}^{T}K_{2i}\tilde{z}_{2i}\\ &-\sum_{k=1}^{3}\tilde{W}_{i,k}^{T}\phi_{i}\left(\hat{Z}_{i}\right)\tilde{z}_{2i,k} + \sum_{k=1}^{3}\epsilon_{i}W_{i,k}^{*T}\phi_{ti}z_{2i,k}\\ &+\sum_{k=1}^{3}\frac{\chi_{i}}{2}\|W_{i,k}^{*}\|^{2} - \sum_{k=1}^{3}\frac{\chi_{i}}{2}\left\|\tilde{W}_{i,k}\right\|^{2} + \frac{1}{2}\|\bar{\epsilon}_{i}\|^{2}.\end{aligned} \tag{12.85}$$

We can then get

$$\begin{aligned}\dot{V}_{3i} \leq& -\sum_{k=1}^{3}\frac{k_{1i,k}k_{ci,k}^{2}z_{1i,k}^{2}}{k_{ci,k}^{2}-\sigma_{i,k}^{2}} - z_{2i}^{T}\left(K_{2i}-I_{3}\right)z_{2i} - z_{2i}^{T}K_{2i}\tilde{z}_{2i}\\ &+\frac{1}{\chi_{i}}\tilde{z}_{2i}^{T}\tilde{z}_{2i}\left\|\phi_{i}\left(\hat{Z}_{i}\right)\right\|^{2} - \sum_{k=1}^{3}\frac{\chi_{i}}{4}\left\|\tilde{W}_{i,k}\right\|^{2}\\ &+\sum_{k=1}^{3}\left(\frac{\chi_{i}}{2}+\frac{1}{2}\epsilon_{i}^{2}\|\phi_{ti}\|^{2}\right)\left\|W_{i,k}^{*}\right\|^{2} + \frac{1}{2}\|\bar{\epsilon}_{i}\|^{2}.\end{aligned} \tag{12.86}$$

Noticing $-z_{2i}^{T}K_{2i}\tilde{z}_{2i} \leq z_{2i}^{T}z_{2i}/2 + \left(K_{2i}\tilde{z}_{2i}\right)^{T}\left(K_{2i}\tilde{z}_{2i}\right)/2$ and substituting (12.72) and (12.83) to (12.86), we further have

$$\begin{aligned}\dot{V}_{3i} \leq& -\sum_{k=1}^{3}\frac{k_{1i,k}k_{ci,k}^{2}z_{1i,k}^{2}}{k_{ci,k}^{2}-\sigma_{i,k}^{2}} - z_{2i}^{T}\left(K_{2i}-\frac{3}{2}I_{3}\right)z_{2i} + \frac{1}{2}\|\bar{\epsilon}_{i}\|^{2}\\ &+\frac{1}{2}\sum_{k=1}^{3}\left(\chi_{i}+\epsilon_{i}^{2}\|\phi_{ti}\|^{2}\right)\left\|W_{i,k}^{*}\right\|^{2} - \sum_{k=1}^{3}\frac{\chi_{i}}{4}\left\|\tilde{W}_{i,k}\right\|^{2}\\ &+\frac{\lambda_{\max}}{2}\left(K_{2i}^{T}K_{2i}+\frac{2l}{\chi_{i}}I_{3}\right)\varepsilon_{i}^{2}h_{1i}^{2}.\end{aligned} \tag{12.87}$$

Then, taking (12.55) into (12.87) gives

$$\dot{V}_{3i} \leq -\vartheta_{i}V_{3i} + C_{i}, \tag{12.88}$$

where ϑ_i and C_i are two constants taking the following forms:

$$\vartheta_i = \min \Big[\lambda_{\min}\left(\mathrm{diag}\left(k_{1i,1}, \ldots, k_{1i,p}\right)\right), \quad \frac{2\lambda_{\min}\left(K_{2i} - \frac{3}{2}I_3\right)}{\lambda_{\max}\left(M_i\right)}, \frac{1}{2}\gamma_i \chi_i \Big], \tag{12.89}$$

$$C_i = \frac{1}{2}\sum_{k=1}^{p}\left(\chi_i + \epsilon_i^2 \|\phi_{ti}\|^2\right)\left\|W_{i,k}^*\right\|^2 + \frac{1}{2}\|\bar{\epsilon}_i\|^2 + \frac{\lambda_{\max}}{2}\left(K_{2i}^T K_{2i} + \frac{2l}{\chi_i}I_3\right)\varepsilon_i^2 h_{li}^2. \tag{12.90}$$

To ensure $\vartheta_i > 0$, we chose the control gains $k_{1i,1}, k_{1i,2}, k_{1i,3}$ and K_{2i} as

$$k_{1i,1}, k_{1i,2}, k_{1i,3} > 0, \tag{12.91}$$

$$\lambda_{\min}\left(K_{2i} - \frac{3}{2}I_3\right) > 0. \tag{12.92}$$

After some manipulations similar to (12.61)–(12.65), we know z_{1i}, z_{2i}, and $\tilde{W}_i$ are UUB, while the bounds are given in (12.75)–(12.77), where D_i=2 (V_{2i} (0) + C_i/ρ_i). Given that $|\sigma_i\,(0)| < k_{ci}$, we have $|\sigma_i\,(t)| < k_{ci}$, $\forall t > 0$. Therefore, the distributed tracking control with output constraints is achieved by using only attitude measurements, which concludes the proof. □

Remark 12.5 For all followers, the closed-loop errors have been proven to remain within a small compact set. From (12.89), (12.90), we know that the increase in the control gains diag $\left(k_{1i,1}, k_{1i,2}, k_{1i,3}\right)$, K_{2i}, the neural network node number l, the high-gain $1/\varepsilon_i$, the adaptive gain γ_i, and the modification term χ_i will contribute to smaller tracking errors because of the decline of C_i/ϑ_i. Additionally, when the measurements are affected by Gaussian white noise, a much smaller ϵ_i can be chosen to deal with the problem. Also, the peaking phenomenons of (12.66) can be fixed utilizing the saturation process in [42].

Remark 12.6 Without angular velocity sensors, we have proposed a distributed output-feedback attitude tracking control algorithm after the DFSEs yield accurate estimations of the leader's attitude and angular velocity. More specifically, only the attitude measurements of the multiple spacecraft are utilized in the controller (12.73) and the adaptive law (12.74). From Remark 12.2, we know the angular velocity measurements are not used in the DFSEs (12.14) and (12.15) either, which provides the proposed output-feedback control algorithm (12.73) a rigorous proof.

12.4 Simulation

Simulation results have been given to validate the feasibility of the proposed approach, where the follower spacecraft are modeled by (12.5) and the inertia matrices are given as

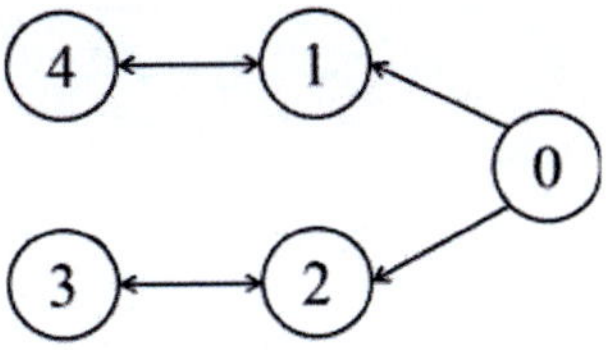

Fig. 12.1 The communication graph among 5 spacecraft

Table 12.2 The parameters for simulation

Parameters	Values	Parameters	Values
β_1	200	k_{ci}	$[0.13, 0.17, 0.15]^T$
β_2	200	k_{1i}	$[0.5, 0.5, 0.5]^T$
K_{2i}	$100I_3$	K_{pi}	$[0.5, 0.5, 0.5]^T$
K_{di}	$100I_3$	$\dot{\sigma}_i$ (0)	$[0, 0, 0]^T$
$\bar{\lambda}_{1i}$	5	$\dot{\sigma}_0^{(1)}$	$0.011 \sin\left(0.09t + \frac{\pi}{2}\right)$
l	12	$\dot{\sigma}_0^{(2)}$	$0.012 \sin\left(0.08t + \frac{\pi}{2}\right)$
ε_i	0.01	$\dot{\sigma}_0^{(3)}$	$0.014 \sin\left(0.1t + \frac{\pi}{2}\right)$
γ_i	4	σ_0 (0)	$[0, 0, 0]^T$
η_r	10	σ_1 (0)	$[0.06, 0.04, -0.07]^T$
μ_r	random $[-1, 1]$	σ_2 (0)	$[0.07, -0.03, 0.09]^T$
$\hat{W}_i$ (0)	0	σ_3 (0)	$[-0.06, 0.04, 0.02]^T$
χ_i	5	σ_4 (0)	$[0.05, -0.02, 0.01]^T$

$$J_1 = \begin{bmatrix} 12 & 0.4 & 0.2 \\ 0.4 & 10 & 0.6 \\ 0.2 & 0.6 & 11 \end{bmatrix}, \quad J_2 = \begin{bmatrix} 14 & 0.2 & 0.4 \\ 0.2 & 12 & 0.4 \\ 0.4 & 0.4 & 10 \end{bmatrix}$$

$$J_3 = \begin{bmatrix} 13 & 0.4 & 0.4 \\ 0.4 & 10 & 0.4 \\ 0.4 & 0.4 & 9 \end{bmatrix}, \quad J_4 = \begin{bmatrix} 16 & 0.6 & 0.2 \\ 0.6 & 14 & 0.4 \\ 0.2 & 0.4 & 12 \end{bmatrix}$$

The network topology is given in Fig. 12.1. It can be seen that follower 1 and follower 2 can directly access the leader states. Follower 3 and follower 4, however, can only obtain information from other followers instead of the leader, which is a rather difficult condition for them to track the dynamic leader.

We consider three different cases. Firstly, we evaluate the control strategy (12.34) for multiple spacecraft systems with attitude constraints by state feedback. Secondly, we examine the control scheme (12.73) for spacecraft with attitude constraints by output feedback. Finally, to present more convincing results, we consider PD control designed as $\tau_i = -K_{pi} z_{1i} - K_{di} \dot{z}_{1i}$. All control parameters are given in Table 12.2.

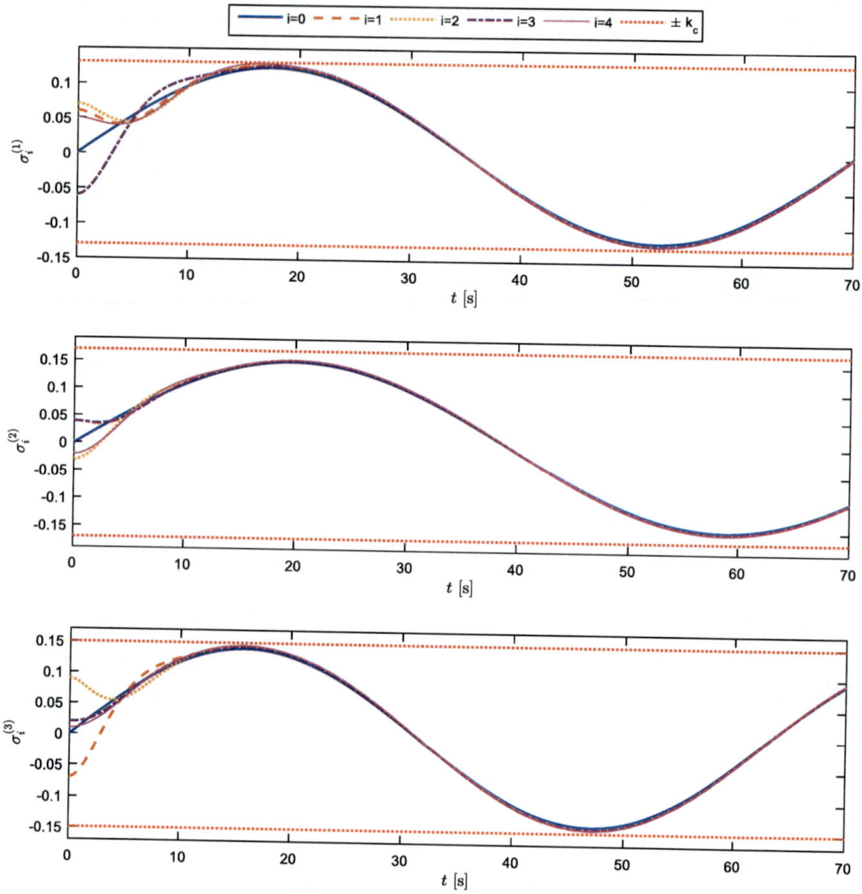

Fig. 12.2 The attitudes of four followers and the leader using (12.34)

The attitude tracking performances of the four followers are shown in Figs. 12.2, 12.3 and 12.4. We can see that the proposed state-feedback control algorithm (12.34), (12.35) and the output-feedback control algorithm (12.73), (12.74) can guarantee satisfying attitude coordination performances. In addition, the attitude constraints $|\sigma_i| \leq k_{ci} = [0.13,\ 0.17,\ 0.15]^T, i = 1, 2, \ldots, 4$, are never violated due to the barrier terms in (12.39) and (12.78). More specifically, the integral BLF will approach infinity while σ_i approaching the constraints k_{ci}. For PD control, as shown in Table 12.2, the values of K_{pi} and K_{di} are identical to the control gains in (12.34) and (12.73). However, from Fig. 12.4, we know the attitude constraints of σ_2 and σ_4 are obviously breached, although the attitudes of followers can roughly track those of the leader. Figures 12.5 and 12.6 demonstrate that the tracking errors remain within a small value close to the origin. Followers 1 and 2 have lower tracking errors as they can access information from the leader directly. Nevertheless, the distributed

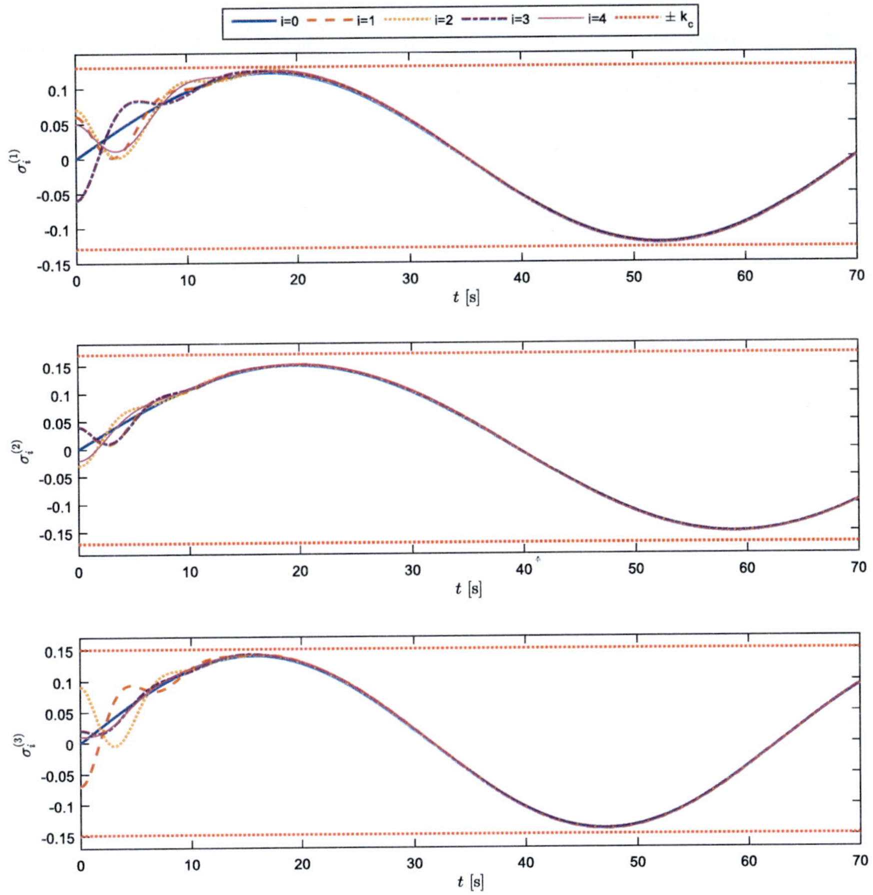

Fig. 12.3 The attitudes of four followers and the leader using (12.73)

state-feedback control algorithm (12.34) requires the full-state feedback to be implemented, which is rather restrictive in practice. The corresponding control torque of each follower is given in Figs. 12.7 and 12.8, which is acceptable to spacecraft systems in practice. Consequently, the control approaches proposed in this chapter are effective and feasible.

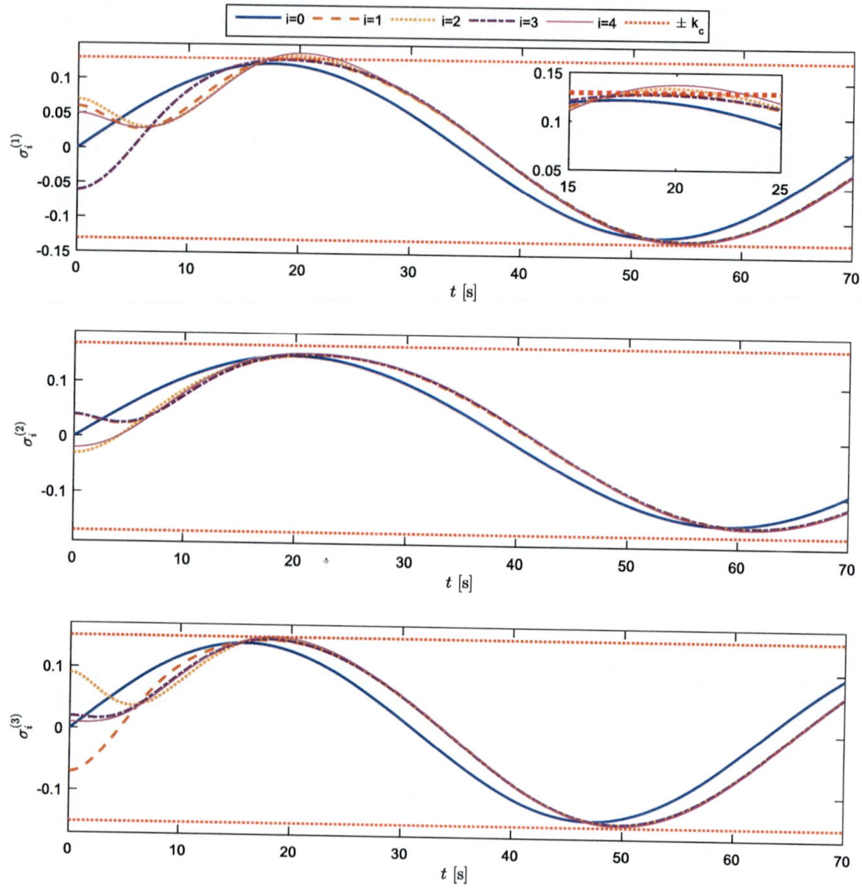

Fig. 12.4 The attitudes of four followers and the leader using PD control

12.5 Conclusion

In this chapter, the distributed attitude coordinated tracking problem has been investigated for multiple spacecraft systems with attitude constraints and model uncertainties. We employ the DFSEs to estimate the accurate attitude and angular velocity of the dynamic leader. To prevent any violation of the attitude constraints, the integral BLF based control algorithms are respectively designed by state and output feedback, while the adaptive NNs are used to approximate model uncertainties of spacecraft systems. Furthermore, we have studied the leaderless synchronization. By Lyapunov theory, the proposed control algorithms have been proven to achieve attitude and angular velocity tracking with a moving leader while never violating the attitude constraints. This result not only provides a new insight into the multiple

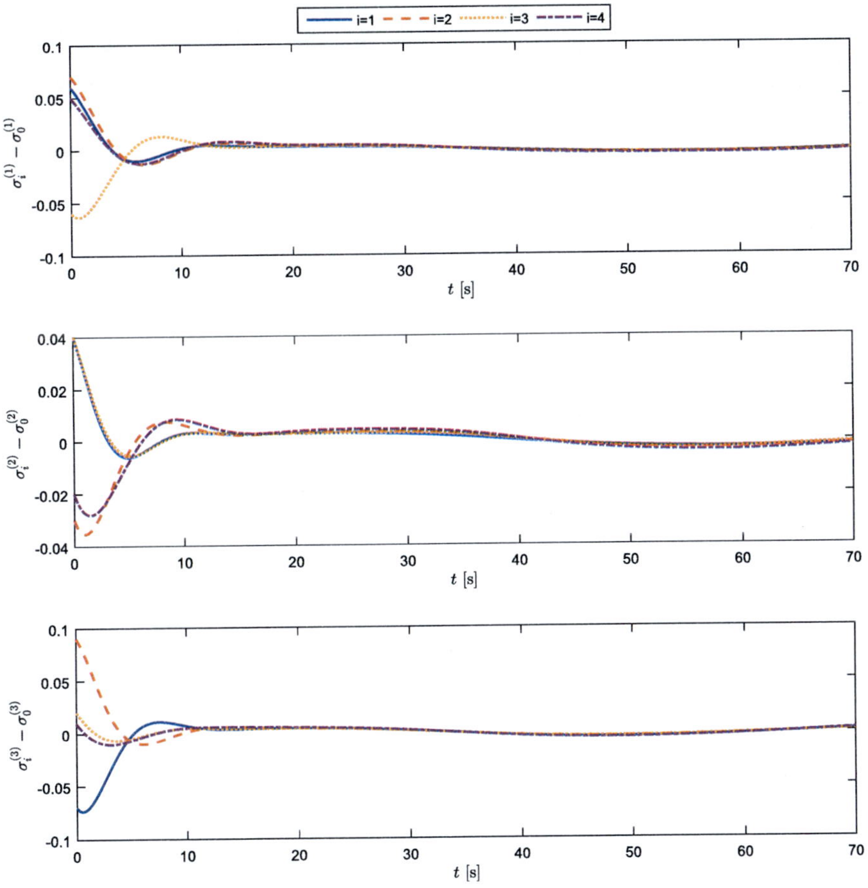

Fig. 12.5 The attitude tracking error of each follower using (12.34)

spacecraft attitude coordination but also has potential benefits in in-orbit services and deep space explorations, since no attitude velocity measurements are required and the attitude constraints are strictly guaranteed. Finally, the simulation results have been given to illustrate the feasibility of our approaches. In the future, more practical situations will be considered, such as the actuator faults, the communication constraints, and the constraints on spacecraft angular velocities.

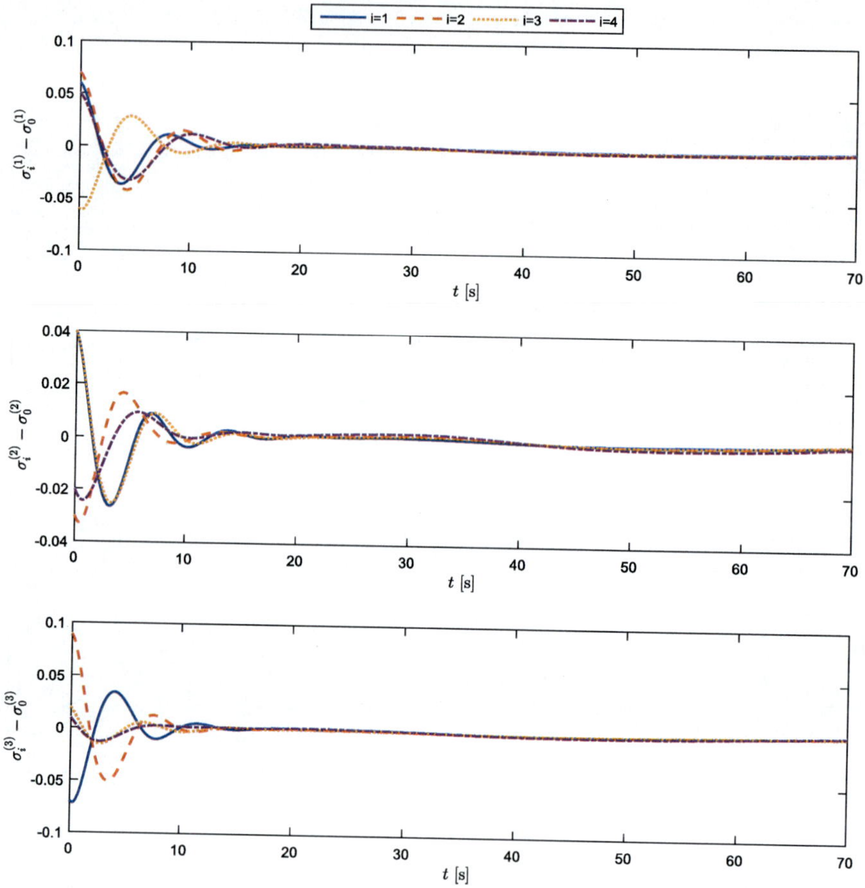

Fig. 12.6 The attitude tracking error of each follower using (12.73)

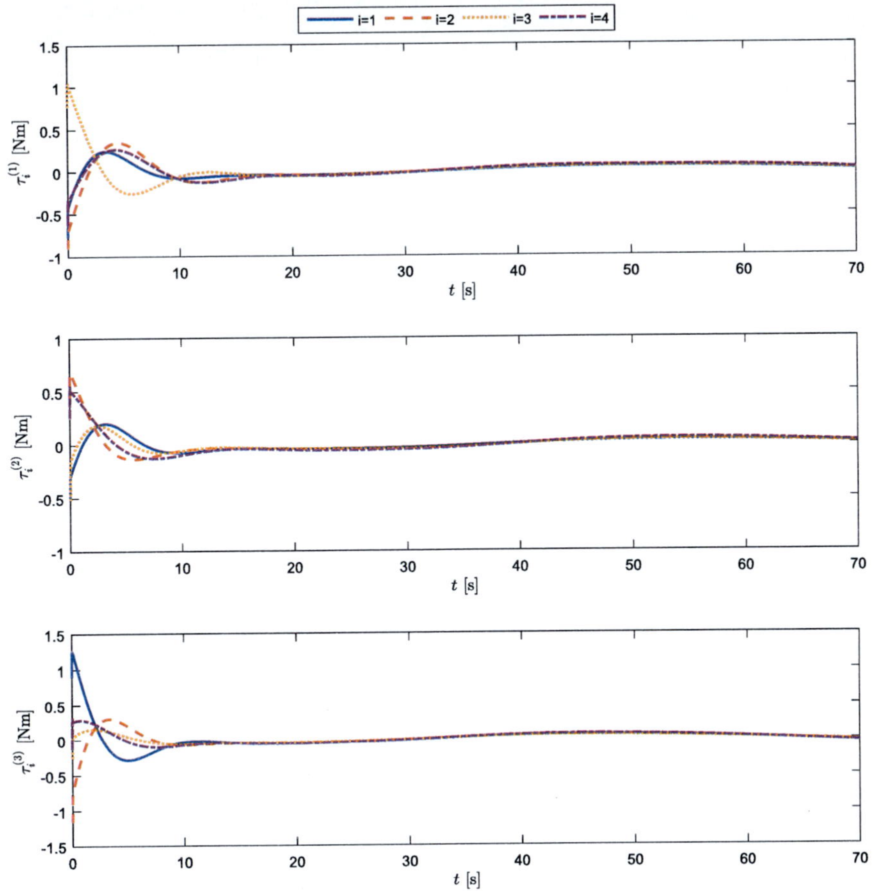

Fig. 12.7 The control torques of four followers using (12.34)

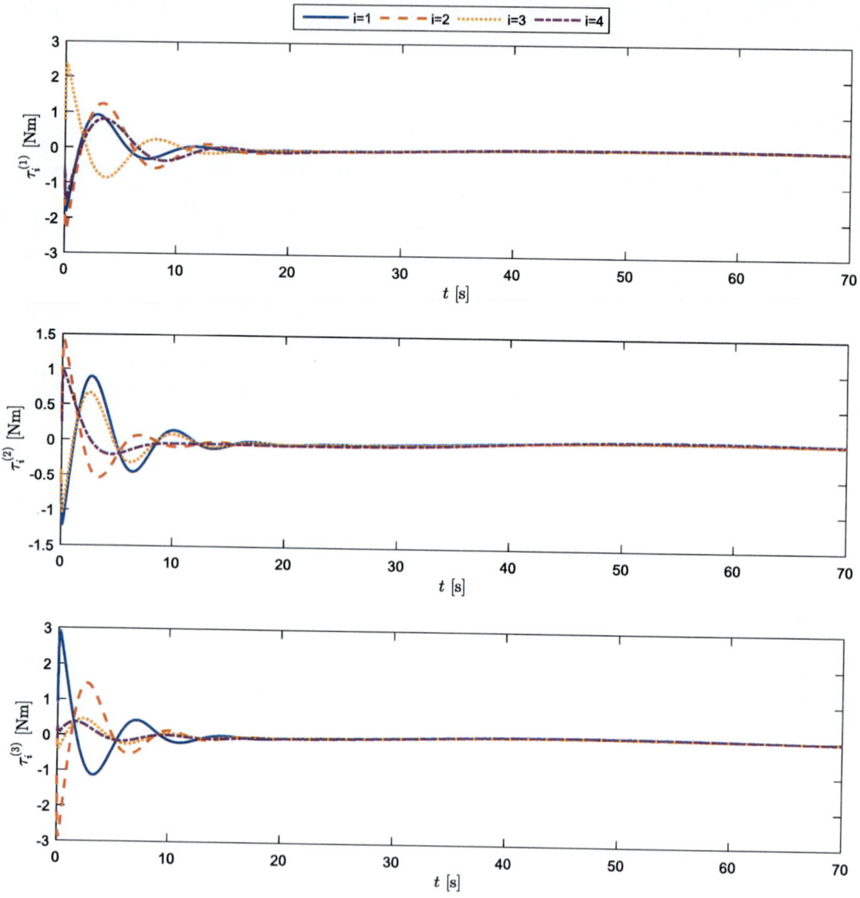

Fig. 12.8 The control torques of four followers using (12.73)

References

1. Lawton, J.R., Beard, R.W.: Synchronized multiple spacecraft rotations. Automatica **38**(8), 1359–1364 (2002)
2. Sarlette, A., Sepulchre, R., Leonard, N.E.: Autonomous rigid body attitude synchronization. Automatica **45**(2), 572–577 (2009)
3. Bai, H., Arcak, M., Wen, J.T.: Rigid body attitude coordination without inertial frame information. Automatica **44**(12), 3170–3175 (2008)
4. Chen, Z., Huang, J.: Attitude tracking and disturbance rejection of rigid spacecraft by adaptive control. IEEE Transactions on Automatic Control **54**(3), 600–605 (2009)
5. Du, H., Li, S., Qian, C.: Finite-time attitude tracking control of spacecraft with application to attitude synchronization. IEEE Transactions on Automatic Control **56**(11), 2711–2717 (2011)
6. Peng, K., Yang, Y.: Leader-following consensus problem with a varying-velocity leader and time-varying delays. Physica A Statistical Mechanics & Its Applications **388**(388), 193–208 (2009)

7. Mei, J., Ren, W., Ma, G.: Distributed containment control for Lagrangian networks with parametric uncertainties under a directed graph. Automatica **48**(4), 653–659 (2012). https://doi.org/10.1016/j.automatica.2012.01.020
8. Jiang, F., Li, J., Baoyin, H., Gao, Y.: Study on relative orbit geometry of spacecraft formations in elliptical reference orbits. Journal of guidance, control, and dynamics **31**(1), 123–134 (2008)
9. Zhang, Z.g., Li, J.f.: Orbit and attitude control of spacecraft formation flying. Applied Mathematics and Mechanics **29**(1), 43–50 (2008)
10. Abdessameud, A., Tayebi, A.: On consensus algorithms for double-integrator dynamics without velocity measurements and with input constraints. Systems & Control Letters **59**(12), 812–821 (2010)
11. Lv, Y., Li, Z., Duan, Z., Chen, J.: Distributed adaptive output feedback consensus protocols for linear systems on directed graphs with a leader of bounded input. Automatica **74**, 308–314 (2016)
12. Cao, Y., Ren, W., Egerstedt, M.: Distributed containment control with multiple stationary or dynamic leaders in fixed and switching directed networks. Automatica **48**(8), 1586–1597 (2012)
13. Yang, Q., Fang, H., Mao, Y., Huang, J.: Distributed tracking for networked Euler-Lagrange systems without velocity measurements. Journal of Systems Engineering and Electronics **25**(4), 671–680 (2014). 10.1109/JSEE.2014.00077
14. Abdessameud, A., Tayebi, A.: Attitude synchronization of a group of spacecraft without velocity measurements. IEEE Transactions on Automatic Control **54**(11), 2642–2648 (2009). 10.1109/TAC.2009.2031567
15. Li, D., Ma, G., He, W., Zhang, W., Li, C., Ge, S.S.: Distributed coordinated tracking control of multiple Euler-Lagrange systems by state and output feedback. IET Control Theory & Applications **11**(14), 2213–2221 (2017)
16. McInnes, C.R.: Large angle slew maneuvers with autonomous sun vector avoidance. Journal of guidance, control, and dynamics **17**(4), 875–877 (1994)
17. Shen, Q., Yue, C., Goh, C.H.: Velocity-Free Attitude Reorientation of a Flexible Spacecraft with Attitude Constraints. Journal of Guidance Control, and Dynamics **40**(5), 1–7 (2017). https://doi.org/10.2514/1.G002129
18. Hablani, H.B.: Attitude commands avoiding bright objects and maintaining communication with ground station. Journal of Guidance, Control, and Dynamics **22**(6), 759–767 (1999)
19. Kjellberg, H.C., Lightsey, E.G.: Discretized constrained attitude pathfinding and control for satellites. Journal of Guidance, Control, and Dynamics **36**(5), 1301–1309 (2013)
20. Wu, C., Xu, R., Zhu, S., Cui, P.: Time-optimal spacecraft attitude maneuver path planning under boundary and pointing constraints. Acta Astronautica **137**(October 2016), 128–137 (2017). https://doi.org/10.1016/j.actaastro.2017.04.004
21. Lee, U., Mesbahi, M.: Feedback control for spacecraft reorientation under attitude constraints via convex potentials. IEEE Transactions on Aerospace and Electronic Systems **50**(4), 2578–2592 (2014). https://doi.org/10.1109/TAES.2014.120240
22. Tee, K.P., Ge, S.S., Tay, E.H.: Automatica Barrier Lyapunov Functions for the control of output-constrained. Automatica **45**(4), 918–927 (2009). https://doi.org/10.1016/j.automatica.2008.11.017
23. He, W., Chen, Y.H., Yin, Z.: Adaptive Neural Network Control of an Uncertain Robot With Full-State Constraints. IEEE Transactions on Cybernetics **46**(3), 620–629 (2016). https://doi.org/10.1109/TCYB.2015.2411285
24. He, W., Huang, H., Ge, S.S.: Adaptive Neural Network Control of a Robotic Manipulator With Time-Varying Output Constraints. IEEE Transactions on Cybernetics **47**(10), 3136–3147 (2017)
25. He, W., Ge, S.S.: Dynamic modeling and vibration control of a flexible satellite. IEEE Transactions on Aerospace and Electronic Systems **51**(2), 1422–1431 (2015). https://doi.org/10.1109/TAES.2014.130804
26. Ren, W., Beard, R.W.: Distributed Consensus in Multi-vehicle Cooperative Control. Communications & Control Engineering **27**(2), 71–82 (2008)

27. Wang, S., Min, H., Liu, Y., Liao, S., Ma, L.: Distributed finite-time attitude dynamic tracking control for multiple rigid spacecraft. IET Control Theory & Applications **9**(17), 2568–2573 (2015). https://doi.org/10.1049/iet-cta.2015.0227
28. Li, S., Du, H., Shi, P.: Distributed attitude control for multiple spacecraft with communication delays. IEEE Transactions on Aerospace and Electronic Systems **50**(3), 1765–1773 (2014). https://doi.org/10.1109/TAES.2014.120572
29. Yang, H., You, X., Xia, Y., Liu, Z.: Nonlinear attitude tracking control for spacecraft formation with multiple delays. Advances in Space Research **54**(4), 759–769 (2014). https://doi.org/10.1016/j.asr.2014.04.026
30. Meng, Z., Ren, W., You, Z.: Distributed finite-time attitude containment control for multiple rigid bodies. Automatica **46**(12), 2092–2099 (2010). https://doi.org/10.1016/j.automatica.2010.09.005
31. Shuster, M.D.: A survey of attitude representations. Navigation **8**(9), 439–517 (1993)
32. Ge, S.S., Lee, T.H., Harris, C.J.: Adaptive Neural Network Control of Robotic Manipulators. Proceedings of SPIE - The International Society for Optical Engineering **4390**, 236–242 (2001)
33. Hu, Q., Zhang, J.: Bounded finite-time coordinated attitude control via output feedback for spacecraft formation. Journal of Aerospace Engineering **28**(5), 4014129 (2014)
34. Bhat, S.P., Bernstein, D.S.: Finite-time stability of continuous autonomous systems. SIAM Journal on Control and Optimization **38**(3), 751–766 (2000)
35. Horn, R., Johnson, C.: Matrix analysis. Cambridge university press, Cambridge UK (1990)
36. Su, Y., Huang, J.: Cooperative output regulation of linear multi-agent systems. IEEE Transactions on Automatic Control **57**(4), 1062–1066 (2011)
37. Kubat, M.: Neural networks: a comprehensive foundation by Simon Haykin, Macmillan, 1994, ISBN 0-02-352781-7. Knowledge Engineering Review **13**(4), 409–412 (1999)
38. Ge, S.S., Wang, C.: Adaptive neural control of uncertain mimo nonlinear systems. IEEE Transactions on Neural Networks **15**(3), 674–692 (2004)
39. Ge, S.S., Hang, C.C., Zhang, T.: Stable adaptive control for nonlinear multivariable systems with a triangular control structure. Automatic Control, IEEE Transactions on **45**(6), 1221–1225 (2000)
40. Behtash, S.: Robust output tracking for non-linear systems. International Journal of Control **51**(6), 1381–1407 (1990)
41. Ge, S.S., Hang, C.C., Lee, T.H., Zhang, T.: Stable adaptive neural network control. Springer, Boston (2002)
42. Khalil, H.K.: Nonlinear systems. Prentice Hall, New Jersey (1996)

Chapter 13
Formation-Containment Control for Networked Spacecraft with Output Feedback

Abstract This chapter addresses the distributed formation-containment (DFC) problem for multiple spacecraft systems with model uncertainties via output feedback in both constant and time-varying formation cases. First, a novel definition of the DFC problem is proposed using a two-layer framework. Since only parts of the followers can acquire the states of the dynamic leader, we design a distributed finite-time sliding-mode estimator (DFSE) to obtain accurate estimations of the desired position and velocity for each spacecraft. Next, to deal with the absence of velocity sensors, we propose two DFC control laws combined with the high-gain observer for the leaders and the followers, respectively, while the time-varying formation in the first layer and the leader-based containment in the second layer can be achieved. Further, the adaptive neural networks (NNs) are applied to deal with the model uncertainties due to their superior approximation capability. The uniform ultimate boundedness of all the state errors can be guaranteed by Lyapunov stability theory. In addition, a unified framework is given which can be transformed to other basic distributed problems. Finally, simulation examples are presented to illustrate the feasibility of the theoretical results.

13.1 Introduction

The multi-spacecraft formation systems can be abstractly described as multi-agent systems (MASs) when many spacecraft form networked systems. Distributed coordinated control strategies for multi-spacecraft formation systems based on the description of the MASs also have been an active area due to its extensive utilization and obvious benefits, which include stronger robustness, less communication requirement and higher efficiency [1–3].

Currently, existing works on MASs mainly focus on the consensus problem, the tracking problem, the formation problem and the containment problem. As for the consensus problem, the follower state is updated using the information from neighbors and reaches a common value as a result. For systems with first-order dynamics, the consensus control has been studied in [4–6], where the consensus can be guaranteed if there exists a directed spanning tree in the communication networks

Y. Gao and D. Li, *Spacecraft Maneuver with Performance Guaranteed*,
https://doi.org/10.1007/978-981-99-4653-2_13

[6]. Practically, a broad class of vehicles requires a second-order dynamic model. Unfortunately, the extension of consensus problems from a class of MASs described by first-order dynamics to those modeled by second-order dynamics is nontrivial because other than the spanning tree condition, the Laplacian matrix generated from the communication topology plays an important role [7–9]. Considering a moving leader, the coordinated tracking problem has also been extensively investigated for systems with first-order [10, 11] and second-order dynamics [12]. In [13, 14], by using the potential field, the formation control problem is studied, where multiple agents move in a desired spatial configuration. Further, the containment control problem is concerned, where each follower is driven to a region formed by multiple leaders [15, 16].

Spacecraft formation flight control tasks often include the relative orbital/position motion and relative attitude control. The relative orbital/position motion and attitude motion of the spacecraft can be unified and abstracted into the Euler-Lagrange Systems (ELSs). For general design, in this chapter, the ELSs are used to describe the multi-spacecraft systems. The ELSs can also be employed to describe a wide class of practical systems such as satellites, underwater robots, and autonomous vehicles [17]. Since the cooperative problems for multiple Euler-Lagrange systems (MELSs) have extensive applications and other potential benefits, we give attention to the distributed control of MELSs. Similar to distributed cooperative control of linear systems, recent works on coordination for MELSs also focus on the problem of leaderless consensus [18–20], tracking a dynamic leader or equivalently a reference [14, 21, 22], formation control [23, 24] and containment control [25].

Previous works on containment control share an assumption that there exists no communication among the leaders. In practice, however, we also require the leaders to achieve a certain formation coordinately. Inspired by the formation problem and the containment problem, an interesting DFC problem arises [26–32], where there exists a communication topology among the leaders. The control objective of DFC problem is to keep the leaders moving in a desired formation and to drive all the followers into the convex hull spanned by the leaders. In [26], under an undirected communication topology, the DFC problem for first-order swarm systems is investigated. Under time-varying delays, the DFC control law is proposed in [27] for second-order MASs. Then, the DFC problem is solved for linear time-invariant MASs [28] and linear time-invariant swarm systems [29] under a directed communication graph. Next, considering coupled heterogeneous linear systems, the DFC problem is studied in [30] with intermittent communications. In [31], without using velocity information, the DFC control algorithm is proposed for second-order MASs under an undirected graph. The DFC problem for multiple unmanned aerial vehicles is studied in [32], in which the controller can be applied to both fixed and switching graphs.

It is worth noticing that the controllers in [27, 29, 32] require the measurements of generalized velocities to be implemented. In practice, however, velocity sensors may not always be accessible because of the rigid constraints on the expense and space. For systems with velocity sensors, however, the measurements of velocity are frequently contaminated with high levels of noise which weakens the system performance [33]. On the other hand, only the linear systems or single-input-single-output systems are considered in [27–32]. In addition, few of the references above consider systems with model uncertainties. Inspired by the aforementioned works, we consider the DFC control by output feedback for multi-spacecraft formation systems described by MELSs under in the presence of model uncertainties. Compared with existing works on the DFC problem [27–32], our study contains the following contributions:

(1) A two-layer DFC control scheme combined with the adaptive neural networks (NNs) is designed, which can compensate for the model uncertainties and enhance the performance of the systems. Using the proposed control scheme, all error signals will remain within a small compact set, and the uniform ultimate boundedness is proved by the Lyapunov direct method. In comparison, most of the existing works consider the linear systems [28, 30, 31], MELSs with known dynamic model [34], or MELSs with the parametric uncertainties, which require the linearity-in-parameters assumption [35, 36]. Moreover, a novel two-layer framework is defined which can be transformed to other basic distributed problems, such as leaderless synchronization problems, formation control problems, coordinated tracking problems, and distributed containment problems.
(2) Compared with the DFC control algorithms [27, 29, 32] using full-state information, the proposed two-layer DFC control is feasible using output feedback when the velocity and acceleration measurements are no longer accessible in practice.
(3) Different from current works considering a constant formation configuration [37, 38], the proposed two-layer DFC control algorithm can be applied to the time-varying formation configuration case by transforming the formation functions, which is more general and practical.
(4) In both layers, since the desired formation configuration or the leaders' information is accessible to only parts of the spacecraft, a distributed finite-time sliding-mode estimator (DFSE) algorithm is proposed to access accurate estimations of the desired information in finite time by using communication topology. Moreover, the control law proposed in this chapter is fully distributed, while the Laplacian matrix or some other global information is required in some other results on MASs such as [39].

The rest of this chapter is organized as follows. In Sect. 13.2, the graph theory, the problem description, the dynamics of the spacecraft systems, and some other technical Lemmas are introduced. The main results are given in Sects. 13.3 and 13.4, where the two-layer DFC control algorithm is proposed combined with the NNs and the high-gain observer. In Sect. 13.5, numerical studies are presented. Finally, we draw a conclusion in Sect. 13.6.

Notations: Throughout this chapter, let $\mathbb{R}$ denote the set of real numbers. Let $I_n \in \mathbb{R}^{n\times n}$ denote the n-dimensional identity matrix. Let $\lambda_{\max}(\cdot)$ and $\lambda_{\min}(\cdot)$ denote the maximum and minimum eigenvalues of a matrix, respectively. Let $\|\cdot\|$ be the Euclidean norm of a vector. Let $\mathrm{diag}(z_1, \ldots, z_n) \in \mathbb{R}^{n\times n}$ be the diagonal matrix, where $z_1, \ldots, z_n$ are the diagonal entries. Let $\mathrm{sgn}(\cdot)$ stand for the signum function.

13.2 Preliminaries and Problem Description

Considering a network of k spacecraft described by (2.34), we utilize a graph group $\mathcal{G} = (V, E)$ to characterize the network topology among the spacecraft, where $V = \{1, \ldots, k\}$ and $E \subseteq V \times V$ stand for the vertex set and the arc set, respectively. An arc $(i, j) \in E$ represents that the spacecraft j has access to the information of the spacecraft i. A path from the vertex i to the vertex j is a sequence of arcs of the form $(i_1, i_2), (i_2, i_3), \ldots$. A subgraph $\mathcal{G}_s = (V_s, E_s)$ of $\mathcal{G}$ is a graph such that $V_s \subseteq V$ and $E_s \subseteq E$. The root is a vertex that has paths to all the other vertices in the graph. The adjacency matrix $\mathcal{A} = \left[a_{ij}\right] \in \mathbb{R}^{k\times k}$ is a nonnegative matrix defined as $a_{ij} > 0$ if $(j, i) \in E$, and $a_{ij} = 0$ otherwise. Moreover, self-arcs are not allowed in this chapter, which means $a_{ii} = 0$. The Laplacian matrix $\mathcal{A}$ is defined as $\mathcal{L} = \left[l_{ij}\right] \in \mathbb{R}^{k\times k}$, where $l_{ii} = \sum_{j=1, j\neq i}^{k} a_{ij}$ and $l_{ij} = -a_{ij}, i \neq j$.

The spacecraft formation described by (2.34) consists of N leader spacecraft, labeled as 1 to N and M follower spacecraft, labeled as $N + 1$ to $N + M$. The corresponding Laplacian matrix $\mathcal{L} \in \mathbb{R}^{(N+M+1)\times(N+M+1)}$ can be written as

$$\mathcal{L} = \begin{bmatrix} 0 & 0_{1\times N} & 0_{1\times M} \\ \mathcal{L}_{L2} & \mathcal{L}_{L1} & 0_{N\times M} \\ 0_{M\times 1} & \mathcal{L}_{F2} & \mathcal{L}_{F1} \end{bmatrix}, \tag{13.1}$$

where $\mathcal{L}_{L1}$ represents the communication topology among the leaders, $\mathcal{L}_{F1}$ represents the communication topology among the followers, and $\mathcal{L}_{F2}$ represents the communication interaction between the leaders and followers. In this chapter, we consider the communication topology among the leaders and the communication topology among the followers are bidirectional. Here, we define $\mathcal{L}_{L2} = [-a_{10}, -a_{20}, \ldots, -a_{N0}]^T$, where $a_{i0} > 0, i = 1, 2, \ldots, N$, if the leader i knows the desired position and velocity of the formation and $a_{i0} = 0$ otherwise. Then, define

$$\mathcal{G}_L = \begin{bmatrix} a_{00} & \cdots & a_{0N} \\ \vdots & \ddots & \vdots \\ a_{N0} & \cdots & a_{NN} \end{bmatrix}. \tag{13.2}$$

Remark 13.1 The spacecraft formation consists of N leaders and M followers, however, the dimension of the corresponding Laplacian matrix $\mathcal{L}$ is $\mathbb{R}^{(N+M+1)\times(N+M+1)}$. The reason is that the communication among the leaders is distributed and only a subset of the leaders know the desired formation state. Then, we manually add a node $a_{i0}, i = 1, 2, \ldots, N$, to describe the ability of the ith leader to access the desired formation position and velocity. Hence, for the convenience of the subsequent control design, we can define a virtual spacecraft 0, whose position and velocity are equal to the desired formation position and velocity.

Assumption 13.1 For the N leaders, $\mathcal{G}_L$ is connected. For the M followers, there exists at least one path from the leader to each follower.

Lemma 13.1 [40] *Each entry of* $-\mathcal{L}_{F1}^{-1}\mathcal{L}_{F2}$, $-\mathcal{L}_{L1}^{-1}\mathcal{L}_{L2}$ *is nonnegative and each row sum of* $-\mathcal{L}_{F1}^{-1}\mathcal{L}_{F2}$, $-\mathcal{L}_{L1}^{-1}\mathcal{L}_{L2}$ *is equal to one.* $\mathcal{L}_{L1} \in \mathbb{R}^{N\times N}$, $\mathcal{L}_{F1} \in \mathbb{R}^{M\times M}$ *are symmetric positive definite.*

Definition 13.1 A two-layer framework is proposed to describe the DFC problem: the leaders' formation layer and the followers' containment layer. In the first layer, the leaders are supposed to achieve a specified formation and move together with a desired reference simultaneously. In the second layer, the followers are supposed to converge into the convex hull spanned by the leaders while maintaining the velocity matching with the leaders. The DFC is accomplished if and only if the control objectives in both layers are achieved.

Remark 13.2 Compared with [27–32], a novel two-layer definition of DCF problem is presented, which has following features:

(1) The communication topology among the leaders is distributed, as such, the configuration of the formation in the first layer is achieved via the cooperative control of the leaders.
(2) The desired formation configuration in the first layer is time-varying (see Remark 13.7), which results in a time-varying containment in the second layer and makes the task more challenging compared with the constant configuration.
(3) The information flows from the first layer to the second layer is unidirectional. In other words, the states of leaders have an influence on the behaviors of followers, but not vice versa, which accords with the practical missions.

Assumption 13.2 The leader is independent of the followers. There exist a nonnegative constant δ_L, such that

$$\|\ddot{q}_0\| \le \delta_L < \infty, \tag{13.3}$$

Remark 13.3 [41] Assumption 13.2 is essential for the reason that the adaptive NNs are only effective in a compact set, which will be detailed later. Moreover, the boundedness of the leader's acceleration is acceptable in practice.

We give other definitions and lemmas needed in this chapter.

Definition 13.2 Let $\mathbb{C}$ be a set in a real vector space $V \subseteq \mathbb{R}^p$. The set $\mathbb{C}$ is convex if, for any x and y in $\mathbb{C}$, the point $(1-t)\,x + ty \in \mathbb{C}$ for any $t \in [0,\ 1]$. The convex hull for a set of points $X \triangleq \{x_1, \ldots, x_n\}$ in V is the minimal convex set containing all points in X. Then, the convex hull of X takes the form

$$\mathrm{Co}\,(X) \triangleq \left\{ \sum_{i=1}^{n} \alpha_i x | x_i \in X,\ \alpha_i \ge 0,\ \sum_{i=1}^{n} \alpha_i = 1 \right\}. \tag{13.4}$$

Lemma 13.2 [42] *For any real numbers $c > 0$ and $\alpha \in (0, 1)$, the origin of a Lyapunov function $V(x)$ is finite-time stable if the following inequality holds*

$$\dot{V}(x) + cV(x)^{\alpha} \leq 0, \tag{13.5}$$

where the settling time is estimated by

$$T(x_0) \leq \frac{V(x_0)^{1-\alpha}}{c(1-\alpha)}. \tag{13.6}$$

13.3 First Layer: Distributed Formation Control Design for the Leaders

In this section, the distributed spacecraft formation problem is considered for the leaders modeled by MELSs with model uncertainties by output feedback. According to Definition 13.1, we will design control algorithms for the control objectives in the first layer.

(1) In the first layer, we firstly propose a DFSE to guarantee that all the leaders can obtain the desired position and velocity in a constant formation configuration case, while the time-varying formation configuration can be achieved by transforming the formation functions in Remark 13.7.
(2) Then, by employing a high-gain observer, the distributed formation control by output feedback is proposed for the leaders, and the NN technique is applied to compensate for the model uncertainties, which can also improve the system performance.

13.3.1 Distributed Finite-Time Sliding-Mode Estimator for Leaders

Inspired by [43], we propose the following DFSE protocol as

$$\dot{\hat{q}}_i = \hat{v}_i - \beta_1 \text{sgn} \sum_{j=0}^{N} a_{ij} \left(\hat{q}_i - \hat{q}_j - \delta_{ij}\right), \tag{13.7}$$

$$\dot{\hat{v}}_i = -\beta_2 \text{sgn} \sum_{j=0}^{N} a_{ij} \left(\hat{v}_i - \hat{v}_j\right), \tag{13.8}$$

where the formation constant δ_{ij} is defined as $\delta_{ij} = \delta_i - \delta_j$, δ_i and δ_j are two constants, β_1 and β_2 are positive constants, $\hat{q}_i$ and $\hat{v}_i, i = 1, \ldots, N$ are the estimations of the desired formation position and velocity for each leader, respectively, and $\hat{q}_0 = q_0$,

$\hat{v}_0 = \dot{q}_0$. It is worth mentioning that q_0 and $\dot{q}_0$ stand for the position and the velocity of the virtual spacecraft 0, respectively, which has been detailed in Remark 13.1. Define

$$Q_L = \left[q_1^T, q_2^T, \ldots, q_N^T\right]^T, \tag{13.9}$$

$$Q_0 = \left[q_0^T, q_0^T, \ldots, q_0^T\right]^T. \tag{13.10}$$

Theorem 13.1 *Let $\beta_1 > 0$ and $\beta_2 > \delta_L$. If Assumptions 13.1 and 13.2 hold, under the DFSEs (13.7) and (13.8), the leaders can acquire the precise estimations of the desired formation position and velocity in finite time, respectively, which means*

$$\hat{q}_i(t) = q_0(t) + \delta_{i0}, \tag{13.11}$$

$$\hat{v}_i(t) = \dot{q}_0(t), \quad i = 1, 2, \ldots, N. \tag{13.12}$$

Proof Let $\tilde{v}_i = \hat{v}_i - \dot{q}_0$ and $\tilde{q}_i = \hat{q}_i - \delta_i - q_0 + \delta_0$, $i = 1, \ldots, N$. Define the vectors $\tilde{v}_L = \left[\tilde{v}_1, \tilde{v}_2, \ldots, \tilde{v}_N\right]^T$ and $\tilde{q}_L = \left[\tilde{q}_1, \tilde{q}_2, \ldots, \tilde{q}_N\right]^T$. Consider the following Lyapunov function according to (13.8).

$$V_1 = \frac{1}{2}\tilde{v}_L^T\left(\mathcal{L}_{L1} \otimes I_p\right)^T \tilde{v}_L. \tag{13.13}$$

According to (13.7) and (13.8), taking the derivative of V_1 yields

$$\begin{aligned}
\dot{V}_1 =& \tilde{v}_L^T(\mathcal{L}_{L1} \otimes I_p)^T \dot{\tilde{v}}_L \\
=& \tilde{v}_L^T\left(\mathcal{L}_{L1} \otimes I_p\right)^T \left[-\beta_2 \text{sgn}\left(\left(\mathcal{L}_{L1} \otimes I_p\right)\tilde{v}_L\right) - \ddot{Q}_0\right] \\
=& \left(\left(\mathcal{L}_{L1} \otimes I_p\right)\tilde{v}_L\right)^T \left[-\beta_2 \text{sgn}\left(\left(\mathcal{L}_{L1} \otimes I_p\right)\tilde{v}_L\right)\right. \\
&\left. + \left({\mathcal{L}_{L1}}^{-1}\mathcal{L}_{L2} \otimes I_p\right)\ddot{Q}_0\right].
\end{aligned} \tag{13.14}$$

According to [44], to any $x \subseteq \mathbb{R}^p$, it holds

$$\|x\|_1 \geq \|x\|_2, \tag{13.15}$$

where $\|x\|_2 = \sqrt{x^T x}$, $\|x\|_1 = \sum_{i=1}^{p} |x_i|$, x_i is ith element of vector x. According to (13.15) and Lemma 13.1, we can get

$$\begin{aligned}
\dot{V}_1 \leq& -\left(\beta_2 - \left\|{\mathcal{L}_{L1}}^{-1}\mathcal{L}_{L2}\right\|\delta_L\right)\left\|\left(\mathcal{L}_{L1} \otimes I_p\right)\tilde{v}_L\right\|_1 \\
\leq& -\left(\beta_2 - \delta_L\right)\left\|\left(\mathcal{L}_{L1} \otimes I_p\right)\tilde{v}_L\right\|_2 \\
\leq& -\left(\beta_2 - \delta_L\right)\sqrt{\tilde{v}_L^T\left(\mathcal{L}_{L1} \otimes I_p\right)^2\tilde{v}_L} \\
\leq& -\left(\beta_2 - \delta_L\right)\lambda_{\min}\left(\mathcal{L}_{L1}\right)\|\tilde{v}\|_2 \\
\leq& -\sqrt{2}\left(\beta_2 - \delta_L\right)\frac{\lambda_{\min}\left(\mathcal{L}_{L1}\right)}{\sqrt{\lambda_{\max}\left(\mathcal{L}_{L1}\right)}}V_1^{\frac{1}{2}}.
\end{aligned} \tag{13.16}$$

Thus, it follows from Lemma 13.2 that $\hat{v}_i \to \dot{q}_0, i = 1, 2, \ldots, N$ in a settling time

$$T_1 = \frac{\sqrt{2}\sqrt{\lambda_{\max}(\mathcal{L}_{L1})}}{(\beta_2 - \delta_L)\lambda_{\min}(\mathcal{L}_{L1})} V_1^{\frac{1}{2}}(0), \tag{13.17}$$

and $\dot{q}_0$ can be utilized to replace $\hat{v}_i$, when $t \geq T_1$.

When $t \geq T_1$, construct the Lyapunov function $V_2 = \frac{1}{2}\tilde{q}_L^T (\mathcal{L}_{L1} \otimes I_p)^T \tilde{q}_L$, and take the derivative of V_2 yields,

$$\begin{aligned}
\dot{V}_2 &= \tilde{q}_L^T (\mathcal{L}_{L1} \otimes I_p)^T \dot{\tilde{q}}_L \\
&= \tilde{q}_L^T (\mathcal{L}_{L1} \otimes I_p)^T \left[-\beta_1 \mathrm{sgn}\left(\left(\mathcal{L}_{L1} \otimes I_p\right) \tilde{q}_L\right)\right] \\
&\leq -\beta_1 \left\| \left(\mathcal{L}_{L1} \otimes I_p\right) \tilde{q}_L \right\|_1 \\
&\leq -\beta_1 \left\| \left(\mathcal{L}_{L1} \otimes I_p\right) \tilde{q}_L \right\|_2 \\
&\leq -\beta_1 \sqrt{\tilde{q}_L^T \left(\mathcal{L}_{L1} \otimes I_p\right)^2 \tilde{q}_L} \\
&\leq -\beta_1 \lambda_{\min}(\mathcal{L}_{L1}) \|\tilde{q}_L\|_2 \\
&\leq -\sqrt{2}\beta_1 \frac{\lambda_{\min}(\mathcal{L}_{L1})}{\sqrt{\lambda_{\max}(\mathcal{L}_{L1})}} V_2^{\frac{1}{2}}.
\end{aligned} \tag{13.18}$$

Next, from Lemma 13.2, we have $\hat{q}_i \to q_0 + \delta_i - \delta_0 = q_0 + \delta_{i0}, i = 1, 2, \ldots, N$ in finite time with a settling time

$$T_2 = T_1 + \frac{\sqrt{2}\sqrt{\lambda_{\max}(\mathcal{L}_{L1})}}{\beta_1 \lambda_{\min}(\mathcal{L}_{L1})} V_2^{\frac{1}{2}}(T_1), \tag{13.19}$$

and $q_0 + \delta_{i0}$ can be used to replace $\hat{q}_i$, when $t \geq T_2$.

Therefore, $\hat{q}_i$ and $\hat{v}_i$, $i = 1, 2, \ldots, N$ can be used to replace the position and velocity of the desired formation when $t \geq T_2$. □

Remark 13.4 Compared with the dynamic estimator in [45], the states of leaders are not used in the DFSEs (13.7) and (13.8), which provides some convenience for us to design the output-feedback control algorithm. More specifically, only the Laplacian matrix is involved in (13.7) and (13.8), which grants the estimators independence from the controller or any system dynamics.

13.3.2 Distributed Formation Control of Leaders by Output Feedback

In this subsection, considering the case with $t \geq T_2$, the distributed spacecraft formation problem for leaders in the first layer without using velocity measurements is considered. To deal with this challenging task, we incorporate a high-gain observer into our distributed adaptive control algorithm, in which the adaptive NNs are employed to approximate the model uncertainties.

Before moving forward, define the following generalized tracking errors as

$$z_{1i} = q_i - \hat{q}_i, \tag{13.20}$$

$$z_{2i} = \dot{q}_i - \alpha_{1i}, \quad i = 1, 2, \ldots, N, \tag{13.21}$$

where the virtual control $\alpha_{1i} = \hat{v}_i - K_{1i} z_{1i}$, and the gain matrix $K_{1i} = K_{1i}^T > 0$.

Then, we have

$$\dot{z}_{1i} = z_{2i} + \alpha_{1i} - \hat{v}_i = z_{2i} - K_{1i} z_{1i}. \tag{13.22}$$

Using (2.34), the derivative of z_{2i} takes the form

$$\dot{z}_{2i} = M_i^{-1}(q_i)\left[\tau_i - C_i(q_i, \dot{q}_i)\dot{q}_i - g_i(q_i)\right] - \dot{\alpha}_{1i}. \tag{13.23}$$

Due to the model uncertainties in $M_i(q_i)$, $C_i\left(q_{i,}\dot{q}_i\right)$ and $g_i(q_i)$, we can not directly use these terms in our controller. To conquer the challenge above, we apply the NN technique to adaptively compensate for the model uncertainties and enhance the control performance, which is detailed in [46, 47]. The following function f_i is introduced:

$$f_i(q_i, \dot{q}_i, \alpha_{1i}, \dot{\alpha}_{1i}) = W_i^{*T} \phi_i(q_i, \dot{q}_i, \alpha_{1i}, \dot{\alpha}_{1i}) + \epsilon_i, \tag{13.24}$$

where $W_i^* \in \mathbb{R}^{l \times p}$ is the ideal constant weight matrix, $l > 1$ is the NN node number, $f_i(q_i, \dot{q}_i, \alpha_{1i}, \dot{\alpha}_{1i}) = g_i(q_i) + C_i(q_i, \dot{q}_i)\alpha_{1i} + M_i(q_i)\dot{\alpha}_{1i}$ represents the model uncertainties and $\phi_i = [\varphi_{i1}, \varphi_{i2}, \ldots, \varphi_{il}]^T \in \mathbb{R}^l$ is the activation function. ϵ_i is the approximation error bounded over the compact set, i.e., $\|\epsilon_i(Z_i)\| \leq \bar{\epsilon}_i$, $\forall Z_i \in \Omega_{z_i}$ with $\bar{\epsilon}_i > 0$ as an unknown constant [48], where $Z_i = \left[q_i^T, \dot{q}_i^T, \alpha_{1i}^T, \dot{\alpha}_{1i}^T\right]^T$.

Here, employing the Gaussian functions, the elements of the activation function are given as

$$\varphi_{ir}(Z_i) = \exp\left[\frac{-(Z_i - \mu_r)^T (Z_i - \mu_r)}{\eta_r{}^2}\right], \quad r = 1, \ldots, l, \tag{13.25}$$

where η_r is the Gaussian function width and μ_r is the receptive field center [49, 50].

Since the velocity measurements are not available, according to the certainty equivalence and separation properties, $\dot{q}_i$ is estimated using a high-gain observer [51], where the unmeasurable state vector z_{2i} is estimated as

$$\hat{z}_{2i} = \frac{\pi_{2i}}{\varepsilon_i} - \alpha_{1i}, \quad i = 1, \ldots, N, \tag{13.26}$$

and the dynamics of π_{2i} takes the form

$$\varepsilon_i \dot{\pi}_{1i} = \pi_{2i}, \tag{13.27}$$

$$\varepsilon_i \dot{\pi}_{2i} = -\bar{\lambda}_{1i} \pi_{2i} - \pi_1 + q_i. \tag{13.28}$$

where the parameter $\bar{\lambda}_1$ is selected s.t., the polynomial $s^2 + \bar{\lambda}_1 s + 1$ is Hurwitz and ε_i can be any small positive constant. Then, we have the following properties [51]:

$$\xi_{2i} = \frac{\pi_{2i}}{\varepsilon_i} - \dot{q}_i = -\varepsilon_i \psi_i{}^{(2)}, \tag{13.29}$$

$$\psi_i = \pi_{2i} + \bar{\lambda}_{1i} \pi_{1i}, \tag{13.30}$$

Hence, the estimation error $\tilde{z}_{2i}$ takes the form

$$\tilde{z}_{2i} = \hat{z}_{2i} - z_{2i} = \frac{\pi_{2i}}{\varepsilon_i} - \dot{q}_i = \xi_{2i}. \tag{13.31}$$

Furthermore, there exist positive constants t_{li}^* and h_{li} [52], such that $\forall t > t_{li}^*$, we have

$$\|\xi_{2i}\| \leq \varepsilon_i h_{li}. \tag{13.32}$$

Remark 13.5 To estimate the velocity information, we have introduced a high-gain observer for each leader. Additionally, π_{k+1}/ε^k will converge asymptotically to $x_1^{(k)}$, and the kth order derivative of x_1, i.e., ξ_k will converge to the origin within a short time as long as x_1 and its kth derivative is bounded. Moreover, an increase in the high-gain $1/\varepsilon_i$ will lead to a smaller convergence time. Therefore, we can say that π_{k+1}/ε^k is an effective observer to estimate output signals to the kth order. It can be noticed that the observer (13.26) is considered with $k = 2$.

We propose the following distributed spacecraft formation control algorithm (13.33) and the adaptive law (13.34) for the first layer using only position measurements

$$\tau_i = -z_{1i} - K_i \hat{z}_{2i} + \hat{W}_i^T \phi_i \left(\hat{Z}_i\right), \tag{13.33}$$

$$\dot{\hat{W}}_{i,k} = -\gamma_i \left(\phi_i \left(\hat{Z}_i\right) \hat{z}_{2i,k} + \sigma_i \hat{W}_{i,k}\right), \tag{13.34}$$

where K_i is the control gain, γ_i, σ_i are positive constants, $\hat{W}_{i,k}$ is the kth column vector of the matrix $\hat{W}_i$ and $\hat{z}_{2i,k}$ is the kth element of the vector $\hat{z}_{2i}$, $k = 1, 2, \ldots, p$, $i = 1, \ldots, N$.

Remark 13.6 We introduce the σ_i modification term in (13.34) to enhance the robustness of the systems [53]. The estimation of $W_i^{*T} \phi_i (Z_i)$ will increase to an excessively high value without the regulation of σ_i, which will eventually induce an instability of the control strategy [54].

Theorem 13.2 *In the first layer, considering the network of spacecraft formation governed by MELSs* (2.34)*, given that the output measurements are accessible, for each compact set* Ω_{i0} *where* $q_i(0)$, $\dot{q}_i(0)$, $\hat{W}_{i,1}(0), \ldots, \hat{W}_{i,k}(0) \in \Omega_{i0}$, $i = 1, \ldots, N$, $k = 1, \ldots, p$, *the tracking errors for the leaders are semi-globally uniformly ultimately bounded (SGUUB) under the distributed control algorithm* (13.33) *with the adaptive law* (13.34)*. The closed-loop error signals* z_{1i}, z_{2i} *and* $\tilde{W}_i$ *will remain, respectively, within the compact sets* $\Omega_{z_{1i}}$, $\Omega_{z_{2i}}$ *and* $\Omega_{\tilde{W}_i}$ *taking the following forms:*

$$\Omega_{z_{1i}} := \left\{ z_{1i} \in \mathbb{R}^{p\times 1} | \, \|z_{1i}\| \leq \sqrt{D_i} \right\}, \tag{13.35}$$

$$\Omega_{z_{2i}} := \left\{ z_{2i} \in \mathbb{R}^{p\times 1} | \, \|z_{2i}\| \leq \sqrt{D_i / \lambda_{\min}(M_i)} \right\}, \tag{13.36}$$

$$\Omega_{\tilde{W}_i} := \left\{ \tilde{W}_i \in \mathbb{R}^{l\times p} | \, \left\|\tilde{W}_i\right\| \leq \sqrt{\gamma_i D_i} \right\}, \tag{13.37}$$

where D_i *is a positive constant which will be detailed later.*

Proof Consider the following Lyapunov function candidate

$$V_{3i} = \frac{1}{2} z_{1i}^T z_{1i} + \frac{1}{2} z_{2i}^T M_i z_{2i} + \frac{1}{2\gamma_i} \sum_{k=1}^{p} \tilde{W}_{i,k}^T \tilde{W}_{i,k}, \tag{13.38}$$

where $\hat{W}_{i,k}$, $W_{i,k}^*$, and $\tilde{W}_{i,k}$, are, respectively, the neural network estimate, ideal value and weight error defined as

$$\tilde{W}_{i,k} = \hat{W}_{i,k} - W_{i,k}^*. \tag{13.39}$$

Substituting (13.23) into the time derivative of V_{3i} yields

$$\begin{aligned} \dot{V}_{3i} =& z_{1i}^T z_{2i} - z_{1i}^T K_{1i} z_{1i} + \frac{1}{\gamma_i} \sum_{k=1}^{p} \tilde{W}_{i,k}^T \dot{\hat{W}}_{i,k} + \frac{1}{2} z_{2i}^T \dot{M}_i z_{2i} \\ &+ z_{2i}^T \left[\tau_i - C_i(q_i, \dot{q}_i) \dot{q}_i - g_i(q_i) \right] - z_{2i}^T M_i \dot{\alpha}_{1i}. \end{aligned} \tag{13.40}$$

Applying Property 2.1, and substituting (13.33), (13.34) into (13.40), we have

$$\begin{aligned} \dot{V}_{3i} =& - z_{1i}^T K_{1i} z_{1i} - z_{2i}^T \left(K_i - \frac{1}{2} I_p \right) z_{2i} - z_{2i}^T K_i \tilde{z}_{2i} \\ &- \sum_{k=1}^{p} \left[\tilde{W}_{i,k}^T \phi_i\left(\hat{Z}_i\right) \hat{z}_{2i,k} + \sigma_i \tilde{W}_{i,k}^T \hat{W}_{i,k} \right] + \frac{1}{2} \|\bar{\epsilon}_i\|^2 \\ &- \sum_{k=1}^{p} z_{2i,k} \left[W_{i,k}^{*T} \phi_i(Z_i) - \hat{W}_{i,k}^T \phi_i\left(\hat{Z}_i\right) \right], \end{aligned} \tag{13.41}$$

where $\tilde{z}_{2i} = \hat{z}_{2i} - z_{2i}$. Note that $\hat{Z}_i$ is the input vector of Gaussian basis function, we have the following equations [55]:

$$\varphi_{ir}(\hat{Z}) = \exp\left[\frac{-\left(\hat{Z}_i - \mu_r\right)^T\left(\hat{Z}_i - \mu_r\right)}{{\eta_r}^2}\right], \tag{13.42}$$

$$r = 1, \ldots, l,$$

$$\phi_i(\hat{Z}_i) = \phi_i\left(Z_i\right) + \epsilon_i \phi_{ti}, \tag{13.43}$$

where ϕ_{ti} is a bounded vector function, ϵ_i and l are positive constants.
From (13.42) and (13.43), we can get

$$\left\|\phi_i\left(\hat{Z}_i\right)\right\|^2 \le l, \tag{13.44}$$

$$\begin{aligned}\hat{W}_{i,k}^T \phi_i\left(\hat{Z}_i\right) &= \tilde{W}_i^T \phi_i\left(\hat{Z}_i\right) + W_{i,k}^{*T}\left(\phi_i\left(Z_i\right) + \epsilon_i \phi_{ti}\right) \\ &= \tilde{W}_i^T \phi_i\left(\hat{Z}_i\right) + W_{i,k}^{*T} \phi_i\left(Z_i\right) \\ &\quad + \epsilon_i W_{i,k}^{*T} \phi_{ti}.\end{aligned} \tag{13.45}$$

Noticing $\tilde{W}_{i,k} = \hat{W}_{i,k} - W_{i,k}^*$, we have

$$\begin{aligned}-\sigma_i \tilde{W}_{i,k}^T \hat{W}_{i,k} &= -\sigma_i \tilde{W}_{i,k}^T\left(\tilde{W}_{i,k} + W_{i,k}^*\right) \\ &\le -\sigma_i\left\|\tilde{W}_{i,k}\right\|^2 + \frac{\sigma_i}{2}\left\|\tilde{W}_{i,k}\right\|^2 + \frac{\sigma_i}{2}\left\|W_{i,k}^*\right\|^2 \\ &\le \frac{\sigma_i}{2}\left(\left\|W_{i,k}^*\right\|^2 - \left\|\tilde{W}_{i,k}\right\|^2\right).\end{aligned} \tag{13.46}$$

Substituting (13.45) and (13.46) to (13.41) yields

$$\begin{aligned}\dot{V}_{3i} = &- z_{1i}^T K_{1i} z_{1i} - z_{2i}^T\left(K_i - \frac{1}{2} I_p\right) z_{2i} - z_{2i}^T K_i \tilde{z}_{2i} \\ &- \sum_{k=1}^{p} \tilde{W}_{i,k}^T \phi_i\left(\hat{Z}_i\right) \tilde{z}_{2i,k} + \sum_{k=1}^{p} \epsilon_i W_{i,k}^{*T} \phi_{ti} z_{2i,k} \\ &+ \sum_{k=1}^{p} \frac{\sigma_i}{2}\left\|W_{i,k}^*\right\|^2 - \sum_{k=1}^{p} \frac{\sigma_i}{2}\left\|\tilde{W}_{i,k}\right\|^2 + \frac{1}{2}\|\bar{\epsilon}_i\|^2.\end{aligned} \tag{13.47}$$

Since

$$\sum_{k=1}^{p} \epsilon_i W_{i,k}^{*T} \phi_{ti} z_{2i,k} \leq \sum_{k=1}^{p} \frac{1}{2} \epsilon_i^2 \left\| W_{i,k}^* \right\|^2 \|\phi_{ti}\|^2 + \frac{1}{2} z_{2i}^T z_{2i}, \tag{13.48}$$

$$-\sum_{k=1}^{p} \tilde{W}_{i,k}^T \phi_i \left(\hat{Z}_i\right) \tilde{z}_{2i,k} \leq \frac{1}{\sigma_i} \tilde{z}_{2i}^T \tilde{z}_{2i} \left\| \phi_i \left(\hat{Z}_i\right) \right\|^2 + \sum_{k=1}^{p} \frac{\sigma_i}{4} \left\| \tilde{W}_{i,k} \right\|^2, \tag{13.49}$$

we can then get

$$\begin{aligned} \dot{V}_{3i} \leq & - z_{1i}^T K_{1i} z_{1i} - z_{2i}^T \left(K_i - I_p\right) z_{2i} - z_{2i}^T K_i \tilde{z}_{2i} \\ & + \frac{1}{\sigma_i} \tilde{z}_{2i}^T \tilde{z}_{2i} \left\| \phi_i \left(\hat{Z}_i\right) \right\|^2 - \sum_{k=1}^{p} \frac{\sigma_i}{4} \left\| \tilde{W}_{i,k} \right\|^2 \\ & + \sum_{k=1}^{p} \left(\frac{\sigma_i}{2} + \frac{1}{2} \epsilon_i^2 \|\phi_{ti}\|^2 \right) \left\| W_{i,k}^* \right\|^2 + \frac{1}{2} \|\bar{\epsilon}_i\|^2. \end{aligned} \tag{13.50}$$

Noticing $-z_{2i}^T K_i \tilde{z}_{2i} \leq z_{2i}^T z_{2i}/2 + (K_i \tilde{z}_{2i})^T (K_i \tilde{z}_{2i})/2$ and substituting (13.31) and (13.44) to (13.50), we further have

$$\begin{aligned} \dot{V}_{3i} \leq & - z_{1i}^T K_{1i} z_{1i} - z_{2i}^T \left(K_i - \frac{3}{2} I_p \right) z_{2i} + \frac{1}{2} \|\bar{\epsilon}_i\|^2 \\ & + \frac{1}{2} \sum_{k=1}^{p} \left(\sigma_i + \epsilon_i^2 \|\phi_{ti}\|^2\right) \left\| W_{i,k}^* \right\|^2 - \sum_{k=1}^{p} \frac{\sigma_i}{4} \left\| \tilde{W}_{i,k} \right\|^2 \\ & + \frac{\lambda_{\max}}{2} \left(K_i^T K_i + \frac{2l}{\sigma_i} I_p \right) \varepsilon_i^2 h_{li}^2 \\ \leq & - \rho_i V_{3i} + C_i, \end{aligned} \tag{13.51}$$

where ρ_i and C_i are two constants taking the following forms:

$$\rho_i = \min \left(2\lambda_{\min}\left(K_{1i}\right), \frac{2\lambda_{\min}\left(K_i - \frac{3}{2} I_p\right)}{\lambda_{\max}\left(M_i\right)}, \frac{1}{2} \gamma_i \sigma_i \right), \tag{13.52}$$

$$\begin{aligned} C_i = & \frac{1}{2} \sum_{k=1}^{p} \left(\sigma_i + \epsilon_i^2 \|\phi_{ti}\|^2\right) \left\| W_{i,k}^* \right\|^2 + \frac{1}{2} \|\bar{\epsilon}_i\|^2 \\ & + \frac{\lambda_{\max}}{2} \left(K_i^T K_i + \frac{2l}{\sigma_i} I_p \right) \varepsilon_i^2 h_{li}^2. \end{aligned} \tag{13.53}$$

To ensure $\rho_i > 0$, we chose the control gains K_{1i} and K_i as

$$\lambda_{\min}(K_{1i}) > 0, \tag{13.54}$$

$$\lambda_{\min}\left(K_i - \frac{3}{2}I_p\right) > 0. \tag{13.55}$$

From the analysis above, we know z_{1i}, z_{2i}, and $\tilde{W}_i$ are SGUUB. For completeness, we provide the details of the proof here.

$$\begin{aligned} V_{3i}(t) \leq & V_{3i}(0)\, e^{-\rho_i t} + \frac{C_i}{\rho_i}\left(1 - e^{-\rho_i t}\right) \\ \leq & V_{3i}(0) + \frac{C_i}{\rho_i}. \end{aligned} \tag{13.56}$$

Thus, we have

$$\|z_{1i}\| \leq \sqrt{D_i}, \tag{13.57}$$

where $D_i = 2\left(V_{3i}(0) + C_i/\rho_i\right)$. Therefore, z_{1i} remains in the compact set $\Omega_{z_{1i}}$. Bounds for z_{2i} and $\tilde{W}_i$ can be given using a similar procedure. On the other hand, according to Theorem 13.3, the proposed DFSEs can, respectively, yield accurate desired formation position and velocity estimations when $t \geq T_2$. As such,

$$\hat{q}_i(t) = q_0(t), \tag{13.58}$$

$$\hat{v}_i(t) = \dot{q}_0(t), \quad i = 1, 2, \ldots, N, \quad t \geq T_2. \tag{13.59}$$

Hence, it is straightforward to present that all the closed-loop error signals z_{1i}, z_{2i}, $\tilde{W}_i$ for leaders in the first layer are SGUUB and the distributed formation control objective for the first layer is achieved using only position measurements, which concludes the proof. □

Remark 13.7 Note that the desired relative formation configuration δ_{ij} between leader i and leader j is a constant in (13.7). To turn our control scheme into a more general one, we can consider the time-varying formation configuration and rewrite the DFSEs and the generalized tracking errors as

$$\dot{\hat{q}}_i = \hat{v}_i - \beta_1 \operatorname{sgn} \sum_{j=0}^{N} a_{ij}\left(\hat{q}_i - \hat{q}_j\right), \tag{13.60}$$

$$\dot{\hat{v}}_i = -\beta_2 \operatorname{sgn} \sum_{j=0}^{N} a_{ij}\left(\hat{v}_i - \hat{v}_j\right), \tag{13.61}$$

$$z_{1i} = q_i - \hat{q}_i - r_i, \tag{13.62}$$

$$z_{2i} = \dot{q}_i - \alpha_{1i}, \tag{13.63}$$

$$\alpha_{1i} = \hat{v}_i + \dot{r}_i - K_{1i} z_{1i}, \quad i = 1, 2, \ldots, N, \tag{13.64}$$

where $r_i(t)$ is the formation function which denotes the time-varying relative distance decomposed from the desired formation configuration. The stability of the closed-loop systems can be achieved by following the proof of Theorems 13.1 and 13.2.

Remark 13.8 The tracking errors of all leaders have been shown to remain within a neighborhood of zero. From (13.52) and (13.53), we can conclude that an increase in the control gain K_{1i}, K_i, the high-gain $1/\varepsilon_i$ and the neural network node number l_i will result in a lower residual error. Additionally, we can select a much smaller ϵ_i if the output measurements are influenced by Gaussian white noise with mean zero. Applying the saturation function, the peaking phenomenon of the observer (13.26) can be avoided utilizing the process in [56].

Remark 13.9 In the absence of velocity sensors, we have designed a distributed formation control algorithm combined with the high-gain observer after the DFSEs yield accurate estimations of the desired formation position and velocity. More specifically, only the position measurements of the networked ELSs are utilized in the controller (13.33) and the adaptive law (13.34). From Remark 13.4, we know the velocity measurements are not used in the DFSEs (13.7) and (13.8) either, which ensures a rigorous proof of the proposed output feedback control schemes.

Remark 13.10 Compared with the tracking control schemes and the formation control schemes with a single leader, the DFC control algorithm proposed in this chapter is more robust for several reasons as follows.

(1) The virtual spacecraft 0 does not really exist, so there is nothing to worry about some unexpected faults or breakdowns.
(2) The formation can always be achieved even if some of the leaders lose efficacy, as long as the connectivity still holds in $\mathcal{G}_L$ formed by the rest of the leaders.

13.4 Second Layer: Distributed Containment Control Design for the Followers

In this section, the distributed containment problem is considered for the follower spacecraft modeled by MELSs with model uncertainties by output feedback. According to Definition 13.1, we will design control algorithms for the control goals in the second layer.

(1) In the second layer, we design a similar DFSE to generate the precise estimation of the weighted average of the leaders' velocities and positions in finite time.

(2) Next, the distributed containment control by output feedback is designed for the followers, which has similar yet different properties compared with the controller in the first layer.

13.4.1 Distributed Finite-Time Sliding-Mode Estimator for Followers

From Theorem 13.2, we know the accelerations of the leaders are bounded. Here, for convenience, define a positive constant δ_F satisfying

$$\|\ddot{q}_i\| \leq \delta_F < \infty, \ i = 1, 2, \ldots, N. \tag{13.65}$$

Next, we propose the following DFSE protocol as

$$\dot{\hat{q}}_i = \hat{v}_i - \beta_3 \text{sgn} \sum_{j=1}^{N+M} a_{ij} \left(\hat{q}_i - \hat{q}_j\right), \tag{13.66}$$

$$\dot{\hat{v}}_i = -\beta_4 \text{sgn} \sum_{j=1}^{N+M} a_{ij} \left(\hat{v}_i - \hat{v}_j\right), \tag{13.67}$$

$$i = N+1, N+2, \ldots, N+M,$$

where β_3 and β_4 are positive constants, $\hat{q}_i$ and $\hat{v}_i$, $i = N+1, \ldots, N+M$ are the estimations of the weighed average of the leaders' position and velocity for each follower, respectively, and $\hat{q}_i = q_i$, $\hat{v}_i = \dot{q}_i$ when $i = 1, \ldots, N$. Define

$$Q_F = \left[q_{N+1}^T, q_{N+2}^T, \ldots, q_{N+M}^T\right]^T, \tag{13.68}$$

$$\begin{aligned} C_F &= \left[c_{N+1}, c_{N+2}, \ldots, c_{N+M}\right]^T \\ &= -\left(L_{F1}^{-1} L_{F2} \otimes I_p\right) Q_L. \end{aligned} \tag{13.69}$$

Theorem 13.3 *Let $\beta_3 > 0$ and $\beta_4 > \delta_F$. If Assumption 13.1 hold, under the DFSEs (13.66) and (13.67), the followers can acquire the precise estimations of the weighed average of the leaders' position and velocity in finite time, respectively, which means*

$$\hat{q}_i \to c_i, \tag{13.70}$$

$$\hat{v}_i \to \dot{c}_i, \ i = N+1, N+2, \ldots, N+M. \tag{13.71}$$

Proof Let $\tilde{v}_i = \hat{v}_i - \dot{c}_i$ and $\tilde{q}_i = \hat{q}_i - c_i$, $i = N+1, \ldots, N+M$. Define the following vectors

$$\tilde{v}_F = \left[\tilde{v}_{N+1}, \tilde{v}_{N+2}, \ldots, \tilde{v}_{N+M}\right]^T, \tag{13.72}$$

$$\tilde{q}_F = \left[\tilde{q}_{N+1}, \tilde{q}_{N+2}, \ldots, \tilde{q}_{N+M}\right]^T. \tag{13.73}$$

Consider the following Lyapunov functions according to (13.66) and (13.67).

$$V_4 = \frac{1}{2}\tilde{v}_F^T(\mathcal{L}_{F1} \otimes I_p)^T \tilde{v}_F, \tag{13.74}$$

$$V_5 = \frac{1}{2}\tilde{q}_F^T(\mathcal{L}_{F1} \otimes I_n)^T \tilde{q}_F. \tag{13.75}$$

Using a similar procedure like (13.14)–(13.19), we have $\hat{v}_i \to \dot{c}_i, \hat{q}_i \to c_i, i = N + 1, N + 2, \ldots, N + M$ in settling times T_3 and T_4, respectively, as follows.

$$T_3 = \frac{\sqrt{2}\sqrt{\lambda_{\max}(\mathcal{L}_{F1})}}{(\beta_4 - \delta_L)\,\lambda_{\min}(\mathcal{L}_{F1})} V_4^{\frac{1}{2}}(0), \tag{13.76}$$

$$T_4 = T_3 + \frac{\sqrt{2}\sqrt{\lambda_{\max}(\mathcal{L}_{F1})}}{\beta_3 \lambda_{\min}(\mathcal{L}_{F1})} V_5^{\frac{1}{2}}(T_3). \tag{13.77}$$

Therefore, $\hat{q}_i$ and $\hat{v}_i$, $i = N + 1, N + 2, \ldots, N + M$ can be used to replace the weighted average of the positions and velocities of the leaders when $t \geq T_4$. □

13.4.2 Distributed Containment Control of Followers by Output Feedback

In this subsection, considering the case with $t \geq T_4$, the distributed containment problem for follower spacecraft in the second layer without using velocity measurements is considered. Similar to the approach in the first layer, we incorporate a high-gain observer into our distributed adaptive control algorithm, while the adaptive NNs are employed to approximate the model uncertainties.

We first define the following generalized tracking errors as

$$z_{1i} = q_i - \hat{q}_i, \tag{13.78}$$

$$z_{2i} = \dot{q}_i - \alpha_{1i}, \quad i = N + 1, N + 2, \ldots, N + M, \tag{13.79}$$

where the virtual control $\alpha_{1i} = \hat{v}_i - K_{2i} z_{1i}$ and the gain matrix $K_{2i} = K_{2i}^T > 0$.

Similarly, employing the adaptive NNs, we can compensate the model uncertainties as follows

$$f_i(q_i, \dot{q}_i, \alpha_{1i}, \dot{\alpha}_{1i}) = W_i^{*T} \phi_i(q_i, \dot{q}_i, \alpha_{1i}, \dot{\alpha}_{1i}) + \epsilon_i, \tag{13.80}$$

where $f_i(q_i, \dot{q}_i, \alpha_{1i}, \dot{\alpha}_{1i}) = -g_i(q_i) - C_i(q_i, \dot{q}_i)\alpha_{1i} - M_i(q_i)\dot{\alpha}_{1i}$, $i = N+1, N+2, \ldots, N+M$ denotes the model uncertainties.

Since the velocity measurements are not available, similar to the design in the first layer (13.26) and (13.32), we utilize the high-gain observer to estimate the unmeasurable state vector

$$\hat{z}_{2i} = \frac{\pi_{2i}}{\varepsilon_i} - \alpha_{1i}, \; i = N+1, N+2, \ldots, N+M, \tag{13.81}$$

and the estimate error $\tilde{z}_{2i}$ is bounded by

$$\|\tilde{z}_{2i}\| \leq \varepsilon_i h_{fi}, \; \forall t > t_{fi}^*, \tag{13.82}$$

where h_{fi} and t_{fi}^* are positive constants.

We then propose the following distributed containment control algorithm (13.83) and the adaptive law (13.84) for the second layer using only position measurements

$$\tau_i = -z_{1i} - K_i\hat{z}_{2i} + \hat{W}_i^T\phi_i\left(\hat{Z}_i\right), \tag{13.83}$$

$$\dot{\hat{W}}_{i,k} = -\gamma_i\left(\phi_i\left(\hat{Z}_i\right)\hat{z}_{2i,k} + \sigma_i\hat{W}_{i,k}\right), \tag{13.84}$$

where K_i is the control gain, γ_i, σ_i are positive constants, $\hat{W}_{i,k}$ is the kth column vector of the matrix $\hat{W}_i$ and $\hat{z}_{2i,k}$ is the kth element of the vector $\hat{z}_{2i}$, $k = 1, 2, \ldots, p$, $i = N+1, N+2, \ldots, N+M$.

Theorem 13.4 *In the second layer, considering the network of spacecraft formation governed by MELSs (2.34), given that the output measurements are accessible, for each compact set* Ω_{i0} *where* $q_i(0)$, $\dot{q}_i(0)$, $\hat{W}_{i,1}(0), \ldots, \hat{W}_{i,k}(0) \in \Omega_{i0}$, $i = N+1, N+2, \ldots, N+M$ $k = 1, \ldots, p$, *the tracking errors for the followers are SGUUB under the distributed control algorithm (13.83) with the adaptive law (13.83). The closed-loop error signals* z_{1i}, z_{2i} *and* $\tilde{W}_i$ *will remain, respectively, within the compact sets* $\Omega_{z_{1i}}$, $\Omega_{z_{2i}}$ *and* $\Omega_{\tilde{W}_i}$ *taking the following forms:*

$$\Omega_{z_{1i}} := \left\{z_{1i} \in \mathbb{R}^{p\times 1} | \, \|z_{1i}\| \leq \sqrt{D_i}\right\}, \tag{13.85}$$

$$\Omega_{z_{2i}} := \left\{z_{2i} \in \mathbb{R}^{p\times 1} | \, \|z_{2i}\| \leq \sqrt{D_i/\lambda_{\min}(M_i)}\right\}, \tag{13.86}$$

$$\Omega_{\tilde{W}_i} := \left\{\tilde{W}_i \in \mathbb{R}^{l\times p} | \left\|\tilde{W}_i\right\| \leq \sqrt{\gamma_i D_i}\right\}, \tag{13.87}$$

Proof A part of the proof is omitted here, please refer to (13.38)–(13.53) for more details. Similarly, we have

$$\|z_{1i}\| \le \sqrt{D_i}, \tag{13.88}$$
$$\|z_{2i}\| \le \sqrt{D_i/\lambda_{\min}(M_i)}, \tag{13.89}$$
$$\left\|\tilde{W}_i\right\| \le \sqrt{\gamma_i D_i}, \tag{13.90}$$

where $D_i = 2\left(V_{6i}(0) + C_i/\rho_i\right)$, V_{6i} is a Lyapunov function chosen identical with V_{3i}. We have two constants ρ_i and C_i, $i = N+1, N+2, \ldots, N+M$ taking the following forms:

$$\rho_i = \min\left(2\lambda_{\min}(K_{2i}), \frac{2\lambda_{\min}\left(K_i - \frac{3}{2}I_p\right)}{\lambda_{\max}(M_i)}, \frac{1}{2}\gamma_i\sigma_i\right), \tag{13.91}$$

$$\begin{aligned} C_i =& \frac{1}{2}\sum_{k=1}^{p}\left(\sigma_i + \epsilon_i^2\|\phi_{ti}\|^2\right)\left\|W_{i,k}^*\right\|^2 + \frac{1}{2}\|\bar{\epsilon}_i\|^2 \\ &+ \frac{\lambda_{\max}}{2}\left(K_i^T K_i + \frac{2l}{\sigma_i}I_p\right)\varepsilon_i^2 h_{fi}^2, \end{aligned} \tag{13.92}$$

in which the control gains K_{2i} and K_i are chosen as

$$\lambda_{\min}(K_{2i}) > 0, \tag{13.93}$$
$$\lambda_{\min}\left(K_i - \frac{3}{2}I_p\right) > 0. \tag{13.94}$$

Next, according to Theorem 13.4, the proposed DFSEs can, respectively, yield precise estimations of the weighed average of the leaders' position and velocity in finite time when $t \ge T_4$. As such,

$$\hat{q}_i(t) = c_i(t), \tag{13.95}$$
$$\hat{v}_i(t) = \dot{c}_i(t), \quad i = N+1, \ldots, N+M, \quad t \ge T_4. \tag{13.96}$$

Hence, it is straightforward to present that all the closed-loop error signals z_{1i}, z_{2i}, $\tilde{W}_i$, $i = N+1, N+2, \ldots, N+M$ for followers in the second layer are SGUUB and the distributed containment control objective for the second layer is achieved using only position measurements, which concludes the proof. □

Remark 13.11 According to the two-layer framework of DFC problem in Definition 13.1, we can transform the DFC problem into 4 basic distributed problems as follows:

(1) Suppose that $M = 0$, $q_0 = \dot{q}_0 = 0$, and $\delta_{ij} = 0$ the DFC problem becomes the leaderless synchronization problem.
(2) Suppose that $M = 0$, the DFC problem becomes the distributed formation control problem.
(3) Suppose that $N = 1$, the DFC problem becomes the distributed coordinated tracking problem.

(4) Suppose that the leaders' desired trajectories can be globally accessed and there is no demand on the formation function δ_{ij}, the DFC problem becomes the distributed containment problem.

Therefore, the two-layer DFC control algorithm proposed in this chapter is a general approach, which can be applied to the above problems in a unified framework.

Remark 13.12 The distributed control scheme (13.83) and adaptive law (13.84) in the second layer are similar to those (13.33), (13.34) in the first layer, however, they have different meanings and play different roles in the DFC problem. The boundedness conclusion derived from Theorem 13.2 in the first layer is the precondition of Theorems 13.3 and 13.4 in the second layer. On the other hand, a leader can only access information from part of the other leaders while a follower can communicate with a subset of both leaders and the other followers. More specifically, the states of leaders have an influence on the behaviors of followers, but not vice versa. In conclusion, the first layer is the foundation of the second layer, and they are not two independent problems but highly coupled.

Remark 13.13 The closed-loop errors in Theorem 13.4 are not effected by those in Theorem 13.2, because the control objective of the second layer is to drive all the followers into the convex hull spanned by the leaders' actual positions. Hence, the containment error for the ith follower should be derived from $q_i(t) - c_i(t)$, $i = N+1, N+2, \ldots, N+M$.

13.5 Simulation

Numerical analysis has been presented to validate the feasibility of the proposed approach. We consider a group of 16 networked spacecraft, where there are 8 leaders labeled as $1, 2, \ldots, 8$ in the first layer and 8 followers labeled as $9, 10, \ldots, 16$ in the second layer. In the Local-Vertical-Local-Horizontal (LVLH) rotating frame, the relative orbital motion of the ith spacecraft can be modeled as (2.34). The spacecraft formation relative orbit motion equations can be rewritten in the form of (2.34).

$$L = \begin{bmatrix} 0 & 0 & 0 & 0 & 0 & 0 & 0 & 0 & 0 & 0 & 0 & 0 & 0 & 0 & 0 & 0 & 0 \\ 0 & 2 & -1 & 0 & 0 & 0 & 0 & 0 & -1 & 0 & 0 & 0 & 0 & 0 & 0 & 0 & 0 \\ 0 & -1 & 2 & -1 & 0 & 0 & 0 & 0 & 0 & 0 & 0 & 0 & 0 & 0 & 0 & 0 & 0 \\ 0 & 0 & -1 & 2 & -1 & 0 & 0 & 0 & 0 & 0 & 0 & 0 & 0 & 0 & 0 & 0 & 0 \\ 0 & 0 & 0 & -1 & 2 & -1 & 0 & 0 & 0 & 0 & 0 & 0 & 0 & 0 & 0 & 0 & 0 \\ 0 & 0 & 0 & 0 & -1 & 2 & -1 & 0 & 0 & 0 & 0 & 0 & 0 & 0 & 0 & 0 & 0 \\ 0 & 0 & 0 & 0 & 0 & -1 & 2 & -1 & 0 & 0 & 0 & 0 & 0 & 0 & 0 & 0 & 0 \\ 0 & 0 & 0 & 0 & 0 & 0 & -1 & 2 & -1 & 0 & 0 & 0 & 0 & 0 & 0 & 0 & 0 \\ -1 & -1 & 0 & 0 & 0 & 0 & 0 & -1 & 3 & 0 & 0 & 0 & 0 & 0 & 0 & 0 & 0 \\ 0 & -1 & 0 & 0 & 0 & 0 & 0 & 0 & 0 & 3 & -1 & 0 & 0 & 0 & 0 & 0 & -1 \\ 0 & 0 & -1 & 0 & 0 & 0 & 0 & 0 & 0 & -1 & 3 & -1 & 0 & 0 & 0 & 0 & 0 \\ 0 & 0 & 0 & -1 & 0 & 0 & 0 & 0 & 0 & 0 & -1 & 3 & -1 & 0 & 0 & 0 & 0 \\ 0 & 0 & 0 & 0 & -1 & 0 & 0 & 0 & 0 & 0 & 0 & -1 & 3 & -1 & 0 & 0 & 0 \\ 0 & 0 & 0 & 0 & 0 & -1 & 0 & 0 & 0 & 0 & 0 & 0 & -1 & 3 & -1 & 0 & 0 \\ 0 & 0 & 0 & 0 & 0 & 0 & -1 & 0 & 0 & 0 & 0 & 0 & 0 & -1 & 3 & -1 & 0 \\ 0 & 0 & 0 & 0 & 0 & 0 & 0 & -1 & 0 & 0 & 0 & 0 & 0 & 0 & -1 & 3 & -1 \\ 0 & 0 & 0 & 0 & 0 & 0 & 0 & 0 & -1 & -1 & 0 & 0 & 0 & 0 & 0 & -1 & 3 \end{bmatrix} \tag{13.97}$$

We consider two different cases for the simulation studies. Firstly, we evaluate the proposed two-layer DFC control laws (13.33) and (13.83) for the spacecraft formation governed by MELSs with model uncertainties, where the desired formation configuration is a constant. Secondly, we modify our control algorithms according to Remark 13.7 and apply them to the DFC case with a time-varying formation configuration.

In the first case, suppose that the group of spacecraft move on a near-circular orbit, whose initial elements are given in Table 13.1. The two-layer network topology among 16 spacecraft is represented in Fig. 13.1, where spacecraft i can access information from spacecraft j if there is an arrow from spacecraft j to spacecraft i, $(i, j = 0, \ldots, 16)$. According to Fig. 13.1, the corresponding Laplacian matrix defined as (13.1) takes the form (see (13.97)). The other parameters for simulation are given in Table 13.2.

Table 13.1 The initial elements of the near-circular orbit

Initial elements	Values
Semi-major axis a	7136 km
Eccentricity e	0.001
Inclination i	60°
Right ascension of ascending node Ω	10°
Periapsis ω	30°
True anomaly υ	0°

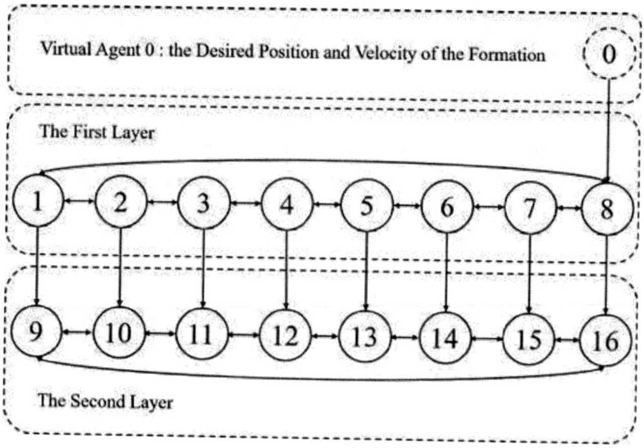

Fig. 13.1 The Network topology among 16 spacecraft

Table 13.2 The parameters for simulation

Parameters	Values	Parameters	Values
m_i	35 kg	β_1	5
ε_i	0.02	β_2	5
$\bar{\lambda}_{1i}$	1	β_3	5
K_i	$0.1I_3$	β_4	5
K_{1i}, K_{2i}	$5I_3$	σ_i	5
γ_i	4	l	256
η_r	10	μ_r	Random $[-1, 1]$
q_0	$\left[t; 200\sin\left(\frac{\pi t}{200}\right); t\right]$ m	$\hat{W}_i(0)$	0
δ_{18}	$[50, -50, 0]^T$ m	$q_1(0)$	$[-75, -125, 0]^T$ m
δ_{21}	$[100, 0, 0]^T$ m	$q_2(0)$	$[75, -125, 0]^T$ m
δ_{32}	$[50, 50, 0]^T$ m	$q_3(0)$	$[125, -75, 0]^T$ m
δ_{43}	$[0, 100, 0]^T$ m	$q_4(0)$	$[125, 75, 0]^T$ m
δ_{54}	$[-50, 50, 0]^T$ m	$q_5(0)$	$[75, 125, 0]^T$ m
δ_{65}	$[100, 0, 0]^T$ m	$q_6(0)$	$[-75, 125, 0]^T$ m
δ_{76}	$[-50, -50, 0]^T$ m	$q_7(0)$	$[-125, 75, 0]^T$ m
δ_{87}	$[0, -100, 0]^T$ m	$q_8(0)$	$[-125, -75, 0]^T$ m
δ_{80}	$[-100, -50, 0]^T$ m	$q_9(0)$	$[-150, -150, 20]^T$ m
$q_{10}(0)$	$[0, -180, 0]^T$ m	$q_{11}(0)$	$[150, -150, 20]^T$ m
$q_{12}(0)$	$[150, 0, 20]^T$ m	$q_{13}(0)$	$[150, 150, 20]^T$ m
$q_{14}(0)$	$[0, 150, 20]^T$ m	$q_{15}(0)$	$[-150, 150, 0]^T$ m
$q_{16}(0)$	$[-180, 0, 0]^T$ m	$\dot{q}_i(0)$	0 m/s

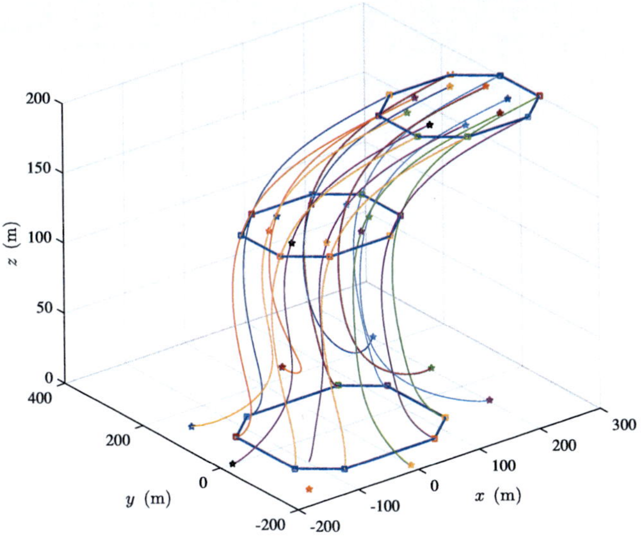

Fig. 13.2 The trajectories of 16 spacecraft with a constant formation configuration

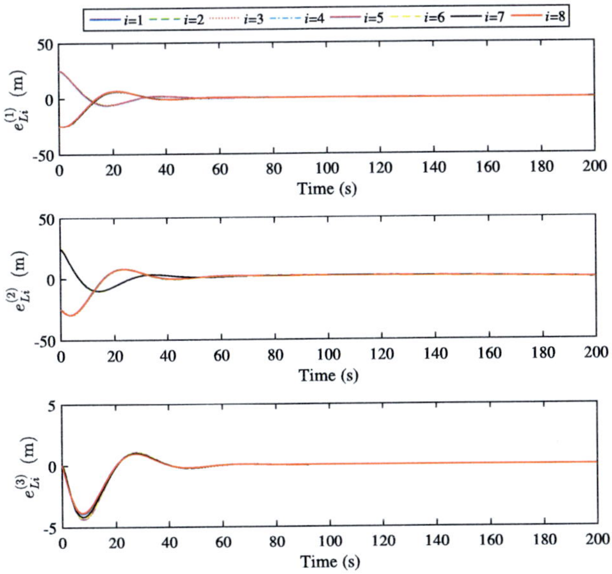

Fig. 13.3 The formation errors of 8 leader spacecraft in the first layer

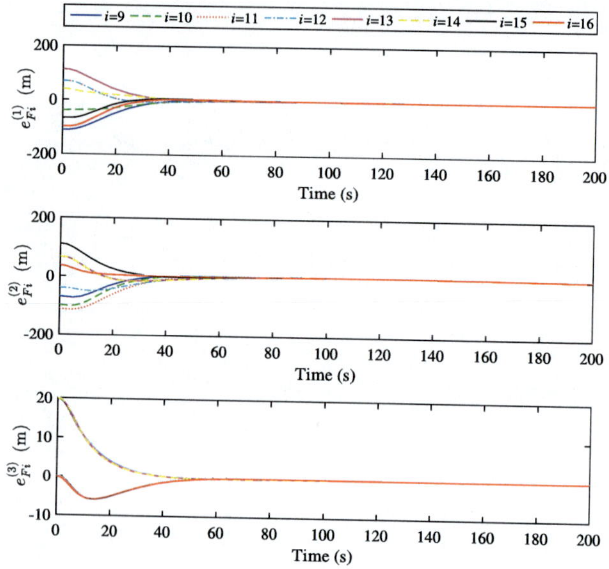

Fig. 13.4 The containment errors of 8 follower spacecraft in the second layer

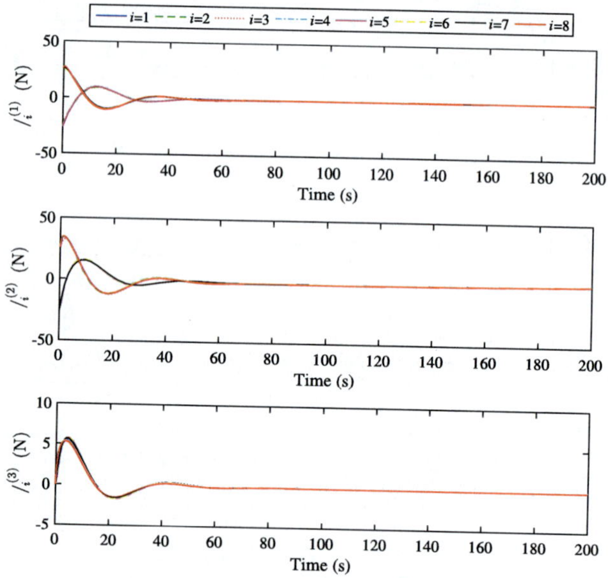

Fig. 13.5 The control forces of 8 leader spacecraft in the first layer

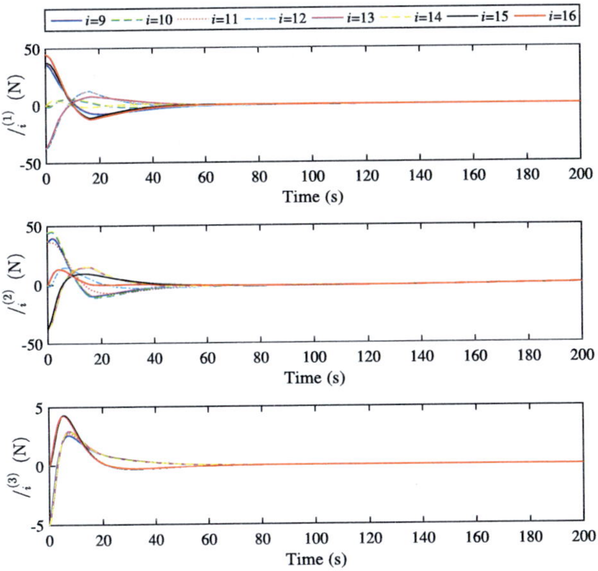

Fig. 13.6 The control forces of 8 follower spacecraft in the second layer

In the first case, the trajectories of 16 spacecraft are given in Fig. 13.2. It can be observed that the proposed two-layer control algorithms (13.33) and (13.83) can guarantee a good performance for the 16 spacecraft. In the first layer, the leaders achieve a constant formation configuration and maintain the performance. In the second layer, the followers converge into the convex hull spanned by the leaders while maintaining the velocity matching with the leaders. Thus, according to Remark 13.7, the DFC is accomplished. Figure 13.3 demonstrates that the formation errors of 8 leader spacecraft in the first layer converge to the origin within 80s. In addition, leaders which can directly receive the information from virtual spacecraft 0 has the smallest formation error. However, other leaders with more communication transmission links may yield higher formation errors. Figure 13.4 shows that the containment errors of 8 follower spacecraft in the second layer converge to the origin within 80s. More specifically, simulation result indicates that the containment errors of followers reduce with the increase in the adaptation gain γ_i. The corresponding control input of each spacecraft is given in Figs. 13.5 and 13.6, which is acceptable to spacecraft in practice. Note that all followers can trace and converge into the specified formation of the leaders, and maintain the containment performance hereafter. Therefore, the group of spacecraft has achieved the two-layer formation-containment objective.

Next, we consider the second case with a time-varying formation configuration to better exam the performance of the proposed two-layer control scheme modified as Remark 13.7, where the time-varying formation function $r_i(t)$ is detailed in Table 13.3 and the other simulation parameters are chosen the same as those in the first case. As can be seen from Table 13.3, the time-varying formation functions

Table 13.3 The time-varying formation function

$r_i(t)$	Values
$r_1(t)$	$\left[-50-\frac{25}{1+e^{-(t-100)}}, -100+\frac{25}{1+e^{-(t-100)}}, \frac{30}{1+e^{-(t-100)}}\right]^T$
$r_2(t)$	$\left[50+\frac{25}{1+e^{-(t-100)}}, -100+\frac{25}{1+e^{-(t-100)}}, \frac{30}{1+e^{-(t-100)}}\right]^T$
$r_3(t)$	$\left[100-\frac{25}{1+e^{-(t-100)}}, -50-\frac{25}{1+e^{-(t-100)}}, -\frac{30}{1+e^{-(t-100)}}\right]^T$
$r_4(t)$	$\left[100-\frac{25}{1+e^{-(t-100)}}, 50+\frac{25}{1+e^{-(t-100)}}, \frac{30}{1+e^{-(t-100)}}\right]^T$
$r_5(t)$	$\left[50+\frac{25}{1+e^{-(t-100)}}, 100-\frac{25}{1+e^{-(t-100)}}, -\frac{30}{1+e^{-(t-100)}}\right]^T$
$r_6(t)$	$\left[-50-\frac{25}{1+e^{-(t-100)}}, 100-\frac{25}{1+e^{-(t-100)}}, \frac{30}{1+e^{-(t-100)}}\right]^T$
$r_7(t)$	$\left[-100-\frac{25}{1+e^{-(t-100)}}, 50+\frac{25}{1+e^{-(t-100)}}, -\frac{30}{1+e^{-(t-100)}}\right]^T$
$r_8(t)$	$\left[-100+\frac{25}{1+e^{-(t-100)}}, -50-\frac{25}{1+e^{-(t-100)}}, -\frac{30}{1+e^{-(t-100)}}\right]^T$

$r_i(t)$, $i = 1, 2, \ldots, N$ are continuous and smooth, which are implementable in the proposed two-layer controller. Figure 13.7 is given to present the trajectories of 16 spacecraft with a time-varying formation configuration. It can be observed that the DFC is achieved, where the formation configuration transforms from an octagon into a cube during the formation flying of 16 spacecraft. The other simulation results of the second case are omitted here.

The above simulation results have proved that the control approach proposed in this chapter is effective and feasible in both constant and time-varying formation cases. It is worth mentioning that only the output information rather than state information is required for the two-layer controllers in both situations, which is more feasible in practical applications.

13.6 Conclusion

In this chapter, the two-layer DFC control schemes have been proposed for multi-spacecraft formation systems with model uncertainties. Under the strict condition that the leaders are neighbor of only parts of the followers, we design DFSEs to estimate the desired positions and velocities for the spacecraft in both layers, respectively. Then, the NNs are used to approximate model uncertainties of the multi-spacecraft formation systems and improve the system performance. Next, when only the position measurements are available, the high-gain observer is developed for each spacecraft to estimate the velocity information. Finally, the numerical examples have been performed to demonstrate the feasibility of the proposed two-layer DFC control

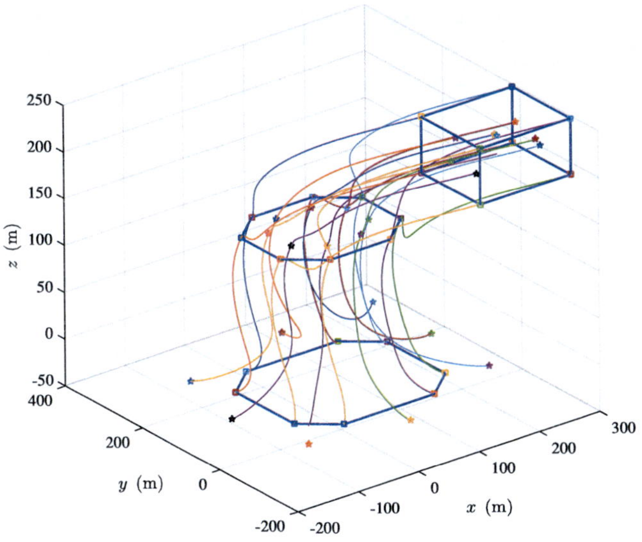

Fig. 13.7 The trajectories of 16 spacecraft with a time-varying formation configuration

algorithms in both the constant formation configuration case and the time-varying formation configuration case.

References

1. Olfati-Saber, R., Fax, J.A., Murray, R.M.: Consensus and cooperation in networked multi-agent systems. Proceedings of the IEEE **95**(1), 215–233 (2007)
2. Bullo, F., Cortés, J., Martinez, S.: Distributed control of robotic networks: a mathematical approach to motion coordination algorithms, vol. 27. Princeton University Press (2009)
3. Cao, Y., Yu, W., Ren, W., Chen, G.: An overview of recent progress in the study of distributed multi-agent coordination. IEEE Transactions on Industrial informatics **9**(1), 427–438 (2012)
4. Jadbabaie, A., Lin, J., Morse, A.S.: Coordination of groups of mobile autonomous agents using nearest neighbor rules. IEEE Transactions on Automatic Control **48**(6), 988–1001 (2003). https://doi.org/10.1109/TAC.2003.812781
5. Olfati-Saber, R., Murray, R.M.: Consensus problems in networks of agents with switching topology and time-delays. IEEE Transactions on Automatic Control **49**(9), 1520–1533 (2004). https://doi.org/10.1109/TAC.2004.834113
6. Ren, W., Beard, R.W.: Consensus seeking in multiagent systems under dynamically changing interaction topologies. IEEE Transactions on Automatic Control **50**(5), 655–661 (2005)
7. Yu, W., Chen, G., Cao, M.: Some necessary and sufficient conditions for second-order consensus in multi-agent dynamical systems. Automatica **46**(6), 1089–1095 (2010)
8. Ren, W., Atkins, E.: Distributed multi-vehicle coordinated control via local information exchange. International Journal of Robust & Nonlinear Control **17**(10-11), 1002–1033 (2007)
9. Mei, J., Ren, W., Chen, J.: Distributed Consensus of Second-Order Multi-Agent Systems With Heterogeneous Unknown Inertias and Control Gains Under a Directed Graph. IEEE Trans-

actions on Automatic Control **61**(8), 2019–2034 (2016). https://doi.org/10.1109/TAC.2015.2480336
10. Cao, Y., Ren, W., Egerstedt, M.: Distributed containment control with multiple stationary or dynamic leaders in fixed and switching directed networks. Automatica **48**(8), 1586–1597 (2012)
11. Peng, K., Yang, Y.: Leader-following consensus problem with a varying-velocity leader and time-varying delays. Physica A Statistical Mechanics & Its Applications **388**(388), 193–208 (2009)
12. Hong, Y., Chen, G., Bushnell, L.: Distributed observers design for leader-following control of multi-agent networks. Automatica **44**(3), 846–850 (2008)
13. Do, K.D.: Bounded controllers for formation stabilization of mobile agents with limited sensing ranges. IEEE Transactions on Automatic Control **52**(3), 569–576 (2007)
14. Cheah, C.C., Hou, S.P., Slotine, J.J.E.: Region-based shape control for a swarm of robots. Automatica **45**(10), 2406–2411 (2009)
15. Notarstefano, G., Egerstedt, M., Haque, M.: Containment in leader–follower networks with switching communication topologies. Automatica **47**(5), 1035–1040 (2011)
16. Ji, M., Ferrari-Trecate, G., Egerstedt, M., Buffa, A.: Containment control in mobile networks. IEEE Transactions on Automatic Control **53**(8), 1972–1975 (2008)
17. Ballard, L., Cao, Y., Ren, W.: Distributed discrete-time coupled harmonic oscillators with application to synchronised motion coordination. IET Control Theory & Applications **4**(5), 806–816 (2010)
18. Chopra, N., Spong, M.W.: Passivity-based control of multi-agent systems. Springer, Berlin, Heidelberg (2006)
19. Hou, Z.G., Cheng, L., Tan, M.: Decentralized robust adaptive control for the multiagent system consensus problem using neural networks. IEEE Transactions on Systems, Man, and Cybernetics, Part B (Cybernetics) **39**(3), 636–647 (2009)
20. Nuno, E., Ortega, R., Basanez, L., Hill, D.: Synchronization of networks of nonidentical Euler-Lagrange systems with uncertain parameters and communication delays. IEEE Transactions on Automatic Control **56**(4), 935–941 (2011)
21. Dong, W.: On consensus algorithms of multiple uncertain mechanical systems with a reference trajectory. Automatica **47**(9), 2023–2028 (2011)
22. Mei, J., Ren, W., Ma, G.: Distributed coordinated tracking with a dynamic leader for multiple Euler-Lagrange systems. IEEE Transactions on Automatic Control **56**(6), 1415–1421 (2011)
23. Pereira, A.R., Hsu, L.: Adaptive formation control using artificial potentials for euler-lagrange agents. IFAC Proceedings Volumes **41**(2), 10788–10793 (2008)
24. Ihle, I.F., Jouffroy, J., Fossen, T.I.: Formation Control of Marine Surface Craft : A Lagrangian Approach. IEEE Journal of Oceanic Engineering **31**(4), 922–934 (2006)
25. Li, J., Ren, W., Xu, S.: Distributed containment control with multiple dynamic leaders for double-integrator dynamics using only position measurements. IEEE Transactions on Automatic Control **57**(6), 1553–1559 (2012)
26. Ferrari-Trecate, G., Egerstedt, M., Buffa, A., Ji, M.: Laplacian sheep: A hybrid, stop-go policy for leader-based containment control. In: International workshop on hybrid systems: Computation and control, pp. 212–226. Springer (2006)
27. Han, L., Dong, X., Li, Q., Ren, Z.: Formation-containment control for second-order multi-agent systems with time-varying delays. Neurocomputing **218**, 439–447 (2016). https://doi.org/10.1016/j.neucom.2016.09.001
28. Dong, X., Li, Q., Ren, Z., Zhong, Y.: Output formation-containment analysis and design for general linear time-invariant multi-agent systems. Journal of the Franklin Institute **353**(2), 322–344 (2016). https://doi.org/10.1016/j.jfranklin.2015.11.004
29. Dong, X., Shi, Z., Lu, G., Zhong, Y.: Formation-containment analysis and design for high-order linear time-invariant swarm systems. International Journal of Robust and Nonlinear Control **25**(17), 3439–3456 (2015)
30. Wang, Y., Liu, X., Xiao, J.: Output formation-containment of coupled heterogeneous linear systems under intermittent communication. Journal of the Franklin Institute **354**(1), 392–414 (2017). https://doi.org/10.1016/j.jfranklin.2016.10.011

31. Zheng, B., Mu, X.: Formation-containment control of second-order multi-agent systems with only sampled position data. International Journal of Systems Science **47**(15), 3609–3618 (2016). https://doi.org/10.1080/00207721.2015.1107148
32. Han, T., Chi, M., Guan, Z.H., Hu, B., Xiao, J.W., Huang, Y.: Distributed Three-Dimensional Formation Containment Control of Multiple Unmanned Aerial Vehicle Systems. Asian Journal of Control **19**(3), 1103–1113 (2017). https://doi.org/10.1002/asjc.1445
33. Wit, C.C.D., Fixot, N.: Adaptive control of robot manipulators via velocity estimated feedback. IEEE Transactions on Automatic Control **37**(8), 1234–1237 (1992). https://doi.org/10.1109/9.151115
34. Fang, H., Chen, J., Wang, X.: Distributed observer-based coordination for multiple Lagrangian systems using only position measurements **8**(November 2013), 2102–2114 (2014)
35. Ghapani, S., Mei, J., Ren, W., Song, Y.: Fully distributed flocking with a moving leader for Lagrange networks with parametric uncertainties. Automatica **67**, 67–76 (2016). https://doi.org/10.1016/j.automatica.2016.01.004
36. Mehrabian, A.R., Khorasani, K.: Constrained distributed cooperative synchronization and reconfigurable control of heterogeneous networked Euler-Lagrange multi-agent systems. Information Sciences **370-371**(5), 578–597 (2016). https://doi.org/10.1016/j.ins.2015.09.032
37. Cao, Y., Ren, W.: Multi-vehicle coordination for double-integrator dynamics under fixed undirected/directed interaction in a sampled-data setting. International Journal of Robust & Nonlinear Control **20**(9), 987–1000 (2010)
38. Ren, W.: Sampled-data discrete-time coordination algorithms for double-integrator dynamics under dynamic directed interaction. International Journal of Control **83**(3), 506–515 (2010)
39. Ren, W., Beard, R.W.: Distributed Consensus in Multi-vehicle Cooperative Control. Communications & Control Engineering **27**(2), 71–82 (2008)
40. Meng, Z., Ren, W., You, Z.: Distributed finite-time attitude containment control for multiple rigid bodies. Automatica **46**(12), 2092–2099 (2010). https://doi.org/10.1016/j.automatica.2010.09.005
41. Li, D., Ma, G., He, W., Zhang, W., Li, C., Ge, S.S.: Distributed coordinated tracking control of multiple Euler-Lagrange systems by state and output feedback. IET Control Theory & Applications **11**(14), 2213–2221 (2017)
42. Bhat, S.P., Bernstein, D.S.: Finite-time stability of continuous autonomous systems. SIAM Journal on Control and Optimization **38**(3), 751–766 (2000)
43. Mei, J., Ren, W., Ma, G.: Distributed containment control for Lagrangian networks with parametric uncertainties under a directed graph. Automatica **48**(4), 653–659 (2012). https://doi.org/10.1016/j.automatica.2012.01.020
44. Horn, R., Johnson, C.: Matrix analysis. Cambridge university press, Cambridge UK (1990)
45. Su, Y., Huang, J.: Cooperative output regulation of linear multi-agent systems. IEEE Transactions on Automatic Control **57**(4), 1062–1066 (2011)
46. Kubat, M.: Neural networks: a comprehensive foundation by Simon Haykin, Macmillan, 1994, ISBN 0-02-352781-7. Knowledge Engineering Review **13**(4), 409–412 (1999)
47. Xu, B., Sun, F.: Composite intelligent learning control of strict-feedback systems with disturbance. IEEE Transactions on Cybernetics **48**(2), 730–741 (2017)
48. He, W., Huang, H., Ge, S.S.: Adaptive Neural Network Control of a Robotic Manipulator With Time-Varying Output Constraints. IEEE Transactions on Cybernetics **47**(10), 3136–3147 (2017)
49. Ge, S.S., Wang, C.: Adaptive neural control of uncertain mimo nonlinear systems. IEEE Transactions on Neural Networks **15**(3), 674–692 (2004)
50. Xu, B., Wang, D., Zhang, Y., Shi, Z.: Dob-based neural control of flexible hypersonic flight vehicle considering wind effects. IEEE Transactions on Industrial Electronics **64**(11), 8676–8685 (2017)
51. Behtash, S.: Robust output tracking for non-linear systems. International Journal of Control **51**(6), 1381–1407 (1990)
52. He, W., Ge, S.S., Li, Y., Chew, E., Ng, Y.S.: Neural network control of a rehabilitation robot by state and output feedback. Journal of Intelligent & Robotic Systems **80**(1), 15–31 (2015)

53. Ioannou, P.A., Sun, J.: Robust adaptive control. Courier Corporation, Massachusetts USA (2012)
54. Xu, B., Yang, C., Pan, Y.: Global neural dynamic surface tracking control of strict-feedback systems with application to hypersonic flight vehicle. IEEE Transactions on Neural Networks and Learning Systems **26**(10), 2563–2575 (2015)
55. Ge, S.S., Hang, C.C., Lee, T.H., Zhang, T.: Stable adaptive neural network control. Springer, Boston (2002)
56. Khalil, H.K.: Nonlinear systems. Prentice hall, New Jersey (1996)

Chapter 14
Formation-Containment Control for Networked Spacecraft with Collision Avoidance

Abstract The distributed formation-containment (DFC) problem under a directed graph is addressed for networked spacecraft systems. First, using a leader-follower framework, the DFC problem is properly defined. For the leader spacecraft and the follower spacecraft, respectively, a DFC control law is next proposed without using velocity information. Based on the artificial potential function, all the spacecraft can achieve the control objective satisfactorily while avoiding collisions with others as well as the obstacles in the environment caused by other close-range spacecraft or on-orbit assembly components. By Lyapunov stability theory, the boundedness of the error signals is guaranteed. Simulations are finally given to show the feasibility of this approach.

14.1 Introduction

The spacecraft in the multi-spacecraft formation flight systems can be abstracted as the agents in the multi-agent systems, which has strong versatility and a large number of theoretical research results. Simultaneously, the multi-agent theory has also brought many supporting control technologies for spacecraft formation flight. For multi-agent systems (MASs), existing works mainly focus on several interesting problems [1–3]: For the consensus problems, the agents update their states via information from neighbors and finally reach a common value [4–6]. Considering a moving leader, works on coordinated tracking problems are extensively studied [7–9]. In [10, 11], the formation control problem is studied, where the agents form a desired spatial configuration. Given a region spanned by several leaders, the containment control problem is addressed to drive the followers into a specific area [12, 13]. On the other hand, most practical systems can be described by Euler–Lagrange Systems (ELSs), such as autonomous vehicles, underwater robots, and spacecraft [14]. Current multi-agent works of multiple Euler–Lagrange systems (MELSs) also focus on the problems of consensus [15–17], tracking [11, 18, 19], formation [20, 21] and containment [22].

Y. Gao and D. Li, *Spacecraft Maneuver with Performance Guaranteed*,
https://doi.org/10.1007/978-981-99-4653-2_14

Based on the containment problem and the formation problem, the DFC problem is investigated in [23–30], where the leaders form a geometric formation shape and the followers move into the area spanned by the leaders. Considering an undirected graph, the DFC problem is studied for first-order swarm systems in [23]. Then, under a directed graph, for linear time-invariant swarm systems [26] and MASs [25], respectively, DFC control laws are proposed. For second-order MASs, DFC control with time-varying delays is investigated in [24]. Next, considering intermittent communications, the DFC problem for coupled heterogeneous linear systems is addressed in [27]. In [28], for a class of second-order MASs under an undirected graph, a DFC control algorithm is proposed using only position measurements. Similarly, the DFC problem is considered in [30] with a stationary formation configuration while no relative velocity information is required. For networked unmanned aerial vehicles, the DFC problem with both fixed and switching graphs is studied in [29]. In [31], under an undirected graph, the DFC problem has been investigated by adopting a two-layer framework and a multilayer framework.

Most approaches mentioned above require generalized velocity information to be implemented. In practice, however, to save cost and space, some spacecraft in formation might not equip velocity sensors. Since the number of spacecraft in a DFC problem may be large, another critical issue is to avoid the collision with each other and with pop-up threats in the environment. Unfortunately, few of the results in the literature considers the obstacle avoidance problem in a spacecraft DFC mission. Concerning obstacle avoidance, output feedback, and model uncertainties simultaneously, the DFC problem poses considerable difficulty and complexity for controller design.

In this chapter, we study DFC control via output feedback for multi-spacecraft formation flight systems under a directed graph with environmental obstacles caused by other close-range spacecraft or on-orbit assembly components and model uncertainties. Compared with existing works [23–30], this approach holds the following contributions. First, a novel spacecraft DFC control law integrated with NNs and artificial potential functions is addressed, which can approximate the model uncertainties and help the close range formation spacecraft avoid collisions with each other and with the obstacles caused by other close-range spacecraft or on-orbit assembly components. By utilizing the proposed controller, all the state errors are semi-globally uniformly ultimately bounded (SGUUB). Compared with the undirected result in Chap. 13, this approach is addressed in a directed graph and gets rid of the discontinuity induced by the estimators introduced in Chap. 13. Different from the results in [24, 26, 29] using full-state measurements, the proposed approach is still feasible without using velocity information. Moreover, compared with existing works concerning a stationary formation configuration [30, 32, 33], this control scheme can tackle a time-varying formation case, which is more general and practical. Last but not least, this chapter deals with systems governed by multi-spacecraft formation flight systems with model uncertainties, while most existing approaches consider linear systems [25, 27, 28], systems with the known dynamic model [34] or systems which has the 'linearity-in-parameters' property [35, 36].

The remainder of this chapter is organized as follows. In Sect. 14.2, the problem under investigation is formulated. Section 14.3 covers the main results of this study, where the DFC control algorithm with collision avoidance via output feedback is proposed. In Sect. 14.4, simulations are given. The conclusion is drawn in Sect. 14.5.

14.2 Preliminaries and Problem Description

In this chapter, the Euler–Lagrange systems are utilized to describe the multi-spacecraft formation flight systems described by (2.34) because of their abstraction of different control objects and the transferability of control algorithms. Consider a group of N leader spacecraft with subset $\mathcal{V}_l = \{1, 2, \ldots, N\}$ and M follower spacecraft with subset $\mathcal{V}_f = \{N+1, N+2, \ldots, N+M\}$ in $\mathbb{R}^p$. The spacecraft are modeled as (2.34).

Here, an assumption on the output measurement of the system (2.34) is addressed.

Assumption 14.1 The state information of the system (2.34) is measurable.

We consider the communication topology as a directed graph. We utilize $\mathcal{G} = (\mathcal{V}, \mathcal{E})$ to describe the communication graph, where $\mathcal{V} = \mathcal{V}_l \cup \mathcal{V}_f$ and $\mathcal{E} \subseteq \mathcal{V} \times \mathcal{V}$ denote the vertex set and the arc set, respectively. Denote $\mathcal{N}_i$ as the in-neighbour set of node i, where $\mathcal{N}_i = \{j : (j, i) \in \mathcal{E}\}$. The adjacency matrix $\mathcal{A} = \left[a_{ij}\right] \in \mathbb{R}^{(N+M)\times(N+M)}$ is defined as $a_{ij} > 0$ if $(j, i) \in \mathcal{E}$, and $a_{ij} = 0$ otherwise. We define $a_{ii} = 0$, and the Laplacian matrix is defined as $\mathcal{L} = \left[l_{ij}\right] \in \mathbb{R}^{(N+M)\times(N+M)}$, where $l_{ii} = \sum_{j=1, j\neq i}^{k} a_{ij}$ and $l_{ij} = -a_{ij}$, $i \neq j$. For convenience, we use a virtual spacecraft 0 to denote the virtual center of desired formation, and only parts of the leaders can access information spacecraft 0.

Thus the corresponding Laplacian matrix in this chapter can be written as

$$\mathcal{L} = \begin{bmatrix} 0 & 0_{1\times N} & 0_{1\times M} \\ \mathcal{L}_{L2} & \mathcal{L}_{L1} & 0_{N\times M} \\ 0_{M\times 1} & \mathcal{L}_{F2} & \mathcal{L}_{F1} \end{bmatrix}, \tag{14.1}$$

where $\mathcal{L}_{L1}$ represents the interaction graph among the leaders, $\mathcal{L}_{L2}$ the information flow from the virtual spacecraft to the leaders, $\mathcal{L}_{F1}$ the interaction graph among the followers, and $\mathcal{L}_{F2}$ the information interaction between the leaders and followers. Let $\mathcal{G}_L$ denote the directed graph associated with $N+1$ spacecraft labeled as 0 to N. Then, an assumption on the graph is made.

Assumption 14.2 There exists a directed spanning tree in $\mathcal{G}_L$, where the root of $\mathcal{G}_L$ is the spacecraft 0. There exists at least one leader has a directed path to each follower.

Based on Assumption 14.2, an useful lemma in a directed graph is recalled.

Lemma 14.1 *Under Assumption 14.2, the following statements hold:*

(1) $\mathcal{L}_{L1}$ *and* $\mathcal{L}_{F1}$ *are non-singular;*
(2) All eigenvalues of $\mathcal{L}_{L1}$ *have positive real parts;*
(3) Each entry of $-\mathcal{L}_{L1}^{-1}\mathcal{L}_{L2}$ *is nonnegative and each row sum of* $-\mathcal{L}_{L1}^{-1}\mathcal{L}_{L2}$ *is equal to one. Similarly, each entry of* $-\mathcal{L}_{F1}^{-1}\mathcal{L}_{F2}$ *is nonnegative and each row sum of* $-\mathcal{L}_{F1}^{-1}\mathcal{L}_{F2}$ *is equal to one.*

Proof The proof is similar to that in [37–39]. □

We propose a spacecraft leader-follower framework to characterize the DFC problem, which is formally defined as follows.

Definition 14.1 Given the initial spacecraft formation $\mathcal{G}(q_i(0)), i \in \mathcal{V}$, and design a DFC control law for spacecraft $i \in \mathcal{V}$ such that $q_i(t) \to q_0(t) + p_i(t)$, $i \in \mathcal{V}_l$, and $q_i(t) \to \sum_{j\in N_i} a_{ij}\left(q_i - q_j\right)$, $i \in \mathcal{V}_f$, respectively with bounded tracking errors while no collision occurs, where $p_i(t)$ is the auxiliary function decomposed from the target formation configuration, i.e., the target relative position between spacecraft 0 and spacecraft i.

14.3 Distributed Formation-Containment Control with Collision Avoidance via Output Feedback

In this section, without using velocity measurements, the DFC problem for collision-free spacecraft is considered. We design our DFC control law with the help of a high-gain observer, where NNs are introduced to tackle model uncertainties. On the other hand, to help the spacecraft avoid collisions, potential functions are also properly designed and applied.

To achieve collision and obstacle avoidance purpose, we first assume that the spacecraft's danger detection range is centered at the spacecraft and the radius is r. Then, we design the potential functions for spacecraft i as

$$P_{i,j}\left(\tilde{q}_{i,j}\right) = \begin{cases} 0, & \left\|\tilde{q}_{i,j}\right\| \geq r, \\ c_{i,j}\ \ln\dfrac{r}{\left\|\tilde{q}_{i,j}\right\|}, & \left\|\tilde{q}_{i,j}\right\| < r, \end{cases} \tag{14.2}$$

$$P_{i,o}\left(\tilde{q}_{i,o}\right) = \begin{cases} 0, & \left\|\tilde{q}_{i,o}\right\| \geq r, \\ c_i\ \ln\dfrac{r-d_{safe}}{\left\|\tilde{q}_{i,o}\right\|-d_{safe}}, & \left\|\tilde{q}_{i,o}\right\| < r, \end{cases} \tag{14.3}$$

where d_{safe} is the required safe distance, $c_{i,j}$ and c_i are positive constant gains, and $c_{i,j} = c_{j,i}$, $\tilde{q}_{i,j} = q_i - q_j$, $(i, j \in \mathcal{V},\ i \neq j)$. Define $\tilde{q}_{i,o} = q_i - q_o$, $o \in O$, and O denotes the set of all obstacles within radius r of spacecraft i. In addition, let H_i represent all spacecraft within the danger range of spacecraft i as

$$H_i = \left\{j \in \mathcal{V}, j \neq i \mid \left\|\tilde{q}_{i,j}\right\| < r\right\}. \tag{14.4}$$

We next define generalized tracking errors as

$$z_{1i} = \begin{cases} \sum_{j \in N_i} a_{ij} \left(q_i - p_i - q_j + p_j\right), \ i \in \mathcal{V}_l, \\ \sum_{j \in N_i} a_{ij} \left(q_i - q_j\right), \ i \in \mathcal{V}_f, \end{cases} \tag{14.5}$$

$$z_{2i} = \dot{q}_i - \alpha_{1i}, \quad i \in \mathcal{V}. \tag{14.6}$$

The virtual control takes the form

$$\begin{aligned}\alpha_{1i} = &\frac{1}{\kappa_i}\beta_i - \frac{1}{\kappa_i}(\Theta_i)^{\dagger}\Bigg(K_{2i} V_{pi} + (\sum_{o \in O} \frac{\partial P_{i,o}}{\partial q_i})^T \frac{\beta_i}{\kappa_i} \\ &+2(\sum_{j \in H_i} \frac{\partial P_{i,j}}{\partial q_i})^T \frac{\beta_i}{\kappa_i}\Bigg),\end{aligned} \tag{14.7}$$

where

$$\Theta_i = z_{1i}^T + \frac{1}{\kappa_i}\sum_{o \in O}\left(\frac{\partial P_{i,o}}{\partial q_i}\right)^T + \frac{2}{\kappa_i}\sum_{j \in H_i}\left(\frac{\partial P_{i,j}}{\partial q_i}\right)^T, \tag{14.8}$$

$$V_{pi} = \frac{1}{2} z_{1i}^T z_{1i} + \sum_{o \in O} P_{i,o} + \sum_{j \in H_i} P_{i,j}, \tag{14.9}$$

$$\beta_i = \begin{cases} \sum_{j \in N_i} a_{ij} \left(\dot{q}_j - \dot{p}_j\right) + \dot{p}_i, \ i \in \mathcal{V}_l, \\ \sum_{j \in N_i} a_{ij} \dot{q}_j, \ i \in \mathcal{V}_f, \end{cases} \tag{14.10}$$

the gain matrix $K_{2i} = K_{2i}^T > 0$, a positive constant $\kappa_i = \sum_{j \in N_i} a_{ij}$, and $(\Theta_i)^{\dagger}$ is the Moore–Penrose inverse of Θ_i defined as

$$\Theta_i (\Theta_i)^{\dagger} = \begin{cases} 0, & \Theta_i = [0, 0, \ldots, 0]^T, \\ 1, & \text{otherwise}. \end{cases} \tag{14.11}$$

Then, we have

$$\begin{aligned}\dot{z}_{1i} =& \kappa_i z_{2i} + \kappa_i \alpha_{1i} - \beta_i \\ =& \kappa_i z_{2i} - (\Theta_i)^{\dagger}\left(K_{2i} V_{pi} + \left(\sum_{o \in O}\frac{\partial P_{i,o}}{\partial q_i}\right)^T \frac{\beta_i}{\kappa_i}\right. \\ &\left. +2\left(\sum_{j \in H_i}\frac{\partial P_{i,j}}{\partial q_i}\right)^T \frac{\beta_i}{\kappa_i}\right).\end{aligned} \tag{14.12}$$

Using (2.34), the derivative of z_{2i} takes the form

$$\dot{z}_{2i} = M_i^{-1}(q_i)\left[\tau_i - C_i(q_i, \dot{q}_i)\dot{q}_i - g_i(q_i)\right] - \dot{\alpha}_{1i}. \tag{14.13}$$

Note that $M_i(q_i)$, $C_i\left(q_i, \dot{q}_i\right)$ and $g_i(q_i)$ cannot be directly used in control design due to model uncertainties. We here introduce adaptive NNs [40] to approximate the model uncertainties as

$$f_i(q_i, \dot{q}_i, \alpha_{1i}, \dot{\alpha}_{1i}) = W_i^{*T}\phi_i(q_i, \dot{q}_i, \alpha_{1i}, \dot{\alpha}_{1i}) + \epsilon_i, \tag{14.14}$$

where $W_i^* \in \mathbb{R}^{l\times p}$ denotes the ideal constant weight matrix, l the NN node number, $f_i(q_i, \dot{q}_i, \alpha_{1i}, \dot{\alpha}_{1i}) = g_i(q_i) + C_i(q_i, \dot{q}_i)\alpha_{1i} + M_i(q_i)\dot{\alpha}_{1i}$ the model uncertainties, and $\phi_i = [\varphi_{i1}, \varphi_{i2}, \ldots, \varphi_{il}]^T \in \mathbb{R}^l$ the activation function. ϵ_i is the bounded approximation error, i.e., $\|\epsilon_i(Z_i)\| \le \bar{\epsilon}_i$, $\forall Z_i \in \Omega_{z_i}$ with $\bar{\epsilon}_i > 0$ as an unknown constant [41], where $Z_i = \left[q_i^T, \dot{q}_i^T, \alpha_{1i}^T, \dot{\alpha}_{1i}^T\right]^T$. The elements of ϕ_i are chosen

$$\varphi_{i\ell}(Z_i) = \exp\left[\frac{-(Z_i - \mu_\ell)^T(Z_i - \mu_\ell)}{{\eta_\ell}^2}\right],\ \ell = 1, \ldots, l, \tag{14.15}$$

where η_ℓ and μ_ℓ are two constant parameters [42].

In the case without velocity measurements, we estimate the immeasurable state z_{2i} using a high-gain observer

$$\hat{z}_{2i} = \frac{\pi_{2i}}{\varepsilon_i} - \alpha_{1i},\ i \in \mathcal{V}, \tag{14.16}$$

and the dynamics of π_{2i} is given as

$$\varepsilon_i\dot{\pi}_{1i} = \pi_{2i}, \tag{14.17}$$

$$\varepsilon_i\dot{\pi}_{2i} = -\bar{\lambda}_{1i}\pi_{2i} - \pi_{1i} + q_i, \tag{14.18}$$

where $\bar{\lambda}_{1i}$ is picked as long as the polynomial $s^2 + \bar{\lambda}_{1i}s + 1$ is Hurwitz, and ε_i is a small positive constant. Then, following property holds [43, 44]:

$$\xi_{2i} = \frac{\pi_{2i}}{\varepsilon_i} - \dot{q}_i = -\varepsilon_i{\psi_i}^{(2)}, \tag{14.19}$$

$$\psi_i = \pi_{2i} + \bar{\lambda}_{1i}\pi_{1i}, \tag{14.20}$$

where ${\psi_i}^{(2)}$ denotes the 2th derivative of ψ_i. We have $\tilde{z}_{2i}$

$$\tilde{z}_{2i} = \hat{z}_{2i} - z_{2i} = \frac{\pi_{2i}}{\varepsilon_i} - \dot{q}_i = \xi_{2i}. \tag{14.21}$$

In addition, there exist positive constants t_i^* and h_i such that

$$\|\xi_{2i}\| \leq \varepsilon_i h_i, \ \forall t > t_i^*. \tag{14.22}$$

The following DFC control law and the adaptive law are formally proposed for the leaders and followers using only position measurements

$$\begin{aligned} \tau_i = & -\kappa_i z_{1i} - K_{1i}\hat{z}_{2i} + \hat{W}_i^T \phi_i\left(\hat{Z}_i\right) \\ & \sum_{o\in O}\frac{\partial P_{i,o}}{\partial q_i} + 2\sum_{j\in H_i}\frac{\partial P_{i,j}}{\partial q_i}, \end{aligned} \tag{14.23}$$

$$\dot{\hat{W}}_{i,k} = -\gamma_i\left(\phi_i\left(\hat{Z}_i\right)\hat{z}_{2i,k} + \sigma_i \hat{W}_{i,k}\right), \tag{14.24}$$

where K_i is the control gain, γ_i, σ_i are positive constants, $\hat{W}_{i,k}$ is the kth column vector of $\hat{W}_i$ and $\hat{z}_{2i,k}$ the kth element of $\hat{z}_{2i}$, $k = 1, 2, \ldots, p$, $i \in \mathcal{V}$.

Define the global closed-loop error signals as

$$z_1 = \left[z_{11}^T, z_{12}^T, \ldots, z_{1(N+M)}^T\right]^T, \tag{14.25}$$

$$z_2 = \left[z_{21}^T, z_{22}^T, \ldots, z_{2(N+M)}^T\right]^T, \tag{14.26}$$

$$\tilde{W} = \left[\tilde{W}_1, \tilde{W}_2, \ldots, \tilde{W}_{N+M}\right]^T, \tag{14.27}$$

where $\tilde{W}_i = \hat{W}_i - W_i^*$, $i \in \mathcal{V}$.

Theorem 14.1 *Concerning the multi-spacecraft formation systems governed by (2.34), under the DFC control algorithms (14.23) and (14.24) and Assumptions 14.1 and 14.2, for each compact set Ω_{i0} where q_i (0), $\dot{q}_i$ (0), $\hat{W}_{i,1}$ (0), ..., $\hat{W}_{i,k}$ (0) $\in \Omega_{i0}$, $k = 1, \ldots, p$, $i \in \mathcal{V}$, the error signals of spacecraft are SGUUB, and z_1, z_2 and $\tilde{W}$ remain within compact sets Ω_{z_1}, Ω_{z_2} and $\Omega_{\tilde{W}}$, respectively:*

$$\Omega_{z_1} := \left\{z_1 \in \mathbb{R}^{p(N+M)} \mid \|z_1\| \leq \sqrt{D}\right\},$$

$$\Omega_{z_2} := \left\{z_2 \in \mathbb{R}^{p(N+M)} \mid \|z_2\| \leq \sqrt{D/\min_{i\in\mathcal{V}}\left(\lambda_{\min}\left(M_i\right)\right)}\right\},$$

$$\Omega_{\tilde{W}} := \left\{\tilde{W} \in \mathbb{R}^{l(N+M)\times p} \mid \left\|\tilde{W}\right\| \max_{i\in\mathcal{V}}\left(\sqrt{\gamma_i D}\right)\right\},$$

where D is a positive constant which will be detailed in the proof.

Proof Consider a Lyapunov function as

$$V = \sum_{i=1}^{N+M} \left[\frac{1}{2} z_{1i}^T z_{1i} + \sum_{o \in O} P_{i,o} + \sum_{j \in H_i} P_{i,j} + \frac{1}{2} z_{2i}^T M_i z_{2i} + \frac{1}{2\gamma_i} \sum_{k=1}^{p} \tilde{W}_{i,k}^T \tilde{W}_{i,k} \right]. \tag{14.28}$$

Based on (14.13), the time derivative of V is given as

$$\begin{aligned} \dot{V} = \sum_{i=1}^{N+M} \Bigg[& \Theta_i \left(\kappa_i z_{2i} - (\Theta_i)^{\dagger} \left(K_{2i} V_{pi} + \sum_{o \in O} \left(\frac{\partial P_{i,o}}{\partial q_i} \right)^T \frac{\beta_i}{\kappa_i} + 2 \left(\sum_{j \in H_i} \frac{\partial P_{i,j}}{\partial q_i} \right)^T \frac{\beta_i}{\kappa_i} \right) \right) + \frac{1}{\gamma_i} \sum_{k=1}^{p} \tilde{W}_{i,k}^T \dot{\tilde{W}}_{i,k} \\ & - z_{2i}^T M_i \dot{\alpha}_{1i} + z_{2i}^T \left(\tau_i - C_i \left(q_i, \dot{q}_i \right) \dot{q}_i - g_i \left(q_i \right) \right) + \frac{1}{2} z_{2i}^T \dot{M}_i z_{2i} \\ & + \sum_{o \in O} \left(\frac{\partial P_{i,o}}{\partial q_i} \right)^T \frac{\beta_i}{\kappa_i} + 2 \left(\sum_{j \in H_i} \frac{\partial P_{i,j}}{\partial q_i} \right)^T \frac{\beta_i}{\kappa_i} \Bigg]. \end{aligned} \tag{14.29}$$

Applying Properties 2.2 and 2.3, and substituting (14.23), (14.24) into (14.29), we have

$$\begin{aligned} \dot{V} = \sum_{i=1}^{N+M} \Bigg[& -K_{2i} V_{pi} - z_{2i}^T \left(K_{1i} - \frac{1}{2} I_p \right) z_{2i} - z_{2i}^T K_{1i} \tilde{z}_{2i} \\ & - \sum_{k=1}^{p} \left[\tilde{W}_{i,k}^T \phi_i \left(\hat{Z}_i \right) \hat{z}_{2i,k} + \sigma_i \tilde{W}_{i,k}^T \hat{W}_{i,k} \right] + \frac{1}{2} \| \bar{\epsilon}_i \|^2 \\ & - \sum_{k=1}^{p} z_{2i,k} \left[W_{i,k}^{*T} \phi_i \left(Z_i \right) - \hat{W}_{i,k}^T \phi_i \left(\hat{Z}_i \right) \right] \Bigg], \end{aligned} \tag{14.30}$$

where $\tilde{z}_{2i} = \hat{z}_{2i} - z_{2i}$. Let $\hat{Z}_i$ denote the input of Gaussian functions, and we have following equations [45]:

$$\varphi_{i\ell}(\hat{Z}_i) = \exp \left[\frac{-\left(\hat{Z}_i - \mu_\ell \right)^T \left(\hat{Z}_i - \mu_\ell \right)}{{\eta_\ell}^2} \right], \tag{14.31}$$

$$\ell = 1, \ldots, l,$$

$$\phi_i(\hat{Z}_i) = \phi_i \left(Z_i \right) + \epsilon_i \phi_{ti}, \tag{14.32}$$

where ϵ_i, is a positive constant and ϕ_{ti} a bounded vector function.

Based on (14.31) and (14.32), we have

$$\left\|\phi_i\left(\hat{Z}_i\right)\right\|^2 \leq l, \tag{14.33}$$

$$\begin{aligned}\hat{W}_{i,k}^T\phi_i\left(\hat{Z}_i\right) &= \tilde{W}_i^T\phi_i\left(\hat{Z}_i\right) + W_{i,k}^{*T}\left(\phi_i\left(Z_i\right) + \epsilon_i\phi_{ti}\right)\\ &= \tilde{W}_i^T\phi_i\left(\hat{Z}_i\right) + W_{i,k}^{*T}\phi_i\left(Z_i\right)\\ &\quad + \epsilon_i W_{i,k}^{*T}\phi_{ti}.\end{aligned} \tag{14.34}$$

Noticing $\tilde{W}_{i,k} = \hat{W}_{i,k} - W_{i,k}^*$, we have

$$\begin{aligned}-\sigma_i\tilde{W}_{i,k}^T\hat{W}_{i,k} &= -\sigma_i\tilde{W}_{i,k}^T\left(\tilde{W}_{i,k} + W_{i,k}^*\right)\\ &\leq -\sigma_i\left\|\tilde{W}_{i,k}\right\|^2 + \frac{\sigma_i}{2}\left\|\tilde{W}_{i,k}\right\|^2 + \frac{\sigma_i}{2}\left\|W_{i,k}^*\right\|^2\\ &\leq \frac{\sigma_i}{2}\left(\left\|W_{i,k}^*\right\|^2 - \left\|\tilde{W}_{i,k}\right\|^2\right).\end{aligned} \tag{14.35}$$

From (14.34) and (14.35), we have

$$\begin{aligned}\dot{V} = \sum_{i=1}^{N+M}\Bigg[&-K_{2i}V_{pi} - z_{2i}^T\left(K_{1i} - \frac{1}{2}I_p\right)z_{2i} - z_{2i}^T K_{1i}\tilde{z}_{2i}\\ &+ \sum_{k=1}^{p}\epsilon_i W_{i,k}^{*T}\phi_{ti}z_{2i,k} - \sum_{k=1}^{p}\tilde{W}_{i,k}^T\phi_i\left(\hat{Z}_i\right)\tilde{z}_{2i,k}\\ &- \sum_{k=1}^{p}\frac{\sigma_i}{2}\left\|\tilde{W}_{i,k}\right\|^2 + \sum_{k=1}^{p}\frac{\sigma_i}{2}\left\|W_{i,k}^*\right\|^2 + \frac{1}{2}\|\bar{\epsilon}_i\|^2\Bigg].\end{aligned} \tag{14.36}$$

Since

$$\sum_{k=1}^{p}\epsilon_i W_{i,k}^{*T}\phi_{ti}z_{2i,k} \leq \sum_{k=1}^{p}\frac{1}{2}\epsilon_i^2\left\|W_{i,k}^*\right\|^2\|\phi_{ti}\|^2 + \frac{1}{2}z_{2i}^T z_{2i}, \tag{14.37}$$

$$-\sum_{k=1}^{p}\tilde{W}_{i,k}^T\phi_i\left(\hat{Z}_i\right)\tilde{z}_{2i,k} \leq \frac{1}{\sigma_i}\tilde{z}_{2i}^T\tilde{z}_{2i}\left\|\phi_i\left(\hat{Z}_i\right)\right\|^2 + \sum_{k=1}^{p}\frac{\sigma_i}{4}\left\|\tilde{W}_{i,k}\right\|^2, \tag{14.38}$$

we then have

$$\begin{aligned}\dot{V} \leq & \sum_{i=1}^{N+M} \Big[-K_{2i} V_{pi} - z_{2i}^T \left(K_{1i} - I_p\right) z_{2i} - z_{2i}^T K_{1i} \tilde{z}_{2i} \\ & - \sum_{k=1}^{p} \frac{\sigma_i}{4} \left\| \tilde{W}_{i,k} \right\|^2 + \frac{1}{\sigma_i} \tilde{z}_{2i}^T \tilde{z}_{2i} \left\| \phi_i \left(\hat{Z}_i \right) \right\|^2 + \frac{1}{2} \|\bar{\epsilon}_i\|^2 \\ & + \sum_{k=1}^{p} \left(\frac{\sigma_i}{2} + \frac{1}{2} \epsilon_i^2 \|\phi_{ti}\|^2 \right) \left\| W_{i,k}^* \right\|^2 \Big]. \end{aligned} \tag{14.39}$$

According to (14.21) and (14.33), (14.39) becomes

$$\begin{aligned}\dot{V} \leq & \sum_{i=1}^{N+M} \Big[-K_{2i} V_{pi} - z_{2i}^T \left(K_{1i} - \frac{3}{2} I_p\right) z_{2i} + \frac{1}{2} \|\bar{\epsilon}_i\|^2 \\ & - \sum_{k=1}^{p} \frac{\sigma_i}{4} \left\| \tilde{W}_{i,k} \right\|^2 + \frac{1}{2} \sum_{k=1}^{p} \left(\sigma_i + \epsilon_i^2 \|\phi_{ti}\|^2 \right) \left\| W_{i,k}^* \right\|^2 \\ & + \frac{\lambda_{\max}}{2} \left(K_{1i}^T K_{1i} + \frac{2l}{\sigma_i} I_p \right) \varepsilon_i^2 h_i^2 \Big] \\ \leq & -\rho V + C, \end{aligned} \tag{14.40}$$

where ρ and C take the form

$$\rho = \min_{i \in \mathcal{V}} \left(\lambda_{\min}\left(K_{2i}\right), \frac{2\lambda_{\min}\left(K_{1i} - \frac{3}{2} I_p\right)}{\lambda_{\max}\left(M_i\right)}, \frac{1}{2} \gamma_i \sigma_i \right), \tag{14.41}$$

$$\begin{aligned} C = & \sum_{i=1}^{N+M} \Big[\frac{1}{2} \sum_{k=1}^{p} \left(\sigma_i + \epsilon_i^2 \|\phi_{ti}\|^2 \right) \left\| W_{i,k}^* \right\|^2 + \frac{1}{2} \|\bar{\epsilon}_i\|^2 \\ & + \frac{\lambda_{\max}}{2} \left(K_{1i}^T K_{1i} + \frac{2l}{\sigma_i} I_p \right) \varepsilon_i^2 h_i^2 \Big]. \end{aligned} \tag{14.42}$$

We choose the following control gains K_{1i} and K_{2i} to guarantee $\rho > 0$:

$$\lambda_{\min}\left(K_{2i}\right) > 0, \tag{14.43}$$

$$\lambda_{\min}\left(K_{1i} - \frac{3}{2} I_p\right) > 0, \; i \in \mathcal{V}. \tag{14.44}$$

Here, we provide the details of the proof to show that z_{1i}, z_{2i}, and $\tilde{W}_i$ are SGUUB. Since

$$V(t) \leq V(0)\, e^{-\rho t} + \frac{C}{\rho}\left(1 - e^{-\rho t}\right) \leq V(0) + \frac{C}{\rho}, \tag{14.45}$$

we can get

$$\|z_1\| \leq \sqrt{D}, \tag{14.46}$$

with $D = 2\left(V(0) + C/\rho\right)$. By a similar procedure, boundedness of z_2 and $\tilde{W}$ can also be given.

We next show the formation tracking error for each spacecraft is bounded. Rewrite (14.5) in a compact form

$$\begin{aligned} z_{1l} &= \left[z_{11}^T, \ldots, z_{1N}^T\right]^T \\ &= \left(\mathcal{L}_{L1} \otimes I_p\right)(q_l - p) + \left(\mathcal{L}_{L2} \otimes I_p\right) q_0, \end{aligned} \tag{14.47}$$

$$\begin{aligned} z_{1f} &= \left[z_{1(N+1)}^T, \ldots, z_{1(N+M)}^T\right]^T \\ &= \left(\mathcal{L}_{F1} \otimes I_p\right) q_f + \left(\mathcal{L}_{F2} \otimes I_p\right) q_l, \end{aligned} \tag{14.48}$$

where $p = \left[p_1^T, \ldots, p_N^T\right]^T$.

According to Assumption 14.2 and Lemma 14.1, considering (14.46) (14.47) and (14.48), we can obtain

$$\begin{aligned} q_l - p + \left(\mathcal{L}_{L1}^{-1}\mathcal{L}_{L2} \otimes I_p\right) q_0 &= \left(\mathcal{L}_{L1}^{-1} \otimes I_p\right) z_{1l} \\ &\leq \sqrt{D}\left\|\mathcal{L}_{L1}^{-1} \otimes I_p\right\|, \end{aligned} \tag{14.49}$$

$$\begin{aligned} q_f + \left(\mathcal{L}_{F1}^{-1}\mathcal{L}_{F2} \otimes I_p\right) q_l &= \left(\mathcal{L}_{F1}^{-1} \otimes I_p\right) z_{1f} \\ &\leq \sqrt{D}\left\|\mathcal{L}_{F1}^{-1} \otimes I_p\right\|. \end{aligned} \tag{14.50}$$

According to Definition 14.1, it is straightforward to show that all the closed-loop errors for leaders and followers are SGUUB. According to Definition 14.1, the DFC control objective is achieved. □

14.4 Simulation

$$L = \begin{bmatrix} 0 & 0 & 0 & 0 & 0 & 0 & 0 & 0 & 0 & 0 & 0 & 0 & 0 & 0 & 0 & 0 & 0 \\ 0 & 1 & 0 & 0 & 0 & 0 & 0 & 0 & -1 & 0 & 0 & 0 & 0 & 0 & 0 & 0 & 0 \\ 0 & -1 & 1 & 0 & 0 & 0 & 0 & 0 & 0 & 0 & 0 & 0 & 0 & 0 & 0 & 0 & 0 \\ 0 & 0 & -1 & 1 & 0 & 0 & 0 & 0 & 0 & 0 & 0 & 0 & 0 & 0 & 0 & 0 & 0 \\ 0 & 0 & 0 & -1 & 1 & 0 & 0 & 0 & 0 & 0 & 0 & 0 & 0 & 0 & 0 & 0 & 0 \\ 0 & 0 & 0 & 0 & -1 & 1 & 0 & 0 & 0 & 0 & 0 & 0 & 0 & 0 & 0 & 0 & 0 \\ 0 & 0 & 0 & 0 & 0 & -1 & 1 & 0 & 0 & 0 & 0 & 0 & 0 & 0 & 0 & 0 & 0 \\ 0 & 0 & 0 & 0 & 0 & 0 & -1 & 1 & 0 & 0 & 0 & 0 & 0 & 0 & 0 & 0 & 0 \\ -1 & 0 & 0 & 0 & 0 & 0 & 0 & -1 & 2 & 0 & 0 & 0 & 0 & 0 & 0 & 0 & 0 \\ 0 & -1 & 0 & 0 & 0 & 0 & 0 & 0 & 0 & 2 & -1 & 0 & 0 & 0 & 0 & 0 & 0 \\ 0 & 0 & -1 & 0 & 0 & 0 & 0 & 0 & 0 & 0 & 2 & -1 & 0 & 0 & 0 & 0 & 0 \\ 0 & 0 & 0 & -1 & 0 & 0 & 0 & 0 & 0 & 0 & 0 & 2 & -1 & 0 & 0 & 0 & 0 \\ 0 & 0 & 0 & 0 & -1 & 0 & 0 & 0 & 0 & 0 & 0 & 0 & 2 & -1 & 0 & 0 & 0 \\ 0 & 0 & 0 & 0 & 0 & -1 & 0 & 0 & 0 & 0 & 0 & 0 & 0 & 2 & -1 & 0 & 0 \\ 0 & 0 & 0 & 0 & 0 & 0 & -1 & 0 & 0 & 0 & 0 & 0 & 0 & 0 & 2 & -1 & 0 \\ 0 & 0 & 0 & 0 & 0 & 0 & 0 & -1 & 0 & 0 & 0 & 0 & 0 & 0 & 0 & 2 & -1 \\ 0 & 0 & 0 & 0 & 0 & 0 & 0 & 0 & -1 & -1 & 0 & 0 & 0 & 0 & 0 & 0 & 2 \end{bmatrix} \tag{14.51}$$

To show the feasibility of the proposed DFC control law (14.23) and (14.24), simulation results have been given by 8 leader spacecraft and 8 follower spacecraft, which are labeled as $1, 2, \ldots, 8$ and $9, 10, \ldots, 16$, respectively. For the ith spacecraft, the relative orbital motion is described by (2.34) and the orbit's initial elements are the same as those in Chap. 13. In Fig. 14.1, The graph is given. Accordingly, the Laplacian matrix is also detailed (see (14.51)). In Tables 14.1 and 14.2, all simulation parameters are presented.

From the trajectories in Fig. 14.2, we know the control laws (14.23) and (14.24) can perform well, while the collisions between the obstacles and the spacecraft are avoided. It shows that the formation is affected by the obstacles, and the configuration of formation is re-formed when the group of spacecraft are away from the obstacles. Figures 14.3 and 14.4 present that the formation errors of 8 leaders and 8 followers converge to the origin within 60s and 80s, respectively, while the corresponding control forces are given in Figs. 14.5 and 14.6. The numerical results above have shown that the proposed control scheme with collision avoidance is effective and feasible.

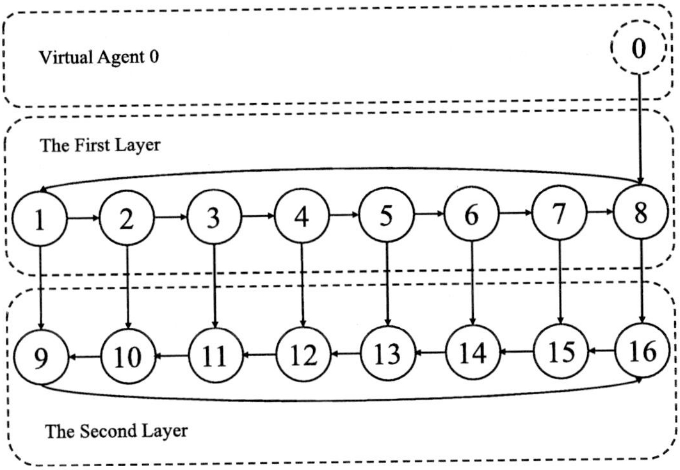

Fig. 14.1 The graph among 16 spacecraft

Table 14.1 The simulation parameters I

Parameters	Values	Parameters	Values
m_i	35 kg	β_1	5
ε_i	0.02	β_2	5
$\bar{\lambda}_{1i}$	1	β_3	5
K_{1i}	$0.1I_3$	β_4	5
K_{2i}	$5I_3$	σ_i	5
γ_i	4	l	256
c_i	10	$c_{i,j}$	10
r	50	d_{safe}	20
η_r	10	μ_r	random $[-1, 1]$
q_0	$\left[t; 200\sin\left(\frac{\pi t}{200}\right); t\right]$ m	$\hat{W}_i(0)$	0
q_{o1}	$[60, 250, 175]^T$ m	q_{o2}	$[145, 40, 40]^T$ m
$q_1(0)$	$[-75, -125, 0]^T$ m	$q_{10}(0)$	$[0, -180, 0]^T$ m
$q_2(0)$	$[75, -125, 0]^T$ m	$q_{11}(0)$	$[150, -150, 20]^T$ m
$q_3(0)$	$[125, -75, 0]^T$ m	$q_{12}(0)$	$[150, 0, 20]^T$ m
$q_4(0)$	$[125, 75, 0]^T$ m	$q_{13}(0)$	$[150, 150, 20]^T$ m
$q_5(0)$	$[75, 125, 0]^T$ m	$q_{14}(0)$	$[0, 150, 20]^T$ m
$q_6(0)$	$[-75, 125, 0]^T$ m	$q_{15}(0)$	$[-150, 150, 0]^T$ m
$q_7(0)$	$[-125, 75, 0]^T$ m	$q_{16}(0)$	$[-180, 0, 0]^T$ m
$q_8(0)$	$[-125, -75, 0]^T$ m	$\dot{q}_i(0)$	0 m/s
$q_9(0)$	$[-150, -150, 20]^T$ m		

Table 14.2 The simulation parameters II

$p_i(t)$	Values
$p_1(t)$	$\left[-50-\frac{25}{1+e^{-(t-100)}}, -100+\frac{25}{1+e^{-(t-100)}}, \frac{30}{1+e^{-(t-100)}}\right]^T$
$p_2(t)$	$\left[50+\frac{25}{1+e^{-(t-100)}}, -100+\frac{25}{1+e^{-(t-100)}}, \frac{30}{1+e^{-(t-100)}}\right]^T$
$p_3(t)$	$\left[100-\frac{25}{1+e^{-(t-100)}}, -50-\frac{25}{1+e^{-(t-100)}}, -\frac{30}{1+e^{-(t-100)}}\right]^T$
$p_4(t)$	$\left[100-\frac{25}{1+e^{-(t-100)}}, 50+\frac{25}{1+e^{-(t-100)}}, \frac{30}{1+e^{-(t-100)}}\right]^T$
$p_5(t)$	$\left[50+\frac{25}{1+e^{-(t-100)}}, 100-\frac{25}{1+e^{-(t-100)}}, -\frac{30}{1+e^{-(t-100)}}\right]^T$
$p_6(t)$	$\left[-50-\frac{25}{1+e^{-(t-100)}}, 100-\frac{25}{1+e^{-(t-100)}}, \frac{30}{1+e^{-(t-100)}}\right]^T$
$p_7(t)$	$\left[-100-\frac{25}{1+e^{-(t-100)}}, 50+\frac{25}{1+e^{-(t-100)}}, -\frac{30}{1+e^{-(t-100)}}\right]^T$
$p_8(t)$	$\left[-100+\frac{25}{1+e^{-(t-100)}}, -50-\frac{25}{1+e^{-(t-100)}}, -\frac{30}{1+e^{-(t-100)}}\right]^T$

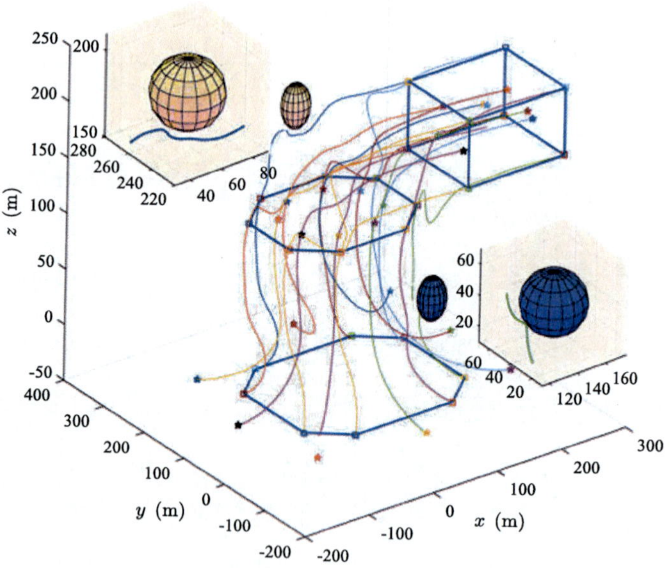

Fig. 14.2 The trajectories of 16 spacecraft

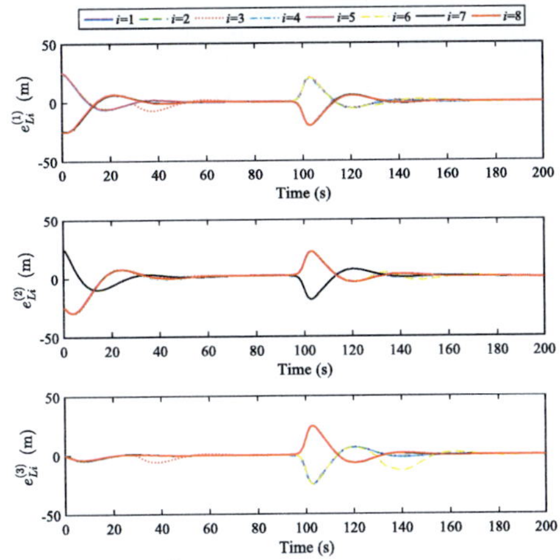

Fig. 14.3 The leader position errors

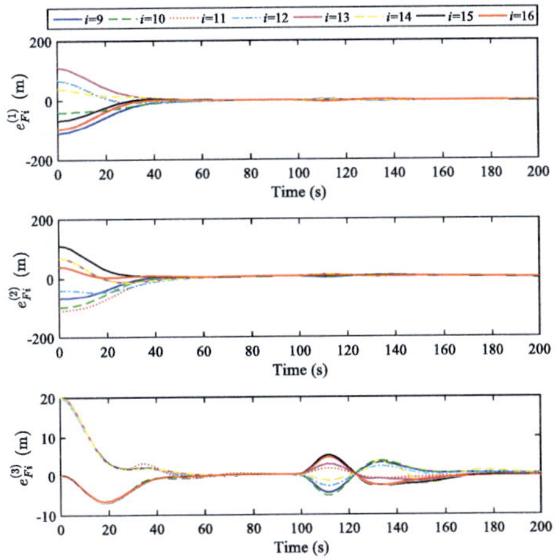

Fig. 14.4 The follower position errors

14.5 Conclusion

In this chapter, under a directed graph, the spacecraft DFC control schemes via output feedback with collision avoidance have been addressed for the multi-spacecraft formation systems. All the spacecraft neither collide with each other nor with obstacles caused by other close-range spacecraft or on-orbit assembly components. Considering the model uncertainties and the lack of velocity measurements, the NNs and the high-gain observer have been employed to design the controller. In the simulation, we have shown the potential application of the proposed approach in the time-varying formation of spacecraft. Practical situations, such as control with communication constraints or state constraints will be considered in future work.

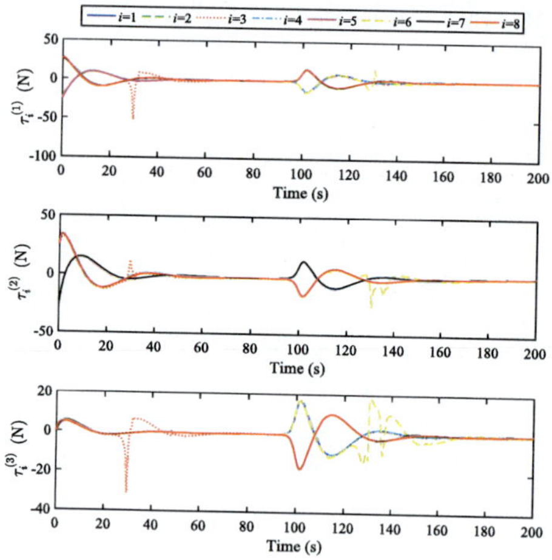

Fig. 14.5 The control inputs of 8 leaders

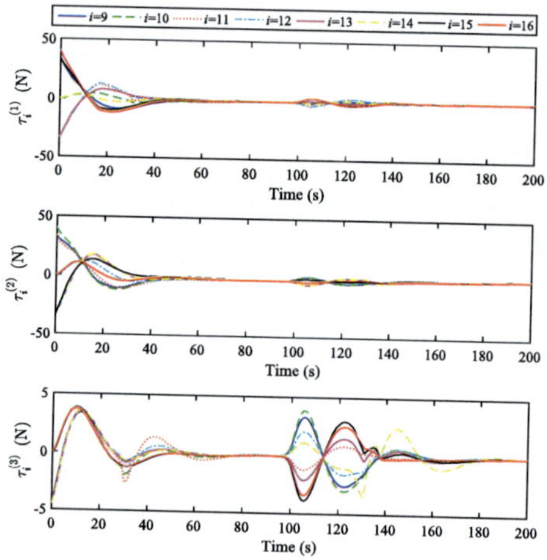

Fig. 14.6 The control inputs of 8 followers

References

1. Olfati-Saber, R., Fax, A., Murray, R.M.: Consensus and Cooperation in Networked Multi-Agent Systems. Proceedings of the IEEE **95**(1), 215–233 (2007)
2. Bullo, F., Cortes, J., Martinez, S.: Distributed Control of Robotic Networks: A Mathematical Approach to Motion Coordination Algorithms. Princeton University Press, Princeton (2009)
3. Cao, Y., Yu, W., Ren, W., Chen, G.: An Overview of Recent Progress in the Study of Distributed Multi-Agent Coordination. IEEE Transactions on Industrial Informatics **9**(1), 427–438 (2012)
4. Jadbabaie, A., Lin, J., Morse, A.S.: Coordination of groups of mobile autonomous agents using nearest neighbor rules. IEEE Transactions on Automatic Control **48**(6), 988–1001 (2003). https://doi.org/10.1109/TAC.2003.812781
5. Olfati-Saber, R., Murray, R.M.: Consensus problems in networks of agents with switching topology and time-delays. IEEE Transactions on Automatic Control **49**(9), 1520–1533 (2004). https://doi.org/10.1109/TAC.2004.834113

6. Ren, W., Beard, R.W.: Consensus seeking in multiagent systems under dynamically changing interaction topologies. IEEE Transactions on Automatic Control **50**(5), 655–661 (2005)
7. Cao, Y., Ren, W., Egerstedt, M.: Distributed containment control with multiple stationary or dynamic leaders in fixed and switching directed networks. Automatica **48**(8), 1586–1597 (2012)
8. Peng, K., Yang, Y.: Leader-following consensus problem with a varying-velocity leader and time-varying delays. Physica A Statistical Mechanics & Its Applications **388**(388), 193–208 (2009)
9. Hong, Y., Chen, G., Bushnell, L.: Distributed observers design for leader-following control of multi-agent networks. Automatica **44**(3), 846–850 (2008)
10. Do, K.D.: Bounded controllers for formation stabilization of mobile agents with limited sensing ranges. IEEE Transactions on Automatic Control **52**(3), 569–576 (2007)
11. Cheah, C.C., Hou, S.P., Slotine, J.J.E.: Region-based shape control for a swarm of robots. Automatica **45**(10), 2406–2411 (2009)
12. Notarstefano, G., Egerstedt, M., Haque, M.: Containment in leader–follower networks with switching communication topologies. Automatica **47**(5), 1035–1040 (2011)
13. Ji, M., Ferrari-Trecate, G., Egerstedt, M., Buffa, A.: Containment control in mobile networks. IEEE Transactions on Automatic Control **53**(8), 1972–1975 (2008)
14. Ballard, L., Cao, Y., Ren, W.: Distributed discrete-time coupled harmonic oscillators with application to synchronised motion coordination. IET Control Theory & Applications **4**(5), 806–816 (2010)
15. Chopra, N., Spong, M.W.: Passivity-based control of multi-agent systems. Springer, Berlin, Heidelberg (2006)
16. Hou, Z.G., Cheng, L., Tan, M.: Decentralized robust adaptive control for the multiagent system consensus problem using neural networks. IEEE Transactions on Systems, Man, and Cybernetics, Part B (Cybernetics) **39**(3), 636–647 (2009)
17. Nuno, E., Ortega, R., Basanez, L., Hill, D.: Synchronization of networks of nonidentical Euler-Lagrange systems with uncertain parameters and communication delays. IEEE Transactions on Automatic Control **56**(4), 935–941 (2011)
18. Mei, J., Ren, W., Ma, G.: Distributed coordinated tracking with a dynamic leader for multiple Euler-Lagrange systems. IEEE Transactions on Automatic Control **56**(6), 1415–1421 (2011)
19. Li, D., Ma, G., He, W., Zhang, W., Li, C., Ge, S.S.: Distributed coordinated tracking control of multiple Euler-Lagrange systems by state and output feedback. IET Control Theory & Applications **11**(14), 2213–2221 (2017)
20. Ihle, I.F., Jouffroy, J., Fossen, T.I.: Formation Control of Marine Surface Craft : A Lagrangian Approach. IEEE Journal of Oceanic Engineering **31**(4), 922–934 (2006)
21. Li, D., Ma, G., He, W., Ge, S.S., Lee, T.H.: Cooperative circumnavigation control of networked microsatellites. IEEE Transactions on Cybernetics **50**(10), 4550–4555 (2019)
22. Li, J., Ren, W., Xu, S.: Distributed containment control with multiple dynamic leaders for double-integrator dynamics using only position measurements. IEEE Transactions on Automatic Control **57**(6), 1553–1559 (2012)
23. Ferrari-Trecate, G., Egerstedt, M., Buffa, A., Ji, M.: Laplacian sheep: A hybrid, stop-go policy for leader-based containment control. In: International workshop on hybrid systems: Computation and control, pp. 212–226. Springer (2006)
24. Han, L., Dong, X., Li, Q., Ren, Z.: Formation-containment control for second-order multi-agent systems with time-varying delays. Neurocomputing **218**, 439–447 (2016). https://doi.org/10.1016/j.neucom.2016.09.001
25. Dong, X., Li, Q., Ren, Z., Zhong, Y.: Output formation-containment analysis and design for general linear time-invariant multi-agent systems. Journal of the Franklin Institute **353**(2), 322–344 (2016). https://doi.org/10.1016/j.jfranklin.2015.11.004
26. Dong, X., Shi, Z., Lu, G., Zhong, Y.: Formation-containment analysis and design for high-order linear time-invariant swarm systems. International Journal of Robust and Nonlinear Control **25**(17), 3439–3456 (2015)

27. Wang, Y., Liu, X., Xiao, J.: Output formation-containment of coupled heterogeneous linear systems under intermittent communication. Journal of the Franklin Institute **354**(1), 392–414 (2017). https://doi.org/10.1016/j.jfranklin.2016.10.011
28. Zheng, B., Mu, X.: Formation-containment control of second-order multi-agent systems with only sampled position data. International Journal of Systems Science **47**(15), 3609–3618 (2016). https://doi.org/10.1080/00207721.2015.1107148
29. Han, T., Chi, M., Guan, Z.H., Hu, B., Xiao, J.W., Huang, Y.: Distributed Three-Dimensional Formation Containment Control of Multiple Unmanned Aerial Vehicle Systems. Asian Journal of Control **19**(3), 1103–1113 (2017). https://doi.org/10.1002/asjc.1445
30. Chen, L.M., Li, C.J., Mei, J., Ma, G.F.: Adaptive cooperative formation-containment control for networked Euler-Lagrange systems without using relative velocity information. IET Control Theory & Applications pp. 1–9 (2017). https://doi.org/10.1049/iet-cta.2016.1185
31. Li, D., Ge, S.S., He, W., Ma, G., Xie, L.: Multilayer formation control of multi-agent systems. Automatica **109**, 108558 (2019). https://doi.org/10.1016/j.automatica.2019.108558
32. Cao, Y., Ren, W.: Multi-vehicle coordination for double-integrator dynamics under fixed undirected/directed interaction in a sampled-data setting. International Journal of Robust & Nonlinear Control **20**(9), 987–1000 (2010)
33. Ren, W.: Sampled-data discrete-time coordination algorithms for double-integrator dynamics under dynamic directed interaction. International Journal of Control **83**(3), 506–515 (2010)
34. Fang, H., Chen, J., Wang, X.: Distributed observer-based coordination for multiple Lagrangian systems using only position measurements **8**(November 2013), 2102–2114 (2014)
35. Ghapani, S., Mei, J., Ren, W., Song, Y.: Fully distributed flocking with a moving leader for Lagrange networks with parametric uncertainties. Automatica **67**, 67–76 (2016). https://doi.org/10.1016/j.automatica.2016.01.004
36. Mehrabian, A.R., Khorasani, K.: Constrained distributed cooperative synchronization and reconfigurable control of heterogeneous networked Euler-Lagrange multi-agent systems. Information Sciences **370-371**(5), 578–597 (2016). https://doi.org/10.1016/j.ins.2015.09.032
37. Mei, J., Ren, W., Ma, G.: Distributed containment control for Lagrangian networks with parametric uncertainties under a directed graph. Automatica **48**(4), 653–659 (2012). https://doi.org/10.1016/j.automatica.2012.01.020
38. Meng, Z., Lin, Z., Ren, W.: Robust cooperative tracking for multiple non-identical second-order nonlinear systems. Automatica **49**(8), 2363–2372 (2013)
39. Meng, Z., Ren, W., You, Z.: Distributed finite-time attitude containment control for multiple rigid bodies. Automatica **46**(12), 2092–2099 (2010). https://doi.org/10.1016/j.automatica.2010.09.005
40. Kubat, M.: Neural networks: a comprehensive foundation by Simon Haykin, Macmillan, 1994, ISBN 0-02-352781-7. Knowledge Engineering Review **13**(4), 409–412 (1999)
41. Funahashi, K.I.: On the approximate realization of continuous mappings by neural networks. Neural networks **2**(3), 183–192 (1989)
42. Ge, S.S., Wang, C.: Adaptive neural control of uncertain mimo nonlinear systems. IEEE Transactions on Neural Networks **15**(3), 674–692 (2004)
43. Behtash, S.: Robust output tracking for non-linear systems. International Journal of Control **51**(6), 1381–1407 (1990)
44. Zhang, B., Jia, Y., Du, J.: Adaptive synchronization control of networked robot systems without velocity measurements. International Journal of Robust and Nonlinear Control **28**(11), 3606–3622 (2018)
45. Ge, S.S., Hang, C.C., Lee, T.H., Zhang, T.: Stable adaptive neural network control. Springer, Boston (2002)

Chapter 15
Formation-Containment Control for Networked Spacecraft in Constrained Space

Abstract This chapter addresses the layered formation-containment (LFC) problem for multi-spacecraft in the constrained space with a directed communication topology. The formation-containment problem is first defined using a layered framework, and a layered distributed finite-time estimator is proposed to acquire the target states for spacecraft in each layer. Based on the proposed framework, the formation configuration and the mechanism of the information flow can be explored and designed naturally. To avoid collisions with borders, obstacles, as well as the other spacecraft in the constrained space, an artificial potential function is designed based on the Dirac delta function. Further, a disturbance observer and adaptive neural networks (NNs) are applied to respectively tackle the external disturbance and the model uncertainties. The desired formation of eachs layer can be achieved while no collision occurs in the constrained space. The semi-global uniform ultimate boundedness of closed-loop errors is guaranteed by Lyapunov stability theory. Simulation results are given to show the effectiveness of the proposed approaches.

15.1 Introduction

The spacecraft in the multi-spacecraft formation flight systems can be abstracted as the agents in the multi-agent systems, which has strong versatility and a large number of theoretical research results. From the control theory perspective, multi-agent system control has been an active area with several topics, such as consensus problems [1–4], tracking problems [5, 6], formation problems [7–10], and containment problems [11–13].

In the formation-containment problems, the leaders achieve a certain formation and the followers move within the safe convex hull spanned by the leaders [14–17]. Further, under an undirected graph, the formation-containment problem is solved by a layered framework in Chap. 13, and the communications among both layers are constructed. As for practical multi-spacecraft close-range formation tasks, a critical issue arises as a large number of spacecraft may carry out a mission in an uncertain constrained space, for example, the vicinity of complex configuration space stations

Y. Gao and D. Li, *Spacecraft Maneuver with Performance Guaranteed*,
https://doi.org/10.1007/978-981-99-4653-2_15

or the vicinity of on-orbit assembly structures, etc. First, the group of spacecraft should avoid collisions among each other. On the other hand, the spacecraft have to avoid collisions with unexpected obstacles or borders caused by other close-range spacecraft or on-orbit assembly components due to the lack of global environmental information. Hence, real-time environment modeling techniques are essential to the spacecraft. Researchers have made many achievements on cooperative control in the constrained space, such as artificial potential filed (APF) [18, 19] and vector filed [20, 21], while few of them can be applied to the problem with spatial borders and global obstacle information is required in advance. To obtain the satisfactory performance for multi-spacecraft control, the unforeseen spatial environment modeling and the complex spatial constraint avoiding are still open problems.

Considering the abovementioned practical issues in the multi-spacecraft formation flight system, we give attention to Euler-Lagrange systems (ELSs) as they can characterize a broad class of practical systems [22–24]. Among them, ELSs can generalize to describe the relative orbital motion, relative attitude motion, and even the related payloads' coordinated motion of the multi-spacecraft formation flight system. For ELSs, another two important issues are inherent model uncertainties and unknown external disturbances. Considering the inherent non-linearity and the model uncertainty in ELSs, adaptive NN-based control is extensively investigated, which provides online approximations of system uncertainties [25–27]. Meanwhile, the disturbance observer (DO) is an effective technique to tackle external disturbances from the environment. Moreover, the DO provides a less conservative scheme compared with most of existing 'worst case based' robust control methods [28–30].

In this chapter, an interesting layered formation-containment (LFC) problem is defined and solved by a unified framework, where a leading layer and a following layer are proposed to describe the behavior of two groups of spacecraft. Inspired by the above mentioned ideas, we consider LFC control for the multi-spacecraft formation systems described by the multiple Euler-Lagrange systems (MELSs) in the presence of complex spatial constraints, model uncertainties, and external disturbances. Compared with our work in Chap. 13, LFC control can tackle a more general case with directed graphs, and new results to handle complex spatial constraints, external disturbances, as well as the layered formation-containment problems are also addressed. In addition, the flexible formation formed by the first layer will help the followers in the second layer to avoid pop-up obstacles and borders, since the leaders will detect the spatial constraints or other pop-up threats first and drive the followers to a safe place through formation changing.

This research has the following contributions. Considering model uncertainties and external disturbances, a novel LFC control law combined with adaptive NNs, APFs, and DOs is designed, which can also help the spacecraft avoid collisions with the borders, the obstacles, as well as each other. The semi-globally uniformly ultimately bounded (SGUUB) property of all error signals is proved by the Lyapunov direct method. Next, utilizing the Dirac delta function and local geometrical information, an artificial potential field of spatial constraints is established, which help the spacecraft to model and avoid complex spatial constraints, such as the obstacles and the borders, while most of the existing works only deal with relatively simple obstacles [18–21]. Last but not least, the undirected result in Chap. 13 requires a connected graph to drive spacecraft to a desired formation shape, but it is not implementable for practical tasks over a directed topology. This approach is addressed under a directed graph as a general extension.

Notations: Let $\mathbb{R}$ denote the set of real numbers. Let $I_n \in \mathbb{R}^{n\times n}$ denote the identity matrix of order n. Let $\lambda_{\max}(\cdot)$ and $\lambda_{\min}(\cdot)$ denote the maximum and minimum eigenvalues of a matrix, respectively. Let $\|\cdot\|$ be the Euclidean norm of a vector and $\operatorname{diag}(z_1, \ldots, z_n) \in \mathbb{R}^{n\times n}$ the diagonal matrix. Let $\operatorname{sgn}(\cdot)$ stand for the signum function.

15.2 Preliminaries and Problem Description

In this chapter, the Euler-Lagrange systems are utilized to describe the multi-spacecraft formation flight systems because of their abstraction of different control objects and the transferability of control algorithms. Consider a group of N leader spacecraft (in the first layer) with subset $\mathcal{V}_l = \{1, 2, \ldots, N\}$ and M follower spacecraft (in the second layer) with subset $\mathcal{V}_f = \{N+1, N+2, \ldots, N+M\}$ in $\mathbb{R}^p$. The ith spacecraft is modeled by (2.35).

Assumption 15.1 According to the real task environment, we assume that the external disturbance $f_{di}(t)$ is continuous and bounded.

Lemma 15.1 [31] *If a continuously differentiable function $f(t)$ is bounded by a positive constant κ as $|f(t)| \le \kappa$, $\forall t \in [0, \infty)$, then $\dot{f}(t)$ is bounded $\forall t \in [0, \infty)$.*

We utilize a directed graph $\mathcal{G} = (\mathcal{V}, \mathcal{E})$ to characterize the communication topology among the spacecraft in both layers, where $\mathcal{E} \subseteq \mathcal{V} \times \mathcal{V}$ and $\mathcal{V} = \mathcal{V}_L \cup \mathcal{V}_F$ denote the arc set and the vertex set, respectively. The adjacency matrix $\mathcal{A} = [a_{ij}] \in \mathbb{R}^{(N+M)\times(n+m)}$ is a nonnegative matrix defined as $a_{ij} > 0$ if $(j, i) \in \mathcal{E}$, and $a_{ij} = 0$ otherwise. We define $a_{ii} = 0$, and the Laplacian matrix is defined as $\mathcal{L} = [l_{ij}] \in \mathbb{R}^{(N+M)\times(N+M)}$, where $l_{ii} = \sum_{j=1, j\neq i}^{N+M} a_{ij}$ and $l_{ij} = -a_{ij}$, $i \neq j$. For convenience, we use a virtual spacecraft 0 to denote the virtual center of desired formation.

In this chapter, the corresponding Laplacian matrix $\mathcal{L} \in \mathbb{R}^{(N+M+1)\times(N+M+1)}$ is given as

$$\mathcal{L} = \begin{bmatrix} 0 & 0_{1\times N} & 0_{1\times M} \\ \mathcal{L}_{L2} & \mathcal{L}_{L1} & 0_{N\times M} \\ 0_{M\times 1} & \mathcal{L}_{F2} & \mathcal{L}_{F1} \end{bmatrix}, \tag{15.1}$$

where $\mathcal{L}_{L1}$ represents the communication topology among the leaders, $\mathcal{L}_{L2}$ represents the information flow from the virtual spacecraft to the leaders, $\mathcal{L}_{F1}$ represents the communication topology among the followers, and $\mathcal{L}_{F2}$ represents the communication interaction between the leaders and followers. Let $\mathcal{G}_L$ denote the directed graph with $N+1$ spacecraft labeled as 0 to N, and the following assumptions are made.

Assumption 15.2 For the first layer, there exists a directed spanning tree in $\mathcal{G}_L$, where the root of $\mathcal{G}_L$ is the virtual spacecraft 0. For the second layer, there exists at least one leader has a directed path to each follower.

Assumption 15.3 Suppose that the acceleration of the virtual spacecraft 0 is bounded by a nonnegative constant δ_L, such that $\|\ddot{q}_0\| \leq \delta_L < \infty$.

A layered framework is proposed to describe the formation-containment problem in the constrained space with two layers, where the leader spacecraft in the first layer move together with a time-varying formation configuration and the follower spacecraft in the second layer move within the convex hull spanned by the leaders.

To solve this problem, in Sect. 15.3, a layered distributed finite-time estimator (LDFE) is first designed: (1) in the first layer, an LDFE is introduced for each spacecraft to estimate the target position and velocity of the formation configuration in finite time; and (2) in the second layer, a similar LDFE is given to estimate the accurate estimation of the weighted average of the leaders' states in finite time. Then, a collision-free LFC control law is designed in Sect. 15.4 to stear all the spacecraft to the target formation based on the estimate from LDFEs. In addition, the flexible formation composed by the first layer will help the followers in the second layer to avoid pop-up obstacles. The LFC problem to be solved in this chapter is formally stated below.

Problem 15.1 Given the initial spacecraft formation $\mathcal{G}(q_i(0))$, $i \in \mathcal{V}$, design an LDFE with $\hat{v}_i$ and $\hat{q}_i$ for spacecraft $i \in \mathcal{V}_L$ such that $\hat{v}_i(t) \to \dot{q}_0(t) + \dot{r}_i(t)$, and $\hat{q}_i(t) \to q_0(t) + r_i(t)$ in finite time, where $r_i(t)$ is the auxiliary function which shows the relative distance decomposed from the target formation configuration of the first layer, i.e., the target relative position between spacecraft 0 and spacecraft i. Similarly, design an LDFE for spacecraft $i \in \mathcal{V}_F$, such that $\hat{v}_i(t) \to \dot{c}_i(t)$, and $\hat{q}_i(t) \to c_i(t)$, where c_i denotes the target containment position for spacecraft i (c_i will be formally introduced in Sect. 15.3). Next, based on the estimates $\hat{v}_i$ and $\hat{q}_i$ from the LDFE, design the LFC control law $\tau_i(t)$ for spacecraft $i \in \mathcal{V}$ such that $q_i(t) \to \hat{q}_i(t)$ and $\dot{q}_i(t) \to \hat{v}_i(t)$ with bounded tracking errors, while no collision with the spatial constraints or other spacecraft occurs.

15.3 Layered Distributed Estimator

In this section, for spacecraft in both layers, we propose the following LDFE protocol (see (15.2a) and (15.3b)), where β_1, β_2, β_3, and β_4 are positive constants, ϑ_i is an auxiliary variable. In addition, we define $\hat{q}_0 = q_0$, $\vartheta_0 = \dot{q}_0$ in (15.2a) and (15.2b), and $\hat{q}_j = q_j$, $\hat{v}_j = \dot{q}_j$ when $j \in \mathcal{V}_L$ in (15.3a) and (15.3b).

$$\text{The First Layer}\begin{cases}\dot{\hat{q}}_i = \hat{v}_i - \beta_1 \text{sgn}\left[\sum_{j\in\mathcal{V}_L} a_{ij}\left(\hat{q}_i - r_i - \hat{q}_j + r_j\right) + a_{i0}\left(\hat{q}_i - r_i - q_0\right)\right], & i\in\mathcal{V}_L, \quad (15.2a)\\ \hat{v}_i = \vartheta_i + \dot{r}_i, & i\in\mathcal{V}_L, \quad (15.2b)\\ \dot{\vartheta}_i = -\beta_2 \text{sgn}\left[\sum_{j\in\mathcal{V}_L} a_{ij}\left(\vartheta_i - \vartheta_j\right) + a_{i0}\left(\vartheta_i - \dot{q}_0\right)\right], & i\in\mathcal{V}_L, \quad (15.2c)\end{cases}$$

$$\text{The Second Layer}\begin{cases}\dot{\hat{q}}_i = \hat{v}_i - \beta_3 \text{sgn}\left[\sum_{j\in\mathcal{V}_F} a_{ij}\left(\hat{q}_i - \hat{q}_j\right) + \sum_{j\in\mathcal{V}_L} a_{ij}\left(\hat{q}_i - q_j\right)\right], & i\in\mathcal{V}_F, \quad (15.3a)\\ \dot{\hat{v}}_i = -\beta_4 \text{sgn}\left[\sum_{j\in\mathcal{V}_F} a_{ij}\left(\hat{v}_i - \hat{v}_j\right) + \sum_{j\in\mathcal{V}_L} a_{ij}\left(\hat{v}_i - \dot{q}_j\right)\right], & i\in\mathcal{V}_F. \quad (15.3b)\end{cases}$$

Before giving the main theorem, we define a positive constant δ_F, which satisfies

$$\|\ddot{q}_i\| \le \delta_F < \infty,\ i \in \mathcal{V}_L, \tag{15.4}$$

as the accelerations of the leaders are bounded, which will be detailed later in Remark 15.2. We then define Q_L, C_F and c_i, $i \in \mathcal{V}_F$ as follows for easy analysis, $Q_L = \left[q_1^T, q_2^T, \ldots, q_N^T\right]^T$, $C_F = -\left(L_{F1}^{-1}L_{F2} \otimes I_p\right) Q_L = \left[c_{N+1}^T, c_{N+2}^T, \ldots, c_{N+M}^T\right]^T$.

Theorem 15.1 *Let $\beta_1 > 0$, $\beta_2 > \delta_L$, $\beta_3 > 0$ and $\beta_4 > \delta_F$. If Assumptions 15.2 and 15.3 hold, the LDFEs (15.2a) and (15.3b) can achieve the estimation object defined in Problem 15.1 in finite time with time instant T_M, i.e., for the first layer we have $\hat{q}_i(t) \to q_0(t) + r_i(t)$, $i \in \mathcal{V}_L$, $\hat{v}_i(t) \to \dot{q}_0(t) + \dot{r}_i(t)$, $i \in \mathcal{V}_L$, and for the second layer we have $\hat{q}_i(t) \to c_i(t)$, $i \in \mathcal{V}_F$, $\hat{v}_i(t) \to \dot{c}_i(t)$, $i \in \mathcal{V}_F$, when $t \to T_M$.*

Proof First, consider the first layer $i \in \mathcal{V}_L$. Define $\tilde{v}_i = \hat{v}_i - \dot{q}_0 - \dot{r}_i$, $\tilde{v}_L^+(t) \triangleq \max_{i\in V_L} \tilde{v}_i(t)$, and $\tilde{v}_L^-(t) \triangleq \min_{i\in V_L} \tilde{v}_i(t)$. We can then get

$$\begin{aligned}\dot{\tilde{v}}_i &= \dot{\vartheta}_i - \ddot{q}_0\\ &= -\beta_2 \text{sgn}\sum_{j=0}^{N} a_{ij}\left(\vartheta_i - \vartheta_j\right) - \ddot{q}_0\\ &= -\beta_2 \text{sgn}\sum_{j=0}^{N} a_{ij}\left(\hat{v}_i - \dot{r}_i - \dot{q}_0 - \hat{v}_j + \dot{r}_j + \dot{q}_0\right) - \ddot{q}_0\\ &= -\beta_2 \text{sgn}\sum_{j=0}^{N} a_{ij}\left(\tilde{v}_i - \tilde{v}_j\right) - \ddot{q}_0,\ i \in \mathcal{V}_L.\end{aligned} \tag{15.5}$$

Analysing the following three cases, we can get the convergence time for $\hat{v}_i(t) \to \dot{q}_0(t) + \dot{r}_i(t)$ in the first layer:

Case L1: $\tilde{v}_L^+(0) \le 0$ and $\tilde{v}_L^-(0) \le 0$. It is trivial to derive $\tilde{v}_L^+(t) = \tilde{v}_L^-(t) = 0$, $\forall t > 0$, if $\tilde{v}_L^+(0) = 0$ and $\tilde{v}_L^-(0) = 0$ [32]. Next, consider the case with $\tilde{v}_L^+(0) \le 0$ and $\tilde{v}_L^-(0) < 0$, where $\tilde{v}_L^+(t) \le 0$, $\forall t > 0$ because $\dot{\tilde{v}}_L^+(t) \le -\beta_2 - \ddot{q}_0 \le -\beta_2 + \delta_L < 0$ when $\tilde{v}_L^+(t) > 0$. Similarly, we know $\tilde{v}^-(t)$ is nondecreasing. Since either $\dot{\tilde{v}}_L^-(t) = -\ddot{q}_0$ or $\dot{\tilde{v}}_L^-(t) > \beta_2 - \delta_L > 0$, it is worth mentioning that when $\tilde{v}_L^-(t) < 0$, $\dot{\tilde{v}}_L^-(t) = -\ddot{q}_0$ only exists at isolated time instants. Here, the proof is given by contradiction. Assume that $\tilde{v}_L^-(t) < 0$ for $T > 0$, and $\dot{\tilde{v}}_L^-(t) = -\ddot{q}_0$ for $t \in [t_1, t_2]$ with $t_1 < t_2 < T$. Therefore, there exists leader j satisfactory $\dot{\tilde{v}}_j(t) = -\ddot{q}_0$ for $t \in [t_1, t_3]$, where $t_1 < t_3 \le t_2$. Then, according to (15.5) and Assumption 15.2, it follows that $\tilde{v}_i(t) = \tilde{v}_j(t) = \tilde{v}_L^-(t) < 0$, $\forall i \in \{0, 1, \ldots, N\}$, for $t \in [t_1, t_3]$, which is contradicted by $\tilde{v}_0(t) = 0$. Thus, except for several isolated instants, we know that $\tilde{v}_L^-(t)$ will keep increasing when $t_1 < T$ with a speed larger than $\beta_2 - \delta_L$. From the above analysis, the convergence time is given by $t \le \frac{|\tilde{v}_L^-(0)|}{\beta_2 - \delta_L}$.

Case L2: $\tilde{v}_L^+(0) \ge 0$ and $\tilde{v}_L^-(0) \ge 0$. Utilizing a similar analysis given in Case L1, we can get the convergence time as $t \le \frac{|\tilde{v}_L^+(0)|}{\beta_2 - \delta_L}$.

Case L3: $\tilde{v}_L^+(0) \ge 0$ and $\tilde{v}_L^-(0) \le 0$. Combine Case L1 and Case L2, and we can derive the convergence time as $t \le \frac{\max\{|\tilde{v}_L^+(0)|, |\tilde{v}_L^-(0)|\}}{\beta_2 - \delta_L}$.

We next study the convergence time for $\hat{q}_i(t) \to q_0(t) + r_i(t)$: Define $\tilde{q}_i = \hat{q}_i - q_0 - r_i$ and $T_{L1} = \frac{\max_{i \in V_L}\{|\tilde{v}_i(0)|\}}{\beta_2 - \delta_L}$. Since $\hat{v}_i(t) = \dot{q}_0(t) + \dot{r}_i$, $\forall t \ge T_{L1}$, we have $\dot{\tilde{q}}_i = -\beta_1 \text{sgn} \sum_{j=0}^{N} a_{ij} \left(\tilde{q}_i - \tilde{q}_j\right)$, $i \in \mathcal{V}_L$.

Then, following the three cases of the first layer, we can derive $\hat{q}_i(t) = q_0(t) + r_i(t)$, for $\forall t \ge T_{L1} + \frac{\max_{i \in V_L}\{|\tilde{q}_i(T_{L1})|\}}{\beta_1}$, which completes the proof. Define $T_{L2} = T_{L1} + \frac{\max_{i \in V_L}\{|\tilde{q}_i(T_{L1})|\}}{\beta_1}$.

Next, consider the second layer $i \in \mathcal{V}_F$. Define $\tilde{v}_i = \hat{v}_i - \dot{c}_i$, $\tilde{v}_F^+(t) \triangleq \max_{i \in V_F} \tilde{v}_i(t)$, and $\tilde{v}_F^-(t) \triangleq \min_{i \in V_F} \tilde{v}_i(t)$. We can then get

$$\dot{\tilde{v}}_i = -\beta_4 \text{sgn} \sum_{j=1}^{N} \left(l_{ij}\tilde{v}_i\right) - \ddot{c}_i, \; i \in \mathcal{V}_F, \tag{15.6}$$

where l_{ij} is the entry of matrix $\mathcal{L}$.

Analysing the following three cases, we can get the convergence time for $\hat{v}_i(t) \to \dot{c}_i$ in the second layer:

Case F1: $\tilde{v}_F^+(0) \le 0$ and $\tilde{v}_F^-(0) \le 0$. Similar to that in the first layer, we have $\tilde{v}_F^+(t) = \tilde{v}_F^-(t) = 0$, $\forall t > 0$, if $\tilde{v}_F^+(0) = 0$ and $\tilde{v}_F^-(0) = 0$. Next, consider the case with $\tilde{v}_F^+(0) \le 0$ and $\tilde{v}_F^-(0) < 0$, and we can get $\tilde{v}_F^+(t) \le 0$, $\forall t > 0$ because

$$l_{ii} \ge -\sum_{j=1, j\neq i}^{N} l_{ij}, \tag{15.7}$$

$$\dot{\tilde{v}}_F^+(t) \le -\beta_4 - \ddot{c}_i \le -\beta_4 + \delta_F < 0, \tag{15.8}$$

when $\tilde{v}_F^+(t) > 0$. Similarly, we know $\tilde{v}_F^-(t)$ is nondecreasing. Since either $\dot{\tilde{v}}_F^-(t) = -\ddot{c}_i$ or $\dot{\tilde{v}}_F^-(t) > \beta_4 - \delta_F > 0$, it is worth mentioning that when $\tilde{v}_F^-(t) < 0$, $\dot{\tilde{v}}_F^-(t) = -\ddot{c}_i$ only exists at isolated time instants. Here, the proof is given by contradiction. Assume that $\tilde{v}_F^-(t) < 0$ for $T > 0$, and $\dot{\tilde{v}}_F^-(t) = -\ddot{c}_i$ for $t \in [t_1, t_2]$ with $0 < t_1 < t_2 < T$. Therefore, there exists leader j satisfactory $\dot{\tilde{v}}_j(t) = -\ddot{c}_i$ for $t \in [t_1, t_3]$, where $t_1 < t_3 \le t_2$. Then, According to (15.3b) and Assumption 15.2, it follows that $\sum_{j=1}^{N+M} l_{ij}\tilde{v}_j = 0$, and $\sum_{j=1}^{N+M} a_{ij}\left(\hat{v}_i - \hat{v}_j\right) = 0$ for $t \in [t_1, t_3]$, which is contradicted by $v_{N+1} \neq v_{N+2} \neq \ldots \neq v_{N+M}$ due to the arbitrariness of v_j, $j \in V_L$. Thus, except for several isolated instants, we know $\tilde{v}_F^-(t)$ will keep increasing when $0 < t_1 < T$ with a speed larger than $\beta_4 - \delta_F$. From the above analysis, the convergence time is given by $t \le \frac{|\tilde{v}_F^-(0)|}{\beta_4 - \delta_F}$.

Case F2: $\tilde{v}_F^+(0) \ge 0$ and $\tilde{v}_F^-(0) \ge 0$. Utilizing a similar analysis given in Case F1, we can get the convergence time as $t \le \frac{|\tilde{v}_F^+(0)|}{\beta_4 - \delta_F}$.

Case F3: $\tilde{v}_F^+(0) \ge 0$ and $\tilde{v}_F^-(0) \le 0$. Combine Case F1 and Case F2, and we can derive the convergence time as $t \le \frac{\max\{|\tilde{v}_F^+(0)|, |\tilde{v}_F^-(0)|\}}{\beta_4 - \delta_F}$.

Define $T_{F1} = \frac{\max_{i\in V_F}\{|\tilde{v}_i(0)|\}}{\beta_4 - \delta_F}$. Similar to the proof of the first layer, the convergence time for $\hat{q}_i \to c_i$ can be derived as

$$t \ge T_{F1} + \frac{\max_{i\in V_F}\{|\tilde{q}_i(T_{F1})|\}}{\beta_3}, \tag{15.9}$$

which completes the proof. For convenience, define $T_{F2} = T_{F1} + \frac{\max_{i\in V_F}\{|\tilde{q}_i(T_{F1})|\}}{\beta_3}$, and $T_M = \max\{T_{L2}, T_{F2}\}$. □

According to Theorem 15.1, we can use $\hat{q}_i$ and $\hat{v}_i$, $i \in \mathcal{V}_L$ to replace the position and velocity of the desired formation, and use $\hat{q}_i$ and $\hat{v}_i$, $i \in \mathcal{V}_F$ to replace the weighted average of the positions and velocities of the leaders when $t \ge T_M$.

Lemma 15.2 *The estimation error* $\tilde{q}_i$, $i \in \mathcal{V}$ *is bounded for* $0 \le t \le \max\{T_{L1}, T_{F1}\}$.

Proof Consider the first layer with $i \in \mathcal{V}_L$. It follows from (15.5) that $|\tilde{v}_i(t)| \le |\tilde{v}_i(0)| + (\delta_L + \beta_2)t$ for $0 \le t \le T_{L1}$. Since $\dot{\tilde{q}}_i = -\beta_1 \text{sgn} \sum_{j=0}^{N} a_{ij}\left(\tilde{q}_i - \tilde{q}_j\right)$, we know $|\tilde{q}_i(t)| \le |\tilde{q}_i(0)| + (|\tilde{v}_i(0)| + \beta_1)t + \frac{(\delta_L+\beta_2)t^2}{2}$ for $0 \le t \le T_{L1}$. Thus, $\tilde{q}_i, i \in \mathcal{V}_L$ is bounded for $0 \le t \le T_{L1}$. Similarly, we have $\tilde{q}_i, i \in \mathcal{V}_F$ is bounded for $0 \le t \le T_{F1}$. □

Lemma 15.3 *Note that $\dot{\hat{q}}_i$ in (15.2a) is discontinuous and $\dot{q}_0 + \dot{r}_i$ is continuous. Still, $\dot{\hat{q}}_i$ can replace $\dot{q}_0 + \dot{r}_i$ or $\hat{v}_i$ when $t \ge T_M$. Similarly, $\dot{\hat{q}}_i$ in (15.3a) can replace $\dot{c}_i$ or $\hat{v}_i$ when $t \ge T_M$.*

Proof The proof is similar to that in Sect. 12.3.1. □

15.4 Layered Formation-Containment Control in Constrained Space

In this section, based on the estimation from the LDFEs, we consider the LFC control of spacecraft formation described by MELSs with model uncertainties and external disturbances in constrained space. The model of the constrained space is first constructed, and the Dirac delta function is used to design an artificial potential function which can gather the potential forces from the spatial constraints. Then, a unified LFC control algorithm is proposed to achieve the LFC control objective with the help of a DO and adaptive NNs.

Define auxiliary variables $z_{1i} = q_i - \hat{q}_i$, and $z_{2i} = \dot{q}_i - \alpha_{1i}$, $i \in \mathcal{V}$, where α_{1i} is the virtual control to be designed. Assume that the danger detection range of each spacecraft is centered at the spacecraft with radius r. To achieve collision avoidance among the spacecraft, we design the potential functions for spacecraft i as

$$P_{i,c}\left(q_{i,j}\right) = \sum_{j \in H_i} c_{i,j} \ln \frac{r - d_{safe}}{\|q_{i,j}\| - d_{safe}}, \tag{15.10}$$

where $q_{i,j} = q_i - q_j$, $(i, j \in \mathcal{V},\ i \neq j)$, $c_{i,j} = c_{j,i}$ are positive constants, d_{safe} is the required safe distance, and H_i represents all spacecraft within the danger range of spacecraft i as $H_i = \left\{j \in \mathcal{V},\ j \neq i \mid \|\tilde{q}_{i,j}\| < r\right\}$.

Next, to avoid collisions with obstacles in the constrained space, we define the constrained space as a point set $\Im = \Im_1 + \Im_2 + \ldots + \Im_R$, where R is the number of continuous regions. Define L_r as the edge of region $\Im_r$. Here, the irregular edges of spatial constraints are considered, where we divide an irregular edge into a number of segments. Inspired by the result in [33], we use Dirac delta function [34] to gather the potential forces from all segments of each edge in the constrained space, and the following potential function is constructed

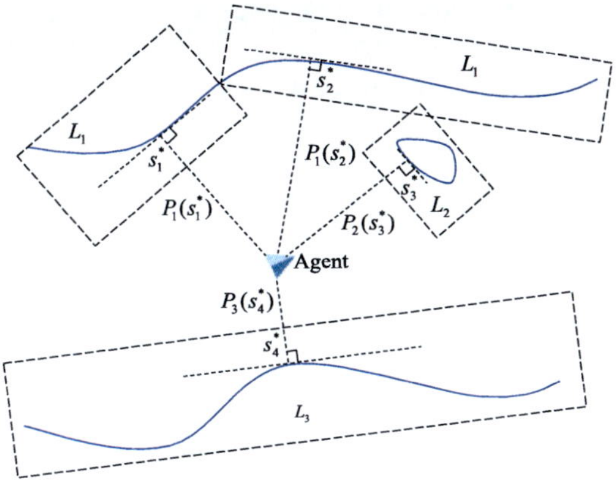

Fig. 15.1 The modelling demonstration of the constrained space

$$P_{i,o}(q_i) = \sum_{x=1}^{R} \underbrace{\int_{q_i^{(1)}-r}^{q_i^{(1)}+r} \cdots \int_{q_i^{(p)}-r}^{q_i^{(p)}+r}}_{p} k_{cx} \frac{\delta(\|q_i - s - kP_x(s)\|)}{\|q_i - s\| - d_{safe}} \underbrace{\mathrm{d}s_1 \cdots \mathrm{d}s_p}_{p}, \tag{15.11}$$

where k is a real number, k_{cx} is a positive constant, $P_x(s)$ represents the perpendicular direction of L_x, and s_i denotes the ith element of vector s, $i \in 1, 2, \ldots, p$. The modelling demonstration of constrained space is given in Fig. 15.1, where $s_1^*, s_2^*, \ldots, s_4^*$ satisfy the following conditions $q_i - s_1^* = k_1 P_1(s_1^*)$, $q_i - s_2^* = k_2 P_1(s_2^*)$, $q_i - s_3^* = k_3 P_2(s_3^*)$, $q_i - s_4^* = k_4 P_3(s_4^*)$, with real numbers $k_1, k_2, \ldots, k_4$.

The virtual control is given as

$$\alpha_{1i} = \hat{v}_i - (\Theta_i)^\dagger \left[K_{2i} V_{pi} + \left(\frac{\partial P_{i,o}}{\partial q_i} + 2\frac{\partial P_{i,c}}{\partial q_i} \right)^T \hat{v}_i \right], \tag{15.12}$$

where $\Theta_i = z_{1i}^T + \left(\frac{\partial P_{i,o}}{\partial q_i}\right)^T + 2\left(\frac{\partial P_{i,c}}{\partial q_i}\right)^T$, and $V_{pi} = \frac{1}{2} z_{1i}^T z_{1i} + P_{i,o} + P_{i,c}$, the gain matrix $K_{2i} = K_{2i}^T > 0$, and $(\Theta_i)^\dagger$ is the Moore-Penrose inverse of Θ_i defined as

$$\Theta_i(\Theta_i)^\dagger = \begin{cases} 0, & \Theta_i = [0, 0, \ldots, 0]^T, \\ 1, & \text{otherwise.} \end{cases} \tag{15.13}$$

Remark 15.1 Notice that $\dot{q}_j$ is only used in (15.12) when spacecraft j is within the danger range of spacecraft i. Thus it is acceptable in practice and not in conflict with the directed communication topology, since the danger range r is rather small compared with the communication range among the directed graph.

Then, according to Lemma 15.3 and (15.12), we have

$$\begin{aligned}\dot{z}_{1i} &= z_{2i} + \alpha_{1i} - \hat{v}_i \\ &= z_{2i} - (\Theta_i)^{\dagger}\left[K_{2i}V_{pi} + \left(\frac{\partial P_{i,o}}{\partial q_i} + 2\frac{\partial P_{i,c}}{\partial q_i}\right)^T \hat{v}_i\right]\end{aligned} \tag{15.14}$$

Using (2.35), the derivative of z_{2i} takes the form

$$\dot{z}_{2i} = M_i^{-1}\left[\tau_i - f_{di} - C_i\dot{q}_i - g_i\right] - \dot{\alpha}_{1i}. \tag{15.15}$$

Note $M_i(q_i)$, $C_i(q_i,\dot{q}_i)$ and $g_i(q_i)$ cannot be directly used in control design due to the model uncertainties. Here, assuming that $M_i(q_i)$ and $C_i(q_i,\dot{q}_i)$ can be divided into the certain parts M_{0i}, C_{0i}, and the uncertain parts ΔM_i, ΔC_i, where $M_i = M_{0i} + \Delta M_i, C_i = C_{0i} + \Delta C_i$. In addition, M_{0i} is a symmetric positive definite matrix, and $(1/2)\dot{M}_{0i} - C_{0i}$ is skew symmetric. We next apply NNs to approximate the model uncertainties [35] as

$$\begin{aligned}f_i(q_i, \dot{q}_i, \alpha_{1i}, \dot{\alpha}_{1i}) &= W_i^{*T}\phi_i(q_i, \dot{q}_i, \alpha_{1i}, \dot{\alpha}_{1i}) + \epsilon_i \\ &= g_i(q_i) + \Delta C_i(q_i, \dot{q}_i)\dot{q}_i + M_i(q_i)\dot{\alpha}_{1i} \\ &\quad + \Delta M_i(q_i)\dot{z}_{2i},\end{aligned} \tag{15.16}$$

where $W_i^* \in \mathbb{R}^{l\times p}$ is the ideal constant weight matrix, l the NN node number, and $\phi_i = [\varphi_{i1}, \varphi_{i2}, \ldots, \varphi_{il}]^T \in \mathbb{R}^l$ is the activation function. ϵ_i is the approximation error bounded over the compact set, i.e., $\|\epsilon_i(Z_i)\| \le \bar{\epsilon}_i$, $\forall Z_i \in \Omega_{z_i}$ with $\bar{\epsilon}_i > 0$ as an unknown constant, where $Z_i = \left[q_i^T, \dot{q}_i^T, \alpha_{1i}^T, \dot{\alpha}_{1i}^T\right]^T$.

The activation functions are chosen as Gaussian functions

$$\varphi_{i\ell}(Z_i) = \exp\left[\frac{-(Z_i - \mu_\ell)^T(Z_i - \mu_\ell)}{\eta_\ell^2}\right], \ \ell = 1, \ldots, l, \tag{15.17}$$

where η_ℓ and μ_ℓ are two constant parameters [25].

To estimate the disturbance, we first define an auxiliary variable as

$$z_{3i} = d_i + \upsilon_i(z_{2i}), \tag{15.18}$$

where $d_i = f_{di} + \epsilon_i$, and $\upsilon_i(z_{2i})$ is a function vector of z_{2i}. Generally, for the easy implementation of the disturbance observer, $\upsilon_i(z_{2i})$ is defined as a linear function of z_{2i}. In the later simulations, $\upsilon_i(z_{2i})$ is chosen as $3z_{2i}$.

Substituting (15.16) into (15.15), we can get

$$\dot{z}_{2i} = M_{0i}^{-1}\left[\tau_i - d_i - C_{0i}\dot{q}_i - W_i^{*T}\phi_i(Z_i)\right]. \tag{15.19}$$

Using (15.19), the derivative of (15.18) is given as $\dot{z}_{3i} = \dot{d}_i + \Im_i(z_{2i}) M_i^{-1}(q_i)$ $\left[\tau_i - d_i - C_{0i}(q_i, \dot{q}_i)\dot{q}_i - W_i^{*T}\phi_i(Z_i)\right]$, where $\Im(z_{2i}) = \frac{\partial v_i(z_{2i})}{\partial z_{2i}^T} \in \mathbb{R}^{p\times p}$.

We next design the following observer to approximate z_{3i}

$$\dot{\hat{z}}_{3i} = \Im_i(z_{2i}) M_i^{-1}(q_i)\left[\tau_i - \hat{d}_i - C_{0i}(q_i, \dot{q}_i)\dot{q}_i\right], \tag{15.20}$$

where $\hat{z}_{3i}$ is the estimation of d_i taking the form as $\hat{d}_i = \hat{z}_{3i} - v_i(z_{2i})$. Then, we have the estimation error and its time derivative as

$$\begin{aligned}
\tilde{z}_{3i} &= \hat{z}_{3i} - z_{3i} = \hat{d}_i - d_i = \tilde{d}_i, \\
\dot{\tilde{z}}_{3i} &= \dot{\tilde{d}}_i = -\dot{d}_i - \Im_i(z_{2i}) M_{0i}^{-1}(q_i)\left(-\tilde{d}_i + W_i^{*T}\phi_i(Z_i)\right).
\end{aligned}$$

Then, the following LFC control algorithms for both layers are formally proposed

$$\begin{aligned}
\tau_i = &- z_{1i} - K_{1i}z_{2i} + \hat{W}_i^T\phi_i(Z_i) + \hat{d}_i \\
&- \frac{\partial P_{i,o}}{\partial q_i} - 2\frac{\partial P_{i,c}}{\partial q_i} + C_{0i}(q_i, \dot{q}_i)\alpha_{1i},
\end{aligned} \tag{15.21}$$

$$\dot{\hat{W}}_{i,k} = -\gamma_i\left(\phi_i(Z_i) z_{2i,k} + \sigma_i\hat{W}_{i,k}\right), \tag{15.22}$$

where K_{1i} is the control gain, γ_i, σ_i are positive constants, $\hat{W}_{i,k}$ denotes the kth column vector of $\hat{W}_i$, and $z_{2i,k}$ stands for the kth element of z_{2i}, $i \in \mathcal{V}$. Define the global closed-loop error signals as $z_1 = [z_{11}^T, z_{12}^T, \ldots, z_{1(N+M)}^T]^T$, $z_2 = [z_{21}^T, z_{22}^T, \ldots, z_{2(N+M)}^T]^T$, $\tilde{W} = [\tilde{W}_1, \tilde{W}_2, \ldots, \tilde{W}_{N+M}]^T$, $\tilde{d} = [\tilde{d}_1^T, \tilde{d}_2^T, \ldots, \tilde{d}_{N+M}^T]^T$.

Theorem 15.2 *In both layers, considering the multi-spacecraft formation systems modeled by MELSs (2.35), under the LFC control algorithm (15.21) and (15.22), the closed-loop errors z_1, z_2, $\tilde{W}$, and $\tilde{d}$ in both layers are SGUUB, which will remain within the following compact sets Ω_{z_1}, Ω_{z_2}, $\Omega_{\tilde{W}}$, and $\Omega_{\tilde{d}}$ respectively:*

$$\Omega_{z_1} = \left\{z_1 \in {}^{p(N+M)} \mid \|z_1\| \le \sqrt{D}\right\}, \tag{15.23}$$

$$\Omega_{z_2} = \left\{z_2 \in {}^{p(N+M)} \mid \|z_2\| \le \sqrt{\frac{D}{\min_{i\in\mathcal{V}} \lambda_{\min}(M_i)}}\right\}, \tag{15.24}$$

$$\Omega_{\tilde{W}} = \left\{\tilde{W} \in {}^{l(N+M)\times p} \mid \left\|\tilde{W}\right\| \le \max_{i\in\mathcal{V}}\left(\sqrt{\gamma_i D}\right)\right\}, \tag{15.25}$$

$$\Omega_{\tilde{d}} = \left\{\tilde{d} \in {}^{p(N+M)} \mid \left\|\tilde{d}\right\| \le \sqrt{D}\right\}, \tag{15.26}$$

as long as for $\forall i \in \mathcal{V}$,

$$\lambda_{\min}(K_{2i}) > 0,\ \lambda_{\min}(K_{1i} - \frac{1}{2}I_p) > 0,$$

$$\lambda_{\min}(\Im_i(z_{2i})M_{0i}^{-1} - \frac{\left\|\Im(z_{2i})M_{0i}^{-1}\right\|^2 - 2\theta}{2\theta}I_p) > 0,$$

where D is a positive constant defined in the following proof.

Proof Consider the following Lyapunov function candidate

$$V = \sum_{i\in\mathcal{V}}\left[\frac{1}{2}z_{1i}^T z_{1i} + \frac{1}{2}\tilde{d}_i^T\tilde{d}_i + P_{i,o} + P_{i,c}\right.$$
$$\left. + \frac{1}{2}z_{2i}^T M_{0i} z_{2i} + \frac{1}{2\gamma_i}\sum_{k=1}^{p}\tilde{W}_{i,k}^T\tilde{W}_{i,k}\right], \tag{15.27}$$

where $\hat{W}_{i,k}$, $W_{i,k}^*$, and $\tilde{W}_{i,k}$, are, respectively, the NN estimation, ideal value, and weight error defined as $\tilde{W}_{i,k} = \hat{W}_{i,k} - W_{i,k}^*$.

Consider that the constrained space is time-invariant. Then, according to Lemma 15.3, take the time derivative of V yielding

$$\dot{V} = \sum_{i\in\mathcal{V}}\left[\left(\frac{\partial P_{i,o}}{\partial q_i}\right)^T\dot{z}_{1i} + 2\left(\frac{\partial P_{i,c}}{\partial q_i}\right)^T(\dot{z}_{1i} + \hat{v}_i) + \tilde{d}_i^T\dot{\tilde{d}}_i\right.$$
$$+ \frac{1}{\gamma_i}\sum_{k=1}^{p}\tilde{W}_{i,k}^T\dot{\tilde{W}}_{i,k} + \frac{1}{2}z_{2i}^T\dot{M}_i z_{2i} - z_{2i}^T M_i\dot{\alpha}_{1i} + z_{1i}^T\dot{z}_{1i}$$
$$\left. + z_{2i}^T\left[\tau_i + d_i - C_0\dot{q}_i - W_i^{*T}\phi_i(Z_i)\right] + \left(\frac{\partial P_{i,o}}{\partial q_i}\right)^T\hat{v}_i\right].$$

Substituting (15.14) and (15.15) into $\dot{V}$ gives

$$\dot{V} = \sum_{i\in\mathcal{V}}\left[\Theta_i\left(z_{2i} - (\Theta_i)^\dagger\left(K_{2i}V_{pi} + \left(\frac{\partial P_{i,o}}{\partial q_i}\right)^T\hat{v}_i\right.\right.\right.$$
$$\left.\left. + 2\left(\frac{\partial P_{i,c}}{\partial q_i}\right)^T\hat{v}_i\right)\right) + \frac{1}{\gamma_i}\sum_{k=1}^{p}\tilde{W}_{i,k}^T\dot{\tilde{W}}_{i,k} + \tilde{d}_i^T\dot{\tilde{d}}_i$$
$$+ \frac{1}{2}z_{2i}^T\dot{M}_i z_{2i} - z_{2i}^T M_i\dot{\alpha}_{1i} + 2\left(\frac{\partial P_{i,c}}{\partial q_i}\right)^T\hat{v}_i$$
$$\left. + z_{2i}^T\left[\tau_i + d_i - C_0\dot{q}_i - W_i^{*T}\phi_i(Z_i)\right] + \left(\frac{\partial P_{i,o}}{\partial q_i}\right)^T\hat{v}_i\right].$$

Applying Property 2.3, and considering (15.21) and (15.22), we have

$$\begin{aligned}\dot{V} =&\sum_{i\in\mathcal{V}}\Big[-z_{2i}^T K_{1i} z_{2i} - \tilde{d}_i^T \Im_i\left(z_{2i}\right) M_{0i}^{-1}\left(q_i\right)\tilde{d}_i \\ &-\sum_{k=1}^{p}\left(\sigma_i \tilde{W}_{i,k}^T \hat{W}_{i,k}\right) - \tilde{d}_i^T \dot{d}_i - K_{2i} V_{pi} + z_{2i}^T \tilde{d}_i \\ &+\tilde{d}_i^T \Im_i\left(z_{2i}\right) M_{0i}^{-1}\left(q_i\right) W_i^{*T} \phi_i\left(Z_i\right)\Big].\end{aligned} \tag{15.28}$$

Noticing $\tilde{W}_{i,k} = \hat{W}_{i,k} - W_{i,k}^*$, we have

$$\begin{aligned}-\sigma_i \tilde{W}_{i,k}^T \hat{W}_{i,k} &= -\sigma_i \tilde{W}_{i,k}^T\left(\tilde{W}_{i,k} + W_{i,k}^*\right) \\ &\leq -\sigma_i\left\|\tilde{W}_{i,k}\right\|^2 + \frac{\sigma_i}{2}\left\|\tilde{W}_{i,k}\right\|^2 + \frac{\sigma_i}{2}\left\|W_{i,k}^*\right\|^2 \\ &\leq \frac{\sigma_i}{2}\left(\left\|W_{i,k}^*\right\|^2 - \left\|\tilde{W}_{i,k}\right\|^2\right).\end{aligned} \tag{15.29}$$

Using average value inequality, we can get $z_{2i}^T\tilde{d}_i \leq (1/2)\, z_{2i}^T z_{2i} + (1/2)\,\tilde{d}_i^T\tilde{d}_i$, and $\tilde{d}_i^T \Im_i M_{0i}^{-1} W_i^{*T}\phi_i\left(Z_i\right) \leq (1/2\theta)\left\|\Im_i M_{0i}^{-1}\right\|^2\left\|\tilde{d}_i\right\|^2 + (\theta/2)\sum_{k=1}^{p}\left\|W_{i,k}^*\right\|^2\left\|\phi_i\left(Z_i\right)\right\|^2$.

According to Assumption 15.1 and Lemma 15.1, we have $\left\|\dot{f}_{di}\right\| \leq \varsigma_{0i}$. Then, we know $\dot{d}_i \leq \varsigma_{0i} + \bar{\epsilon}_i$, and $-\tilde{d}_i^T\dot{d}_i \leq (1/2)\,\tilde{d}_i^T\tilde{d}_i + (1/2)\left(\varsigma_{0i} + \bar{\epsilon}_i\right)^2$.

Substituting the above inequality to (15.28) yields

$$\begin{aligned}\dot{V} =&\sum_{i\in\mathcal{V}}\Big[-z_{2i}^T\left(K_{1i} - \frac{1}{2}I_p\right) z_{2i} - \tilde{d}_i^T \Im_i\left(z_{2i}\right) M_{0i}^{-1}\left(q_i\right)\tilde{d}_i \\ &+\frac{\left\|\Im\left(z_{2i}\right) M_{0i}^{-1}\left(q_i\right)\right\|^2}{2\theta}\left\|\tilde{d}_i\right\|^2 + \frac{\theta}{2}\sum_{k=1}^{p}\left\|W_i^*\right\|^2\left\|\phi_i\left(Z_i\right)\right\|^2 \\ &+\tilde{d}_i^T\tilde{d}_i + \frac{1}{2}\left(\varsigma_{0i} + \bar{\epsilon}_i\right)^2 - K_{2i} V_{pi} \\ &\sum_{k=1}^{p}\left(\frac{\sigma_i}{2}\left(\left\|W_{i,k}^*\right\|^2 - \left\|\tilde{W}_{i,k}\right\|^2\right)\right)\Big],\end{aligned} \tag{15.30}$$

and we have

$$
\begin{aligned}
\dot{V} =& \sum_{i\in\mathcal{V}} \Bigg[-z_{2i}^T \left(K_{1i} - \frac{1}{2} I_p\right) z_{2i} - K_{2i} V_{pi} \\
& - \tilde{d}_i^T \left(\Im_i\left(z_{2i}\right) M_{0i}^{-1} - \frac{\left\|\Im\left(z_{2i}\right) M_{0i}^{-1}\right\|^2 - 2\theta}{2\theta} I_p \right) \tilde{d}_i \\
& + \sum_{k=1}^{p} \left(\left(\frac{\theta \|\phi_i\left(Z_i\right)\|^2 + \sigma_i}{2} \right) \|W_{i,k}^*\|^2 \right) \\
& - \sum_{k=1}^{p} \left(\frac{\sigma_i}{2} \tilde{W}_{i,k}^T \tilde{W}_{i,k} \right) + \frac{1}{2} (\varsigma_{0i} + \bar{\epsilon}_i)^2 \Bigg] \\
\leq & - \rho V + C,
\end{aligned}
\tag{15.31}
$$

where ρ and C take the forms

$$
\begin{aligned}
\rho = \min_{i\in\mathcal{V}} & \Bigg(\lambda_{\min}\left(K_{2i}\right), \frac{2\lambda_{\min}\left(K_{1i} - \frac{1}{2} I_p\right)}{\lambda_{\max}\left(M_i\right)}, \gamma_i \sigma_i, \\
& \lambda_{\min} \left(\Im_i\left(z_{2i}\right) M_{0i}^{-1} - \frac{\left\|\Im\left(z_{2i}\right) M_{0i}^{-1}\right\|^2 - 2\theta}{2\theta} I_p \right) \Bigg), \\
C = \sum_{i\in\mathcal{V}} & \Bigg(\sum_{k=1}^{p} \left(\left(\frac{\theta \|\phi_i\left(Z_i\right)\|^2 + \sigma_i}{2} \right) \|W_{i,k}^*\|^2 \right) \\
& + \frac{1}{2} (\varsigma_{0i} + \bar{\epsilon}_i)^2 \Bigg).
\end{aligned}
$$

To ensure $\rho > 0$, we choose the control gains K_{1i} and K_{2i} as

$$
\lambda_{\min}\left(K_{1i} - \frac{1}{2} I_p\right) > 0, \ \lambda_{\min}\left(K_{2i}\right) > 0, \ i \in \mathcal{V}, \tag{15.32}
$$

and $\Im_i\left(z_{2i}\right)$ satisfies

$$
\lambda_{\min} \left(\Im_i\left(z_{2i}\right) M_{0i}^{-1} - \frac{\left\|\Im\left(z_{2i}\right) M_{0i}^{-1}\right\|^2 - 2\theta}{2\theta} I_p \right) > 0.
$$

Here, we provide the details of the proof to show that z_{1i}, z_{2i}, $\tilde{W}_i$, and $\tilde{d}$ are SGUUB. Since $V(t) \leq V(0) e^{-\rho t} + (C/\rho)\left(1 - e^{-\rho t}\right) \leq V(0) + C/\rho$, we have $\|z_1\| \leq \sqrt{D}$, where $D = 2\left(V(0) + C/\rho\right)$. Therefore, z_1 remains in the compact set Ω_{z_1}. Bounds for z_2, $\tilde{W}$, and $\tilde{d}$ can be given using a similar procedure. Based on Theorem 15.1, when $t \geq T_M$, LDFEs can generate accurate desired formation position and velocity estimations, respectively. Hence, the layered formation-containment control objective is achieveds which concludes the proof. □

Remark 15.2 From the analysis above, we know the states of the leaders in the first layer are bounded. Thus, the boundedness property of leaders' accelerations (15.4) can be utilized as a precondition Case F1 in Theorem 15.1.

Remark 15.3 Based on Lemma 15.2 and Theorem 15.2, it is known that $\tilde{q}_i$ and z_{1i}, $i \in \mathcal{V}$ are bounded during $0 \leq t \leq T_M$. Therefore, the states q_i will remain bounded when $t \leq T_M$, since $z_{1i} = q_i - \hat{q}_i$.

15.5 Simulation

Numerical results are given to validate the full power of the proposed approach. 16 networked spacecraft are considered, of which the relative orbital motion and the initial elements of the near-circular orbit are the same as those of the simulations in Chap. 13. We evaluate the control algorithms (15.21) and (15.22) for networked spacecraft in the constrained space with model uncertainties and external disturbances, where the desired formation configuration is time-varying.

The layered interaction graph among 16 spacecraft is shown in Fig. 15.2. The other parameters for simulation are given in Table 15.1, where $q_0 = \left[200\cos\left(\frac{\pi t}{100}\right),\ 200\sin\left(\frac{\pi t}{100}\right),\ 2t\right]^T$. The external disturbances are chosen as $f_{di} = [\sin(t) + 1,\ \sin(t) + \frac{1}{2},\ 2\cos\left(\frac{1}{2}t\right) + \frac{1}{2}]^T$. The detailed initial states of spacecraft and the smooth formation functions $r_i(t)$ are omitted here.

The trajectories of spacecraft are given in Fig. 15.3. In the constrained space, as shown in Fig. 15.3, a spatial border is at $x = 250$, two sphere obstacles with 90m radius are respectively at $[0, 0, 300]^T$ and $[0, 0, 700]^T$. It can be observed that the LFC control algorithms can guarantee a satisfactory performance when external disturbances and model uncertainties exist. From the side view of trajectories in Fig. 15.4, we know the collisions with the border, the obstacles, as well as the spacecraft, are avoided successfully. The formation in the first layer is affected by the obstacles and the border, and the configuration of formation is re-formed when the group of spacecraft are away from the obstacles, while the followers in the second can safely converge into the leaders' convex hull.

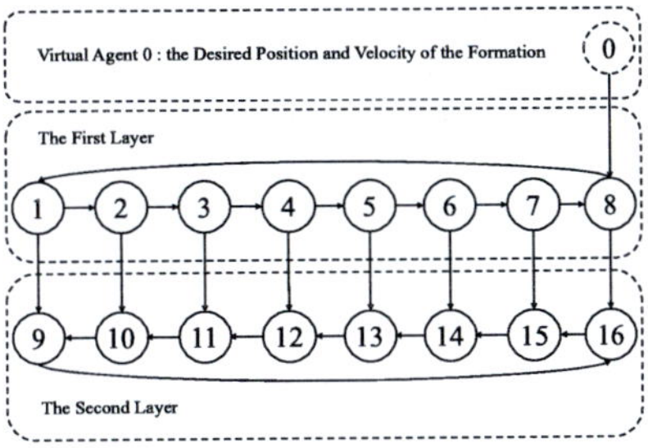

Fig. 15.2 The layered communication graph

Table 15.1 The parameters for simulation

Parameters	Values	Parameters	Values
m_i	35 kg	β_1	5
$\hat{d}_i\ (0)$	0	β_2	5
$\upsilon_i\ (z_{2i})$	$3z_{2i}$	β_3	5
K_{1i}	$0.1I_3$	β_4	5
K_{2i}	$5I_3$	σ_i	5
γ_i	4	l	256
k_{cx}	100	$c_{i,j}$	10
r	100	d_{safe}	10
η_r	10	μ_r	random $[-1, 1]$
C_{0i}	$(1/2)\ C_i$	M_{0i}	$(1/2)\ M_i$

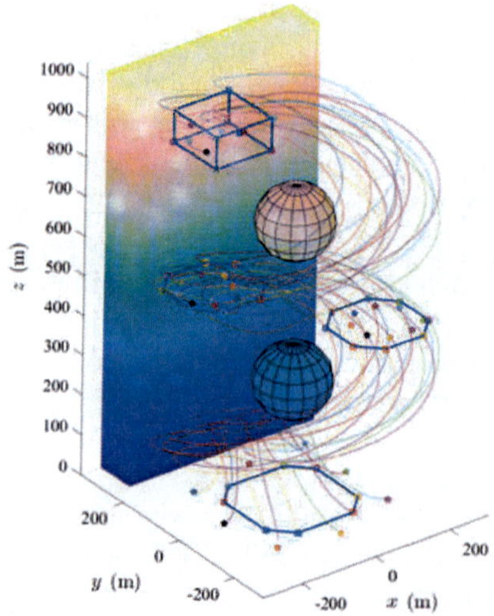

Fig. 15.3 The trajectories of 16 spacecraft in the constrained space (3D view)

In fact, the adjustable formation shape formed by the first layer will help the followers avoid pop-up obstacles and borders, since the leaders will detect the spatial constrains first and drive the followers to a safe place through formation changing. Figure 15.5 shows that the formation errors of 8 leaders converge to the origin within 100s. Figure 15.6 shows that the tracking errors of 8 followers converge to the origin within 80s. In Figs. 15.7 and 15.8, the corresponding control inputs of spacecraft are shown, which are acceptable for spacecraft in practice. The numerical results above have shown that the proposed LFC control scheme is effective and feasible in the

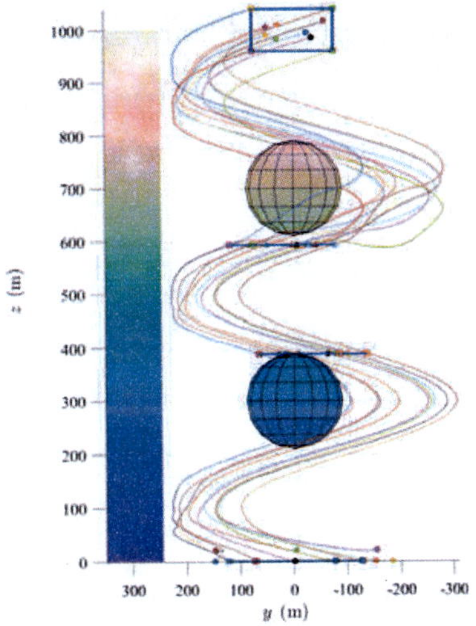

Fig. 15.4 The trajectories of 16 spacecraft in the constrained space (side view)

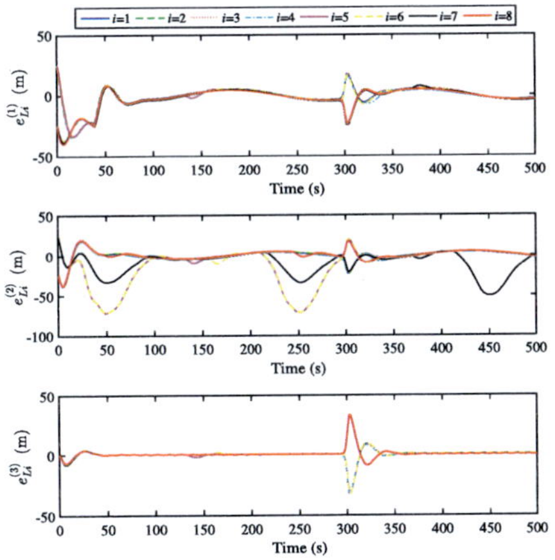

Fig. 15.5 The formation errors of 8 leaders in the first layer

constrained space with a time-varying formation case. (These simulation examples are for demonstration purpose only, which aims to show the control performance and may not match practical spacecraft missions).

15.6 Conclusion

In this chapter, based on a layered framework, we have proposed a layered formation-containment control algorithm in the constrained space for multi-spacecraft formation systems described by MELSs with external disturbances and model

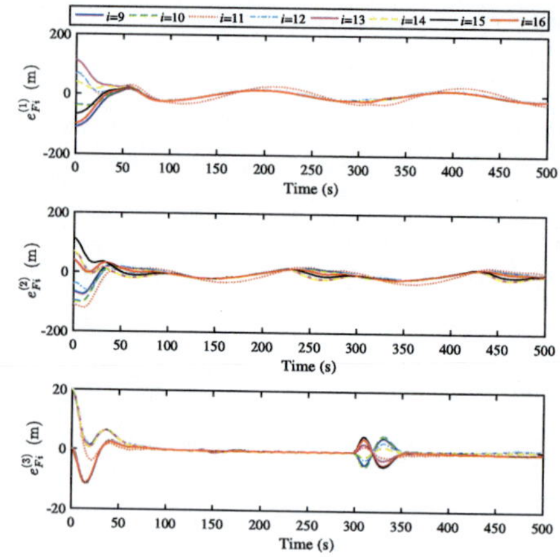

Fig. 15.6 The tracking errors of 8 followers in the second layer

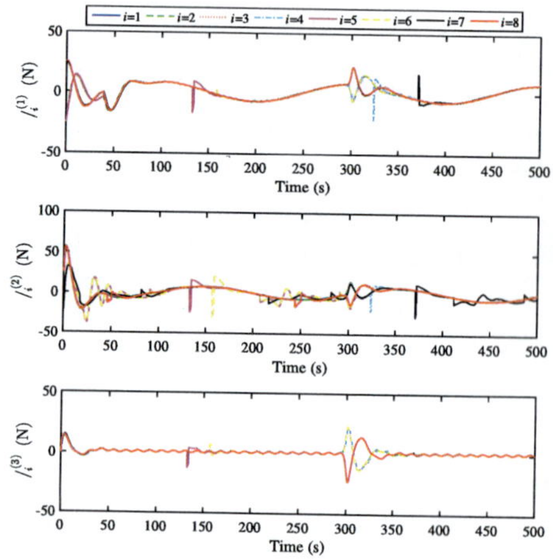

Fig. 15.7 The control forces of 8 leaders in the first layer

uncertainties. Under a directed communication graph, an LDFE has been proposed to help the spacecraft acquire their desired positions and velocities. Using the Dirac delta function, an artificial potential field has been constructed to guide spacecraft in both layers to a safe place while avoiding collisions with spatial constraints or each other. Then, a DO and NNs have been properly designed to approximate external disturbances and model uncertainties, while improving the robustness and the performance of multi-spacecraft formation systems. The approaches in this chapter not only provide a new insight into the multi-spacecraft formation problems but also have potential benefits in practical applications. The numerical examples have shown the feasibility of the proposed approach in the constrained space with a spacecraft formation case.

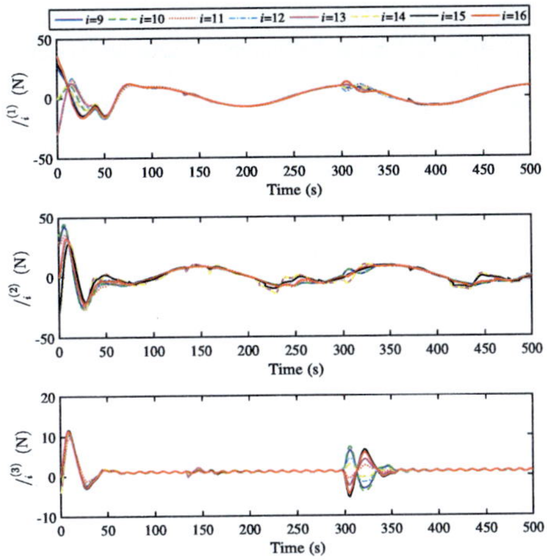

Fig. 15.8 The control forces of 8 followers in the second layer

References

1. Jadbabaie, A., Lin, J., Morse, A.S.: Coordination of groups of mobile autonomous agents using nearest neighbor rules. IEEE Transactions on Automatic Control **48**(6), 988–1001 (2003). https://doi.org/10.1109/TAC.2003.812781
2. Olfati-Saber, R., Murray, R.M.: Consensus problems in networks of agents with switching topology and time-delays. IEEE Transactions on Automatic Control **49**(9), 1520–1533 (2004). https://doi.org/10.1109/TAC.2004.834113
3. Ren, W., Beard, R.W.: Consensus seeking in multiagent systems under dynamically changing interaction topologies. IEEE Transactions on Automatic Control **50**(5), 655–661 (2005)
4. Xie, D., Xu, S., Zhang, B., Li, Y., Chu, Y.: Consensus for multi-agent systems with distributed adaptive control and an event-triggered communication strategy. IET Control Theory & Applications **10**(13), 1547–1555 (2016)
5. Muscio, G., Pierri, F., Trujillo, M.A., Cataldi, E., Antonelli, G., Caccavale, F., Viguria, A., Chiaverini, S., Ollero, A.: Coordinated Control of Aerial Robotic Manipulators: Theory and Experiments. IEEE Transactions on Control Systems Technology (2017)
6. Peng, K., Yang, Y.: Leader-following consensus problem with a varying-velocity leader and time-varying delays. Physica A Statistical Mechanics & Its Applications **388**(388), 193–208 (2009)
7. Beard, R.W., Lawton, J., Hadaegh, F.Y.: A coordination architecture for spacecraft formation control. IEEE Transactions on control systems technology **9**(6), 777–790 (2001)
8. Do, K.D.: Bounded controllers for formation stabilization of mobile agents with limited sensing ranges. IEEE Transactions on Automatic Control **52**(3), 569–576 (2007)
9. Cheah, C.C., Hou, S.P., Slotine, J.J.E.: Region-based shape control for a swarm of robots. Automatica **45**(10), 2406–2411 (2009)
10. Sakai, D., Fukushima, H., Matsuno, F.: Leader–follower navigation in obstacle environments while preserving connectivity without data transmission. IEEE Transactions on Control Systems Technology **26**(4), 1233–1248 (2017)
11. Notarstefano, G., Egerstedt, M., Haque, M.: Containment in leader–follower networks with switching communication topologies. Automatica **47**(5), 1035–1040 (2011)
12. Rong, L., Lu, J., Xu, S., Chu, Y.: Reference model-based containment control of multi-agent systems with higher-order dynamics. IET Control Theory & Applications **8**(10), 796–802 (2014)

13. Cui, G., Xu, S., Ma, Q., Li, Z., Chu, Y.: Command-filter-based distributed containment control of nonlinear multi-agent systems with actuator failures. International Journal of Control **91**(7), 1708–1719 (2018)
14. Han, L., Dong, X., Li, Q., Ren, Z.: Formation-containment control for second-order multi-agent systems with time-varying delays. Neurocomputing **218**, 439–447 (2016). https://doi.org/10.1016/j.neucom.2016.09.001
15. Dong, X., Shi, Z., Lu, G., Zhong, Y.: Formation-containment analysis and design for high-order linear time-invariant swarm systems. International Journal of Robust and Nonlinear Control **25**(17), 3439–3456 (2015)
16. Wang, Y., Liu, X., Xiao, J.: Output formation-containment of coupled heterogeneous linear systems under intermittent communication. Journal of the Franklin Institute **354**(1), 392–414 (2017). https://doi.org/10.1016/j.jfranklin.2016.10.011
17. Han, T., Chi, M., Guan, Z.H., Hu, B., Xiao, J.W., Huang, Y.: Distributed Three-Dimensional Formation Containment Control of Multiple Unmanned Aerial Vehicle Systems. Asian Journal of Control **19**(3), 1103–1113 (2017). https://doi.org/10.1002/asjc.1445
18. Khatib, O.: Real-time obstacle avoidance for manipulators and mobile robots. The international journal of robotics research **5**(1), 90–98 (1986)
19. Lee, L.F., Krovi, V.: A standardized testing-ground for artificial potential-field based motion planning for robot collectives. In: Proceedings of the Performance Metrics for Intelligent Systems Workshop, Gaithersburg (2006)
20. Rezaee, H., Abdollahi, F.: A decentralized cooperative control scheme with obstacle avoidance for a team of mobile robots. IEEE Transactions on Industrial Electronics **61**(1), 347–354 (2014)
21. Panagou, D., Kumar, V.: Cooperative visibility maintenance for leader–follower formations in obstacle environments. IEEE Transactions on Robotics **30**(4), 831–844 (2014)
22. Kallu, K.D., Jie, W., Lee, M.C.: Sensorless reaction force estimation of the end effector of a dual-arm robot manipulator using sliding mode control with a sliding perturbation observer. International Journal of Control, Automation and Systems **16**(3), 1367–1378 (2018)
23. Ballard, L., Cao, Y., Ren, W.: Distributed discrete-time coupled harmonic oscillators with application to synchronised motion coordination. IET Control Theory & Applications **4**(5), 806–816 (2010)
24. Li, P., Xu, S., Chu, Y., Chen, W., Zhang, Z.: Finite-time leader-following rendezvous for Euler–Lagrange multi-agent systems with an uncertain leader. Transactions of the Institute of Measurement and Control **40**(6), 1766–1775 (2018)
25. Ge, S.S., Wang, C.: Adaptive neural control of uncertain mimo nonlinear systems. IEEE Transactions on Neural Networks **15**(3), 674–692 (2004)
26. He, W., Dong, Y.: Adaptive fuzzy neural network control for a constrained robot using impedance learning. IEEE transactions on neural networks and learning systems **29**(4), 1174–1186 (2018)
27. Li, D., Ma, G., He, W., Zhang, W., Li, C., Ge, S.S.: Distributed coordinated tracking control of multiple Euler-Lagrange systems by state and output feedback. IET Control Theory & Applications **11**(14), 2213–2221 (2017)
28. Chen, M., Ge, S.S.: Adaptive neural output feedback control of uncertain nonlinear systems with unknown hysteresis using disturbance observer. IEEE Transactions on Industrial Electronics **62**(12), 7706–7716 (2015)
29. Guo, Q., Zhang, Y., Celler, B.G., Su, S.W.: Backstepping control of electro-hydraulic system based on extended-state-observer with plant dynamics largely unknown. IEEE Transactions on Industrial Electronics **63**(11), 6909–6920 (2016)
30. Zhang, Z., Xu, S.: Observer design for uncertain nonlinear systems with unmodeled dynamics. Automatica **51**, 80–84 (2015)
31. Li, Z., Su, C.Y., Wang, L., Chen, Z., Chai, T.: Nonlinear Disturbance Observer-Based Control Design for a Robotic Exoskeleton Incorporating Fuzzy Approximation. IEEE Transactions on Industrial Electronics **62**(9), 5763–5775 (2015). https://doi.org/10.1109/TIE.2015.2447498
32. Cao, Y., Ren, W., Meng, Z.: Decentralized finite-time sliding mode estimators and their applications in decentralized finite-time formation tracking. Systems & Control Letters **59**(9), 522–529 (2010)

33. Ge, S.S., Liu, X., Goh, C.H., Xu, L.: Formation tracking control of multiagents in constrained space. IEEE Transactions on Control Systems Technology **24**(3), 992–1003 (2016)
34. Dirac, P.A.M.: The principles of quantum mechanics. 27. Oxford university press (1981)
35. Kubat, M.: Neural networks: a comprehensive foundation by Simon Haykin, Macmillan, 1994, ISBN 0-02-352781-7. Knowledge Engineering Review **13**(4), 409–412 (1999)

Chapter 16
Conclusion

Abstract The research in this book provides a series of high-precision constrained control methods based on performance guaranteed and explicit settling time for the more complex control problems of spacecraft on-orbit service missions. These include the general performance guaranteed control method, command-filter-based control method, optimization control method, cooperative game control method, and formation control method. Applying these control methods can make the relative motion of the on-orbit service spacecraft have higher control accuracy and better performance during missions. The research in this book is expected to provide the theoretical basis and the advanced design approaches for the relative motion and general motion control of the on-orbit service spacecraft under high-precision and multi-constraint conditions. These methods and algorithms are expected to have significant theoretical value and application prospects in aerospace engineering.

For the maneuvering, rendezvous, docking, and on-orbit operation tasks of the on-orbit service spacecraft, the service spacecraft needs to perform relative motion according to the geometric characteristics and motion characteristics of the target spacecraft, and these characteristics will bring the actual kinetic *physical constraint* conditions to the mission design and the control design. Facing complex moving targets or docking interfaces, the relative motion control task of the service spacecraft becomes the key to the success of the on-orbit service mission, and the control algorithm design considering the constraints is crucial. Correspondingly, the algorithms that consider state or output constraints can provide performance guaranteed and safe motion regions for the relative motion of the on-orbit service spacecraft. Simultaneously, these algorithms can also provide high-precision control responses for the motion and pointing tasks of the general spacecraft. For the high-precision control response requirements, it is necessary to consider the *performance constraint* conditions of the closed-loop system response.

This book studies and analyzes the problem of the precise and constrained relative motion control of the on-orbit service spacecraft relative to the target spacecraft. Starting from the abstract physical constraints and performance constraints, this research moves towards a motion control technology route that considers performance guaran-

Y. Gao and D. Li, *Spacecraft Maneuver with Performance Guaranteed*,
https://doi.org/10.1007/978-981-99-4653-2_16

teed. A series of control algorithms based on the performance guaranteed techniques are proposed. These algorithms describe the authors' recent research achievements in the fields of performance guaranteed control and on-orbit service spacecraft control from three aspects: control, game, and cooperation. In Part I, performance guaranteed control for single spacecraft, for a single service spacecraft relative to the target spacecraft, this book proposes: the spacecraft performance guaranteed asymptotic control with the time parameters, the spacecraft performance guaranteed finite-time control, the spacecraft performance guaranteed control with improved transient performance, the spacecraft performance guaranteed control with the command filter, the spacecraft performance guaranteed finite-time control with the command filter, and the spacecraft performance guaranteed optimal control with adaptive dynamic programming. In Part II, performance guaranteed games for multiple spacecraft, for the cooperative service spacecraft and the target spacecraft, this book proposes: the performance guaranteed non-zero-sum games for multiple spacecraft and the combined spacecraft. In Part III, performance guaranteed cooperation for spacecraft formation, for the multi-spacecraft or abstract network systems, this book introduces: the distributed coordinated control algorithms with full-state constraints and with output constraints; the formation-containment control algorithms for networked spacecraft with output feedback, with collision avoidance, and in constrained space. These algorithms realize the state constraints and formation configuration constraints and guarantees of spacecraft formation systems, respectively.

Applying the algorithms presented in this book, the service spacecraft's closed-loop system state or tracking error is constrained by performance guaranteed conditions constructed from the performance functions. The fundamental objective of the performance guaranteed control is to make the actual time-domain response of the closed-loop system state satisfy these complex constraints. The time-domain response of the closed-loop system satisfies the expected performance constraint parameters, including predefined maximum settling time, convergence rate, overshoot, steady-state error, input constraints, formation configuration constraints, etc. The behavior of the closed-loop system meets design expectations. Simultaneously, considering a variety of practical scenarios, this book provides the finite-time stability algorithm designs, the describing motion trajectory algorithm design, the algorithm design considering input saturation, the optimal control algorithm design, the game algorithm design, the multi-spacecraft constraints algorithm design, and the multi-spacecraft formation-containment algorithm design. These designs give the algorithms in this book practical application value for different scenarios. The rationality and effectiveness of the algorithms proposed in this book are verified through rigorous theoretical proof and comparative simulation analysis.

This book's research objects, design ideas, and control theory methods are logical and rich. The research objects in the three parts of this book are progressive. In Part I, the performance guaranteed control algorithms for the single service spacecraft are designed. The relative motion of the service spacecraft to the target spacecraft is mainly based on tracking control. Moreover, these tracking controls can also be considered as the tracking control of a single spacecraft on the desired trajectory. In Part II, considering the scenario where multiple spacecraft cooperate, based on the non-zero-sum game theory, the spacecraft performance guaranteed cooperative control strategies suitable for two or more are designed. These strategies make the control input of each spacecraft cooperatively optimal, and all control inputs in these strategies reach Nash equilibrium. In Part III, more cooperative behaviors of spacecraft are considered, and the constrained control algorithms and formation-containment control algorithms of the spacecraft formation are designed.

Based on the performance guaranteed methods, the control methods of the three parts in this book are diverse. From the perspective of stability, the stability of the closed-loop systems with performance guaranteed is enriched, including stability, asymptotic stability, uniformly ultimately bounded, finite-time stability, and finite-time uniformly ultimately bounded, etc. From the perspective of the control technology, this book studies performance guaranteed control technique, ratio persistence property, command filter backstepping, adaptive dynamic programming technique, non-zero-sum game theory, multi-spacecraft consistency, formation-containment control technique, etc. From the perspective of the performance indicators, the maximum settling time of the closed-loop system, the finite settling time of the closed-loop system, convergence rate, overshoot, steady-state error, control input constraints, and cooperative game performance indicators are considered.

On the one hand, the writing of this book summarizes the authors' research results in recent years. On the other hand, it also expects to provide innovative ideas and methods for spacecraft control, especially in occasions subject to actual physical constraints or high control precision requirements. Based on the algorithm designs in this book, engineering designers can derive control schemes and algorithms suitable for spacecraft engineering according to the practical task requirements. Based on the theoretical demonstration, this book also hopes to provide theoretical designers with algorithm design ideas that guarantee performance and constraint states.

Furthermore, based on this book, there are many derived issues worthy of further study in the future, including but not limited to the following: the estimation of the steady-state boundaries and their relation to strict constraints; the control margins under the constraints and the tightening and relaxation of the constraints; further optimization of the controlled trajectory of the closed-loop system, etc.

The authors will continue to work on the performance guaranteed control and the constrained precise motion of the spacecraft.